Handbook of Plastics Test Methods

The first edition of this book was prepared for the Plastics Institute by G. C. Ives, J. A. Mead and M. M. Riley.

The revision for this second edition was undertaken by the following staff of the Rubber and Plastics Research Association of Great Britain Limited:

J. H. Beesley
R. P. Brown
D. Hands
S. W. Hawley
D. I. James
M. L. Lomax
R. H. Norman
B. J. Wain

Handbook of Plastics Test Methods

Second edition

Edited by
R. P. Brown

George Godwin Limited
in association with
The Plastics and Rubber Institute

First published in Great Britain 1971 by
Iliffe Books, an imprint of
the Butterworth Group

Second edition published 1981 by
George Godwin Limited
the book publishing subsidiary of
The Builder Group
1–3 Pemberton Row, Fleet Street,
London EC4P 4HL
in association with
The Plastics and Rubber Institute

British Library Cataloguing in Publication Data

Handbook of plastics test methods.—2nd ed.
1. Plastics—Testing—Standards
I. Brown, R. P.
II. Ives, G. C. Handbook of plastics test methods.
668.4′197 TP1122

ISBN 0-7114-5618-6

Printed and bound in Great Britain by
The Pitman Press, Bath.

Foreword

by Professor A. W. Birley
Director, Institute of Polymer Technology, Loughborough University of Technology

'Is it necessary to test plastics?' is a question that is often posed—and one which must find an affirmative answer, since millions of tests are carried out each day. The purpose of such tests is to measure properties, so it is pertinent to define the significance of these. Properties are the quantitative response of a material to stimuli applied externally; for example, if a component is loaded, the material reacts by deforming or breaking, the relative properties being 'modulus' and 'strength'.

Properties are measured for various reasons, with two main purposes: as a basis for design or for predicting product performance; or for quality control. Over the whole range of property measurement, however, the results obtained need to be accepted and understood by others in the field so that there is economy of effort on the part of those providing data for design, and also a proper basis for commercial activity.

Testing of plastics may yield information on basic properties, as in tests for tensile strength and for apparent permittivity, or may be a limited evaluation of the behaviour of a particular artefact, as in testing the ability of a dustbin to withstand a defined drop without fracture. Between the two are hybrid tests such as those for impact strength, which are obviously related to some feature of behaviour but which cannot be used directly in design. There is a clear distinction between, on the one hand, the properties of materials and, on the other, the response of products made from these materials in simulated or idealised use conditions. Obviously the latter are of crucial importance; however, it is not practical or economic to restrict testing to finished products.

Since the response of plastics to a wide variety of tests is of such universal interest, the question of the standardisation of test procedures attracts much attention. Over the last 10–15 years there has been much progress in standardisation, particularly at the international level, and the present authors have been deeply involved in this work. They are thus peculiarly well qualified to comment on the standardisation of test methods. However, beyond this is the need to set the test methods in context—to provide a background of appreciation of the restrictions imposed in standard tests so as to ensure that reproducible and comparable results can be obtained in different laboratories. This is the main purpose of this book, which certainly achieves this end. It is to the credit of the Plastics and Rubber Institute that they have appreciated the need for a revised edition of this work.

Missing from the international standards literature, and from that of many national bodies, is a classified index of test methods. Although I doubt if this is a prime objective of the present volume, it does serve this purpose and will, for this reason, be appreciated by many technologists in industry who are responsible for specifications but are not familiar with the standards literature.

Loughborough
May 1981

A. W. BIRLEY

Contents

Preface

When *Handbook of Plastics Test Methods* was first published in 1971 for the Plastics Institute (now the Plastics and Rubber Institute) it was quickly accepted as the standard work on the subject and has remained in that position to the present time. In the preface, G. C. Ives asked the question, 'Is there any need for a handbook of plastics test methods?' The success of the book leaves no doubt that the answer is 'Yes', and we believe that there will continue to be a need.

In the decade or so prior to 1971 there had been drastic changes and extensive additions to the standard methods for testing plastics. Since 1971, development has continued and in particular International Standards have assumed increasing prominence, so that it is inevitable that a revision of the Handbook is needed. Unfortunately, the original authors were not in a position to undertake this work and members of the testing staff at the Rubber and Plastics Research Association were happy to answer the call from the PRI to undertake this important task.

In making the revision we have had no wish to depart from the original style and scope: since the book deals with physical testing, chemical analysis has been excluded completely but polymer characterisation tests, which are perhaps on the borderline of the two disciplines, have been retained. Similarly, such tests as thermal ageing, which are normally thought of as physical although they may induce chemical change, are included.

There has been no question of including elastomers as there exists a recent volume on that subject (*Physical Testing of Rubbers*) which may be considered complementary to the present work. Apparatus requirements are discussed as an intrinsic part of test methods but attention is also drawn to the *RAPRA Guide to Rubber and Plastics Test Equipment*, which gives information on the sources of supply of test equipment for both plastics and rubbers. Both of these works are to be found cited as references. Tests on cellular materials have been excluded on the basis that this distinct class of materials should be treated separately, both rubbers and plastics being considered together. Some tests on finished and semi-finished products have been included but this section could not be exhaustive because such tests are too specialised for general treatment. Similarly, tests restricted to specific shapes or forms, or to a single polymer, have only been accorded, at best, a passing reference. The testing of compounding ingredients and aids to processing—solvents, plasticisers, lubricants, stabilisers, pigments and the like—has not been covered except in so far as their evaluation logically forms part of the testing of a plastics material.

The aim of the book is to present an up to date account of plastics testing procedures; within the limitations of scope discussed above, it is intended to be comprehensive in covering all the tests in common, and sometimes not so common, use. Inevitably, most of the tests are the standard ones, intended primarily for quality control use and often of a somewhat arbitrary nature. However, where appropriate, the requirements for obtaining meaningful design data are discussed. The subject of plastics testing could have been treated in a number of ways but it is hoped that the one selected will be found logical and convenient to two classes of reader in particular:

(1) the person in industry directly involved with testing plastics;
(2) the student of plastics technology.

In addition, it is hoped that the book will be of value to those indirectly involved with the evaluation of plastics, such as design engineers.

In the original Handbook the national standards considered were those of the United Kingdom, the United States of America, and Germany, while International (ISO) Standards were not described in detail because their practical value emerges only when they have been adopted nationally. Since that time, considerable advance has been made in aligning national standards with those of ISO, which have therefore become increasingly important. Consequently, it has been found convenient in many cases to describe the international method and then consider national deviations from this. As regards national standards, the emphasis has been placed on British Standards, together with American (USA) and German methods, as these are those most frequently encountered by the authors and are certainly among the most widely used and technically advanced. Generally, test methods peculiar to particular commercial companies have not been considered at all. The trend towards the adoption of International Standards is reducing the regrettable differences between the standards of different countries and organisations, and it is perhaps appropriate to make a plea for the adoption of recognised standards without modification where there is really no strong technical reason for change.

Technical literature does not stand still—between writing and publication it is inevitable that some standards will have been revised. To counteract this as far as possible, the likely trends in test methods have been estimated from the draft standards in circulation and the known activities of the various committees. It should be emphasised that the latest edition of a standard should be called up for all contractual arrangements.

With changes over the years in emphasis on particular areas of testing, the subdivision of subjects in the first edition has been modified in a few places. The layout of a book on testing is bound to be to some extent subjective but we can only hope that the majority of readers find our choice logical and practical. The first chapter, which G. C. Ives described as 'a gentle introduction to the subject', has been retained and covers aspects of testing that are common to the individual topics dealt with in the rest of the book.

R. P. B.

RAPRA
Shawbury
April 1981

Acknowledgements

On undertaking this work the authors of the revision were quickly very appreciative of the sound base that the original authors had given us. We are also most grateful to the Rubber and Plastics Research Association for their permission to carry out the task and their support for the project.

The chapter on Standards and Standards Organisations has been taken with little modification from *Physical Testing of Rubbers* (1979) by kind permission of Applied Science Publishers Limited.

Many of the figures have been taken from other works and we are grateful for permission to use them. Illustrations supplied by companies in industry are acknowledged individually. Figures from British Standards, as cited, are reproduced by permission of the British Standards Institution, 2 Park Street, London W1A 2BS, from whom complete copies of the relevant standards can be obtained. Those illustrating ASTM test methods are copyright American Society for Testing and Materials, 1916 Race Street, Philadelphia, PA 19103, and are reprinted or adapted with permission. The following figures have kindly been supplied by the Rubber and Plastics Research Association of Great Britain Limited: 10.2, 11.10, 13.4, 13.5, 13.8, 13.26, 15.1–15.8, 17.1, 18.7, 20.4.

Thanks are due to colleagues at the Rubber and Plastics Research Association and friends in industry for helpful advice and criticism. Finally, we thank the ladies who typed the text which constituted a considerable addition to their normal work load.

PUBLISHER'S NOTE

Chapter One

Introduction

1.1 PHILOSOPHY

Why test plastics? For that matter, why test anything? Why not rely on experience and good workmanship, backed up by sound judgment? It cannot be denied that testing costs money and it will be seen throughout what follows that practically every test has limitations, the data it yields being applicable only with discretion to the everyday conditions of service.

When dealing specifically with plastics materials the questions above are already part answered by the very case history of the industry, or rather the lack of it! The plastics in use today are very often not precisely the same as those available ten years ago, even if the polymer is basically the same—and there may be refinements of processing which have led, for instance, to a more pure product—the plasticiser has perhaps changed or a new stabiliser system has found favour. New materials are continually being introduced and—of great consequence—plastics are being used in more and more new applications, and frequently more critical applications, than hitherto. Thus, for many circumstances there is not much experience upon which to rely and this makes it very difficult to promote the use of plastics in, for instance, structural applications where the designer may insist upon a guaranteed 50 year performance.

Testing may thus be undertaken in order to assess the performance of a material in relation to the duty which it is to perform, i.e. to establish suitability for purpose; the mechanical properties will be measured and correlated with the calculated stresses, the electrical properties with the known circuitry, and so on. Judicious use of such measurements ideally will obviate the inspired guess, or gross wastage because of the necessity to 'play safe', and in this way the necessary experience will be assembled with a minimum of headaches.

Not that the plastics industry can aspire ultimately to an idyllic future, free from tiresome and profit-reducing testing, any more than the more traditional industries today. All men and machines are fallible and liable to vary in performance for a variety of reasons; the things they produce will alter in quality correspondingly. Thus we find the second principal reason for testing: as a control of quality. Under this broad heading come regular quality control schemes, production batch testing by random sample, purchasers' acceptancy tests and control sample examination in the case of dispute. With the trend towards ever increasing consumer protection legislation, the level of quality control testing is likely to become greater rather than less.

If it is accepted that we must test, then surely that is the end of the matter. Why write a whole book on the unfortunate necessity? After all, everyone knows that 'tensile strength' is the resistance of a material to breaking by stretching, so where can there be room for argument? Well, tensile strength is a relatively straightforward example: most people would have the same broad picture in mind in discussing the tensile strength of a material. However, the measured strength will depend on a number of conditions of sample preparation and test procedure. More generally, there are many cases where different interpretations can be put on one property name or where a number of names apply to the same property, and always the result is dependent on the details of how the test is carried out. In consequence, the test will require a lengthy description; tensile strength serves as an illustration of the factors involved.

Let us suppose that a new moulding composition has been offered, of unknown origin and with a plain statement on its container:

'Tensile strength: 8000 force units/area'—say, 8000*X*.

It is a penny per unit weight cheaper than the material we are at present using, which is satisfactory for our application; we will assume for convenience that this application is one (highly fictional!) where only tensile forces come into play and that there are no other considerations to bother us (for the perspicacious—the specific gravities are the same too!). Our current material also has, we have been told by our suppliers, a tensile strength of 8000*X*. Why should we not change materials and save money—the same strength for a penny a kilo cheaper?

In fact one good reason for looking into the matter further has already become evident. Moulding compositions are usually in the form of granules, pellets, chips or powders, but tensile strength is determined on a moulding of some form or other; therefore we must produce the appropriate test specimens and we must ensure that, although small, they are representative of the product as a whole; in other words, our sampling procedure must be sound. Any moulder or fabricator knows just how much the appearance and strength of a moulding can be altered simply by varying the cycle applied to a given moulding composition. Thus our two values of 8000*X* can only be compared if the conditions of any preheat, moulding temperature, cycle time, etc., are comparable or, better still, if the figures are realistic optima which can be obtained using conditions (especially of time) that are commercially feasible. (Obviously a material which only attains an adequate tensile strength after an uneconomic cycle cannot be considered.)

So far we have achieved the moulding of our test pieces and already we have run into two good reasons why the quoted tensile strengths might not be comparable: however, we have hardly started yet! Most plastics materials are affected by quite small changes in temperature; the stiffening of some types of highly plasticised PVC is a classic example at low temperatures, but many other materials, especially in the thermoplastics class, are affected in a similar if not so spectacular way. How can we know that our two alternative materials were both tested at the same temperature? Perhaps the cheaper product was examined at a lower temperature—and the tensile strength of plastics generally shows a negative coefficient with respect to temperature change! A limited number of plastics are influenced by the relative humidity (or measure of moisture content) of the surrounding air; cellulose acetate is a classic example. If our materials are like cellulose acetate then perhaps the cheaper was tested in a drier condition, which again will tend to give it a higher tensile strength.

For the mechanically minded, it is not difficult to envisage that shape of the test specimen may be important; the stress distributions in the two shapes shown in Figure 1.1 are quite different because (b) has undesirable 'stress raisers' at the sharp 'shoulders'. It is not difficult to imagine that these tend to weaken the specimen and this effect may be so significant as to cause (b) to break at a lower load than (a), thus yielding a lower tensile strength. Can we be certain that our two alternative materials were examined using comparable test specimens?

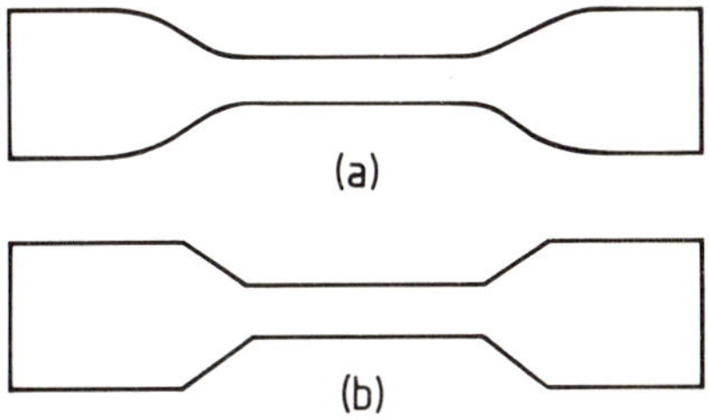

Figure 1.1 Possible shapes for tensile test pieces

Again, many materials behave rather differently according to how quickly they are stressed; the classical example is 'bouncing putty', a silicone composition which behaves like conventional putty when 'worked' (slow stressing), but which rebounds like an elastic solid when dropped on to the floor (quick stressing). Although this particular response to the time factor may be unique, in general plastics materials react to changes in speed of stressing and therefore will give different strengths when the rate of increase of load of tensile specimens is altered. Were our two materials tested under identical stressing conditions?

There are even more influences which can affect the measured value of tensile strength of a moulding composition, but enough has now been said to make the point that the bald statement 'Tensile strength: 8000*X*' is meaningless without detailed qualification of how the figure was obtained. Almost without exception, analogous doubts could be raised for the other properties commonly quoted with abandon.

1.2 GENERAL TESTING PROBLEMS

Throughout this book, when dealing with practically every test method described, it will be necessary to mention pitfalls specific to that technique and peculiar limitations of the data so obtained. There are, however, a number of problems or pitfalls which are pertinent to testing in general and it is desirable and convenient to deal with them at the beginning.

1.2.1 The Need for Standards

From the example of tensile strength discussed in §1.1 it can be seen that, even if a property is quoted with all the experimental test details to put the data into correct perspective, the situation is still far from satisfactory. Two materials, to be compared as before, may have (known) tensile strengths measured by different (known) techniques; we may know the qualitative effects of the changes of temperature, loading rate, etc., between the two techniques, but our knowledge at the moment is certainly not generally sufficient to be able to reduce both sets of data to a common and comparable level or 'norm'—not even sufficient to reduce one variable, still less a whole host of them which as likely as not interact. Thus, knowledge of the test conditions will probably indicate that the data are not comparable, which is a somewhat negative contribution and means that we must ourselves test the one material in a manner identical with the details of moulding, etc., cited for the other; even then we may not feel absolutely confident of our ability to interpret correctly the conditions given for the first set of tests and so we may decide to test both materials by our own preferred method, just for safety.

How can we overcome this sorry state of affairs? Quite simply and obviously by insisting that everyone uses the same methods of test. In this way supplier and user can be guaranteed to speak the same language, thus facilitating trade by giving some definite meaning to sales contracts. This is not only a national matter, for if agreement on methods can be achieved continentally or, better still, universally, then international trade also benefits. The subject of standards, specifications and standards organisations is dealt with in detail in Chapter 2 together with the units used to express the results.

1.2.2 Preparation of Test Pieces

Reference has already been made to the influence on test results of fabrication conditions; when specified, these must be followed rigorously and quoted with the property data obtained. There are further complications, however, particularly in any test where the specimens are examined in other than the complete moulded form. If they have to be cut, sawn, drilled, routed, etc., from sheet or large mouldings, care must be taken to avoid excess heating which can change the degree of cure of, for instance, a phenol formaldehyde moulding or laminated sheet, and may advance it sufficiently to give specimens of different physical properties. Another possible effect is the driving off of absorbed moisture which again can materially affect certain characteristics, particularly electrical insulating properties. Severe overheating tends to degrade the plastics polymer, with detrimental effects in this case, but again yielding unrepresentative test data.

Many properties are influenced by the condition of the surface of the specimen. Notches and scores act as 'stress raisers' and weaken mechanical test specimens, sometimes seriously. Most plastics are hydrophobic or water-resistant and it is not surprising that the finishing of specimens for water absorption tests must be very precisely specified. The importance of polishing the surface when assessing the light transmission characteristics of a plastics material is obvious; surface fissures upset resistance to high voltage ('electric strength'); and so on. Methods of preparing test pieces are discussed in Chapter 3.

By no means all plastics articles have identical properties along their principal axes, in other words there are some which are anisotropic. It is easy to envisage this in a fabric-reinforced laminate because woven fabrics are usually themselves rather stronger in the warp direction than the weft, unless a special weave is incorporated. However, materials that are more homogeneous with respect to composition can also be anisotropic. An injection-moulded bar with the gate at one end is likely to show pronounced orientation of the polymer molecules in the length direction unless special precautions are taken, and the tensile strength, for instance, will be higher along the bar than across it. It is therefore important to state the direction of testing when relevant or, preferably, to examine properties in two, or in some cases three, orthogonal directions.

The influence of preconditioning and test atmospheres has been briefly mentioned and is dealt with more fully in Chapter 4. Quite apart from these ambient influences, however, the very age of a test specimen may be important, even if it has been stored carefully at normal ambient temperatures. Quite apart from the obvious hazards of degradation of light-unstable materials by sunlight, or the hypothetical case of a material so unstable as to be attacked by the oxygen of the air, rather more subtle effects must be considered. Some materials crystallise very slowly at room temperature and, since polymer properties often differ significantly in the crystalline and amorphous forms, it is necessary to test at equilibrium or when the rate of crystallisation is so slow as to be without significance. Similarly, the process of plasticiser–polymer gelation may proceed slowly, even at room temperature, after the normal processing cycle of mixing and pressing, so that plasticised PVC, for instance, must be examined only after some specified minimum time.

1.2.3 Size and Shape of Test Pieces

It might be thought that, since most properties are reduced to units of length, area or volume to yield the basic data of the material, the precise size of test specimens cannot be of importance—it should all come out in the calculation. In actual fact, official test methods generally specify dimensions very closely and one reason for this, in the case of moulding materials, is not hard to see. The properties of a given material moulded in thin section and in very thick section may be quite different. In a thermosetting material the degree of cure is likely to be less in the thicker section, while, with both thermosets and thermoplastics, the extent of locked-in strain is greater in the thicker specimen because of the slower cooling of the centre with respect to the skin. Clearly the measured properties may vary between thin and thick mouldings as a result of effects other than the ratio of the cross-sectional areas.

There is a host of other reasons, however; while it is not appropriate to enumerate them at this early stage, a few illustrations serve as a warning. Reverting to the example of water absorption, which we have seen is mainly a surface phenomenon, it might be thought that logically the property would be calculated per unit surface area. In fact very often it is stated as a percentage w/w! Thus, rigorous control of test piece size is necessary so that the surface area is standard and the fictitious percentages may be compared. The mechanical property of bending (flexural) strength is calculated from the force required to break a bar at its mid-point, the breadth and thickness of the bar and the distance between the outer supports (Figure 1.2):

$$\text{bending strength} = \frac{3FS}{2bt^2} \tag{1.1}$$

Why, then, worry about the precise values of S, b and t as long as they are measured accurately? In fact the classical bending formula above only holds in the case of ideal three-point bending for certain ratios of S/t. Thus, whatever the absolute merits of the property, for comparison of data the test specimens must be essentially identical.

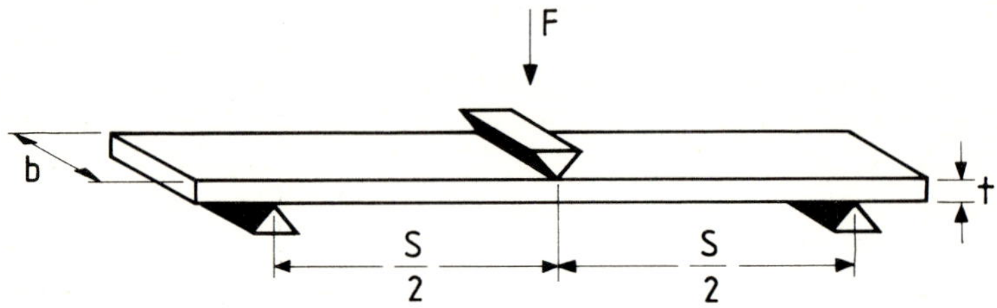

Figure 1.2 Flexural strength measurement—general arrangement

Electrical strength is calculated from the failure voltage (breakdown voltage) of a flat specimen and thickness, often in kV/cm. The thickness must be carefully controlled, however, as this calculated value is by no means itself independent of thickness (Figure 1.3).

These are just typical examples of the influence of test specimen shape on measured properties; there are many more (and many more outside influences

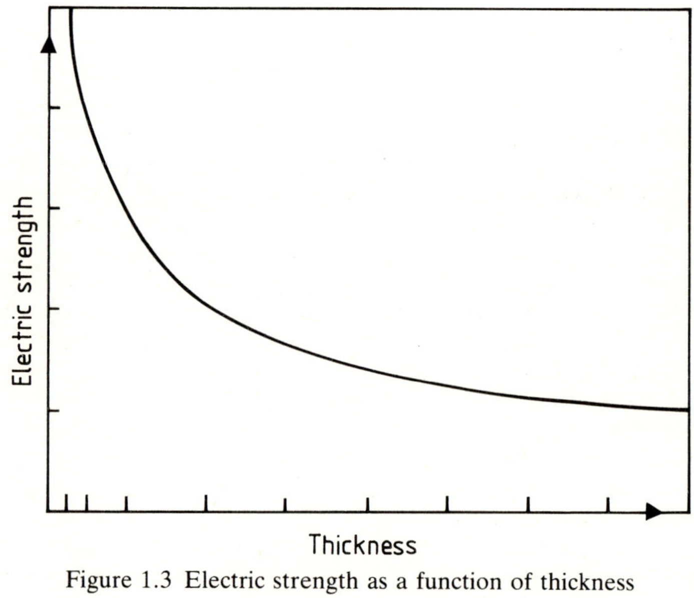

Figure 1.3 Electric strength as a function of thickness

which affect the properties considered above), but these serve to illustrate the extreme care that is necessary in checking all instructions concerning the execution of any given test.

1.2.4 Presentation of Data

For quality control work it is normal to use charts or graphs to display data on a continuing basis so that trends and any unusual departure from the expected levels can be seen at a glance. There is also quite an art in the general manner of recording and reporting test data. Obviously sufficient details of test conditions and other variables must be given. Most up-to-date test method standards give a comprehensive list of the information which must be recorded: do not leave out an apparently trivial item such as the date—you are bound to regret it when the data are referred to a year later. Some indication of the reliability of the results should be given, perhaps by quoting the individual test values or the calculated standard deviation as well as the mean (§1.3).

Reports should be as brief as possible, commensurate with the recording of all the relevant facts. Tabulation is desirable when comparisons are the main object of the work. Opinions differ on the ideal layout of a technical report, but one recommended order of sections is as follows:

Summary
(Contents)
1. Introduction
2. Materials/products examined
3. Test method details
4. Results
5. Conclusions
6. Recommendations
(Appendices)

The summary is for the busy managing director who has not the necessary time to digest all the report; it is not identical with the conclusions because the summary is a précis of the report as a whole. It should not, therefore, contain any data or opinions which are not in the body of the report. The introduction essentially gives the reasons why the work was done and the various happenings which led up to it. It is always important to record precisely what was examined, not just in terms of a type description, but also batch numbers and other relevant data. Sections 3 and 4 are self-explanatory and so also is Section 5, a statement of the conclusions to be drawn from the work in the light of the data of the report and any other information available from experience. A section giving recommendations, for example for further investigations, if often desirable. The managing director (or anyone else for that matter) will not be pleased if the general standard of English is poor!

If the work being reported contains, for example, a mass of numerical data, it is advisable to collect this together into an appropriate number of appendices suitably cross-referred in the text, so that the main body of the report keeps within reasonable proportions; Section 4 can then be restricted to summary tables of mean values, for instance. Graphical representation of results should always be considered as a concise way of reporting data and one which will often assist the drawing of the correct conclusions. Graphs are usually best placed with the appendices.

In certain cases, a section of bibliography or literature cited should be included, between Section 6 and any appendices. In large reports, a contents list immediately after the summary is very useful.

While it is not desirable to dispense with individual style, numbers should be presented so that clarity, elegance and uniformity take precedence. As a simple example, a quarter of a metre should be written 0·25 m and not ·25 m. For guidance on the presentation of numerical values attention is drawn to BS 1957 (1953).[1] The units used should be stated correctly. Care should be taken that symbols, signs and abbreviations are unambiguous and conform to accepted practice. ISO 1000 (1973)[2] and BS 5555 (1976)[3] give information on SI units; symbols and quantities are covered by the ISO 31 series[4] and the BS 1991 series.[5] A glossary of terms and common abbreviations used in the plastics industry is contained in BS 1755[6] and abbreviations for polymers and compounding ingredients in BS 3502 (1967)[7] and BS 4589 (1970).[8]

1.2.5 Limitations of Test Results

When we have taken all these precautions—prepared the test pieces correctly, followed the prescribed method to the letter, reported all the facts clearly without mixing our units—what is the value of the data obtained? Essentially the figures for strength, resistance, etc., derived from our measurements relate only to conditions which simulate precisely those under which we performed our tests, and strictly only to the particular sample tested. There are two aspects to the value, or rather the significance, of our results. Firstly, there is significance in the statistical sense (that dreaded subject which must inevitably rear its head): the result is useless unless we know its significance in terms of the extent to which the sample is representative of the material and what reliance can be placed on the result, taking account of experimental error and material variation. The essential subject of statistics will be approached gently in the next section. The second aspect of significance is the relevance of the result in terms of material or product performance.

It has already been pointed out, and will be further amplified throughout the book, that the property value obtained may vary according to the test method used; furthermore, as a result particularly of pressures of economics and time, user conditions can rarely approach those of normal laboratory test practice. The test data obtained from the routine type of test, while admirable for quality control and perhaps as an indication of service performance if interpreted carefully, will therefore rarely give the designer the values upon which to base his calculations. The largest gap between most published property data and performance behaviour lies in the time-scale: just how long will a given component withstand a certain stress without failing? Trouble-free service over a period of years may be essential, yet tests have lasted but a few seconds or minutes. (One of the greatest drawbacks in the young plastics industry is this lack of case history, though in some traditional industries the necessity of designing more competitively has raised some doubts as to the validity of many time-honoured canons!) Accelerated tests can often provide very useful guides and frequently are the only solution to such demands. However, it must always be remembered that, to produce this very acceleration, some test variable or variables have had to be intensified, for example the temperature raised, the nature of the environment changed or the frequency of stressing increased. These necessary changes may in themselves induce effects which would never occur at the usual ambient temperature, etc., and thus misleading data may result.

It cannot be emphasised too forcefully that all measured properties should be most critically assessed to establish their true significance and applicability. Conversely, any test intended to provide data for purposes other than quality control must be carried out under conditions as near as possible to those in service. Where accelerated tests are used, the dependence of the measured property on different levels of the accelerated parameters must be known if the results are to be extrapolated to service conditions.

1.3 THE CONCEPT OF STATISTICS

'A judicious man looks at statistics, not to get knowledge, but to save himself from having ignorance foisted on him' (Carlyle).

1.3.1 Introduction

No responsible person even remotely connected with science or technology should be without a working knowledge of statistical principles; anyone ignorant of the basic ideas of this subject runs the risk or undertaking his work inefficiently and of being unable to draw the correct conclusion from the results he obtains. An example in plastics testing illustrates the second point quite simply.

Suppose that two laminated sheet materials A and B of equal thickness are to be compared for strength (§8.7). The edgewise Izod values are determined in the usual way, in quintuplicate, with the following results:

	A	B
	1·65 J	1·09 J
	1·37 J	1·48 J
	1·41 J	1·38 J
	1·26 J	1·26 J
	1·60 J	1·39 J
Mean:	1·46 J	1·32 J

On the basis of the two mean values, A may be selected as the material with the better impact strength; however, reflection indicates that the picture may not be so clear-cut. Impact strength tests on laminated sheet, by this method, invariably show a high scatter of results (the spread of individual values) and this is certainly true of the figures above. It is instructive to compare the ranges of the sets of values:

A: 1·26—1·65 J
B: 1·09—1·48 J

There is seen to be a region, 1·26—1·48 J, within which a test figure taken at random might be placed in either set of figures; three out of five of the individual values for A fall in this region and four out of five for B! The essential difference, therefore, lies in two high figures, 1.60 and 1.65, for A and one low figure, 1.09, for B. Are these the all-important values, pointing out the relative merits of the two materials, or are they the 'odd men out', being outside the common region as a result of some chance happening—say, test pieces cut from a fabric-reinforced laminate in such a way that they contain one more or one less thread than the average? We can tell from these figures whether there is likely to be a real or significant difference between the two sets of data only by statistically analysing them to find out whether there is a sound reason for thinking that all ten figures have been drawn from one set of data (i.e. there is no difference between the materials), rather than from two different sets (i.e. A is more impact-resistant than B). The greater part of the study of statistics is to learn to assess the magnitude of chance error in relation to the particular effect under examination.

It is not appropriate to include here a treatise on statistics, but it is worth while to outline some of the basic tools and concepts and make reference to relevant textbooks and standards. A fairly substantial account of statistics in terms of polymer testing is given in a complementary volume, *Physical Testing of Rubbers*.[9] General statistical textbooks are not, of course, written with particular reference to plastics testing. Two volumes edited by Davies[10,11] are good examples of the many comprehensive books available. Extremely good accounts of basic statistical techniques can be found in standards, notably BS 2846, Parts 1–6[12] (ISO equivalents, 2602, 2854, 3207, 3301, 3494 . . .).

1.3.2 Variability

All measurements are subject to variability. We need to know the sources of variability and make a reliable estimate of its magnitude. From this information we may then judge the reliability of our results and hence their significance.

The sources of variability are numerous. In testing a sheet of plastic for tensile strength it is necessary to take into account the intrinsic variability of the material, arising from the fact that it is not perfectly homogeneous, and the variability due to the testing procedure, including test piece preparation, machine accuracy and operator error. This sheet is only a sample taken from the total population of sheets which could be made from a batch of material. Hence, if more than one sheet is tested we introduce variability due to variations in moulding. Different batches may show variations because of changes in the ingredients or the mixing process. Different testing machines, different operators and different laboratories may all introduce further variability.

The tester is always trying to reduce the variability of his results by carefully standardising preparation and testing procedures, frequent calibration of the apparatus and training of operators. Distinction should be made between (a) the random variation due, for example, to real material variations and (b) bias or systematic error due, for example, to a machine being wrongly calibrated.

For many properties the distribution of results will be normal or Gaussian so that, with a sufficient number of results, a plot of frequency (of the measured value) against the value itself will follow the form shown in Figure 1.4.

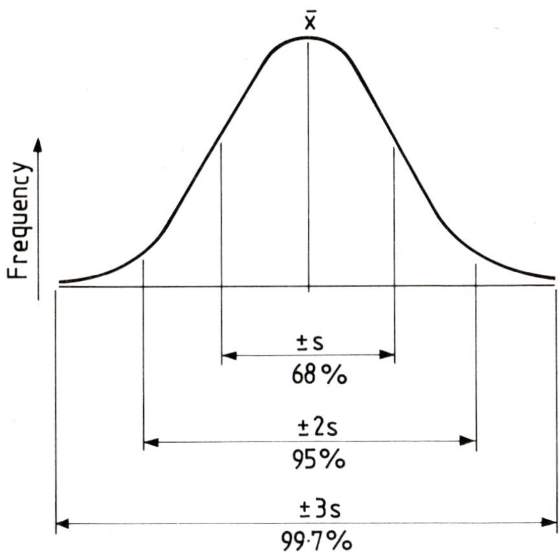

Figure 1.4 Normal frequency distribution curve

The arithmetic mean $\bar{x}$ of a set of n values is defined by

$$\bar{x} = \frac{1}{n}\sum_{i=1} x_i \tag{1.2}$$

i.e. the sum of all the values divided by the number of readings.

Other measures of central tendency are the median, which is the middle value when the results are arranged in ascending or descending order, and the mode, which is the most frequently occurring value. The mean, median and mode are coincident for a normal distribution but not for a skew distribution.

The most useful measure of dispersion or variability is the standard deviation, s, which is calculated from

$$s = \sqrt{\left[\sum_{i=1} (x_i - \bar{x})^2 \Big/ (n - 1)\right]} \tag{1.3}$$

In Figure 1.4 it can be seen that approximately 68 per cent of all values lie within the range $\pm s$ of the mean, 95 per cent within $\pm 2s$ of the mean and 99.7 per cent within $\pm 3s$.

It is often more convenient to use the coefficient of variation, v, which is the standard deviation expressed as a percentage of the mean, i.e.

$$v = \frac{s}{\bar{x}} \times 100 \text{ per cent} \tag{1.4}$$

The variance is the standard deviation squared.

Unexpectedly high or low results sometimes occur, however well the test is organised. Such results should not be rejected unless some physical reason for their occurrence, for example a flaw in the test piece, can be determined. Even a flawed test piece would not be rejected if it was typical of the product or material.

1.3.3 Significance

In the example of impact strength values given in §1.3.1 a difference in mean results for two sheets was recorded. The results would be said to be 'significantly different' if it could be shown that the odds were heavily against the difference being due to chance, i.e. there was a high probability that the difference was real. It is usual to make statistical tests at the 5 per cent or 1 per cent level (i.e. less than 1 in 20 or 1 in 100 probability of the difference being due to chance). Significance is dependent on the magnitude of the difference and also on the amount of variability, so that for a test showing a wide scatter of results a relatively large difference between means would be needed to be proved significant. In the same way the significance or the reliability which can be attributed to a mean result depends on the degree of scatter of results about the mean. It follows that no reliability can be placed on a result without a knowledge of the variability.

The usual method of expressing the precision of an estimate is by calculating the confidence limits, i.e. the limits within which there is a given probability that the true value will lie.

The confidence limits of the mean are given by $\bar{x} \pm (ts/n)$ where s is the standard deviation, n the number of test pieces and t is a factor which varies with the degree of confidence and the value of n. Published tables of values of t are available.[9–12]

When there are two sets of results and it is of interest to know whether their means are significantly different, the least significant difference is calculated. Details of the calculation are given, for example, in the works referred to above.

The significance of the components of variability arising from difference sources can be isolated and assessed by the technique of analysis of variance. Accounts of this technique can be found in statistics textbooks. In, for example, a comparison of natural and artificial ageing results it is of interest whether there is a significant relationship between the two: i.e., do they correlate? Correlation is assessed by regression analysis which consists of estimating the best 'fitting' line through the plotted pairs of two variables. Again, the necessary calculations are covered in the references given above.

1.3.4 Design of Experiments

It should by now be apparent that it is virtually essential to have some grasp of the basic principles of statistics in order to make sensible judgments about one's own or other people's test results. However, no amount of clever analysis will compensate for poor experiment design and planning. The usefulness of a series of tests is increased and the difficulty of analysing the results eased if the experiment is planned carefully and properly in the first place.

Much of experiment planning is common sense. The object is to eliminate as many unwanted sources of variability as possible while not missing any of the important factors which should be studied. A simple example of common sense is to test aged and unaged samples on the same occasion to avoid variability due to different operators on different days.

It is not surprising that statistical principles play as great a part in experiment design as they do in the analysis of results. The test must be repeated often enough to give the required significance of results. The differences in materials or methods that are under investigation must not be masked by changes in uncontrolled variables. The experiment plan must be such that statistical analysis can be sensibly carried out on the results, i.e. the experiment and the method of analysis are planned together.

Each experiment must be designed to suit the particular circumstances and requirements but there is a considerable number of 'standard' approaches which can be used. Once again the reader is referred to works giving a fuller coverage of the subject.[9,11,13,14]

1.4 QUALITY CONTROL

1.4.1 General Considerations

The manufacturer needs some form of quality control:

(a) in order to produce a consistent product, satisfy his customers and meet contractual obligations;
(b) in order to monitor the manufacturing process and to help to identify, and thus rectify, faults at the earliest possible stage.

The frequency and comprehensiveness of quality testing must be governed, to a large extent, by commercial considerations. The only way to be absolutely certain that nothing below standard gets past inspection is to examine 100 per cent of the product, which may be feasible if a simple visual inspection is adequate. At the other end of the scale, if the only usable test is destructive in nature the cost is very high indeed!

In general, it is necessary to take samples from production, and perhaps from deliveries of incoming materials, and test these on the basis that they are representative of the total quality. Sampling is of great importance in all testing work and is given further attention below. For routine quality control, samples must be taken repetitively in time and there is need for a long-term sampling plan and a continuous method for assessing the results.

Quality control or quality assurance is concerned with maintaining the quality of products to set standards. This embraces the control of incoming materials, the control of manufacturing processes and the inspection and testing of the final product. Hence, the quality engineer is concerned with testing, specifications, statistical control schemes, inspection and the monitoring of production methods. Quality control is a discipline in which testing plays but a part, and is dealt with in depth in various specialist works, e.g. Juran *et al.*[15] (this is very comprehensive), Enrick[16] and Hopper.[17] Ultimately it is human fallibility which limits the control of quality and Drury and Fox[18] have considered the subject from this angle. Reliability testing is covered by McCormick[19] and control chart analysis is the subject of an ASTM manual.[20] A number of very useful British Standards have been published in the field of quality control.[21–31]

A particularly useful method of recording production property data is by the use of quality control charts. For example, with any one property regularly checked, a graph is constructed, with the property as ordinate and time of sampling as abscissa. This visual aid allows almost immediate identification of trends or gross variations. The greatest value, however, is derived when the scheme has been operating long enough for reliable estimates to be made of the grand mean (the average figure derived from all measurements to date) and the standard deviation of results (a statistical measure of the overall variation between samples). From the latter and from published tables the limits can be calculated between which 95 per cent, 99·9 per cent, etc., of all data should lie if everything is under control. A horizontal line is drawn on the quality control chart at the value of the grand mean and pairs of lines are drawn, one either side of the mean line, to indicate the 95 per cent, etc., limits (the 95 per cent lines, for example, are called the one-in-forty limits because, on average, not more than one in forty measurements should stray above or below them). With newly acquired data plotted day by day the charts can be used to distinguish a real production failing from the chance low result that is within normal statistical variability.

A number of variants may be operated along these general lines; for example, the range of results for each set of readings, such as the values pertaining to a particular shift, may be plotted as ordinates and limits calculated to indicate whether the variability within, say, each shift is under control or a particular shift or other fixed period is exhibiting excessive variation.

The selection of the actual tests to employ for quality control must be a compromise between scientific interest and economic necessity. If the control can be exercised adequately with a non-destructive test or tests, only the actual cost of the latter is relevant for no product is actually destroyed. This, unfortunately, is rarely the case. It is, however, often possible to reduce to a reasonable minimum the number of different types of test carried out regularly, even though full specification tests must be carried out at less frequent intervals.

1.4.2 Sampling

The whole point of efficient sampling is to select small quantities which are truly representative of the much larger whole. The majority of standard test pieces are quite small, weighing about 50 g or less; usually five is the maximum number used in any one property measurement. Therefore the combined mass of test pieces is very small in relation, say, to the contents of the normal sized sack, perhaps 25 or 50 kg, and far smaller in relation to the production batch from which the selected sack was taken for sampling and testing. Similarly, small test pieces from a large sheet represent a very small fraction of the total area, and the sheet is only one taken from perhaps hundreds produced.

The ordinary fabric-filled phenol-formaldehyde moulding powder is an example of a mixture of 'fines' and coarse agglomerates; in the vibration of transportation the particles tend to segregate, the finer ones to the bottom. There are devices for taking representative samples over the whole depth, for instance a type of soil tester which consists of a half cylinder, open down its length and fitted with steps. Another type, specified at one time for sampling cellulose acetate moulding materials, is a hollow cylinder with an Archimedean screw running down the centre.

A general technique for sampling powders, pellets and granules is that of 'quartering'. The sack contents, for instance, are tipped on to a clean flat surface to form a conical pile, which is then flattened from the centre outwards and divided into quarters. A pair of opposite quarters is discarded and the remainder is formed into a new conical pile, by taking shovelfuls alternately from each quarter; care must be taken in emptying the shovel that the materials run uniformly down the surface of the cone. The quartering process is repeated as many times as are necessary to yield a sample of the requisite size.

There is an ASTM guide to sampling plastics.[32]

Sheets, film, rod, tube and pipe should be sampled randomly by division into a numbers of smaller areas or lengths as appropriate; a table of random numbers is useful in selecting the portions for test without bias. Alternatively, for some purposes it is advantageous to use a particular experiment plan to select the position of test pieces.[9] If the sheet is not isotropic the direction of test pieces relative to the axes of the sheet is important.

The selection of products from production should be randomised and, in particular, care must be taken that the sampling procedure is not biased by the sampling time coinciding with some regular occurrence, for example a shift change. The number of test pieces taken obviously affects the statistical significance of the results, but in a continuous quality control scheme it may be preferable to use fewer test pieces at each sampling time and sample more often. A number of product or material specifications now include sampling techniques appropriate to the particular product.

1.4.3 Calibration

Testing laboratories are as much in need of quality control as is the production unit. The test procedures must be controlled, the operators trained and the equipment calibrated. Good test method standards go a long way to ensure that the test procedures are carried out reproducibly but, particularly in the areas of test piece preparation and conditioning, there is often need to supplement the standards with in-house standard procedures which should be written down, not passed on verbally. Operators should be examined for competence in performing each test and a formal record kept of their training and examination.

It is self-evident that apparatus should be correctly calibrated. This is generally appreciated for such items as the force scales of tensile machines and often the manufacturer offers a calibration service. There is also a British Standard[33] dealing specifically with the requirements and accuracy of such machines. It is the little things that get overlooked: the calibration of dial gauges and thermometers, the check that pointers have not been bent or that dirt is not causing undue friction. Many of these seem trivial but if errors from such sources were not present these would not be the poor inter-laboratory agreement so often witnessed in practice. Errors seem to arise from the most unlikely sources and a good defence is a suspicious mind and a passion for detail. A good insurance is a formal maintenance and calibration scheme for all apparatus, including written records of all checks and measurements and clear labelling of the apparatus as to its state of calibration.

REFERENCES

1. BS 1957 (1953) *Presentation of Numerical Values.*
2. ISO 1000 (1973) *SI units and Recommendations for the Use of their Multiples and of Certain Other Units.*
3. BS 5555 (1976) *SI Units and Recommendations for the Use of their Multiples and of Certain Other Units.*
4. ISO 31 (in several parts) *Quantities and Units.*
5. BS 1991 (in several parts) *Letters Symbols, Signs and Abbreviations.*
6. BS 1755 (in several parts) *Glossary of Terms Used in the Plastics Industry.*
7. BS 3502 (in several parts) *Schedule of Common Names and Abbreviations for Plastics and Rubbers.*
8. BS 4589 (1970) *Abbreviations for Rubber and Plastics Compounding Materials.*
9. Brown, R. P. (1979) *Physical Testing of Rubbers*, Applied Science, Chapter 3.
10. Davies, O. L. (ed.) (1957) *Statistical Methods in Research and Production*, Oliver & Boyd.
11. Davies, O. L. and Goldsmith, P. L. (Eds.) (1972) *The Design and Analysis of Industrial Experiments*, Oliver & Boyd.
12. BS 2846 Parts 1–6, *Guide to the Statistical Interpretation of Data.*
13. Cochran, W. G. and Cox, G. M. (1956) *Experimental Designs*, Wiley.
14. Cox, D. R. (1958) *Planning of Experiments*, Wiley.
15. Jurna, J. M., Gryna, F. M. and Bingham, R. S. (1975) *Quality Control Handbook* 3rd edn, McGraw-Hill.
16. Enrick, N. L. (1962) *Quality Control*, Machinery Publishing.
17. Hopper, A. G. (1969) *Basic Statistical Quality Control*, McGraw-Hill.
18. Drury, C. G. and Fox J. G. (1975) *Human Reliability in Quality Control*, Taylor & Francis.
19. McCormick, C. (1977) *The Statistical Reliability of Industrial Products*, Continental Carbon Company.
20. ASTM Manual STP 150 (1976) *Presentation of Data and Control Chart Analysis.*
21. BS 600 (1935) *Application of Statistical Methods to Industrial Standardisation and Quality Control.*
22. BS 1313 (1947) *Fraction Defective Charts for Quality Control.*
23. BS 2564 (1955) *Control Chart Technique when Manufacturing to a Specification with Special Reference to Articles Machined to Dimensional Tolerances.*
24. BSI Publication PD 6452 (1970) *Guide to Inspection Procedures.*
25. BS 2635 (1955) *Drafting Specifications Based on Limiting the Number of Defectives Permitted in Small Samples.*
26. BS 4891 (1972) *A Guide to Quality Assurance.*
27. BS 5179 Parts 1–3 (1974) *Guide to the Operation and Evaluation of Quality Assurance Systems.*
28. BS 4778 (1979) *Glossary of General Terms Used in Quality Assurance.*
29. BS 6000 (1972) *Guide to the Use of BS 6001.*
30. BS 6001 (1972) *Sampling Procedures and Tables for Inspection by Attributes.*
31. BSI Drafts for Development DD 10—DD 16, *Guides to Reliability of Engineering Equipment.*
32. ASTM D1898 (1968) *Recommended Practice for Sampling of Plastics.*
33. BS 5214 Part 1 (1975) *Tensile, Flexural and Compression Machines.*

Chapter Two

Standards and Standards Organisations

2.1 STANDARDS—TEST METHODS AND SPECIFICATIONS

2.1.1 Introduction

It does not take much imagination to realise that if there were no standardised test methods trade would be severely impaired, progress would be stunted and chaos would ensue. Fortunately, technologists appear to have a strong sense of order and from the early days of the industry have supported the development of standard procedures and the use of these in product specifications; not that the effort put in should make us complacent, as our standards for plastics are by no means complete.

To avoid misunderstanding over terminology it is as well to note that all the British Standards Institution documents are called 'standards' and the word 'specification' is reserved for those standards which specify minimum requirements for materials or products. Other types of standard are 'methods of test', 'glossaries of terms' and 'codes of practice.' It follows that a specification may refer to several methods of test and that a commercially written specification can refer to nationally standardised test methods.

In terms of trade it is ultimately specifications that are important, with test methods acting as building blocks. For this reason it has made chronological sense for the test method standards to be developed first and indeed this has generally been the case in practice. Now that test methods are fairly well developed at the national and international level it can be argued that rather more of the effort available should be put into specifications, especially in the current economic climate. However, it will be apparent throughout this book that considerable activity in test method standardisation is still taking place and that there is room for improvement. Nevertheless there is no case for the development of standard test methods which are of academic interest only and unlikely to be generally used. Some of the problems facing the writer of performance specifications have been discussed by Manning,[1] for example, and discussion of the state of standardisation in the plastics industry by Reinhart[2] is still relevant.

2.1.2 Test Methods

In this book we are concerned with methods of test and only indirectly with specifications. Leaving aside for the moment the various sources of standard test methods, one can recognise different styles or types of published methods. This is not a matter of accident but rather one of progression, the most obvious yardstick being the number of options left open to the user. In the simplest case a particular apparatus is specified, one set of mandatory test conditions given and no choice allowed as to the parameters to be reported; this is the form in which the specification writer needs a test method. Unfortunately for those who want a quiet life, many national and international test methods have become rather more complex. This is partially a result of compromise but more importantly because the measurements being described are not intrinsically simple and the method will be required for a number of different purposes and probably for many different end-products. The specification user must therefore select the particular conditions which best suit his individual purposes. In practice he frequently fails to do this either because he omitted to read the standard carefully enough or because his understanding of it was somewhat limited. As more advanced concepts are being introduced into test method standards so there is an increase in the practice of including explanatory notes, which unfortunately do not always achieve their desired aim.

We can conveniently distinguish three different purposes for which a standard test method is used: (a) purely for quality control; (b) as a performance requirement; (c) for development purposes. In the first case the prime consideration is that precisely the same procedure is always used and also that this procedure is relatively simple and rapid. The test conditions may be completely arbitrary but one set of conditions and one set only is required. If the test is intended, apart from a quality control function, to be a measure of the performance of the product then test conditions will be chosen which have some relevance to the product end use. For development work it is highly probable that a series of conditions will be wanted in the hope that data of use in designing future products will be realised. Committees preparing standard test methods have all these possibilities in mind, with the result that the user of the standard needs to understand the subject sufficiently well to make an intelligent selection of conditions to suit his particular purpose.

The following example may not stand up to too close an inspection but serves to illustrate the point. If a test for resistance to liquids is considered, one would expect a quality control procedure to involve one liquid at one temperature for a relatively short time. The liquid might be a standard fuel such as Liquid B as referred to in ISO Draft Proposal 175, *Determination of the Effects of Liquid Chemicals*, and the test might involve 24 h exposure with weight change being measured. If the testing was intended to have a performance function then a liquid met in service would be used, for example commercial petrol, and testing continued long enough for equilibrium absorption to be reached. Apart from weight change, other relevant physical properties would be measured before and after exposure. For development purposes testing would be further extended to cover a number of fluids, each tested at several temperatures. An international or national standard would attempt to cater for these and other possibilities and would hence include a choice of measuring procedure, test temperature, duration of exposure, properties to be monitored and test liquids. Preferred test parameters might be indicated for use when there were no outside factors influencing the choice.

This is not the place to discuss in any detail what should or should not be included in standard test methods or how they should be written. The quality and the style of those in current existence varies very considerably but it is possible to detect certain general trends in recent years. Standards have become more involved as more factors which cause variability are identified and control of these is specified. At the same time some apparatus has been specified in a more general way, stating what its performance must be without restricting its design or construction to any particular form. This can only be done when all the important parameters have been identified. Incidentally, one way in which causes of experimental variability are uncovered is by means of controlled inter-laboratory comparisons undertaken by standards committees. Results of these are rarely published, but, for example, details of some studies in the USA on rubbers were given at an ASTM symposium.[3] Standards also become more complicated as the underlying principles of the property being measured become better understood and as more meaningful results are demanded by product designers. It is very much a matter of opinion as to whether we at present have taken this progress too far or not far enough.

2.2 ORGANISATIONS PRODUCING STANDARDS

Generally, the sources of standards form three groups:

International organisations
National organisations
Individual companies

Although, in terms of trade, it is commercial specifications which are most important, it appears proper and logical to discuss these groups in descending order of scope, i.e. from the international downwards. In practice, of course, a new test method usually proceeds in the opposite direction from humble beginnings in particular laboratories via national recognition to international status, by that time having become much modified.

2.2.1 International Standards

The ultimate state of unity would be for all countries to be using the same standards. This would obviously be of great value in smoothing the course of international trade and make it easier for technologists to exchange technical information. It is also a very ambitious concept that the countries of the world can compromise on their national procedures and overcome the very great difficulties of language in a field where language is the most important tool of trade.

The International Standards Organisation
In most fields, including plastics, the principal body attempting to achieve the ideal of international agreement is the International Standards Organisation (ISO) which is hence our most important organisation in the standards field. The ISO in its present form is not a very old organisation, having been formed in

1946 after previous attempts at setting up this sort of body had met with little success.

The work of ISO is administered by a permanent central secretariat which has headquarters at 1, rue de Varembe, 1221 Geneva 20. It has as members more than 60 national standards bodies, one body per country. Apart from central committees concerned with planning, certification, etc., the technical work of ISO is carried out by technical committees each relating to a particular area of industry. The secretariat of each technical committee is held by a member country and each member may join any committee either as a participating (P) member or observing (O) member. The P members have voting rights at committee meetings. The choice of P or O member depends on the country's interest and the finance available.

The committee for plastics is TC61 with the United States of America holding the secretariat; rubber is covered by TC45. Most ISO technical committees operate with an infrastructure of subcommittees, workings groups and task groups, although in TC61 subcommittees have been introduced only relatively recently. TC61 normally meets once per year, member countries acting as hosts more or less in rotation. The delegates to the technical committee are decided by the national standards body.

There are at present nine subcommittees in TC61 as listed below, the secretariat for each being taken by a member country.

SC1 Terminology
SC2 Mechanical strength properties
SC4 Burning behaviour
SC5 Physical–chemical properties
SC6 Ageing, chemical and environmental resistance
SC7 Preparation of test specimens
SC9 Specifications for materials for moulding and extrusion
SC10 Cellular materials
SC11 Semi-finished and finished products

Within each subcommittee there is a number of working groups and/or task groups, each with a leader to take care of the details of various aspects of their subject. For example, in SC2 there is a task group for impact tests, another for tensile properties and so on. The subcommittees of TC61 were until recently called working groups and it must be expected that the infrastructure of the technical committee will undergo further change and development. At present, it can be seen that the subcommittees of interest to physical testing are TCs 2,4,5,6, and 7 but other subcommittees have of course an interest in tests to be included in specifications, particularly specialised product-oriented methods.

The order of progress towards an International Standard is that, after consideration at task or working group and subcommittee level, a document is proposed to the plenary session of the technical committee for circulation to members for postal vote as a draft proposal (DP). If it is approved, balloting and commenting take place and the votes and comments are considered at the next meeting of the technical committee. If agreement is reached the revised document is again circulated, this time to all ISO member countries, as a draft international standard (DIS). The comments and votes on this are considered in the following year and if approved at the plenary session the final revision is sent to the central secretariat for approval by the ISO council and publication as an International Standard. It is immediately obvious that this process is slow. Although certain short cuts are possible, it is not unusual for a document to pass through a second draft proposal or second draft standard stage if agreement proves difficult.

ISO standards were first published in 1972; before that time 'ISO Recommendation' was the title used. It is not obligatory for the ISO standards to be incorporated into a national system but obviously the whole aim is a little defeated if this is not done. The BSI takes a very positive attitude in this direction, following the dictum 'Do it once, do it right, do it internationally'. Wherever possible British Standards at least agree with ISO standards technically and the aim is to reproduce them verbatim. To become an ISO standard requires that a document receives the approval of 75 per cent of the members casting a vote. This fairly high percentage helps to ensure that unsatisfactory documents do not get through. Normally, if the UK gives a positive vote at ISO that standard is then adopted as regards technical content into the BSI system.

It is difficult enough to reach agreement on a standard procedure within one country and the problems internationally are considerably greater, not being helped by language difficulties—the official languages of ISO are English, French and Russian. Therefore the slow pace of production of an ISO standard is hardly surprising. Differences between national documents and ISO methods are often not a matter of disagreement, but simply that the two time-scales are different and things have got out of step. However, there are encouraging signs that progress towards more complete rationalisation is being achieved. There are at present some 156 ISO standards published in the plastics field. These are listed in the *ISO Year Book* and additions during the year are published in the British Standards publication *BSI News*. In the UK ISO standards can be obtained from the BSI and in other countries through the national standards body.

Other International Standards

In the electrical field the International Electrotechnical Commission (IEC) performs the same function as ISO. The work of this body is of interest where plastics are used in electrical insulation, etc. As regards electrical test methods for plastics, both ISO and BSI adopt the basic procedures and principles standardised by IEC.

Britain has particular interest in the more limited scope of European standardisation.[4] The European Committee for Standardisation (CEN) was founded in 1960 and comprises the national standards bodies of EEC and EFTA countries. CENELEC is the equivalent body in the electrical field. To many people the concept of European Standards is an unnecessary complication, it being argued that there is no need for any activity in between ISO and the national bodies. However, the work of CEN is likely to assume importance where documents are drawn up at the specific request of the EEC in its programme for the removal of technical trade barriers. Such standards will be used in EEC directives which, if ratified, are binding on the community members. CEN standards are voluntary, being adopted only by approving countries.

There are also fundamental differences in the voting and approval systems in CEN and EEC and it is apparent that those concerned with standardisation will have to watch the European developments carefully if unsatisfactory standards and legislation are not to be forced upon us.

There are many other international organisations concerned with standards and a list of these has been provided by the BSI[5] which makes clear the confusing abbreviations of titles in use (see also Appendix D to this chapter). Unfortunately changes are always taking place and such lists can quickly become out of date.

2.2.2 National Standards

Although generally each country has one principal standards organisation which provides the official membership of ISO, other organisations can be issuing standards at national level. It is usual to include government departments in this category. It is not practical, and indeed not necessary, to consider here the national standards bodies of all countries; a list of ISO members is given as Appendix A. The operations of the British Standards Institution (BSI) will be described in some detail, which, apart from being of particular interest to those trading in Britain, serves to illustrate how the process of generating standards at a national level can be undertaken. It must not be assumed that other countries operate in even roughly the same format. It should be noted, however, that the BSI is one of the longest established and most highly rated of national standards bodies.

The British Standards Institution
The BSI was formed in 1901 and has now developed to the point where it covers an astonishing range of subjects from virtually all branches of industry. Apart from its main function of producing standards it also operates a quality assurance establishment at Maylands Avenue, Hemel Hempstead. From here the BSI's Kitemark certification scheme is operated and much of the testing associated with this carried out.

The BSI receives government support but also raises a significant part of its income from membership fees and sales of publications. The Hemel Hempstead centre is financially self-supporting. Membership is open to virtually anyone, various categories of organisations being defined for the purpose of computing membership fees. It is in fact rather difficult to keep up to date on standards matters without being in membership. Details of new standards, amendments and articles on standards matters generally are published monthly in *BSI News* which is circulated to members. There are also annual editions of the *British Standards Year Book* which lists all the British Standards available and is issued to members and may be purchased by non-members. It is appropriate to mention here BS O (1975)[6] *A Standard for Standards* which gives in considerable detail an account of BSI structure, procedure and editorial practice.

Structure. The permanent staff of BSI, administrative, technical and editorial, operate under a Director General, with the overall responsibility for policy being vested in an executive board. Matters of policy relating to convenient groups of industries (e.g. building, engineering) are dealt with by Divisional

Councils and within each of these divisions is a number of Industry Standards Committees. Of particular interest to us is the Plastics Industry Standards Committee PLC/– which operates within the umbrella of the Chemical Divisional Council.

We now have an introduction to the numbering system used for BSI Committees. The PL refers to 'plastics' and the C to 'chemical'. (The Codes of Practice Council committees have a four-letter code.) PLC/– has responsibility for authorising the initiation of standards projects and for final approval of drafts before publication. The preparation of standards is carried out by technical committees and subcommittees reporting to the industry committee. Hence we have, for example, 'PLC/7 Styrene plastics' and 'PLC/9 Plastics pipes and fittings.' The BSI supplies the secretariat for technical committees and the members are nominated by industry, government departments and research associations. The industry representatives are usually nominated by trade associations and not the individual company, although individual experts are sometimes co-opted.

Testing of plastics is the concern of PLC/17 which currently has five active panels. Panels rather than subcommittees are used purely for administrative convenience. These *ad hoc* panels, despite being almost off the end of the numbering system, are very important and PLC/17 would get very bogged down without them. PLC/36 is concerned with the presentation of plastics design data and hence formulates and suggests test procedures suitable for the generation of such data.

Other committees which should be mentioned in the context of testing are the PLC/RUC series which are joint committees between the plastics and rubber industries. PLC/RUC/10 deals with fire tests, PLC/RUC/6 with accuracy of test machines and PLC/RUC/9 with electrical tests. Cellular materials are also covered by means of PLC/RUC committees. Unfortunately systems are never as simple as we would like; specialised tests may be considered in product committees and not all products containing plastics are covered in PLC committees. For example, coated fabrics, including testing, are the subject of FBT/13. Lastly, the subcommittee PLC/–/1, reporting to the industry committee, exists to coordinate international work and hence is the prime link with ISO affairs.

Preparation of British Standards. Consideration of an initial draft is carried out in technical committee, subcommittee or panel as appropriate. This initial draft may have come from one of a number of sources; for example, it may be based on work carried out by one of the bodies represented on the committee. The draft works upwards to the main committee, being presented there as a private circulation. When agreement is reached the document is circulated as a draft British Standard (DC) to industry for comment. These drafts have no official standing and nowadays are always prepared after careful consideration of the position in ISO. The comments are considered by the technical committee and the amended document is passed to the industry committee for approval. The BSI editorial department then vets the document for consistency of layout, etc., before it goes for printing. This procedure is generally efficient and quicker than the ISO procedure but it must be admitted that unfortunate delays do occur, often as a result of too great a pressure of work on the BSI resources.

The technical committee also has the responsibility of keeping its published standards up to date, which for a major revision would mean a complete re-issue and would be dealt with in the same manner as a new standard. Relatively minor changes are dealt with by amendment slips, the publication of which, as of new standards and revisions, is announced in *BSI News*.

Each standard is given a separate number although there may be more than one part to a standard, each issued separately. The year of publication is added so that different editions of the same standard can be recognised. The *British Standards Year Book* gives a complete list of standards in numerical order, but unfortunately not grouped into subjects. For a complete picture of the up-to-date situation it is necessary to consult the year book and the subsequent issues of *BSI News*, remembering that the year book may be 18 months out of date.

Other British National Standards

Standards or specifications issued by individual companies are not considered to be of national status, however large or multi-national the concern may be. Specifications issued by local authorities and nationalised industries would be in the same bracket. The Rubber and Plastics Research Association, the British Plastics Federation and the Plastics and Rubber Institute do not issue standards.

There are, however, certain independent organisations with national influence which have issued standards that may in part have plastics interest: for example, the Shoe and Allied Trades Research Association and the Society of Motor Manufacturers and Traders, whose standards are now being absorbed into the BSI system.

Government departments, although contributing a great deal to the work of BSI, produce a large amount of their own standardisation. The reasons for this are really similar to those which apply to individual companies—they are unable to wait for the BS system or they have specific requirements unique to themselves. This latter reason applies particularly to the armed forces. Unfortunately the various standards issued by the government departments are rather confusing to outsiders, as indeed are the departments themselves, which appear to undergo frequent change of title or scope. Sources of government standards are given in Appendix B.

USA Standards

American standards, particularly those of ASTM, are widely used in many parts of the world and indeed many companies adopt wholesale methods from ASTM under their own name. The national standards system in the USA differs in many respects from the British; in particular, the organisation which publishes the standards of most interest, the American Society for Testing and Materials (ASTM), is not the official national standards body having ISO membership. That function is fulfilled by the American National Standards Institute (ANSI). Sources of other USA standards of interest are given in Appendix C.

American National Standards Institute (ANSI). ANSI is the premier USA standardisation body and is the counterpart of BSI in being the official ISO representative. It was until a few years ago known as the American Standards Association. ANSI does not itself write standards but approves as American

Standards those produced by ASTM and other similar organisations. Not all ASTM standards are approved in this way and approval may take place years after the introduction of the standard by ASTM.

American Society for Testing and Materials (ASTM). Apart from work at ISO, it is ASTM that most people in the polymer industry think of as representing American standards. ASTM has a membership drawn from similar sources to those of BSI and operates through more than 120 technical committees which in turn have a subcommittee structure.

ASTM standards can be obtained individually but are more usually seen as ASTM books, each being a collection of standards covering a particular subject or related group of subjects. The books are revised annually and although some standards remain unchanged for years there is always a significant amount of new or revised matter. Hence it is advisable to use only the current edition. Although this is rather expensive it is easier for the user than keeping British Standards up to date by studying *BSI News*. There are currently some 50 parts to the annual book of ASTM standards, those concerning plastics in particular being Part 34 *Pipes and fittings*, Part 35 *General test methods* and Part 36 *Materials*. The ASTM is active in the technical field apart from purely producing standards. It organises conferences and publishes numerous books and reports as well as the journal *Standardization News*.

German Standards

The reason for including a mention of German Standards as opposed to those of France, Russia or any other country in an English language book is that DIN standards are those most commonly met with in Britain after BS and ASTM.

The principal German standards organisation, the Deutscher Normenausschuss (DNA), was founded in 1917. It operates rather similarly to the BSI in that it has subscription-paying members from industry, trade associations, etc., and is independent of individual pressures. It has well over 100 technical committees and some 2000 subcommittees, membership of which is honorary with representatives from all interested areas.

The DIN series of standards, of which there are many thousands, is grouped into sections numerically. For example the 530 series deals with materials testing and the 210 series with rubber and plastics. The standards are published separately and are all listed in the *DIN Standards Year Book*. Many DIN standards are also published in English and Spanish and a few in French. There is a separate annual publication which lists those standards which have been translated. The DNA also publishes the journal *DIN-Mitteilungen*.

The use of DIN standards is generally voluntary unless they are incorporated into legislation. The DIN mark on a product signifies that it complies with the relevant DIN standard and manufacturers can be forbidden to use the mark if they do not stick to the rules. The mark is hence similar in concept to the BSI Kitemark, although the policing system appears to be less rigorous.

Other National Standards

It is not possible to discuss all of the national standards bodies but it is necessary to be able to identify the source of a standard from the abbreviations, such as BS and DIN, which are used. Taking the whole world and including government standards, the total becomes enormous and very confusing. The list given in Appendix D is by no means exhaustive but should prove useful.

Which national standards are most in use in any laboratory will depend on the company's trading partners. Information is available from the national bodies listed in Appendix A.

2.2.3 Company Standards

There must be literally millions of company standards in existence. Although they have relatively little significance in a national or international sense, they are the basis of many commercial contracts and hence are perhaps the most important standards of all. Among their number are some of the best examples of standardisation and also some of the worst; sadly, the worst appear very frequently.

Using a commercial standard is like using any standard: the user must be careful that he has the lastest edition and that he has read it very carefully and missed none of the detail. A common fault in commercial standards is that rather a lot of detail is missing, for example there may be insufficient information in a test method to ensure that it is carried out correctly. In this case, all that one can do is to talk to the originator of the standard.

It would save a great deal of pain and confusion if those writing commercial specifications would wherever possible use published standard test methods, preferably those of ISO. Special tests are often needed, but there is no point in inventing one's own procedure for a straightforward test which has been well standardised. Perhaps the origin of a lot of the trouble is that in some cases those writing specifications are not well versed in standardisation outside their own organisation and that many engineers have a poor understanding of plastics and their properties.

2.3 UNITS

It should be unnecessary to state that SI units will be used—that is assumed to be the case. The universal adoption of SI units virtually eliminates the need to include here a section on units because there is no question of conversions or explanations of obscure systems. All is not quite perfect, however, and it seems appropriate to make reference to relevant information.

The basic reference is to ISO 1000 (1973)[7] and its British equivalent BS 5555 (1976)[8] which detail all the units, multiples and sub-multiples to be used. PD 5686 (1972)[9] is very similar but with some extra information, while BS 3763 (1970)[10] sets out the basic features of SI. Certain special considerations apply in any particular industry and both ISO committees TC61 and TC45 have actively considered the subject, but their documents listing special cases relevant to plastics and rubber respectively are not generally available at present. The British Standards for conversion factors are BS 350 Parts 1 and 2 (1962)[11] and PD 6203 (1967).[12]

APPENDIX A: NATIONAL STANDARDS BODIES—ISO MEMBERS

Member Bodies

Albania/Albanie (BSA)
Byroja e Standarteve
Prane Komisionit te Planit te Shtetit
Tirana

Algeria/Algérie (INAPI)
Institut Algérien de Normalisation et de Propriété Industrielle
5, rue Abou Hamou Moussa
BP 1021—Centre de Tri
Alger

Australia/Australie (SAA)
Standards Association of Australia
Standards House
80–86 Arthur Street
North Sydney—NSW 2060

Austria/Autriche (ON)
Österreichisches Normungsinstitut
Leopoldsgasse 4
Postfach 130
A–1021 Wien 2

Bangladesh (BDSI)
Bangladesh Standards Institution
3-DIT (Extension) Avenue
Motijheel Commercial Area
Dacca 2

Belgium/Belgique (IBN)
Institut Belge de Normalisation
Av. de la Brabançonne, 29
B-1040 Bruxelles

Brazil/Brésil (ABNT)
Associacäo Brasileira de Normas Téchnicas
Av. 13 de Maio, n° 13–2° andar
Caixa Postal, 1680
CEP: 20.000—Rio de Janeiro-RJ

Bulgaria/Bulgarie (DKC)
State Committee for Standardization at the Council of Ministers
21, 6th September Str.
Sofia

Canada (SCC)
Standards Council of Canada
International Standardization Branch
Meadowvale Corporate Centre
2000 Argentia Road, Suite 2–401
Mississauga, Ontario
L5N 1V8

Chile/Chili (INN)
Instituto Nacional de Normalización
Matias Cousino 64—6° piso
Casilla 995—Correo 1
Santiago

China
China Association for Standardization
PO Box 820
Peking

Colombia/Colombie (ICONTEC)
Instituto Colombiano de Normas Técnicas
Carrera 37 No. 52–95
PO Box 14237
Bogotá

Cuba (NC)
Comité Estatel de Normalización
5ta nr 306 Entre Cyd Vedado
Zona Postal 4
La Habana

Cyprus
Cyprus Organization for Standards and Control of Quality
Ministry of Commerce and Industry
Nicosia

Czechoslovakia/Tchécoslovaquie (CSN)
Uřad pro normalizaci Měřeni
Václavské Náměsti 19
113 47 Praha 1

Denmark/Danemark (DS)
Dansk Standardiseringsraad
Aurehøjvej 12 and 15
Postbox 77
DK-2900 Hellerup

Egypt, Arab Rep. of/Egypte, Rép. arabe d' (EOS)
Egyptian Organization for Standardization
2 Latin America Street
Garden City
Cairo

Ethiopia/Éthiopie (ESI)
Ethiopian Standards Institution
PO Box 2310
Addis Ababa

Finland/Finlande (SFS)
Suomen Standardisoimisliitto r.y.
PO Box 205
SE-00121 Helsinki 12

France (AFNOR)
Association Française de Normalisation
Tour Europe
Cedex 7
92080 Paris La Defense

Germany/Allemagne (DIN)
DIN Deutsches Institut für Normung
Burggrafenstrasse 4–10
Postfach 1107
D-1000 Berlin 30

Ghana (GSB)
Ghana Standards Board
PO Box M.245
Accra

Greece/Grèce (ELOT)
Hellenic Organization for Standardization
Didotou 15
Athens 144

Hungary/Hongrie (MSZH)
Magyar Szabványügyi Hivatal
Postafiók 24
1450 Budapest 9

India/Inde (ISI)
Indian Standards Institution
Manak Bhavan
9 Bahadur Shah Zafar Marg
New Delhi 110001

Indonesia/Indonésie (YDNI)
Yayasan Dana Normalisasi Indonesia
Indonesian Institute of Sciences
Jalan Teuku Chik Ditiro No. 43
PO Box 250
Jakarta

Iran (ISIRI)
Institute of Standards and Industrial Research of Iran
Ministry of Industries and Mines
PO Box 2937
Teheran

Iraq/Irak (IOS)
Iraqi Organization for Standards
Planning Board
PO Box 13032
Baghdad

Ireland/Irlande (IIRS)
Institute for Industrial Research and Standards
Ballymun Road
Dublin-9

Israel/Israel (SII)
Standards Institution of Israel
42 University Street
Tel Aviv 69977

Italy/Italie (UNI)
Ente Nazionale Italiano di Unificazione
Piazza Armando Diaz 2
120123 Milano

Ivory Coast/Côte d'Ivoire (BIN)
Bureau Ivoirien de Normalisation
BP 1318
Abidjan

Jamaica/Jamaïque (JBS)
Jamaican Bureau of Standards
6 Winchester Road,
PO Box 113
Kingston 10

Japan/Japon (JISC)
c/o International Standards Office
Agency of Industrial Science and Technology
33rd Mori Bldg.
3–8–21 Toranomae
Minato-ku
Tokyo 105

Kenya (KEBS)
Kenya Bureau of Standards
PO Box 54974
NHC House
Harambee Avenue
Nairobi

Korea, Dem. P. Rep. of/Corée, Rép. dém. p. de (CSK)
Committee for Standardization of the Democratic People's Republic of Korea
Pyongyang

Korea, Rep. of /Corée, Rép. de (KBS)
Bureau of Standards
Industrial Advancement Administration
Yongdeungpo-Dong
Yongdeungpo-Ku
Seoul

Lebanon/Liban (LIBNOR)
Institut Libanais de Normalisation
BP 195144
Beyrouth

Libya
Libyan Standards and Specifications Office
Department of Industrial Organization
Secretariat of Industry
Tripoli

Malaysia/Malaisie (SIRIM)
Standards and Industrial Research Institute of Malaysia
Lot 10810, Phase 3, Federal Highway
PO Box 35, Shah Alam
Selangor

Mexico/Mexique (DGN)
Dirección General de Normas
Tuxpan No. 2
Mexico 7, DF

Morocco/Maroc (SNIMA)
Service de Normalisation Industrielle Marocaine
Direction de l'Industrie
Ministère du Commerce, de l'Industrie, des Mines et de la Marine Marchande
Rabat

Netherlands/Pays-Bas (NNI)
Nederlands Normalisatie-instituut
Polakweg 5
PO Box 5810
2280 HV Rijswijk ZH

New Zealand/Nouvelle-Zélande (SANZ)
Standards Association of New Zealand
Private Bag
Wellington

Nigeria (NSO)
Nigerian Standards Organisation
Federal Ministry of Industries
11 Kofo Abayomi Road
Victoria Island
Lagos

Norway/Norvège (NSF)
Norges Standardiseringsforbund
Haakon VII's gt. 2
N-Oslo 1

Pakistan (PSI)
Pakistan Standards Institution
39 Garden Road, Saddar
Karachi-3

Philippines (PS)
Philippines Bureau of Standards
TML Commercial Bldg
100 Quezon Avenue
Quezon City
PO Box 3719
Manila

Poland/Pologne (PKNIM)
Polski Komitet Normalizacji i Miar
Ul. Elektoralna 2
00–139 Warszawa

Portugal (IGPAI)
Direccàs General de Qualidade
Repartiçäo de Normalizaçäo
Rua Jose L'sterao, 83-A
Lisboa-1

Romania/Roumaine (IRS)
Institutul Român de Standardizare
Casuta Postala 6214
Bucarest 1

Saudi Arabia/Arabie Saoudite (SASO)
Saudi Arabian Standards Organization
Airport Street
PO Box 3437
Riyadh

Singapore/Singapour (SISIR)
Singapore Institute of Standards and Industrial Research
179 River Valley Road
PO Box 2611
Singapore 6

South Africa, Rep. of/Afrique du Sud, Rép. d' (SABS)
South African Bureau of Standards
Private Bag X191
Pretoria 0001

Spain/Espagne (IRANOR)
Instituto Nacional de Racionalización y Normalización
Serrano 150
Madrid 6

Sri Lanka (BCS)
Bureau of Ceylon Standards
53 Dharmapala Mawatha
Colombo 3

Sudan/Soudan (SSD)
Standardization and Quality Control Department
Ministry of Industry and Mining
PO Box 2184
Khartoum

Sweden/Suède (SIS)
SIS—Standardiseringskommissionen i Sverige
Tegnergatan 11
Box 3 295
S-103 66 Stockholm

Switzerland/Suisse (SNV)
Association Suisse de Normalisation
Kirchenweg 4
Postfach
8032 Zurich

Thailand/Thaïlande (TISI)
Thai Industrial Standards Institute
Department of Science
Ministry of Industry
Rama VI Street
Bangkok 4

Turkey/ Turquie (TSE)
Turk Standardlari Enstitüsü
Necatibey Caddesi 112
Bakanliklar
Ankara

United Kingdon/Royaume-Uni (BSI)
British Standards Institution
2 Park Street
London WIA 2BS

USA (ANSI)
American National Standards Institute
1430 Broadway
New York, NY 10018

USSR/URSS (GOST)
Gosudarstvennyj Komitet SSSR po Standartam
Soveta Ministrov SSSR
Leninsky Prospekt 9
Moskva 117049

Venezuela (COVENIN)
Comisión Venezolana de Normas Industriales
Av. Boyacà (Cota Mil)
Edf. Fundación La Salle, 5° piso
Caracas 105

Viet Nam, Socialist Republic of/République socialiste du (TCVN)
Departement de Normalisation
Comité d'Etat des Sciences et Techniques
39, rue Tràn Hung Dao
Hanoi

Yugoslavia/Yougoslavie (JZS)
Jugoslovenski Zavod za Standardizaciju
Slobodana Penezića-Krcuna br. 35
Pošt. Pregr. 933
11000 Beograd

Correspondent Members

Barbados/Barbade
Barbados National Standards Institution (BNSI)
'Flodden', Culloden Road
St Michael

Bolivia/Bolivie
Dirección General de Normas y Technologia
Ministerio de Industria, Comercio y Turismo
Casilla 4430, Piso 9
La Paz

Cameroon/Cameroun
Service de la Normalisation
Direction de l'Industrie
Ministère de l'Economie et du Plan
BP 1604
Yaounde

Ecuador/Équateur
Instituto Ecuatoriano de Normalización
Casilla 3999
Av. Universitaria 784
Quito

Hong Kong
Hong Kong Standards and Testing Centre
Eldex Industrial Building.
12th floor, Unit A
21 Ma Tau Wei Road, Hung Hom
Kowloon

Iceland/Islande
Industrial Development Institute
Skipholt 37
Reykjavik

Jordan/Jordanie
Directorate of Standards
Ministry of Industry and Trade
PO Box 2019
Amman

Kuwait/Koweït
Standards and Metrology Department
Ministry of Commerce and Industry
Post Box No. 2944
Kuwait

Liberia
Ministry of Commerce, Industry and Transportation
Division of Standards
Monrovia

Madagascar
Ministère du Développement Rural et de la Réforme Agraire
Direction de l'Agriculture
Service du Contrôle des Qualités et du Conditionnement
BP 1.316
Antananarivo

Malawi
Malawi Bureau of Standards
PO Box 946
Blantyre

Mauritius/Île Maurice
Mauritius Standards Bureau
Ministry of Commerce and Industry
Reduit

Oman
Directorate General for Specifications and Measurements
Ministry of Commerce and Industry
PO Box 550
Muscat

Syria/Syrie
Industrial Testing and Research Centre
(Syrian Standards Organization)
PO Box 845
Damascus

Trinidad and Tobago/Trinité-Et-Tobago
Trinidad and Tobago Bureau of Standards
Room 318, Salvatori Building
Frederick Street
PO Box 288
Port of Spain

Tunisia/Tunisie
Ministère de l'Économie Nationale
Tunis

APPENDIX B: SOURCES OF BRITISH GOVERNMENT STANDARDS

CS and TS Specifications

DMQA Materials Quality Assurance Directorate
Headquarters Building
Royal Arsenal East
London SE18 6TD

DGW(N) Requests from MOD(N) authorities should be forwarded to:

DGW(N)
Naval Ordnance Service Directorate
MOD(N) Ensleigh
Bath BA1 5AB

Defence Standards

D STAN D Stan
13, First Avenue House
High Holborn
London WC1V 6HE

Defence Specifications and DTD Specifications

Her Majesty's Stationery Office
PO Box 569
London SE1 9NH

There are HMSO branches in other cities, certain booksellers act as HMSO agents and there are HMSO agents in other countries.

D Eng R.D. Specifications

Directorate General Engines
AX Registry, Procurement Executive
Ministry of Defence
St Giles Court
London WC2H 8LD

D.G. Ships Procurement Specifications

D.G. Ships
MOD(PE) Ships Dept, Section D
191 Block G, Foxhill
Bath BA1 5AB

APPENDIX C: SOME SOURCES OF USA STANDARDS

ASTM Standards

American Society for Testing and Materials
1916 Race Street
Philadelphia, Pa 19103

SAE Standards

Society of Automotive Engineers
2, Pennsylvania Plaza
New York, NY10011

Federal and Mil Standards

Specifications and Consumer Information Distribution Branch
Washington Navy Yard, Bldg. 197
Washington, DC 20407

CS Standards

Commodity Standards
National Bureau of Standards
Institute for Applied Technology
Washington, DC 20234

SPI Standards

Society of Plastics Industry
250, Park Avenue
New York, NY 10017

NEMA Standards

National Electrical Manufacturers Association
155, E. 44th Street
New York, NY 10017

APPENDIX D: LETTERS USED TO DESIGNATE STANDARDS—COUNTRY OF ORIGIN

ABNT	Brazil
AISI	USA
ANSI	USA
API	USA
AS	Australia
ASME	USA
AWS	USA
BDS	Bulgaria
BDSS	Bangladesh
BNS	Barbados
BS	United Kingdom
BST	Sweden
CAN	Canada
CAS	Central Africa
CEE	International
CEI	Italy
CEMA	Canada
CGA	Canada
CGSB	Canada
CISPR	International
CKS	South Africa
CNS	Taiwan
COPANT	Pan America
COVENIN	Venezuela
CRS	Costa Rica
CSN	Czechoslovakia
CUNA	Italy
CYS	Cyprus
DEMKO	Denmark
DGN	Mexico
DGNT	Boliva
DIN	Germany FR
DS	Denmark
ELOT	Greece
ES	Ethiopia
E.S.	Arab Republic of Egypt
EURONORM	Europe
FED.	USA
GB	China
GOST	USSR
GS	Ghana
I	Portugal
ICAITI	Central America
ICONTEC	Colombia
IEC	International
IEEE	USA
INANTIC	Peru
INDITECNOR	Chile
INEN	Ecuador
IOS	Iraq
IRAM	Argentina
IRS	India
IS	India
I.S.	Eire
ISIRI	Iran
ISO	International
IST	Iceland
JIS	Japan
JS	Jamaica
JSS	Jordan
JUS	Yugoslavia
KEMA	Netherlands
KS	South Korea
K.S.	Kenya

KSS	Kuwait
LS	Lebanon
LSS	Libya
MBS	Malawi
MI	Hungary
MNC	Sweden
MS	Malaysia
MSZ	Hungary
NBN	Belgium
NBS	USA
NEMA	USA
NEMKO	Norway
NEN	Netherlands
NF	France
NI	Indonesia
NIS	Nigeria
NM	Morocco
NORVEN	Venezuela
NP	Portugal
NPR	Netherlands
NS	Norway
NSRDS	USA
NVS	Norway
NZS	New Zealand
ONORM	Austria
OS	Oman
OVE	Austria
PN	Poland
PS	Philippines

SAA	Australia
SABS	South Africa
SAE	USA
SAS	Saudi Arabia
SEMKO	Sweden
SEV	Switzerland
SFS	Finland
SI	Israel
SLS	Sri Lanka
SN	Switzerland
SS	Sweden
S.S.	Singapore
SSS	Syria
STAS	Rumania
TGL	Germany DR
TIS	Thailand
TS	Turkey
TTS	Trinidad & Tobago
UL	USA
ULC	Canada
UNE	Spain
UNEL	Italy
UNI	Italy
UNIT	Uruguay
UTE	France
VDE	Germany FR
VIS	Sweden
VSM	Switzerland
ZS	Zambia

REFERENCES

1. Manning, J. R. (1972) Paper 1, *Conference on Testing of Polymers for Service*, Institute of Physics, October.
2. Reinhart, F. W. (1963) *Society of Plastics Engineers Journal,* **24**, April, 100.
3. Stiehler, R. D. (1974) Symposium Paper, STP No. 553, ASTM.
4. Anon (1975) *BSI News*, February.
5. Anon (1972) *BSI News*, March.
6. BS O Parts 1–4 (1975) *A Standard for Standards.*
7. ISO 1000 (1973) *SI Units and Recommendations for the Use of their Multiples and of Certain Other Units.*
8. BS 5555 (1976) *SI Units and Recommendations for the Use of their Multiples and of Certain Other Units.*
9. BS PD 5686 (1978) *The Use of SI Units.*
10 BS 3763 (1970) *The International System of Units.*
11. BS 350 Part 1 (1974) *Basis of Tables*; Part 2 (1962) *Detailed Conversion Tables.*
12. BS PD 6203 (1967) *Additional Tables for SI Conversion.*

Chapter Three

Preparation of Test Pieces

3.1 INTRODUCTION

In Chapter 6 it is mentioned that the degree of crystallinity can be estimated from density measurements, the density of the amorphous and the crystalline forms generally being significantly different. Since the degree of crystallinity in a test piece depends on the manner in which it has been prepared—temperature, pressure, profile of the mould cavity, rate of cooling, etc.—it therefore follows that the magnitude of this apparently most simple and straightforward of properties, the mass per unit volume, varies according to the history of the test piece under examination. This comment applies to practically every property measured on the massive polymer. Reference is made in other chapters to the influences of test piece shape and preconditioning and to these can be added such considerations as mechanical surface finish and, for certain properties, cleanliness of surface.

While for the purposes of assessing merit in a specific end use the condition of material under examination should simulate realism, for the purposes of a standard test the conditions should approach, as it were, idealism—that is, history, shape, condition, etc., should be carefully defined. Much effort is wasted on producing standard test data, even within the limited confines of the value of such data, because of insufficient attention to the preparation of the test pieces.

The subject as a whole has been discussed by Cohen.[1]

By and large the examination of plastics 'raw materials' is outside the scope of this book but, in passing, it is worth commenting on the necessity of, for instance, predrying hygroscopic powders or flake before making up solutions for viscosity measurement. Similarly, moulding granules may need predrying, e.g. of urea-formaldehyde or nylon, if satisfactory test mouldings are to be obtained.

Many tests are normally performed on specially moulded test pieces; in fact all the tests usually described as being on moulding materials are really on test pieces moulded from the latter, with the exception of the few which may be broadly described as checking the 'mouldability' of the material—particle size, bulk factor, volatile content, cup flow and melt behaviour generally (see Chapters 5, 7 and 16).

3.2 MIXING AND MOULDING

3.2.1 General Considerations

Nearly all of the common plastics moulding materials (both thermosets and thermoplastics) arriving at the physical test laboratory are in the form of granules or powder which may be directly moulded to test pieces or test sheets from which test pieces can be machined. As such, very little mixing is necessary prior to moulding but for one or two thermoplastic materials, notably PVC and ABS, it is essential to premill the granules to sheet form prior to compression moulding test sheets if consistent and meaningful results are to be obtained. This is recognised in ISO 293 (1974)[2] *Compression Moulding Test Specimens of Thermoplastics Materials*, Appendix B of BS 2571 (1963)[3] *Flexible PVC Compounds*, Method 901A of BS 2782 (1977)[4] (which is identical to ISO 293), ASTM D3463 (1975)[5] *Compression Molding Test Specimens of Rigid Acrylonitrile–Butadiene–Styrene (ABS) Plastics* and ASTM D3010-71 (1977)[6] *Preparing Compression-molded Test Sample Plaques of Rigid Poly(vinyl chloride) Compounds*.

In all of these a two roll mill is specified but the ISO and BS 2782 test methods leave the temperature conditions of milling open, the test sample to be mixed at a 'suitable' temperature. The ASTM methods are much more precise, not only specifying the temperature (182° ± 2°C for rigid PVC and 165° to 170°C for ABS) but also the roll clearance and rolling bank size together with an indication of the time on the mill. Appendix B of BS 2571 also gives an indication of roll temperature and milling time.

In some cases where liquid resins are used to make test sheets, premixing of the liquid resin system is necessary prior to casting (e.g. preparing cast poly(methyl methacrylate) sheet, or using a polyester resin to prepare a low-pressure laminated test sheet). Such techniques are not often encountered and usually refer the user to the manufacturer of the resin for instructions. Such is the case in ISO 1268 (1974)[7] *Preparation of Glass Fibre Reinforced, Resin Bonded, Low Pressure Laminated Plastics or Panels for Test Purposes* (Methods 920A to 920C of BS 2782 (1977) are identical to ISO 1268).

When considering the moulding of test pieces or test sheets the most important decision is to settle which method of moulding is appropriate, for the orientation induced in an injection moulded test piece may yield quite different test results from those from a relatively non-orientated compression moulded or extruded test piece. Hayes[8] illustrates this well with some bending strength measurements on $5 \times \frac{1}{2} \times \frac{1}{4}$ in (127 × 12·5 × 7 mm) bars of toughened polystyrene, compression moulded in one case and injection moulded from one end in the other (Table 3.1). In the same paper, the effect of degree of orientation is demonstrated by measuring the same property on bars produced at different moulding temperatures and pressures, the greatest degree of orientation resulting from the lowest temperature and highest pressure (Table 3.2).

Similar studies have been made *inter alia* by Budesheim and Knappe,[9] Horsley *et al.*,[10] Bryant and Hulse,[11] Bossu *et al.*,[12] Hogberg,[13] Morgan and Vale,[14] Williams and Mighton,[15] Dasch,[16,17] Koda[18] and Malac.[19] Hogberg, for instance, examined impact strength measured on bars in which, in one case, flow had occurred in the direction of the bar and, in the other, the direction of flow was perpendicular thereto. Four styrene–acrylonitrile copolymers were ex-

Table 3.1 Comparison of Bending Strengths of Compression and Injection Moulded Toughened Polystyrene (after Hayes[8])

	Bending strength		Deflection at break	
	(lbf/in^2)	(MPa)	(in)	(mm)
Compression moulded	6 570	45·30	0·71	18·0
Injection moulded	8 170	56·33	>1·00	>25·4

Table 3.2 Breaking Strengths at Different Moulding Temperatures and Pressures (after Hayes[8])

Temperature	Pressure		Breaking strength	
(°C)	(lbf/in^2)	(MPa)	(lbf/in^2)	(MPa)
180	10 000	69·0	8 170	56·33
200	8 000	55·0	7 430	51·23
220	5 000	34·0	7 460	51·45
240	4 500	31·0	7 010	48·33

amined and the ratios of the values were between $2\frac{1}{2}$ and 3 to 1, in favour of the bars where flow had been along their length. Amongst the effects studied by Williams and Mighton was the influence of moulding conditions on the stability of mouldings. Koda's study, on polycarbonates, embraced abrasion resistance, hardness, density and heat shrinkage measurements.

The subject, in relation to injection moulding, has been comprehensively reviewed by Whisson.[20] Wintergerst[21] has endeavoured to anneal test pieces of high impact polystyrene back to a condition of 'basic strength' from the injection moulded condition.

On the other hand, moulding conditions may have a more macroscopic effect on the material; a filled or reinforced composition can yield rather different data depending on whether the filler is subject to fracture and whether the process is rigorous enough to cause it. Work reported by Spiwak[22] well demonstrates these points. A range of thermosetting materials was examined, using compression moulded test pieces and two sets prepared by transfer moulding, the size of one gate (*B*) being twice that of the other (*A*). The figures shown in Table 3.3 have been selected from Spiwak's work. A somewhat similar exercise has been described by Elmer and Harrington using phenolic materials in an interlaboratory study.[23]

Panov[24] investigated the effect of test piece size on the tensile strengths of a range of plastics materials and found that the strengths decreased with increasing cross-sectional area—over 20 per cent decrease was observed for polystyrene and unplasticised PVC by increasing the area from 25 mm^2 to 400 mm^2.

Parts of a book edited by Ogorkiewicz[25] deal with the effect of processing variables on the properties of thermoplastics.

Very recently Crawford and co-workers[26] have commenced a comprehensive study of the effect of injection moulding parameters and their influence on

Table 3.3 Impact and Bending Strengths of Test Pieces made of Different Materials and by Different Moulding Techniques (after Spiwak,[22] courtesy Society of Plastics Engineers Inc.)

	Compression moulded		Impact strength Transfer moulded (gate size A)		Transfer moulded (gate size $B = 2 \times A$)	
	(ft lbf/in)	(J/mm)	(ft lbf/in)	(J/mm)	(ft lbf/in)	(J/mm)
PF with short cotton fibre	0·44	0·023	0·45	0·024	0·44	0·023
PF with long asbestos fibre	2·31	0·123	1·23	0·066	1·17	0·062
DAP with short acrylic fibre	0·64	0·034	0·66	0·035	0·66	0·035
DAP with long glass fibre	3·68	0·196	0·68	0·036	0·88	0·047
			Bending strength			
	(lbf/in^2)	(MPa)	(lbf/in^2)	(MPa)	(lbf/in^2)	(MPa)
PF with short cotton fibre	9 200	63·5	9 930	68·5	9 210	63·5
PF with long asbestos fibre	12 100	83·5	6 400	44·0	7 280	50·0
DAP with short acrylic fibre	10 160	70·0	11 590	80·0	9 940	68·5
DAP with long glass fibre	14 800	102·0	10 220	20·5	8 210	56·5

mechanical properties. They describe the design of a mould to produce a set of injection mouldings whose properties can be related generally to product performance.

3.2.2 Standard Procedures

Until comparatively recently it has been traditional to include moulding conditions in the specification for the material under test, certainly in British Standards, although ISO and ASTM have had for some years a small number of published documents dealing exclusively with the moulding of test pieces. During the past 10 years a number of international and national standards on the moulding of test pieces have been published; material specifications, as they are revised, tend to refer to these methods rather than include them in appendices, as has been the practice hitherto.

ISO Standards

There are six ISO Standards relating to the moulding of test pieces. ISO 293 (1974)[2] lays down the general principles to be followed when compression moulding test pieces of thermoplastics materials. The exact conditions are not given because they are considered to be part of the specification for the material under test, or to be agreed between purchaser and supplier.

ISO 294 (1975)[27] deals with injection moulding of test pieces of thermoplastics and, like ISO 293, only lays down general principles to be followed. Although a warning is given under 'Scope' that many factors can influence the characteristics of injection moulded test pieces, no details are given and one is referred to the material specification for exact moulding conditions.

ISO 295 (1974)[28] is intended to serve as a basis for ensuring comparability of results between test houses when moulding thermosetting materials. It includes three annexes:

A. Preparation of test specimens from phenolic moulding materials.
B. Preparation of test specimens from aminoplastic moulding materials.
C. Preparation of test specimens from polyester and epoxy resin moulding materials.

These annexes give recommended moulding conditions for the various materials.

ISO 2557 (1976)[29] *Amorphous Thermoplastic Moulding Materials—Preparation of Test Specimens with a Defined Level of Shrinkage* is somewhat complementary to ISO 293 and ISO 294. It lays down procedures for preparing test pieces with a defined level of shrinkage as measured by finding the longitudinal shrinkage of test pieces by a standardised (annealing) process. Moulding conditions are adjusted to achieve the required level of shrinkage as specified by the material specification or as agreed between the parties concerned.

ISO 1268 (1974)[7] *Preparation of Glass Fibre Reinforced, Resin Bonded, Low Pressure Laminated Plates or Panels for Test Purposes* specifies the procedure for the preparation of test sheets by bonding glass cloth or mats with low-pressure thermosetting resins. Three methods are given: Method A for glass reinforcement impregnated with liquid resin moulded in a press, Method B for moulding pre-pregs and Method C for hand lay-up. The basic design of suitable moulds is given and detailed instructions are included for the hand lay-up method.

ISO 3167 (1975)[30] describes how a Type 1 tensile test specimen of ISO R527 (see Chapter 8), can, with only simple machining, be made suitable for a variety of other tests. Emphasis is placed on strict control of moulding conditions and reference is made to ISO 2557 (see above) and to ISO 2818 (see below). The use of such multipurpose test pieces is attractive for a busy test laboratory where one set of moulded test pieces can be produced under identical conditions followed by machining to yield test pieces suitable for a wide range of physical tests. ISO 3167 proposes that such test pieces may be used for tensile, flexural, compressive, impact, thermal properties, environmental stress cracking and certain flammability properties.

British Standards

At the present time British Standards on moulding test pieces closely follow the ISO practice. Thus:

BS 2782, Method 901A (1977)[4] is the same as ISO 293
BS 2782, Method 902A (1977)[31] is the same as ISO 295
BS 2782, Method 910A (1977)[32] is the same as ISO 294

BS 2782, Methods 920A, 920B and 920C (1977)[33] are the same as those given in ISO 1268.

In addition some British Standards for material include procedures for the preparation of test pieces, notably Appendix B of BS 2571 (1963)[3] *Flexible PVC Compounds*, BS 771 (1980)[34] *Phenolic Moulding Materials*, BS 3126 (1959)[35] *Toughened Polystyrene Moulding Materials* and BS 1493 (1967)[36] *General Purpose Polystyrene Moulding Materials*.

USA Standards

As mentioned earlier, the ASTM standards for the moulding of test pieces are more comprehensive and detailed than the ISO standards. There are no fewer than twelve standards which are related to the moulding of test pieces.

ASTM D647-68 (1977)[37] on design of moulds for test pieces, gives details of compression and injection tools including fully dimensioned drawings. It is referred to in the other ASTM methods for moulding test pieces.

ASTM D958-72 (1977)[38] covers two optional procedures for finding the temperature condition of moulds either by a thermometer or by a pyrometer.

ASTM D956-51 (1977)[39] on compression moulding of AMC test pieces is roughly equivalent to ISO 295 but is much more precise in terms of mould design and moulding conditions.

ASTM D796-65 (1977)[40] on compression moulding of PMC test pieces is, again, roughly equivalent to ISO 295 but more precise.

ASTM D2292-68 (1977)[41] (styrene–butadiene moulding and extrusion materials) uses a simple picture-frame flash mould similar to that given in ISO 293.

ASTM D3027-72 (1977)[42] (allyl moulding compounds) calls for use of the positive moulds described in D647 and closely specifies moulding conditions.

ASTM D3463 (1975)[5] (rigid ABS plastics) gives precise details of moulding test pieces and has already been mentioned above in connection with premilling.

Similarly, ASTM D3010 (1971)[6] (test sample plaques of rigid PVC compounds) closely specifies the procedure to be followed for moulding test pieces of rigid PVC, including milling prior to moulding to ensure homogeneous test pieces.

ASTM D1130 (1969)[43] *Injection Molding Test Specimens of Thermoplastic Materials* is similar to ISO 294. ASTM 1897 (1972), *Injection Molding Test Specimens of Thermoplastic Molding and Extrusion Materials*, is a more precise version of D1130 and allows the use of screw injection machines. It gives more details of moulding cycles than D1130 but these have been based on interlaboratory tests using polystyrene, toughened polystyrene and polycarbonate only. In time, D1130 will probably be superseded by D1897 as there would appear to be no need to have two standards dealing with the same subject.

The preparation of test pieces of thermosetting materials by injection moulding is covered in ASTM D3419 (1975).[44] This standard specifies the type of machine and mould to be used and gives useful guidance on moulding temperatures. Similarly, ASTM D1896-72 (1977)[45] lays down procedure for moulding test pieces by transfer moulding and specifies radio-frequency preheating. Again, recommended moulding temperatures and cycles are given.

German Standard
DIN 53451 (1972)[46] on compression moulded test specimens of thermosetting plastics is broadly similar to ISO 295.

3.3 CUTTING FROM SHEET

When the sample to be tested has already been moulded to some 'massive form', for example calendered sheet, laminate, casting or large commercial moulding, test pieces of the appropriate form have to be cut from the sample.

In many ways this is a more complex problem than moulding specimens directly, for the machining of test specimens is a difficult operation to standardise, especially in terms of the quality of the finishing. No simple recourse to using a stainless steel mould with mirror finish is possible here; yet many properties are profoundly influenced by the presence of surface flaws such as nicks and scratches.

The point is well illustrated by reference to the apparently simple operation of cutting out specimens from soft thin film with a die. Patterson[47] gives the results of a correlation exercise in tensile testing of 1·5 mil polyethylene, 1 mil polyester and 4 mil polycarbonate (1 mil = 0·001 in or 0·025 mm) undertaken by four laboratories using four different techniques: die cut, manual razor cut, hand-driven rotary razor and shear cutter.

Yield strength data showed the least scatter—10 per cent or less covering all four techniques and laboratories. However, much greater divergencies were found with ultimate tensile strength (30 per cent or more) and elongation at break (up to approximately 75 per cent), with the laboratories being relatively consistent and the techniques accounting for the majority if not all the variation. Die cut specimens were clearly the worst in this exercise, though this should not be taken as a total indictment of the technique; indeed, although in this work simple rectangular specimens were examined, many more complex shapes could not be conveniently prepared by any other technique. More importantly, the lesson to be learned is of the necessity to ensure that cutting dies are in good sharp condition and 'clean' cuts are made, for example by a clicking press.

At one time standards did not specify or even recommend methods of cutting and machining test pieces from sheet and other products, but since 1970 this omission has been substantially rectified. For example, ISO R1184 (1970),[48] BS 2782 Methods 326A to 326C (1977)[49] and ASTM D882 (1979),[50] draw attention to the need for care in preparing test pieces for the tensile testing of plastics films and the ISO and BS standards suggest suitable methods of cutting the test pieces. Similarly, BS 2782 Methods 320A to 320F (1976),[51] for tensile properties, have a section on preparation of test pieces. The current revision of ISO R527 (1966)[52] (DIS 527) *Tensile Properties* by ISO TC61 also has a section giving guidance on preparation of test pieces.

It is often necessary to cut rectangular blanks from sheet samples which are subsequently machined to test pieces of specified form and size (§3.4). Such test piece blanks are cut using hand saws, circular saws and diamond impregnated discs. With care and careful hand finishing by file and abrasive paper it is possible to produce finished test pieces from such sawn blanks which are acceptable for many tests, but some form of machining using a properly designed machine speeds up the preparation of test pieces and generally leads to more consistent test results.

3.4 MACHINING FROM SHEET AND FINISHED PRODUCTS

For flexible, semi-flexible and some of the 'softer' rigid plastics sheet and finished products (up to 1·5 mm thick), test pieces can be cut by using a sharp and properly maintained die cutter and clicking press. This method cannot be used satisfactorily for the more rigid materials and recourse must then be made to machining. Sometimes, too, the product when tested is too thick and then one or both surfaces have to be machined to reduce the thickness to the required level.

Both milling and grinding have been found suitable and standard machine tools may be used provided they can be operated at a high enough speed and be fitted with suitable cutters. Tungsten carbide or diamond tipped milling cutters give the longest life but high-grade high-speed steel cutters may be used with many materials and will give an acceptable service life. Tool rotation speeds vary according to the material being machined and are best found by trial and error but generally lie within the range 8000 to 30 000 r.p.m.

The technique of grinding (or buffing) plastics is not so widely practised as for rubber where it is quite commonly used for reducing the thickness of rubber test pieces. When grinding plastics it is important to use the right grade of abrasive wheel; open grit wheels give the best results but in the absence of specific information the advice of the abrasive wheel manufacturer should be sought both as to the correct abrasion grade and the optimum speed of rotation of the wheel.

Standard machine tools and surface grinders can be used for making rectangular bar test pieces; the production of curved test pieces (e.g. dumbbells) requires some form of copying milling machine or router. Many variations have been used but all essentially consist of a motor-driven cutting tool guided by a pin following a template of the required form.

When machining some plastics, a coolant is necessary if a satisfactory surface finish is to be obtained. A jet of compressed air is suitable in many cases, but if a liquid coolant is used care must be taken to ensure that it does not affect the plastics material being machined.

There is much room for experiment in the machining of plastics before precise details of machining conditions can be laid down for any given material, but the need for standardised procedures has been recognised and a start has been made by the publication of ISO 2818 (1974),[53] *Preparation of Test Specimens by Machining*. This gives machining conditions which have been found suitable for thermoplastics and thermosets where the thickness does not exceed 10 mm and also includes diagrams for a manually guided copy milling machine and a tubular abrasive machine tool. ISO 2818 does not offer any recommendation about reducing the thickness of a test piece blank or sheet if it is greater than 10 mm, and it is considered that this is a weakness of the standard. BS 2782 Method 930A (1976)[54] is identical to ISO 2818.

Apparatus and materials for the preparation of test pieces have been discussed by Brown,[55] who also mentions the techniques available for machining the notches in impact test pieces—an important subject because in some plastics small variations in notch geometry can cause large variation in impact test results. Shaping or milling is recommended and hand operated notch cutters are now available commercially.

Finally, we return to the subject of reducing the thickness of sheet where this is necessary to produce test pieces of the specified dimensions. The specification for the material under test, or the test method should be consulted before cutting because some tests (e.g. BS 2782 Method 320E (1976)[56] for tensile properties) requires that one face of the manufactured sheet should be left intact, while others (e.g. BS 2782 Method 351A (1977)[57] for Charpy impact resistance) specify that both surfaces shall be machined uniformly to give the required thickness.

REFERENCES

1. Cohen, L. A. (1967) in *Testing of Polymers* vol. 3, Scmiditz, J. V. and Brown, W. E. (Eds.), Interscience, page 15.
2. ISO 293 (1974) *Compression Moulding Test Specimens of Thermoplastics Materials.*
3. BS 2571 (1963) *Flexible PVC Compounds.*
4. BS 2782, Method 901A (1977) *Compression Moulding Test Specimens of Thermoplastics Materials.*
5. ASTM D3463 (1975) *Compression Molding Test Specimens of Rigid Acrylonitrile–Butadiene–Styrene (ABS) Plastics.*
6. ASTM D3010-71 (1977) *Preparing Compression-molded Test Sample Plaques of Rigid Poly(vinyl chloride) Compounds.*
7. ISO 1268 (1974) *Preparation of Glass Fibre Reinforced Resin Bonded, Low Pressure Laminated Plastics or Panels for Test Purposes.*
8. Hayes, R. (1954) *Transactions of the Plastics Institute*, London, **27** (49), 219.
9. Budesheim, H. and Knappe, W. (1959) *Kunststoffe*, **49** (6), 257.
10. Horsley, R. A., Lee, D. J. A. and Wright, P. B. (1959) in *The Physical Properties of Polymers*, SCI Monograph No. 5, Society of the Chemical Industry, page 63.
11. Bryant, K. C. and Hulse, G. (Eds.) (1955) *Plastics Progress: Papers and Discussions at the British Plastics Convention 1955*, Iliffe.
12. Bossu, B., Chatan, M., Dubois, P. and Rougeaux, J. (1959) in *International Symposium on Plastics Testing Standards*, STP No. 247, ASTM, page 67.
13. Hogberg, H, (1959) in *International Symposium on Plastics Testing Standards*, STP No. 247, ASTM, page 95.
14. Morgan, D. E. and Vale, C. P. (1959) in *The Physical Properties of Polymers*, SCI Monograph No. 5, Society of the Chemical Industry, page 169.
15. Williams J. L. and Mighton, J. W. (1953) in *Symposium on Plastics Testing—Present and Future*, STP No. 132, ASTM, page 32.
16. Dasch, J. (1967) *Kunststoffe*, **57** (2), 117.
17. Dasch, J. (1968) *Kunststoffe*, **58** (11), 769.
18. Koda, H. (1968) *Journal of Applied Polymer Science*, **12** (10), 2257.
19. Malac, J. (1969) *Journal of Applied Polymer Science*, **13** (8), 1767.
20. Whisson, R. R. (1967) *Technical Review 42*, RAPRA.
21. Wintergerst, S. (1967) *Kunststoffe*, **57** (3), 188.
22. Spiwak, L. (1963) *Society of Plastics Engineers Journal*, **19** (6), 557.

23. Elmer, C. and Harrington, E. C. Jr (1960) *ASTM Bulletin*, 249 (October), 35.
24. Panov, P. (1967) *Plaste und Kautschuk*, **14** (7), 491.
25. Ogorkiewicz, R. M. (Ed.) (1969) *Thermoplastics: Effects of Processing*, Iliffe.
26. Crawford, R., Klewpatinond, V. and Benham, P. P. (1978) *Plastics and Rubber: Processing*, December, 133.
27. ISO 294 (1975) *Injection Moulding Test Specimens of Thermoplastic Materials.*
28. ISO 295 (1974) *Compression Moulding Test Specimens of Thermosetting Materials.*
29. ISO 2557/2 (1979) *Amorphous Thermoplastic Moulding Materials—Preparation of Test Specimens with a Defined Level of Shrinkage.*
30. ISO 3167 (1975) *Preparation and Use of Multipurpose Test Specimens.*
31. BS 2782 Method 902A (1977) *Compression Moulding Test Specimens of Thermosetting Materials.*
32. BS 2782 Method 910A (1977) *Injection Moulding Test Specimens of Thermoplastic Materials.*
33. BS 2782 Methods 920A to 920C (1977) *Preparation of Glass Fibre Reinforced, Resin Bonded, Low Pressure Laminated Plates or Panels for Test Purposes.*
34. BS 771 (1980) *Phenolic Moulding Materials.*
35. BS 3126 (1959) *Toughened Polystyrene Moulding Materials.*
36. BS 1493 (1967) *General Purpose Polystyrene Moulding Materials.*
37. ASTM D647-68(1977) *Design of Molds for Test Specimens of Plastics Molding Materials.*
38. ASTM D958-72 (1977) *Determining Temperature of Standard ASTM Test Molds for Test Specimens of Plastics.*
39. ASTM D956-51 (1977) *Compression Molding Test Specimens of Amine Molding Compounds.*
40. ASTM D796-65 (1977) *Compression Molding Test Specimens of Phenolic Molding Compounds.*
41. ASTM D2292-68 (1977) *Compression Molding Test Specimens of Styrene–Butadiene Molding and Extrusion Materials.*
42. ASTM D3027-72 (1977) *Compression Molding Test Specimens of Alkyl Molding Compounds.*
43. ASTM D1130-63 (1969) *Injection Molding Test Specimens of Thermoplastic Materials.*
44. ASTM D3419 (1975) *In-line Screw Injection Molding Test Specimens from Thermosetting Compounds.*
45. ASTM D1896-72 (1977) *Transfer Molding Test Specimens of Thermosetting Compounds.*
46. DIN 53451 (1972) *Plastics—Preparation of Compression Moulded Test Specimens of Thermosetting Moulding Materials.*
47. Patterson, C. D. Jr (1964) *Materials, Research and Standards*, **4** (4), 159.
48. ISO R1184 (1970) *Plastics–Determination of Tensile Properties of Films.*
49. BS 2782 Methods 326A to 326C (1977) *Determination of Tensile Strength and Elongation of Plastics Films.*
50. ASTM D882 (1979) *Tensile Properties of Thin Plastic Sheeting.*
51. BS 2782 Methods 320A to 320F (1976) *Tensile Strength, Elongation and Elastic Modulus.*
52. ISO R527 (1966) *Determination of Tensile Properties.*
53. ISO 2818 (1974) *Preparation of Test Specimens by Machining.*
54. BS 2782 Method 930A (1976) *Preparation of Test Specimens by Machining.*
55. Brown, R. P. (Ed.) (1979) *Guide to Rubber and Plastics Test Equipment,* RAPRA.
56. BS 2782 Method 320E (1976) *Tensile Properties of Plastics.*
57. BS 2782 Method 351A (1977) *Determination of Charpy Impact Resistance of Rigid Plastics and Ebonite (Charpy Impact Flexural Test).*

Chapter Four

Conditioning and Test Atmospheres

4.1 INTRODUCTION

Plastics materials, generally, are affected by ambient conditions and, in the normal test laboratory, these variables may be limited to varying temperature and humidity; hazards of irradiation, be they ultraviolet from sunlight, gamma rays from nuclear power or heat from rocket propulsion, are more logically treated under the general heading, 'Heat Ageing and Environmental Resistance' (see Chapters 16 and 17).

Sensitivity to the effects of temperature is primarily a consequence of the fact that most present-day plastics materials are based on polymers built upon a backbone chain consisting solely or predominantly of covalently bonded carbon atoms. Crosslinking, secondary intermolecular forces and crystallinity can influence reaction to temperature quite significantly, but by comparison with most other 'materials of construction' (in the broadest sense of the phrase) plastics are *temperature sensitive*. A number of them contain hydroxyl groups, e.g. cellulose acetate, polyvinyl acetate and polyvinyl alcohol, while others contain amino groups, such as the nylons and casein, and these groups have a distinct affinity for water molecules, such as are present in air of any finite moisture content or humidity; the water acts as a plasticiser and brings about a change of properties. One of the most spectacular cases is the increase in impact strength of about twenty-fold which may be achieved by allowing dry Nylon 6 to absorb moisture to saturation. Even if the polymer itself is not subject to the influences of temperature or humidity, the other materials present in the product may well be—particularly to humidity. Wood flour or chopped rag filler and cotton, paper or glass fibre reinforcement are well known examples.

It will be obvious, therefore, that to obtain results of any meaning, from the point of view of defining the physical condition of the material when tested and providing data comparable with others obtained similarly, test pieces should be subjected to standard preconditioning to bring them into an equilibrium state with a specified atmosphere. The need for preconditioning (hereafter 'conditioning') clearly depends on the nature of the materials under test; in many cases control of humidity is unimportant because the materials are to all intents and purposes completely hydrophobic.

4.2 STORAGE

The properties of vulcanised rubber change most rapidly immediately after vulcanisation and later, assuming that no accelerating influences are present, the changes become so slow as to be negligible over a period of, say, a few weeks. Hence it is desirable that a minimum period is allowed between vulcanisation and testing. Consequently most standard test methods for rubbers require that no test shall be made less than 16 h after vulcanisation and any sample preparation, and that the maximum time between vulcanisation and testing shall be 28 days (90 days in the case of tests on products).

Similar considerations do not apply to most plastics materials which, by and large, do not show significant changes in properties with storage after moulding; however, a notable exception to this applies in the case of plasticised PVC. Walter[1] found that the elastic properties of PVC gels changed over a period of 2 years storage at ordinary temperatures, which change could not be related to plasticiser loss. Heap and Norman[2] investigated the changes in BS softness after moulding plasticised PVC and concluded that the mechanism was not due to loss of plasticiser nor to laminar inhomogeneity. They suggested a hypothesis of droplet separation but the precise mechanism is still not fully understood. This phenomenon is recognised in the case of BS softness measurements (see Chapter 8) on flexible PVC by specifying a seven-day conditioning period between moulding the test piece and carrying out the test.

Common sense and good housekeeping dictate that samples of moulding powder should be stored in cool and dry conditions prior to moulding and the same also applies to samples of semi-finished and finished products. With the exception mentioned above, test methods for plastics do not usually specify permissible maximum and minimum storage times prior to testing. However some product specifications offer recommendations on this subject. For example, BS 771 (1980)[3] for phenolic moulding materials recommends that tests should be commenced within 24 h of preparing test pieces and BS 1763 (1975),[4] for calendered PVC sheeting, recommends that the sample should be wound on to a tube of adequate diameter and requires that this sample be stored so that it is out of contact with other plastics materials. In any case it is advisable to take note of the storage time and conditions of storage for further reference in the event of anomalous test results being obtained.

4.3 CONDITIONING

4.3.1 Definitions

In the past there have been wide variations in conditioning procedures because national standards naturally tend to simulate average ambient weather conditions of the country in question. This is not only to facilitate the actual conditioning and testing operations but also to produce data relevant to the likely environment of use of the plastics material. In general, too, there has been a tendency for each specification for each different material to dictate its own, often different, conditions. Internationally, standardisation has been even more difficult; 'ambient' in North Canada is rather different from 'ambient' in the

Middle East. Nevertheless agreement has been reached for a limited number of alternative standard conditions to be applied internationally.

Before discussing the various standard conditions it is worth defining the terms used. The preferred unit of temperature is the centigrade degree (°C) although the SI system specifies kelvin (K). The water content of the air is invariably specified as relative humidity (r.h.) and this is defined in BS 1339 (1965)[5] as 'the ratio of actual vapour pressure to the saturation vapour pressure over a plane liquid water surface at the same (dry bulb) temperature, expressed as a percentage'.

Further definitions are given in BS 1339 (1965)[5] and ASTM E41 (1971)[6].

4.3.2 Standard Procedures

ISO Standards

The ISO standard on conditioning most referred to is ISO 291 (1977).[7] This specifies four atmospheres as follows:

23°C	50% r.h.	Pressure 86 to 106 kPa	(recommended atmosphere)
27°C	65% r.h.	Pressure 86 to 106 kPa	(for tropical countries)

and, where humidity has no influence on the properties being examined, either of the above used with uncontrolled relative humidity.

Two levels of tolerance are allowed in the standard. 'Normal' requires ±2°C on temperature and ±5 per cent on r.h., while 'close tolerance' requires ±1°C on temperature and ±2 per cent on r.h. Periods of conditioning are as stated in the appropriate international standard, but where this is not so ISO 291 specifies a minimum conditioning period of 88 h for the atmospheres with controlled relative humidity and a minimum of 4 hours for the others. The Annex to ISO 291 points out that for certain materials these conditioning periods do not allow equilibrium to be reached and suggests that, in such cases, the test piece either be dried at high temperature or be conditioned at 23°C, 50 per cent r.h. until equilibrium is reached. To achieve this it is recommended to condition to constant mass (within 0·1 per cent) for two determinations made at an interval of d^2 weeks where d is the thickness of the test piece (mm). Both of these recommendations have disadvantages: drying out completely at high temperature often results in mechanical properties quite different from those obtained at moisture equilibrium, while conditioning to equilibrium at 23°C 50 per cent r.h. can be time-consuming, extending to weeks or even to months.

These problems have been recognised in respect of polyamides 66, 610 and 6 by the publication of ISO 1110 (1975)[8] which offers a procedure for accelerated conditioning of these materials by heating test pieces between 7 h and 7 days at 95°C to 100°C in aqueous potassium acetate solutions of specified concentrations depending on polymer type and thickness of test piece. The disadvantages of the method are (a) that one has to know the type of polyamide being tested before the method can be applied with confidence, (b) there is no guarantee that mechanical properties obtained will be the same as those obtained by conditioning at 23°C 50 per cent r.h. for a long period and (c) electrical tests on accelerated conditioned test pieces may not be valid because of residual salt in the test pieces. The United Kingdom disapproved of this conditioning procedure and consequently it does not appear as a method in BS 2782.

ISO 291 makes reference to ISO 554 (1976)[9] which offers the choice of six standard atmospheres (Table 4.1). It must be noted that ISO 554 makes no allowance for the use of uncontrolled humidity, yet ISO 291 does so; since ISO 554 is the general standard for conditioning for all ISO standards there would appear to be some inconsistency which needs attention by ISO.

Table 4.1 ISO 554[9] Standard Atmospheres for Conditioning
Each condition has a permitted choice of two tolerences:

(a) Ordinary: ±2°C for temperature and ±5% r.h.
(b) Reduced: ±1°C for temperature and ±2% r.h.

Designation	Temperature (°C)	Relative humidity (%)	Pressure (kPa)	Remarks
23/50	23	50	between 86 and 106	Recommended atmosphere
27/65	27	65	between 86 and 106	For tropical countries
20/65	20	65	between 86 and 106	Used in certain fields of application

British Standards
Until comparatively recently there has been no one standard conditioning procedure for plastics; individual material specifications had to be consulted for details of the levels of temperature and humidity to be used in each case. Even in the earlier editions of BS 2782, each test method carried its own instructions on conditioning. The rectification of this state of affairs is now well advanced. The later edition of BS 2782 (1975)[10] carries a Foreword, General Introduction and Instructions which include the conditioning procedure of ISO 291; a new Part O, Method 010[11] which includes *Standard Atmospheres for Conditioning and Testing*, is at an advanced stage. This is identical to ISO 291.

USA Standards
ASTM E171-63 (1972)[12] specifies the same conditioning atmospheres as ISO 554 and also gives a list of conditioning procedures for various materials.

The ASTM standard specific for the conditioning of plastics is ASTM D618-61 (1971)[13] which goes into the subject very thoroughly. The following definitions are given:

Standard Laboratory Temperature. A temperature of 23°C with a normal tolerance ±2°C and a closer tolerance ±1°C if required.

Standard Laboratory Atmosphere. An atmosphere having a temperature of 23°C (tolerances as above) and a relative humidity of 50 per cent, the latter with a normal tolerance of ±5 per cent and a closer tolerance of ±2 per cent when required.

Room Temperature. A temperature in the range 20°–30°C.

The standard lays down a useful shorthand for identifying the type of conditioning used. A first number indicates the number of hours of conditioning, a second the temperature and a third the relative humidity; the last mentioned may be replaced in certain circumstances by words indicating special treatments. Thus:

(1) 96/23/50	96 h at 23°C and 50% r.h.
(2) 48/50/Water	48 h at 50°C in water
(3) 48/50 + 96/23/50	48 h at 50°C followed by 96 h at 23°C and 50% r.h.

For the actual test condition, the conditioning data as above are followed by a colon, then a capital T, a number indicating the test temperature, and after it another number indicating the relative humidity if controlled, e.g.

96/23/50: T—35/90	Condition as (1) above, and test at 35°C and 90% r.h.

ASTM D618 lays down minimum conditioning times according to the thickness of test specimens and for the most commonly used atmosphere (t/23/50), t is 40 h for specimens 7 mm or under, and 88 h for specimens above 7 mm. This is known as *Procedure A*, but there are five other alternatives to meet the needs of rapidity or to simulate special atmospheres or use environments. Corresponding instructions for test conditions are given.

As in British Standard practice, there is no ASTM specification for accelerated conditioning of polyamides. ASTM D789 (1973)[14] for nylon moulding materials requires the test for flexural modulus of elasticity (the only mechanical property specified) to be determined on test pieces moulded and immediately stored in watertight containers prior to the test which is carried out at 23°C ± 1°C and 50 per cent ± 2 per cent r.h.

German Standards

The most important DIN standard on conditioning, to which most German test methods for plastics refer, is DIN 50014 (1970).[15] This specifies a conditioning atmosphere of 23°C ± 2°C, 50 per cent ± 5 per cent r.h. at a pressure of 800 to 1060 mbar.

DIN 50005 (1975)[16] lists a selection of climates for conditioning and testing of plastics and other electrical insulating materials.

DIN 53714 (1978)[17] offers a procedure for the accelerated conditioning of polyamide test pieces. This procedure is more precise than that of ISO 1110 in that it requires the test pieces to be brought to a specified water content by treatment in water-saturated air in a closed vessel for a few days at a temperature not exceeding 40°C, followed by a further 4 weeks in saturated air sealed in polythene bags and, finally, by a further 4 weeks in a standard atmosphere of 23°C and 50 per cent r.h.

4.4 TESTING CONDITIONS

The object of conditioning is to bring the test piece as nearly as possible into equilibrium with a standard atmosphere so as to eliminate one source of varia-

Table 4.2 Preferred Sub-normal and Elevated Temperatures according to ISO 3205[19] and ASTM D618[13]

ISO 3205 Preferred test temperature* (°C)		ASTM D618 Preferred test temperature (°C)	
−269, −196, −161	Tolerance depends on equipment	− 70, − 55, − 40, − 25, 0	±2
− 70, − 55, − 40, − 25, − 10	±3	35	±1
0, 5, 25, 40, 55, 70, 85, 100	±2	50, 70, 90, 105, 120, 130, 155, 180	±2
105, 125, 150, 175, 200	±3	200, 225, 250, 275, 300	±3
225, 250, 275, 300, 350	±5	325	±4
400, 450, 500, 600, 700	±10	350	±5
800, 900, 1000	±15	400	±6
		450	±8
		500	±10
		600	±12

* ISO 3205 also has an Annex which lists temperatures used in particular fields, culled from documents prepared by various ISO Technical Committees. Thirty-two temperatures are quoted in the Annex, ranging from 165°C to 950°C.

tion in test results either within one test using replicate test pieces or between test laboratories. It is therefore desirable that the test is carried out in a test atmosphere identical to that used for conditioning. ISO 291 (1977)[7] states that this should be the case unless otherwise specified but goes on to say that in all cases the test shall be made immediately after the removal of test specimens from the conditioning enclosure. The sister standard for rubber, ISO 471 (1977),[18] allows for testing in a less rigorous atmosphere than that used for conditioning where such procedure does not affect the result. This recognises the fact that it is more difficult, and costly, to maintain a test room at constant temperature and (particularly) humidity than a small enclosure. No sure concession is made in ISO 291 but it is common practice in the plastics testing laboratory only to control temperature, carrying out the test immediately after removal from the conditioning cabinet. This is sound practice for most plastics materials but it is *not* satisfactory to follow this procedure if the test piece has been conditioned at elevated or subnormal temperatures unless the test piece is very bulky and the test can be made extremely rapidly.

British Standard[10] and ASTM[13] practices follow that of ISO.

The majority of testing is carried out in one of the normal standard atmospheres but ISO 3205 (1976)[19] and ASTM D618–61 (1971)[13] list preferred subnormal and elevated temperatures. The British Standard[11] practice follows that of ISO 3205. It will be noted from Table 4.2 that ISO 3205 and ASTM D618 do not entirely agree either in individual temperatures or in tolerances. These standards do not specify precisely the conditioning periods to be used with the sub-normal and elevated temperatures. ASTM D618 merely states that test pieces should be maintained at the test temperature for a time not less than that required to ensure thermal equilibrium and not more than 5 h prior to testing, but no advice is given on how the time to reach equilibrium can be estimated. Tables of approximate times required to reach equilibrium in both air and liquid media have been given[20] for a wide range of temperatures and various test piece geometries, and these are produced in full as an appendix at the end of this chapter.

For further information on non-ambient temperature testing the reader is referred to §16.9.

4.5 APPARATUS FOR CONDITIONING

It is not feasible, in a small section of this book, to describe the construction of conditioning chambers and design of temperature and humidity controlled test rooms; this is a science in itself which, particularly in the case of the test room or laboratory, is best left to the expert. However, considering the lack of any reference to apparatus in standards such as ISO 291, it is worth discussing briefly certain aspects of the subject.

4.5.1 Air-conditioned Rooms

It is difficult to run a plastics physical testing laboratory without air conditioning of the room in terms of temperature. Nevertheless large numbers of laboratories do not have this facility, presumably on the grounds of cost. However, if the room is reasonably well isolated by doors, does not have excessive window area and is of moderate size the cost of installing self-contained heating and cooling units is surprisingly low. Complete air conditioning of both temperature and humidity is inevitably much more expensive and is not necessary unless the majority of physical testing is to be carried out on moisture-sensitive materials such as cellulose acetate and polyamides. Even then, use of suitable enclosures (§4.5.2) can be quite adequate and give satisfactory results.

As mentioned earlier it is necessary to consult an expert when considering the installation of air conditioning but it should be noted that there is a British Standard covering the design of controlled atmosphere laboratories.[21]

4.5.2 Enclosures

There has been a tendency to treat conditioning chambers in the 'do-it-yourself' manner—a suitable box, a dish of salt solution, a circulation fan, a low wattage heater with controller, a thermometer and some form of dial hygrometer all slung together. This is not to decry salt solutions as a means of controlling humidity; they can provide a cheap and accurate method if used properly and particularly if the salt solution has a large surface area. The often overlooked factors are in the design of the chamber: lack of thermal insulation, doors which when opened immediately cause an almost complete interchange of atmosphere with that of the surrounding room, shelves and corners which prevent adequate air circulation, fans which do not move the air sufficiently, heaters which radiate to test pieces and hygrometers which do not indicate r.h. correctly.

Two types of humidity cabinet are in common use, salt-tray cabinets and moisture-injection types, which are covered by BS 3718 (1964)[22] and BS 3898 (1965),[23] respectively. The salt-tray cabinet is by far the simpler of the two types, being essentially a temperature-controlled enclosure in which the humidity is controlled by the use of saturated salt solutions. Despite their simplicity, much care is required in their design and operation if accurate conditions are to be realised and much useful detail is to be found in BS 3718.

ISO R483 (1966),[24] which refers specifically to plastics, is similar to BS 3718 but additionally describes the use of glycerol solutions to replace saturated salt solutions. ASTM E104-51 (1971)[25] covers the same ground as the other two standards and also describes the use of aqueous sulphuric acid solutions as well as glycerol and salt solutions. It is, however, of less use than BS 3718 and ISO R483 because the 'usual' plastics conditioning temperature of 23°C is not included. At the condition of most interest, i.e. 23°C, 50 per cent r.h., sodium dichromate is just within the tolerance of ±5 per cent r.h. but no salt is listed which will give 50 per cent ± 2 per cent r.h. ISO R483 claims that a glycerol solution will achieve this tolerance if its refractive index is maintained at 1·444 ± 0·002.

The injection type of humidity cabinet uses a humidity sensing device to control the injection of moisture into the cabinet from a reservoir. The humidity sensing device is often a wet and dry bulb hygrometer with mercury contact thermometers. Humidity levels are more readily changed with this type of equipment and some have the facility to cycle both humidity and temperature, so extending the range of tests which can be carried out. BS 4864 (1973)[26] describes enclosures with a greater operating range than those of BS 3898 but this type of apparatus is more pertinent to accelerated ageing than to conditioning as defined in this chapter.

4.5.3 Thermometers

The ordinary mercury-in-glass thermometer as covered by BS 593 (1974)[27] is in such common use that it is taken for granted. In practice, variability of test results obtained at a set temperature can often be traced to the misuse of thermometers, which should be calibrated frequently, inspected for separation of mercury and immersed to the correct depth. The worst errors occur with low-temperature thermometers (not covered by BS 593) and particular care needs to be taken when conditioning or testing at sub-zero temperatures. ISO R653 (1968)[28] and ISO R654 (1968)[29] are International Standards for thermometers and these now have British Standard equivalents in BS 5074 (1974)[30] for precision thermometers and BS 1704 (1951)[31] for general purpose thermometers.

Apart from the ordinary liquid-in-glass thermometer, there are many other types of temperature-measuring instrument and these require just as much care in use. Useful descriptions of the various types may be found in *The Instrument Manual*[32] and techniques of calibration together with a bibliography are given in an HMSO publication.[33]

4.5.4 Hygrometers

The most useful standard hygrometer is the wet and dry bulb type which should be used in conditions where air is circulating around the instrument at a velocity not less than 3 m/s. Hygrometric tables for use with wet and dry bulb hygrometers are published in BS 4833 (1972)[34] which also contains a bibliography; however, probably the most useful reference is an HMSO publication[35] giving general guidance on the use of hygrometers.

In recent years a large number of different types of hygrometer have become available commercially and many of these are described in *The Instrument Manual*.[32] Simple 'mechanical' types of hygrometer, such as those which depend on the change of dimensions of hair or paper, can be useful; they are small in size and relatively inexpensive but they require frequent calibration and attention if their accuracy is to be maintained.

4.5.5 Apparatus for Elevated and Sub-normal Temperatures

Generally, conditioning at elevated or sub-normal temperatures indicates that the test will be carried out at that same temperature and hence in the same enclosure as is used for conditioning. Under these circumstances the conditioning enclosure usually forms part of the testing apparatus and is therefore likely to take one of many forms, depending on the test in question. Comments on the types of enclosure available are given by Brown[36] and the requirements for particular tests will be discussed in the relevant sections of this book.

4.6 MECHANICAL CONDITIONING

Brown[37] has drawn attention to the need for mechanical conditioning when performing certain dynamic mechanical tests on rubbers. This is because it has been shown that the stress–strain curves of vulcanised rubbers containing filler are permanently damaged when they are deformed.

There does not appear to be a need for the mechanical conditioning of plastics, judging by the lack of any published standards on this subject. However, in any mechanical test involving repeated stressing of the test piece, the test results should be observed carefully to see whether mechanical conditioning is necessary to achieve meaningful results from the material under test.

APPENDIX: THERMAL EQUILIBRIUM IN CONDITIONING

If a product is used at non-ambient temperatures, it is logical that it should be tested under these same service conditions. However, elevated or sub-normal temperature testing inevitably requires more complex, and hence more expensive apparatus and normally takes a considerably longer time to carry out. In this respect the conditioning period is a crucial consideration and is often the limiting factor to efficient testing. This is of course of particular importance in quality control testing where both time and money may be of paramount importance.

If the conditioning time used is too short the effect of temperature change on the material will be underrated, and furthermore the variability of the test results is liable to increase. Alternatively, if excessive periods of conditioning are allowed, not only is the test made more expensive but it may become more difficult to distinguish time-dependent effects on the material. Generally technologists have used a rule-of-thumb method to arrive at conditioning times, frequently biased by financial considerations. Their difficulty has been the almost complete lack of available information and the complication of working with a multitude of different test piece geometries. Relatively few standards are available in which conditioning times are specified and the derivations of the figures quoted are never given. For example, BS 903 Part A2 *Tensile Stress Strain Properties of Rubber* states: 'If the test is to be carried out at a temperature other than 20°C the test pieces shall be conditioned at the test temperature immediately prior to testing for a period sufficient to reach substantial temperature equilibrium'.

What are required are the times for the centre of the test piece to come within the tolerance allowed at the temperature in question, calculated for a variety of materials, each tested in liquid and gaseous media over a range of temperature. Furthermore the figures will be required for all the test piece geometries in normal use. The absence of these figures or, in some cases, substitution of inspired guesses at the true values, has been the result of both lack of basic thermal data and the difficulty of deriving the required information. Thermal property data are certainly still rather sparse but considerable advance has been made in the technique of calculating temperature distributions during heating or cooling.

In this report the methods described by Hands[38] have been extended and applied to standard test pieces for rubbers and plastics subjected to step temperature changes in conditioning chambers and baths (Tables 1–3 below).

Test Pieces and Conditions

To make individual calculations for every test piece in current use would clearly be impractical. Fortunately nearly all test pieces used for rubber and plastics fall into three basic geometries. These are cylinders, characterised by diameter and length, flat sheets, characterised by thickness, and flat strips. A test piece may be considered to approximate to an infinite sheet when both its length and width are greater than 4 times its thickness, or to an infinite strip when its length only is greater than 4 times its thickness. The principal form of test piece which does not obviously fall into one of these three geometries is the dumbbell used in tensile testing. However for our purpose dumbbells may be considered as flat strips. For most purposes it is the central parallel part of a dumbbell which is of importance and this width may be taken as the strip width. If the tab ends of the dumbbell are to be considered as well then the overall width is used.

Generally, when a measurement at a non-ambient temperature is to be made, the test piece, at room temperature, is introduced into a cabinet preset at the test temperature. It must be expected that the introduction of a test piece will alter the temperature of the cabinet to some extent but in theory at least the test piece is subjected to a step change of temperature. If time is required for the cabinet to regain its set temperature then this must be considered additional to the times given in the tables. The equilibrium times given have been calculated for step changes from 20°C, and normal variations of room temperature around 20°C may be neglected. The range from −50°C to +250°C was chosen to cover the test temperatures most commonly specified.

Tolerances allowed on temperatures of test are commonly ±1°C or ±2°C. Consequently the equilibrium times calculated are those to reach within 1°C of the set temperature, which should be adequate for all normal circumstances.

The majority of temperature-controlled cabinets used for testing use either air or a liquid as the heat transfer medium, although occasionally test pieces may be in contact with metal platens. Separate figures are quoted in the tables for both air and liquid media. Although a variety of liquids is used, their heat transfer properties have been considered to be very similar.

Although it is essential that the test piece is given sufficient time to reach equilibrium, in practice the conditioning time is not critical to the nearest minute and consequently all times in the tables have been rounded up to the next highest multiple of five minutes. For intermediate temperature changes the time for the next highest temperature should be used.

Cylinders (Table 1, page 58)

Cylindrical test pieces are frequently used in compression stress–strain and compression set tests and also for some resilience and abrasion tests. In the table cylinders have been characterised by diameter and length, specific sizes having been chosen to correspond with currently used standard test pieces. For example the 28·7 mm × 12·7 mm cylinder is that specified in BS 903 part A4.

Flat Sheets (Table 2, page 60)
Flat sheets refer to test pieces where both length and width are large compared with thickness. This covers, for example, specimens for falling weight impact strength measurements and sheets used for hardness and indentation tests. Because the thickness used is often not closely specified, results are given for a range of thicknesses up to 25 mm; for any intermediate thickness the time given for the next thickest sheet in the table should be used.

Flat Strips (Table 3, page 62)
This is the most important table because the majority of specimens for mechanical tests are of this form. There are, for example, the flexural tests and many impact tests and, the largest group of all, the tensile tests. Dumbbells, although shaped, are essentially of strip form and may be treated for most purposes as strips of the same width as the central parallel portion. For example, the type 1 dumbbell of BS 903 Part A2 (Fig. 301.9 of BS 2782) would be taken as a 6 mm wide strip. Also, ring test pieces may be considered as strips with dimensions equal to the cross-sections of the ring. Results are given for a range of widths up to 25 mm. Above this the strip could be treated as a flat sheet. At each width a range of thicknesses are tabulated, which includes some specific to a particular test. Again, for any intermediate dimensions the time for the next largest strip should be used. It should also be noted that the terms 'width' and 'thickness' are completely interchangeable.

Assumed Thermophysical Properties

	Crystalline plastic	*Amorphous plastic*	*Rubber*
Thermal conductivity			
cal/cm s °C	8×10^{-4}	4×10^{-4}	5×10^{-4}
(J/cm s °C)	($33{\cdot}48 \times 10^{-4}$)	($16{\cdot}74 \times 10^{-4}$)	($20{\cdot}92 \times 10^{-4}$)
Thermal diffusivity			
cm^2/s	$1{\cdot}2 \times 10^{-3}$	$0{\cdot}9 \times 10^{-3}$	$1{\cdot}0 \times 10^{-3}$

Assumed values for surface heat transfer coefficient
in air: 5×10^{-4} cal/cm^2 s °C ($20{\cdot}92 \times 10^{-4}$ J/cm^2 s °C)
in oil: $1{\cdot}8 \times 10^{-2}$ cal/cm^2 s °C ($7{\cdot}53 \times 10^{-2}$ J/cm^2 s °C)

REFERENCES

1. Walter, A. T. (1954) *Journal of Polymer Science*, **13**, 217.
2. Heap, R. D. and Norman, R. H. (1966) *Research Report No. 149*, RAPRA.
3. BS 771 (1980) *Phenolic Moulding Materials.*
4. BS 1763 (1975) *Flexible, Unsupported, Thin PVC Sheeting.*
5. BS 1339 (1965) *Definitions, Formulae and Constants relating to the Humidity of the Air.*
6. ASTM E41–63 (1971) *Standard Definition of Terms Relating to Conditioning.*
7. ISO 291 (1977) *Plastics—Standard Atmospheres for Conditioning and Testing.*

8. ISO 1110 (1975) *Polyamide 66, 610 and 6—Accelerated Conditioning of Test Specimens.*
9. ISO 554 (1976) *Standard Atmosphere for Conditioning and/or Testing—Specifications.*
10. BS 2782 (1975) *Methods of Testing Plastics, Foreword, General Introduction and Instructions.*
11. BS 2782 Part 0, Method 010 (draft 1978) *Foreword, General Introduction and Instructions. Standard Atmospheres for Conditioning and Testing, Preferred Test Temperatures, List of Published Methods 1979.*
12. ASTM E171-63 (1972) *Standard Atmospheres for Conditioning and Testing Materials.*
13. ASTM D618-61 (1971) *Conditioning Plastics and Electrical Insulating Materials for Testing.*
14. ASTM D789 (1973) *Nylon Injection Molding and Extrusion Materials.*
15. DIN 50014 (1970) *Testing of Materials, Structural Components and Equipment, Standard Atmospheres for Conditioning and Testing of Plastics, Rubber, Elastomers, Paint, Varnishes and Similar Coating Materials as well as Electrical Insulating Material.*
16. DIN 50005 (1975) *Testing of Plastics and Other Electrical Insulating Material; Selection of Climates for Preconditioning, Conditioning and Testing of Specimens.*
17. DIN 53714 (1978) *Testing of Plastics; Accelerated Conditioning of Test Specimens of Polyamides.*
18. ISO 471 (1977) *Standard Atmospheres for the Conditioning and Testing of Rubber Test Pieces.*
19. ISO 3205 (1976) *Preferred Test Temperatures.*
20. Brown, R. P. and Hands, D. (1973) *RAPRA Members Journal*, April.
21. BS 4194 (1967), *'Design Requirements and Testing of Controlled Atmosphere Laboratories'.*
22. BS 3718 (1964) *Laboratory Humidity Ovens* (*Non Injection Type*).
23. BS 3898 (1965) *Laboratory Humidity Ovens* (*Injection type*).
24. ISO R483 (1966) *Plastics—Methods for Maintaining Constant Relative Humidity in Small Enclosures by means of Aqueous Solutions.*
25. ASTM E104-51 (1971) *Maintaining Constant Relative Humidity by means of Aqueous Solutions.*
26. BS 4864 (1973) *Design and Testing of Enclosures for Environmental Testing.*
27. BS 593 (1974) *Laboratory Thermometers.*
28. ISO R653 (1968) *Long Solid Stem Thermometers for Precision Use.*
29. ISO R654 (1968) *Short Solid Stem Thermometers for Precision Use.*
30. BS 5074 (1974) *Short and Long Solid Stem Thermometers for Precision Use.*
31. BS 1704 (1951) *General Purpose Thermometers.*
32. Miller, T. J. (Ed.) (1971) *The Instrument Manual*, United Trade Press.
33. Barber, C. R. (1971) *The Calibration of Thermometers*, HMSO.
34. BS 4833 (1972) *Hygrometer Tables for use in the Testing or Operation of Environmental Enclosures.*
35. Hickman, M. J. (1970) *Measurement of Humidity*, Notes on Applied Science No. 4, HMSO.
36. Brown, R. P. (1973) *RAPRA Guide to Rubber and Plastics Test Equipment*, RAPRA.
37. Brown, R. P. (1979) *The Physical Testing of Rubbers*, Applied Science, page 66.
38. Hands, D. (1971) *RAPRA Technical Review No. 60*, July.

Diameter (mm)	Height (mm)	Temperature (°C)	Time (min) to 1°C off equilibrium: Crystalline plastic Air	Crystalline plastic Oil	Amorphous plastic Air	Amorphous plastic Oil	Rubber Air	Rubber Oil
64	38	−50	135	60	130	80	130	75
		0	100	45	95	65	95	60
		50	115	50	105	70	105	65
		100	140	60	130	85	130	80
		150	155	65	145	90	145	85
		200	165	70	155	95	155	90
		250	170	75	160	100	160	90
40	30	−50	85	30	75	40	75	35
		0	60	25	55	35	55	30
		50	70	25	60	35	60	30
		100	85	30	75	45	75	35
		150	95	35	85	45	85	40
		200	100	35	90	50	90	45
		250	105	40	90	50	95	45
37	10·2	−50	40	10	35	10	35	10
		0	30	10	25	10	25	10
		50	35	10	25	10	30	10
		100	40	10	35	10	35	10
		150	45	10	35	10	40	10
		200	50	10	40	15	40	10
		250	50	10	40	15	45	15
32	16·5	−50	50	15	45	20	45	15
		0	40	10	30	15	35	15
		50	45	15	35	15	35	15
		100	55	15	45	20	45	20
		150	60	15	50	20	50	20
		200	65	20	50	20	55	20
		250	65	20	55	25	55	20
29	25	−50	60	20	50	25	50	20
		0	45	15	40	20	40	15
		50	50	15	40	20	45	20
		100	60	20	50	25	55	25
		150	70	20	55	25	60	25
		200	70	25	60	30	65	25
		250	75	25	65	30	65	25
28·7	12·7	−50	40	10	35	15	35	10
		0	30	10	25	10	25	10
		50	35	10	30	10	30	10
		100	45	10	35	15	35	15
		150	50	10	40	15	40	15
		200	50	15	40	15	45	15
		250	55	15	40	15	45	15

Diameter (mm)	Height (mm)	Temperature (°C)	Time (min) to 1°C off equilibrium: Crystalline plastic		Amorphous plastic		Rubber	
			Air	Oil	Air	Oil	Air	Oil
25	20	−50	50	15	40	20	40	15
		0	35	10	30	15	30	15
		50	40	10	35	15	35	15
		100	50	15	40	20	45	15
		150	55	15	45	20	45	20
		200	60	15	50	20	50	20
		250	60	15	50	20	50	20
25	8	−50	30	5	25	10	25	5
		0	20	5	20	5	20	5
		50	25	5	20	5	20	5
		100	30	5	25	10	25	5
		150	35	5	25	10	30	10
		200	35	10	30	10	30	10
		250	35	10	30	10	30	10
25	6·3	−50	25	5	20	5	20	5
		0	20	5	15	5	15	5
		50	20	5	15	5	20	5
		100	25	5	20	5	20	5
		150	30	5	20	5	25	5
		200	30	5	25	5	25	5
		250	30	5	25	5	25	5
13	12·6	−50	25	5	20	5	20	5
		0	20	5	15	5	15	5
		50	20	5	15	5	20	5
		100	30	5	20	10	20	5
		150	30	5	25	10	25	10
		200	30	5	25	10	25	10
		250	35	5	25	10	25	10
13	6·3	−50	20	5	15	5	15	5
		0	15	5	10	5	10	5
		50	15	5	15	5	15	5
		100	20	5	15	5	15	5
		150	20	5	15	5	20	5
		200	25	5	20	5	20	5
		250	25	5	20	5	20	5
9·5	9·5	−50	5	5	15	5	15	5
		0	5	5	10	5	10	5
		50	5	5	10	5	15	5
		100	5	5	15	5	15	5
		150	5	5	15	5	20	5
		200	5	5	15	5	20	5
		250	5	5	20	5	20	5

Thickness (mm)	Temperature (°C)	Time (min) to 1°C off equilibrium Crystalline plastic Air	Oil	Amorphous plastic Air	Oil	Rubber Air	Oil
25	−50	115	80	145	100	135	90
	0	80	65	105	85	95	75
	50	90	70	120	90	110	80
	100	115	80	150	100	140	90
	150	130	85	165	105	155	95
	200	135	85	180	110	160	100
	250	140	90	185	115	170	105
15	−50	60	30	80	40	70	35
	0	40	25	55	30	50	30
	50	45	30	65	35	60	30
	100	60	30	80	40	75	35
	150	65	35	90	40	80	40
	200	70	35	95	40	85	40
	250	75	35	100	45	90	40
10	−50	35	15	50	20	45	15
	0	25	15	35	15	30	15
	50	30	15	40	15	35	15
	100	40	15	50	20	45	20
	150	40	15	55	20	50	20
	200	45	15	60	20	55	20
	250	45	20	60	20	55	20
8	−50	30	10	40	15	35	10
	0	20	10	30	10	25	10
	50	25	10	30	10	30	10
	100	30	10	40	15	35	10
	150	35	10	45	15	40	10
	200	35	10	45	15	40	15
	250	35	15	50	15	45	15
5	−50	20	5	25	5	20	5
	0	15	5	20	5	15	5
	50	15	5	20	5	20	5
	100	20	5	25	5	20	5
	150	20	5	25	5	25	5
	200	20	5	30	5	25	5
	250	20	5	30	10	25	5
3	−50	10	5	15	5	15	5
	0	10	5	10	5	10	5
	50	10	5	15	5	10	5
	100	10	5	15	5	15	5
	150	15	5	15	5	15	5
	200	15	5	20	5	15	5
	250	15	5	20	5	15	5

Thickness (mm)	Temperature (°C)	Time (min) to 1°C off equilibrium: Crystalline plastic		Amorphous plastic		Rubber	
		Air	Oil	Air	Oil	Air	Oil
2	−50	10	5	10	5	10	5
	0	5	5	10	5	10	5
	50	5	5	10	5	10	5
	100	10	5	10	5	10	5
	150	10	5	10	5	10	5
	200	10	5	15	5	10	5
	250	10	5	15	5	10	5
1	−50	5	5	5	5	5	5
	0	5	5	5	5	5	5
	50	5	5	5	5	5	5
	100	5	5	5	5	5	5
	150	5	5	5	5	5	5
	200	5	5	5	5	5	5
	250	5	5	10	5	5	5
0·2	−50	5	5	5	5	5	5
	0	5	5	5	5	5	5
	50	5	5	5	5	5	5
	100	5	5	5	5	5	5
	150	5	5	5	5	5	5
	200	5	5	5	5	5	5
	250	5	5	5	5	5	5

Width (mm)	Thickness (mm)	Temperature (°C)	Time (min) to 1°C off equilibrium: Crystalline plastic Air	Crystalline plastic Oil	Amorphous plastic Air	Amorphous plastic Oil	Rubber Air	Rubber Oil
25·4	12·7	−50	50	10	40	15	45	15
		0	35	10	30	10	30	10
		50	40	10	35	15	35	10
		100	55	10	40	15	45	15
		150	60	15	45	15	50	15
		200	60	15	50	15	50	15
		250	65	15	50	20	55	15
	10	−50	45	10	35	10	35	10
		0	30	5	25	10	25	10
		50	35	10	30	10	30	10
		100	45	10	35	10	35	10
		150	50	10	40	10	40	10
		200	50	10	40	15	40	10
		250	55	10	40	15	45	10
	9·5	−50	40	10	30	10	35	10
		0	30	5	25	10	25	10
		50	35	10	25	10	30	10
		100	40	10	35	10	35	10
		150	45	10	35	10	40	10
		200	50	10	40	10	40	10
		250	50	10	40	10	40	10
	6·5	−50	30	5	25	5	25	5
		0	20	5	15	5	20	5
		50	25	5	20	5	20	5
		100	30	5	25	5	25	5
		150	35	5	25	5	30	5
		200	35	5	25	5	30	5
		250	40	5	30	10	30	5
	5·0	−50	25	5	20	5	20	5
		0	20	5	15	5	15	5
		50	20	5	15	5	15	5
		100	25	5	20	5	20	5
		150	30	5	20	5	20	5
		200	30	5	20	5	25	5
		250	30	5	25	5	25	5
	3·0	−50	15	5	10	5	15	5
		0	10	5	10	5	10	5
		50	15	5	10	5	10	5
		100	15	5	10	5	15	5
		150	20	5	15	5	15	5
		200	20	5	15	5	15	5
		250	20	5	15	5	15	5

Width (mm)	Thickness (mm)	Temperature (°C)	Time (min) to 1°C off equilibrium: Crystalline plastic Air	Crystalline plastic Oil	Amorphous plastic Air	Amorphous plastic Oil	Rubber Air	Rubber Oil
25·4	2·0	−50	10	5	10	5	10	5
		0	10	5	5	5	10	5
		50	10	5	10	5	10	5
		100	10	5	10	5	10	5
		150	15	5	10	5	10	5
		200	15	5	10	5	10	5
		250	15	5	10	5	10	5
	1·0	−50	5	5	5	5	5	5
		0	5	5	5	5	5	5
		50	5	5	5	5	5	5
		100	5	5	5	5	5	5
		150	10	5	5	5	5	5
		200	10	5	5	5	5	5
		250	10	5	5	5	5	5
15·0	15·0	−50	45	10	35	15	35	10
		0	35	10	25	10	30	10
		50	35	10	30	10	30	10
		100	45	10	35	15	40	10
		150	50	10	40	15	40	15
		200	55	15	40	15	45	15
		250	55	15	45	15	45	15
12·7	12·7	−50	35	10	30	10	30	10
		0	25	5	20	10	25	10
		50	30	10	25	10	25	10
		100	40	10	30	10	30	10
		150	40	10	35	10	35	10
		200	45	10	35	10	35	10
		250	45	10	35	10	40	10
	10·0	−50	35	5	25	10	25	10
		0	25	5	20	5	20	5
		50	25	5	20	5	20	5
		100	35	5	25	10	30	10
		150	35	10	30	10	30	10
		200	40	10	30	10	30	10
		250	40	10	30	10	35	10
	9·5	−50	30	5	25	10	25	10
		0	25	5	20	5	20	5
		50	25	5	20	5	20	5
		100	35	5	25	10	25	10
		150	35	10	30	10	30	10
		200	40	10	30	10	30	10
		250	40	10	30	10	35	10

Width (mm)	Thickness (mm)	Temperature (°C)	Time (min) to 1°C off equilibrium: Crystalline plastic Air	Crystalline plastic Oil	Amorphous plastic Air	Amorphous plastic Oil	Rubber Air	Rubber Oil
12·7	6·5	−50	25	5	20	5	20	5
		0	20	5	15	5	15	5
		50	20	5	15	5	15	5
		100	25	5	20	5	20	5
		150	30	5	20	5	25	5
		200	30	5	25	5	25	5
		250	30	5	25	5	25	5
	5·0	−50	20	5	15	5	15	5
		0	15	5	10	5	15	5
		50	15	5	15	5	15	5
		100	20	5	15	5	20	5
		150	25	5	20	5	20	5
		200	25	5	20	5	20	5
		250	25	5	20	5	20	5
	3·2	−50	15	5	10	5	15	5
		0	10	5	10	5	10	5
		50	15	5	10	5	10	5
		100	15	5	10	5	15	5
		150	15	5	15	5	15	5
		200	20	5	15	5	15	5
		250	20	5	15	5	15	5
	3·0	−50	15	5	10	5	10	5
		0	10	5	10	5	10	5
		50	10	5	10	5	10	5
		100	15	5	10	5	10	5
		150	15	5	10	5	15	5
		200	15	5	15	5	15	5
		250	20	5	15	5	15	5
	2·0	−50	10	5	10	5	10	5
		0	10	5	5	5	5	5
		50	10	5	5	5	10	5
		100	10	5	10	5	10	5
		150	10	5	10	5	10	5
		200	15	5	10	5	10	5
		250	15	5	10	5	10	5
	1·0	−50	5	5	5	5	5	5
		0	5	5	5	5	5	5
		50	5	5	5	5	5	5
		100	5	5	5	5	5	5
		150	10	5	5	5	5	5
		200	10	5	5	5	5	5
		250	10	5	5	5	5	5

Width (mm)	Thickness (mm)	Temperature (°C)	Time (min) to 1°C off equilibrium: Crystalline plastic Air	Crystalline plastic Oil	Amorphous plastic Air	Amorphous plastic Oil	Rubber Air	Rubber Oil
6·35	12·7	−50	25	5	20	5	20	5
		0	20	5	15	5	15	5
		50	20	5	15	5	15	5
		100	25	5	20	5	20	5
		150	30	5	20	5	25	5
		200	30	5	20	5	25	5
		250	30	5	25	5	25	5
	10·0	−50	25	5	15	5	20	5
		0	15	5	15	5	15	5
		50	20	5	15	5	15	5
		100	25	5	20	5	20	5
		150	25	5	20	5	20	5
		200	25	5	20	5	20	5
		250	30	5	20	5	25	5
	6·5	−50	20	5	15	5	15	5
		0	15	5	10	5	10	5
		50	15	5	10	5	15	5
		100	20	5	15	5	15	5
		150	20	5	15	5	15	5
		200	25	5	15	5	20	5
		250	25	5	20	5	20	5
	5·0	−50	15	5	15	5	15	5
		0	15	5	10	5	10	5
		50	15	5	10	5	10	5
		100	15	5	15	5	15	5
		150	20	5	15	5	15	5
		200	20	5	15	5	15	5
		250	20	5	15	5	15	5
	3·0	−50	15	5	10	5	10	5
		0	10	5	10	5	10	5
		50	10	5	10	5	10	5
		100	15	5	10	5	10	5
		150	15	5	10	5	10	5
		200	15	5	10	5	10	5
		250	15	5	10	5	10	5
	2·0	−50	10	5	10	5	10	5
		0	10	5	5	5	5	5
		50	10	5	5	5	5	5
		100	10	5	10	5	10	5
		150	10	5	10	5	10	5
		200	10	5	10	5	10	5
		250	10	5	10	5	10	5

Width (mm)	Thickness (mm)	Temperature (°C)	Time (min) to 1°C off equilibrium: Crystalline plastic, Air	Crystalline plastic, Oil	Amorphous plastic, Air	Amorphous plastic, Oil	Rubber, Air	Rubber, Oil
6·35	1·52	−50	10	5	5	5	5	5
		0	5	5	5	5	5	5
		50	5	5	5	5	5	5
		100	10	5	5	5	5	5
		150	10	5	5	5	10	5
		200	10	5	10	5	10	5
		250	10	5	10	5	10	5
	1·0	−50	5	5	5	5	5	5
		0	5	5	5	5	5	5
		50	5	5	5	5	5	5
		100	5	5	5	5	5	5
		150	5	5	5	5	5	5
		200	10	5	5	5	5	5
		250	10	5	5	5	5	5
4·0	12·7	−50	20	5	15	5	15	5
		0	15	5	10	5	10	5
		50	15	5	10	5	10	5
		100	20	5	15	5	15	5
		150	20	5	15	5	15	5
		200	20	5	15	5	15	5
		250	20	5	15	5	20	5
	10·0	−50	15	5	15	5	15	5
		0	15	5	10	5	10	5
		50	15	5	10	5	10	5
		100	15	5	15	5	15	5
		150	20	5	15	5	15	5
		200	20	5	15	5	15	5
		250	20	5	15	5	15	5
	6·5	−50	15	5	10	5	10	5
		0	10	5	10	5	10	5
		50	10	5	10	5	10	5
		100	15	5	10	5	10	5
		150	15	5	15	5	15	5
		200	20	5	15	5	15	5
		250	20	5	15	5	15	5
	5·0	−50	15	5	10	5	10	5
		0	10	5	10	5	10	5
		50	10	5	10	5	10	5
		100	15	5	10	5	10	5
		150	15	5	10	5	10	5
		200	15	5	10	5	15	5
		250	15	5	10	5	15	5

Width (mm)	Thickness (mm)	Temperature (°C)	Time (min) to 1°C off equilibrium: Crystalline plastic Air	Crystalline plastic Oil	Amorphous plastic Air	Amorphous plastic Oil	Rubber Air	Rubber Oil
4·0	3·0	−50	10	5	10	5	10	5
		0	10	5	5	5	5	5
		50	10	5	5	5	10	5
		100	10	5	10	5	10	5
		150	10	5	10	5	10	5
		200	15	5	10	5	10	5
		250	15	5	10	5	10	5
	2·0	−50	10	5	5	5	5	5
		0	5	5	5	5	5	5
		50	10	5	5	5	5	5
		100	10	5	5	5	10	5
		150	10	5	10	5	10	5
		200	10	5	10	5	10	5
		250	10	5	10	5	10	5
	1·0	−50	5	5	5	5	5	5
		0	5	5	5	5	5	5
		50	5	5	5	5	5	5
		100	5	5	5	5	5	5
		150	5	5	5	5	5	5
		200	5	5	5	5	5	5
		250	5	5	5	5	5	5

Chapter Five

Processability Tests

5.1 INTRODUCTION

The majority of the properties of plastics discussed in this book relate to finished articles. While these are important to users, there are a number of tests which are applied to plastics compositions in the unmoulded or pre-cast state. Certain of the properties of such 'raw materials' are of paramount importance to the processor and fabricator if he is to predict his machinery requirements rather than pursue an entirely *ad hoc* approach. Again, there are tests designed for application to polymers prior to their formulation into moulding compositions. This chapter considers tests which can be grouped under the general heading of 'processability' and which range from the gel time of a polyester resin to a study of the rheological (flow) behaviour of a polymer in the molten state.

5.2 VISCOELASTIC FLOW BEHAVIOUR

Study of the flow behaviour of molten polymers is essential to the moulder or fabricator if he is to establish the feasibility of any intended process involving melting. If the test is ideal it will forecast a variety of properties such as thermal stability, ease of processing (is the machine powerful enough?) and quality of mouldings. The manufacturer of plastics compositions will likewise be interested in these subjects in order to give his customers an adequate technical service, for he can exercise a considerable degree of control over processability by altering parameters such as average molecular weight and molecular weight distribution of polymer and by incorporating various types and quantities of plasticisers, lubricants and stabilisers. Efficient and reliable laboratory-scale test procedures for processability are essential if experimental work is to be kept within reasonable bounds during the early stages of development of a new polymer or moulding compound. Ultimately, however, when a 'short list' has been prepared, there is still no real substitute for trying out the most promising compositions on a factory scale, using the actual processing equipment to be employed in manufacture.

Molten polymers are viscoelastic materials and hence the study of their flow behaviour is not without complication. In an ideal or Newtonian liquid, to which water at normal temperatures approaches very closely, there is a linear relationship between shear rate (rate of deformation) du/dr and shearing stress τ

such that

$$\tau = \eta \frac{du}{dr} \tag{5.1}$$

where η is a constant known as the coefficient of viscosity, or simply the *viscosity*; this constant, as is well known, is markedly dependent on temperature. Application of the concept to flow of a Newtonian liquid through a capillary of radius r and length l leads to the familiar Poiseuille formula:

$$\frac{Pr}{2l} = \eta \frac{4Q}{\pi r^3} \tag{5.2}$$

where P is the pressure on the liquid and Q the volume rate of flow. It is unfortunate that this formula does not apply to a great number of liquid systems, including polymer melts, all of which are non-ideal. Some liquids change their flow behaviour with time, for example one may 'thicken up' after stirring, another 'thin down'. Even ignoring these complications and considering only time-stable phenomena, there are still a number of possibilities, some of which are illustrated in Figure 5.1.

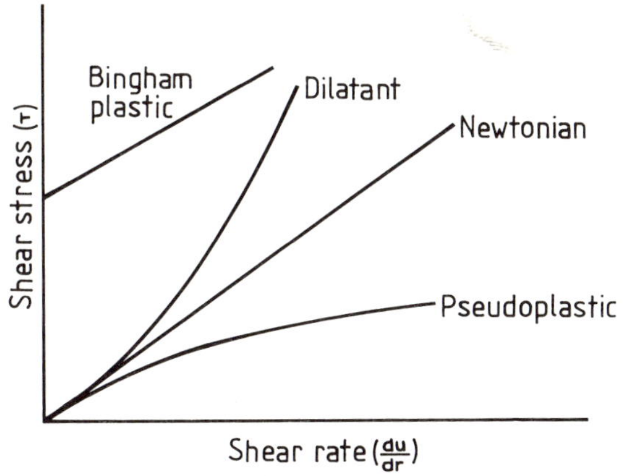

Figure 5.1 Flow relationships for various types of liquid

Many polymer melts follow the 'pseudoplastic' pattern and hence the relationship between shear rate and shear stress is not linear. It follows that, since the common processing techniques for thermoplastics are associated with shear rates between about 100 s^{-1} and 100 000 s^{-1}, there is really no substitute for a comprehensive study of shear stress against shear rate over a range of the latter embracing all likely processing techniques throughout a suitable temperature range. The majority of standard flow tests employ only one (low) shear rate, either fixed by the machine used or dictated in part by the flow behaviour of the material under examination. Clearly, such a one-point observation cannot yield much information for general processability and, although the test may be useful for quality control measures, it can be misleading even in this context since the precise shapes of all pseudoplastic curves are not identical (Figure 5.2).

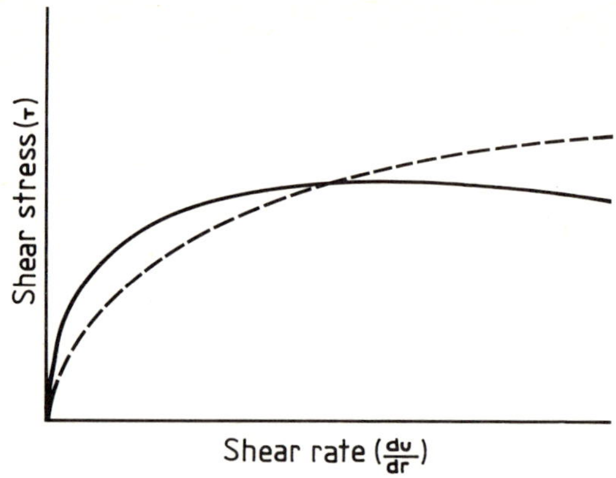

Figure 5.2 Different pseudoplastic curves

Detailed consideration of the flow behaviour of molten polymers is a science in itself: Bernhardt,[1] for instance, studies the subject in relation to common processing techniques, while Cogswell[2] provides a comprehensive study of modern concepts of the subject as a whole.

So far we have only considered materials whose flow characteristics, if not time-independent, are such that no change of state is involved. Thermoplastics fit this description, except when heating to fluidise has been carried on for so long that significant decomposition has taken place. When, however, we turn to thermosets, where a fluid mass sets to the solid state while still hot and during processing, it can well be imagined that study of such an unstable system is very complicated.

During the past ten years there have been increasing studies of the rheological properties of polymer melts and their relationship to processability. For example, Brydson[3] has outlined the relationship for a number of the common processing techniques for plastics; Paul[4] has suggested that measurement using a variable torque rheometer, capillary or orifice rheometer and oscillating disc rheometer can characterise thermosetting materials for injection moulding and, perhaps, for other processes; and Wissbrun[5] has outlined the utility and limitations in the application of rheology to polymer processing with particular reference to development of suitable grades of acetal resin for extrusion blow moulding. Even so, it is regrettable that we still do not have a single standard test method for determining the melt viscosity of polymers at shear rates approaching those of the common processing techniques, which would be useful for a realistic prediction of processability.

5.3 FLOW TESTS FOR THERMOPLASTIC MATERIALS

5.3.1 Melt Flow Index

The determination of melt flow index (MFI) is one of the most widely used tests, though it suffers from the disadvantages discussed above. (In the ISO, BS and ASTM tests for MFI values in the range 0·3 to 20, the corresponding shear rates are 1 to 50 s^{-1}.) The test exists in a number of variants according to polymer type and molecular weight. Although widely used, the test has limitations; while it is of particular value as a quality control test on thermoplastics having relatively low melt viscosities, it does not necessarily indicate uniformity of other properties of polymers of the same type but made by other manufacturers, nor is MFI a fundamental polymer property.

In Method 105C of BS 2782 (1970),[6] three variants are described corresponding to ISO R292 (1967).[7] These two standards refer specifically to 'The Determination of Melt Flow Index of Polyethylene and Polyethylene Compounds'. ISO R1133 (1969)[8] and the later BS method 720A (1979),[9] which replaces 105C, uses the same principle of test but carries a broader variation of test procedures to cover other thermoplastics. ASTM D1238 (1979)[10] and DIN 53735 (1977)[11] differ from ISO R1133 in respect of the number of different loading conditions permitted; these will be detailed later.

The principle employed is that of determining the rate of flow of molten polymer through a closely defined extrusion plastometer: Figure 5.3 shows a cross-section of the important parts.

The cylinder is of hardened steel and is fitted with heaters, lagged and controlled for operation at the required temperature (between 125°C and 300°C; all ±0·5°C); the dimensions of the cylinder, particularly its radius, are closely specified (see Poiseuille formula above: although, as explained, it does not apply

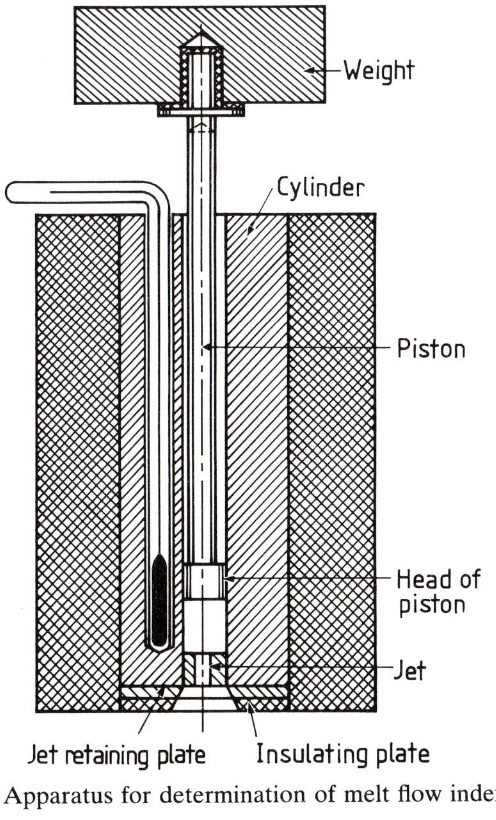

Figure 5.3 Apparatus for determination of melt flow index: BS 2782

to polymer melts, often a power law of the type $\tau = K[(du/dr)^n]$ does so). The piston is of steel and the diameter of its head is 0·075 ± 0·015 mm less than that of the internal diameter of the cylinder (which is 9·5 mm). The die (or 'jet') has an internal diameter of 2·095 mm ± 0·005 mm or 1·180 mm ± 0·005 mm (depending on procedure used) and is made of hardened steel. All surfaces of the apparatus which come into contact with the molten polymer are highly polished. Weights are provided so that the piston can be loaded to give a total weight of between 325 g and 21 600 g, according to the procedure used.

The first requisite is to ensure that the cylinder, piston and jet are scrupulously clean by treatment with hot solvents and wiping with lint-free cloths (*abrasives must not be used*). The cylinder is maintained at the test temperature for 15 minutes and then charged with between 4 and 8 g of sample (within a period not exceeding 1 minute) and the unloaded piston reinserted. Four to six minutes later the weight is added and molten material begins to extrude through the die. The rate of extrusion is measured by cutting off the extrudate at suitable time intervals (given in the standard). Several such 'cut-offs' are taken up to 30 minutes after insertion of the sample and these must all be taken when the piston head is between 50 mm and 20 mm above the upper end of the die (as marked by scribed lines on the piston). The first cut-off and any containing bubbles are discarded; the remainder, at least three, are weighed (to nearest 0·001 g) and the average mass is calculated. The maximum and minimum values of the individual weighings must be within specified values of the average, otherwise the result is discarded and a further test performed on a fresh portion of the sample.

The melt flow rate (or melt flow index) is calculated as the weight (g) extruded in a reference time as specified in the standard (usually 600 s). In the report of results the temperature of test and piston loading must be quoted.

ISO R1133 is currently under revision and Method 720A of BS 2782 has been modelled on the revised version of ISO R1133.

ISO R292 allows three standard procedures, (A), (B) and (C), which correspond to Procedures (4), (1) and (5) respectively of ISO R1133. The differences in the three procedures are simply different die sizes and reference times, the temperature remaining the same at 190°C. It should be noted that ASTM D1238 and DIN 53735 allow only one size of die and therefore procedures (B) of ISO R292 and (1) of R1133 cannot be employed in these standard test methods.

ISO R1133 incorporates seventeen procedural variations depending on temperature, die size, piston loading and reference times (120 to 600 s). ASTM D1238 allows twenty conditions of test (A to T) depending on temperature and piston load, while DIN 53735 suggests twenty conditions based on variations of temperature and piston loads.

When using the test method for polymers other than the polyolefins, caution is necessary in selecting the appropriate conditions. ISO R1133, ASTM D1238 and DIN 53735 do not always agree on the suggested procedures. For example, ISO R1133 and ASTM D1238 mention a temperature of 200°C and a piston loading of 5000 g for determining the melt flow rate of ABS, whereas DIN 53735 offers two alternatives of 200°C with a loading of 21 000 g or 220°C with a loading of 10 000 g.

ASTM D3364–74 (1979)[12] provides an extension of the ASTM D1238 test for the measurement of flow rates of poly(vinyl chloride) and rheologically unstable thermoplastics. The apparatus differs only in respect of the die which has a 120° angle of entry and is about 3 times as long as for D1238. In addition it is recommended that the temperature control be maintained at ±0·1°C. A cylinder temperature of 175°C and piston loading of 20 000 g is specified but may be changed to 5000 g for PVC compounds having high flow rates (in excess of 10 g/10 min). ASTM D3364 is claimed as being useful for quality control tests on PVC compound having a wide range of melt viscosities. Measurements are made at shear rates close to 1 s^{-1}. However, for rigid PVC compound D1238 is recommended at 190°C with a piston load of 21 600 g.

5.3.2 Rossi-Peakes Test

Method 105A of BS 2782 (1970)[6] describes a test for determining the flow temperature of thermoplastic moulding material, defined as that temperature at which a thermoplastic moulding material flows a specified distance under a specified pressure when tested in a standard apparatus.

The apparatus is specified in terms of all essential dimensions, etc., and illustrated by reference to the Rossi-Peakes flow tester (Figure 5.4).

To carry out the test, the apparatus is assembled and the temperature adjusted according to the material under examination. A pellet of the test material, of size to match the charge chamber, is introduced therein and the pressure applied immediately. The length of flow after 2 min is determined and then the material is removed and the operation repeated with a fresh pellet. Two pellets are tested at each of three temperatures, at all of which the flow length is between 13 and 38 mm and where at least one measurement is above 25 mm and one below. The results are plotted and the flow temperature is read off where the length of flow is 25·4 mm.

It is possible that Method 105A will be withdrawn when the current revision of BS 2782 is complete, but it is mentioned because it is still used for cellulose acetate.

In ASTM D569–59 (1976),[13] two alternative procedures are offered, one as above and the other where the degree of flow is measured at a specified temperature. In this latter method, a pressure of 3·45, 6·89 or 10·35 MPa is used and the degree of flow is measured at the prescribed temperature.

5.3.3 Other Flow Tests

Although the melt flow index and Rossi-Peakes tests are the only ones generally standardised by ISO, BSI or ASTM for thermoplastics there are many others, most of which suffer from the same disadvantages as those described above, namely that they use operating conditions, especially shear rates, far removed from practice. As far back as 1952 Hayes[14] described two extrusion plastometers and Gray *et al.*[15] in 1968 developed an apparatus where change in apparent viscosity with temperature could be followed continuously and, it was claimed, the minimum and maximum processing temperatures could be predicted. Again, the apparatus suffered from determination at low shear rates.

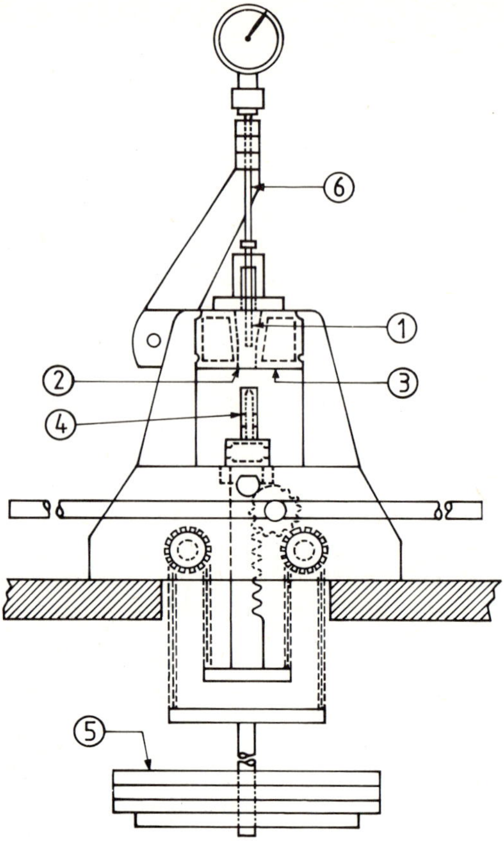

Figure 5.4 Rossi-Peakes flow tester: BS 2782
The orifice (1) is highly polished and runs in a split cone which is clamped in the heater block (3). A thermometer well is inserted into a hole drilled into one half of the split cone. The charge chamber (2) is situated below the orifice and a heated ram (4) is so positioned that it forces material in the chamber up into the orifice. (5) is a system for applying a pressure of 10·3 MPa to the ram. The rate of flow is measured by (6), consisting of a follower rod and indicator exerting a specified pressure

A useful *ad hoc* method of comparative testing is a spiral disc mould where a fair length of flow channel can be provided by a conveniently sized mould fitted to an injection machine. Campbell and Griffiths[16] describe one where the effective length is 77 in (1·956 m), in a mould of overall dimensions 12 in × 9 in (305 × 230 mm). The flow length of the material into the mould, which is vented to atmosphere at its inner end, is determined and related to the test conditions of cycle time, temperatures of mould and barrel, pressure, etc.

During the past 20 years the torque rheometer has found wide use for studying the behaviour of plastics materials. The apparatus consists of a small mixing head, similar in many respects to a miniature internally heated mixer of the Banbury type, driven by a motor via a torque-measuring device. Various sizes of mixing head are available, all fitted with a feed chute and loaded ram. The mixing chamber is fitted with a thermocouple for recording graphically against time as the test proceeds. The mixing head is usually heated by oil circulation and the temperature is controlled to within 1°C. Provision is made to vary the rotor speed, modern instruments having a speed range between 10 and 200 r.p.m.

In use, the weighed charge of plastics material, in powder or granule form, is introduced into heated chamber with the motor running, the load is applied to the feed chute ram and the chart recorder(s) are started. In this way material change with time under dynamic conditions can be determined and this provides useful data, for example, as to what happens during the compounding of a plastics material.

The torque rheometer is available commercially; both the Brabender and Hampden/RAPRA instruments are well known. Some torque rheometers can be fitted with a small instrumented extruder in place of the mixing head, thus adding versatility to the machine.

McCabe[17,18] has described rheological measurements with the Brabender instrument, Schramm[19] the measurement of fusion rate of rigid PVC dry blend and Matthan[20] has evaluated the variables of Brabender Plastograph. Paul[21,22] has described the Hampden/RAPRA machine and has shown how it may be used with PVC. Allen and Williams[23] have used a Brabender instrument to predict polymer processing properties and use of their technique can give relative melt viscosity, stability and moulding temperatures of a range of thermoplastics.

In recent years much work has been described with specially designed pieces of apparatus aimed at bridging the gap between laboratory tests and factory operations. Various workers have attached measuring instruments to actual extruders, injection moulding machines and rheometers, and facilities for studying the effects of a wide range or rates of shear have been described.[24] Often small-scale laboratory extruders, etc., are instrumented and these are available commercially. One such apparatus which has found use in the laboratory is the extrusion capillary rheometer. This is basically a miniature ram extruder fitted with a capillary die and instrumented so that melt temperature and pressure behind the die can be monitored. Speed of extrusion can be varied so that the properties of a plastics melt can be determined under a wide range of shear rates. Dies with different lengths and angles of entry are available to permit study of the effect of die design on output and quality of extrudate.

Several extrusion capillary rheometers available commercially are capable of operating at shear rates up to 10 000 s^{-1}.[25] One commercially available type of extrusion rheometer is shown in Figure 5.5.

Two properties of polymers which are of considerable interest to processors are die swell and melt fracture. Die swell is simply defined as the degree of swelling of a polymer melt as it issues from a die. It is important for the design of extrusion dies and as a quality control parameter for polymers used in critical extrusion applications (for example, extrusion blow moulding of bottles). Melt fracture is manifested by a curious 'sharkskin' effect seen on the surface of extrudates and clearly must be avoided if extruded products of acceptable quality are to be produced. There is no standardised test method for either of these properties but it is possible to use an extrusion rheometer to investigate

Figure 5.5 Davenport extrusion capillary rheometer
(Courtesy Daventest Ltd)

both melt fracture and die swell. Melt fracture may be examined by finding the temperature and rate of extrusion at which it just commences, while die swell can be determined by comparison of the diameter of extrudate with the die diameter. At best, such methods serve as useful quality control tests only, and great care is needed in interpreting the results to predict performance on full-scale processing equipment. Some control laboratories use the ordinary MFI apparatus (BS 2782, Method 720A) to determine the die swell at the same time as the normal MFI test is carried out.

5.4 FLOW TESTS FOR THERMOSETTING MATERIALS

The tests mentioned in §5.3 have all been designed, basically, for thermoplastics materials, although mention has been made in §5.2 that some workers have used, for example, capillary rheometers to assess the processability of thermosetting materials. The Rossi-Peakes test (§5.3.2) has been used for thermosets and Von Meysenburge[26] has described a modified form of this test. A number of tests designed to determine the flow properties of thermosetting materials have now been standardised.

5.4.1 'Cup Flow' Test

A test designed specifically for thermosetting moulding materials is the so-called 'cup flow' determination. Method 105B of BS 2782 (1970)[6] (to be revised and issued as Method 720B) describes the UK practice and limits applicability to phenolic and alkyd moulding materials, excluding 'fast-curing' ones. The cup flow mould is shown in Figure 5.6.

Needless to say, in a flow test such as this the surfaces of the inner male and female members must be smooth and highly polished. The general form and dimensions (including steam coring) are very carefully specified.

The temperature of the mould is laid down as 163° ± 1°C and the thrust on the mould 100 ± 5 kN. The closing speed of the mould when empty must be 130 mm in 4 to 5 s. The mass of moulding material to use is found by trial and error so that a flash of between 2 and 2·5 g is obtained in the test. The mould is opened just sufficiently to charge the weighed-out sample of material, at 10°C to 40°C, with the aid of a scoop and the mould is closed again immediately; the time

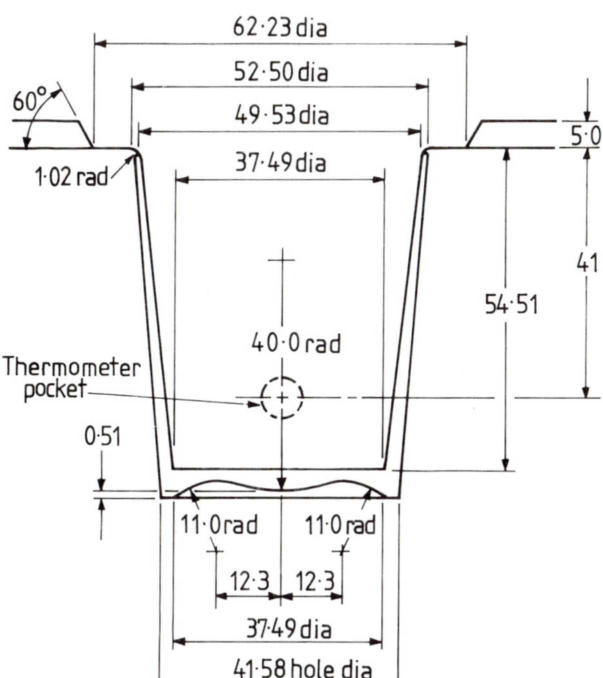

Figure 5.6 Detail of cup flow mould: BS 2782—dimensions in millimetres

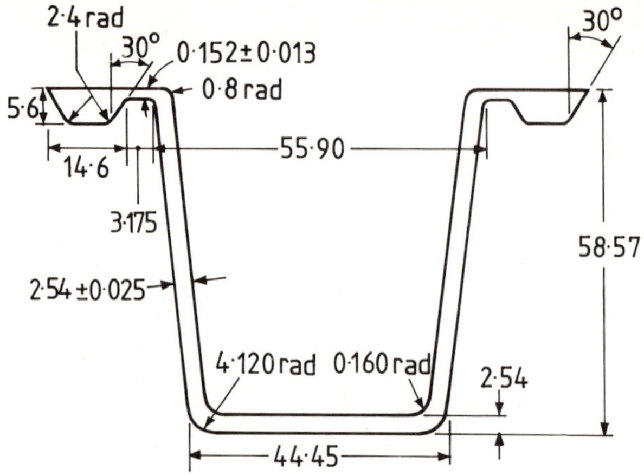

Figure 5.7 Detail of cup flow mould: ASTM D731—dimensions in millimetres
All surfaces are highly polished to No. 2 microfinish (SPI–SPE standard for mould finish). Rockwell C-58 steel. Tolerances on dimensions are ±0·025 mm except as noted

between dropping the powder into the mould and the registration of full pressure must be between 5 and 10 s.

The time of flow is measured in seconds as that between the first registration of pressure on the hydraulic gauge of the press and the instant the flash is seen to cease moving. This is the 'cup flow' of the material.

The technique described in ASTM D731–67 (1976),[27] to determine moulding index, follows much the same pattern although the mould cavity differs in design (Figure 5.7).

An elaborate method is laid down for obtaining the appropriate weight of material to use. For materials of impact strength under 27 J/m (by ASTM D256 Method A (Chapter 8)) a cup with flash of thickness 0·15 to 0·20 mm is moulded, the flash removed and the cup weighed; this weight, multiplied by 1·1, is the charge to use in the test, except for materials of impact strength 27 J/m and above when flash of thickness 0·51 to 0·66 mm is produced and the cup weight multiplied by 1·05. The preferred temperature is 150°, 155° or 165°C (all ±1°C), according to the material under test. A load is first determined which produces a cup of the required flash thickness and then the next lowest load is selected, from a table of eight specified alternatives, for test purposes. The time of flow in seconds is recorded from the instant the hydraulic gauge registers an applied load of 454 kg to when the prescribed flash thickness has been reached.

The result is expressed as the 'moulding index', the minimum force required to produce the necessary flash, with a subscript of the closing time in seconds, e.g. 6800_{18} kgf for a 6800 kg minimum load and 18 s flow time.

DIN 53465 (1963)[28] also uses a 'cup flow' type of mould and expresses the results as 'closing time' of thermosetting moulding materials.

5.4.2 Spiral Flow Test

A test designed for low-pressure thermosetting moulding compounds (moulding pressures less than 6·9 MPa) is described in ASTM D3123-72 (1979).[29] This utilises a transfer moulding press fitted with a specified spiral flow mould and the test is conducted at a mould temperature of 150° ± 3°C and transfer pressure of 6·9 MPa. After cure the moulding is removed and its length is read off to the nearest 6·35 mm which is reported as the spiral flow length of the material under test.

5.4.3 Disc Flow Test

Method 105D of BS 2782 (1970)[6] describes a method for determining the flow of synthetic resin impregnated glass fabric. This method is currently under review and may be withdrawn when the current revision of BS 2782 is complete. Nevertheless it is thought worth while to detail the test method. At least four pieces of the sample, each 100 mm square, carefully superimposed to form a rectangular pack, comprise the test piece. This is weighed (to nearest 0·01 g) and placed between stainless steel plates, not less than 125 mm square and approximately 1·6 mm thick, that have been treated with an appropriate type of release agent. The assembly is then placed between the platens of a heated press and pressure is applied within 10 s and for not less than 15 min. Temperatures and pressures are according to the following schedules:

Resin	*Temperature* (°C ± 5°C)	*Pressure* (MPa ± 10%)
Polyester	120	1·4
Epoxy	160	1·4
Silicone	175	1·4 or 6·9

The laminate is then removed from the press and:

(a) if the flash can be cleanly removed from it, this is done and the laminate is weighed (W_2);
(b) if the flash cannot be removed cleanly, a piece 50 mm ± 1 mm square is cut from the central area of the laminate and weighed (W_3).

The 'flow' of the test piece is then calculated as follows: if W_1 is the original weight of the test piece then, if (a) above has been followed,

$$\text{flow} = 100 \frac{(W_1 - W_2)}{W_1} \quad \text{per cent} \tag{5.3a}$$

or, if (b) above has been followed,

$$\text{flow} = 100 \frac{(W_1 - 4W_3)}{W_1} \quad \text{per cent} \tag{5.3b}$$

5.5 VISCOSITY—STANDARD TEST METHODS

Mention has already been made (§5.1) of the difficulties of determining meaningful melt viscosities of polymers in relation to their processing characteristics. The molecular weight and molecular weight distribution of a polymer have a marked effect on the processibility of polymers; determination of molecular weight by solution viscosity methods is common and is dealt with in Chapter 6. The viscosity of resins, plastisols and organosols (dispersion of PVC resin in plasticiser), which is of interest to the processor handling the materials, is commonly determined by the use of cone and plate, double cone, coaxial cylinder and rotating spindle viscometers. ISO 3219 (1972)[30] specifies a method for determining the viscosity of polymers in the liquid state with a rotational viscometer in which the theoretical shear rate employed in the test is defined. This standard shows how to calculate the theoretical shear rate from the dimensions and rotational frequency but cautions the user that the equations only give results near the true shear rate of Newtonian liquids if the instrument characteristics agree with the standard. BS 2782 Method 730B (1978)[31] is identical to ISO 3219 and DIN Preliminary Standard 53788 (1974)[32] is based on ISO 3219.

It is worth mentioning at this point that the SI unit of viscosity, which is now being increasingly used in viscometry, is the Pascal second (Pa s). The Pascal second is equivalent to the Newton second per square metre (Ns/m^2) which is equal to 10^3 cP (centipoise).

The Brookfield instrument is a well known type of rotational spindle viscometer and is the subject of several standard test methods for polymers in the liquid state.

ISO 2555 (1974)[33] states the method of test for the determination of Brookfield RV viscosity of resins in the liquid state or as an emulsion or dispersion. It specifies the use of the Brookfield RV Model viscometer (RVF, RVF-100 or RVT), chosen according to the product to be tested and the desired experimental accuracy. By selection of the appropriate instrument type it is possible to measure viscosities in the range 0·02 to 8000 Pa s. This standard has two appendices, Annex A and Annex B. Annex A sets out the general application of the method to resins in the liquid state and Annex B gives the principle, description and characteristics of Brookfield viscometers. This standard only details the method of using the viscometer and requires reference to material standards for details of spindle size, speed of operation and temperature of test. As such it does little more than duplicate the manufacturer's instruction manual which comes with the instrument. During its passage through ISO it met with some opposition, including that of the UK, and consequently does not appear as a method in BS 2782.

ASTM D2393-68 (1972)[34] specifies the use of the Brookfield viscometer for finding the viscosity of epoxy resins and related components and ASTM D1824-66 (1972)[35] uses the same apparatus for finding the apparent viscosity of plastisols and organosols at low shear rates.

ASTM D1823-66 (1979)[36] offers a standard method of test for determining the apparent viscosity of plastisols and organosols at high shear rates by the Castor-Severs viscometer. The test method specifies the Castor-Severs A100 extrusion rheometer but does not describe the apparatus, which is gas-operated. Again, the method appears to duplicate the instruction manual provided with the instrument and does not give any indication of the shear rates obtainable. DIN Draft Standard E54801 (1977)[37] refers to the determination of viscosity of PVC paste by a Severs viscometer and is similar to ASTM D1823.

5.6 GELATION TIMES OF THERMOSETTING MATERIALS—STANDARD TEST METHODS

The time required for a thermosetting liquid resin to thicken after addition of catalysts and accelerators is clearly of importance to users of these products since, *inter alia*, the gelation (usually shortened to gel) time governs the size of batch of activated resin a processor can mix to satisfy his production line without waste caused by the usage time exceeding the pot life. Most of the standardised methods depend upon stirring the resin until resistance is felt, the gel time being quoted in minutes between the start of the test and the point at which a specified resistance to agitation is obtained.

ISO 2535 (1974)[38] describes the procedure for unsaturated polyester resins at 25°C using a reference initiator and accelerator. The end point is found by a rotating (1–2 mm/min) glass rod 6 mm in diameter. Suspension of the rod is by a torsion drive so that when the viscosity of the resin reaches 50 Pa s an indication is given and the time can be taken. The result is expressed as 'gel time at 25°C'.

ASTM D2471 (1979)[39] describes a standard test method for gel time and peak exothermic temperature of reacting thermosetting resins. This uses wooden probes 2·4 mm in diameter manipulated by hand every 15 s during the test. At the same time a thermocouple is inserted in the resin to indicate the temperature change throughout the test. The method permits the use of mechanical gel time meters if the sample is large enough, but it is claimed that these do not give results consistent with hand stirring. With the hand probe method the end point is recorded when the reacting resin no longer adheres to the end of a clean probe. Unless otherwise specified, the temperature of test is 23·0° ± 1·0°C. BS 2782 (1970)[40] includes four tests for gel time:

Method 111A	Gelation time at 130°C of phenolic resins for reinforced low-pressure mouldings
Method 111B	Gelation time at 150°C of phenolic resins for reinforced low-pressure mouldings
Method 111C	Gelation time at 82°C of polyester resin (manual determination)
Method 111D	Gelation time of polyester and epoxide resins with gel timer

Replacements for these are already in draft form as:

Method 835A (replaces 111A and 111B)	Determination of gelation time of phenolic resins for reinforced low pressure mouldings
Method 835B (replaces 111C)	Determination of gelation time of polyester resin (manual method)
Method 835C (replaces 111D)	Determination of gelation time of polyester and epoxide resins with gel timer

In addition there will be three new test methods:

Method 835D	Determination of gelation time at 170°C of solid resins and

	laminate bonding sheet material (equivalent to IEC publication 249–3)
Method 835E	Determination of gelation time of thermosetting resins using a hot plate
Method 835F	Determination of gelation time of resin used in laminations

The method using a gel timer (111D or 835C) relies on a vertically reciprocating disc coupled to a trip device which stops a timer when the force on the disc during a downward stroke exceeds 160 ± 4 mN.

All of the BS 2782 tests are carried out at elevated temperatures and the manual methods (i.e. 835A and 835B) employ glass probes.

It is surprising to find that, although the UK approved ISO 2535, this method has not yet been published as a method in BS 2782.

Before concluding this section, mention is made of a method of test for viscosity of highly viscous resins and pastes. ASTM D2730-68 (1979)[41] is a standard test method for sag flow of highly viscous resins. It is offered for quality control and acceptance tests for highly viscous materials and pastes. The principle of the test is to fill horizontal channels of various depths (from 1·6 mm to 19·1 mm) with the test material, then, after turning the channels to the vertical plane, measuring (in mm) the amount of movement of the material out of the channel after 30 minutes. The result is expressed as the length of sag flow for each depth of channel.

5.7 OTHER TESTS

Three further properties of interest to the processor of plastics materials are melting point, particle size and apparent density and bulk factor.

Melting point is of particular interest when processing those plastics and resins which have a fairly sharp change of state, e.g. certain of the polyamides (nylons) and novolac resins. A number of methods have been standardised and these are dealt with in Chapter 16.

The particle size of powders, granules or pellets may be of profound importance for it influences the packing density of a moulding powder, and hence the charge in a fully positive mould, and the gelation characteristics of a paste-making polymer, to take two common examples.

Test methods for determining particle size are dealt with in Chapter 7, which also deals with the measurement of apparent density and bulk factor. All of these properties are of importance to the machine designer as well as the processor because they have to be taken into account when designing, for example, screws for injection machines and extruders. Furthermore, having designed a machine to operate with optimum output and quality with a material of certain bulk factor and particle size, it is then in the processor's interest to use routine tests for these properties to control the quality of different batches of moulding powder.

There are undoubtedly other properties which are of interest to the processor but, except for shrinkage which is dealt with in Chapter 16 and those described above, it is considered that they are so restricted in application as to be outside the scope of this book.

REFERENCES

1. Bernhardt, E. C. (1959) *Processing of Thermoplastic Materials*, Reinhold.
2. Cogswell, F. N. (1981) *Polymer Melt Rheology: A Guide for Industrial Practice*, Godwin/Plastics and Rubber Institute.
3. Brydson. J. A. (1981) *Flow Properties of Polymer Melts* 2nd Edn, Godwin/Plastics and Rubber Institute.
4. Paul, K. T. (1975) *Joint Conference of British Society of Rheology and Rubber Institute, Loughborough, 1975*, page 49.
5. Wissbrun, K. F. (1977) *Polymer News*, **4** (2), 55.
6. BS 2782 (1970) Method 105A *Flow Temperature of Thermoplastic Moulding Material*; Method 105B *Cup Flow of Phenolic and Alkyd Moulding Materials*; Method 105C *Melt Flow Index of Polythene and Polythene Compounds*; Method 105D *Flow of Synthetic Resin Impregnated Glass Fabric*.
7. ISO R292 (2nd Edn) (1967) *Plastics—Determination of the Melt Flow Index of Polyethylene and Polyethylene Compounds*.
8. ISO R1133 (1969) *Plastics—Determination of the Melt Flow Rate of Thermoplastics*.
9. BS 2782 Method 720A (1979) *Determination of Melt Flow Rate of Thermoplastics*.
10. ASTM D1238 (1979) *Flow Rates of Thermoplastics by Extrusion Plastimeter*.
11. DIN 53735 (1977) *Determination of the Melt Flow Index of Thermoplastics*.
12. ASTM D3364-74 (1979) *Flow Rates for Poly(vinyl chloride) and Rheologically Unstable Thermoplastics*.
13. ASTM D569-59 (1976) *Measuring the Flow Properties of Thermoplastic Molding Materials*.
14. Hayes, R. (1952) *Chemical Industry*, **44,** 1069.
15. Gray, T. F. Jr, Holford, T. G. and Combs, R. L. (1968) *Society of Plastics Engineers Journal*, **24** (9), 35.
16. Campbell, G. and Griffiths, L. I. (1956) *Plastics Progress 1955*, Iliffe, page 259.
17. McCabe, C. C. (1960) *Transactions of the Society of Rheology*, **IV,** 335.
18. McCabe, C. C. (1960) *Chemistry in Canada*, **12** (10), 44.
19. Schramm, G. (1965) *International Plastics Engineering*, **5** (12), 420.
20. Matthan, J, (1968) *RAPRA Research Report No. 165*.
21. Paul, K. T. (1972) *RAPRA Bulletin*, February.
22. Paul, K. T. (1973) *RAPRA Members Journal*, November.
23. Allen, E. O. and Williams, R. F. (1971) *Proceedings of SPE Annual Technical Conference 1971*, page 587.
24. Ballam, R. L. and Brown, J. J. *Capillary Rheometer*, Instron Engineering Company, Canton, Mass.
25. RAPRA (1979) *RAPRA Guide to Rubber and Plastics Test Equipment*, page 18.
26. Von Meysenburge, C. M. (1959) in *International Symposium of Plastics Testing Standards*, STP No. 247, ASTM.
27. ASTM D731-67 (1976) *Molding Index of Thermosetting Molding Powder*.
28. DIN 53465 (1963) *Determination of the Closing Time of Thermosetting Moulding Materials; Cup Flow Test Method*.
29. ASTM D3123-72 (1978) *Spiral Flow of Low Pressure Thermosetting Molding Compounds*.
30. ISO 3219 (1972) *Determination of Viscosity of Polymers in Liquid, Emulsified or Dispersed State with a Rotational Viscometer Working at a Defined Shear Rate*.
31. BS 2782 Method 730B (1978) *Determination of the Viscosity of Polymers in the Liquid, Emulsified or Dispersed State using a Rotational Viscometer Working at a Defined Shear Rate*.
32. DIN 53788 (1974) *Determination of the Viscosity at a Rate of Shear of Polymer and Copolymer Aqueous Dispersions*.

33. ISO 2555 (1974) *Resins in the Liquid State or as Emulsions or Dispersions. Determination of Brookfield RV Viscosity.*
34. ASTM D2393-68 (1979) *Viscosity of Expoxy Resins and Related Components.*
35. ASTM D1824-66 (1972) *Apparent Viscosity of Plastisols and Organosols at Low Shear Rates by Brookfield Viscometer.*
36. ASTM D1823-66 (1979) *Apparent Viscosity of Plastisols and Organosols at High Shear Rates by Caster-Severs Viscometer.*
37. DIN E54801 (1977) *Determination of Apparent Viscosity at High Rates of Shear of Polyvinyl Chloride (PVC) Paste by Severs Capillary Viscometer.*
38. ISO 2535 (1974) *Unsaturated Polyester Resins—Measurement of Gel Time at 25°C.*
39. ASTM D2471 (1979) *Gel Time and Peak Exothermic Temperature of Reacting Resins.*
40. BS 2782, Methods 111A to 111D (1970) and Methods 835A to 835F *Determination of Gel Times.*
41. ASTM D2730-68 (1979) *Sag Flow of Highly Viscous Resins.*

Chapter Six

Polymer Characterisation

6.1 INTRODUCTION

The most profound influence on the properties of a polymer is exerted by the chemical constitution of the polymer molecule itself. However, this handbook is concerned with physical testing and accordingly it will be assumed that the chemical constitution is known, either by reason of awareness of the starting materials used or by detailed analysis. After this, properties are influenced to varying degrees (a) by molecular weight (i.e. size of molecules) and (b) by crystallinity (i.e. relative positioning of molecules one to another)—though the range of values of the latter that may be encountered is very dependent on the chemical nature of the polymer. Since the result of polymerisation is invariably not a number of polymers of identical molecular weight, but rather polymers of a range of molecular weights, the influence of molecular weight on properties may be due to the average value and/or the range of values in the sample under consideration.

Generally speaking, crystallinity influences the following parameters:

Density (Chapter 7) Rigidity (Chapter 8) Short-term tensile strength (Chapter 8) Softening point (Chapter 16) Solvent resistance (Chapter 17)	Positive correlation
Permeability to gases (Chapter 18) Shock (impact) resistance (Chapter 8)	Negative correlation

Molecular weight influences the following:

Melt viscosity and processibility (Chapter 5)
Shock (impact) resistance (Chapter 8)
Loadbearing properties (long-term) (Chapter 11)

It must emphasised that these are broad generalisations and by no means exhaustive; thus, as far as molecular weight is concerned, the comments apply to the normal commercial range of polymer size and not at low degrees of polymerisation where the influence of the variable is very significant indeed.

6.2 CRYSTALLINITY

Methods which may be used to measure the degree of crystallinity are:

X-ray diffraction
Nuclear magnetic resonance
Infra-red absorption
Density

X-ray diffraction, the fundamental and most widely employed technique, and NMR are dealt with fully elsewhere[1–13] and are beyond the scope of this book.

Infra-red techniques are applicable to certain polymers, e.g. polypropylene, where the absorptions of the crystalline and amorphous regions differ significantly. It is not a direct measurement and the relationship of absorption, or ratio of absorptions at two frequencies (one selected as markedly dependent on crystallinity and the other essentially independent thereof, to act as a test variable reducing quotient) with crystallinity has to be established by calibration.[14,15]

The most simple method is by density measurement, which again is dependent on accurate calibration in the first place. It is not difficult to envisage that the more ordered (crystalline) the packing of polymer molecules the greater will be the density of the mass polymer and usually the relationship between density and percentage crystallinity is linear. Nevertheless there are exceptions: with poly(4-methyl pentene-1), for instance, the crystalline polymer is less dense than the amorphous variety.

The measurment of density is described in Chapter 7: any method accurate to 0·0001 g/cm^3 and appropriate to the test specimen in question should be suitable.

6.3 MOLECULAR WEIGHT

6.3.1 General Considerations

Unless the molecular weights of all the polymer species in a given sample are identical, i.e. it is a mono-disperse system where the average value has a unique meaning, the expression of average molecular weight can be made in several different ways.

The number average molecular weight, $\bar{M}_n$, results from 'counting' the number of molecules present and is obtained from such methods as end-group analysis, electron microscopy and the measurement of colligative properties such as osmotic pressure, lowering of freezing point, elevation of boiling point and depression of vapour pressure. Light-scattering techniques yield a weight average molecular, $\bar{M}_w$, while solution viscosity measurements result in an average, $\bar{M}_\alpha$, generally somewhere between the two. Higher still is *z*-average, $\bar{M}_z$, approximately that obtained by sedimentation equilibrium (ultracentrifuge). Progressing through $\bar{M}_n$, $\bar{M}_\alpha$, $\bar{M}_w$ to $\bar{M}_z$, the average becomes more dependent on the higher molecular weight species present in the sample.

Mathematically,

$$\bar{M}_n = \frac{\sum_i N_i M_i}{\sum_i N_i} \quad (6.1)$$

$$\bar{M}_\alpha = \left(\frac{\sum_i N_i (M_i^{\alpha+1})}{\sum_i N_i M_i} \right)^{1/\alpha} \quad (6.2)$$

$$\bar{M}_w = \frac{\sum_i N_i (M_i^2)}{\sum_i N_i M_i} \quad (6.3)$$

$$\bar{M}_z = \frac{\sum_i N_i (M_i^3)}{\sum_i N_i M_i^2} \quad (6.4)$$

where α is derived from the Mark-Houwink formula, $[\eta] = KM^\alpha$ (§6.3.2)
N_i is the number of molecules of molecular weight M_i.

The techniques available for measuring molecular weights may be classified as primary, wherein molecular weight is derived directly from the method, and secondary, wherein molecular weight is obtained from a calibration curve derived from standards of known molecular weight, usually of narrow molecular weight distribution, i.e. $\bar{M}_n \simeq \bar{M}_w \simeq \bar{M}_z$.

The primary methods include end-group analysis, microscopy, ultracentrifuge, light-scattering and the colligative techniques of measuring changes in osmotic pressure, vapour pressure, freezing point and boiling point. Of these, the most frequently encountered are osmotic and vapour pressure measurements, to yield $\bar{M}_n$, and light-scattering, to yield $\bar{M}_w$, because of their wide range of molecular weight and relative simplicity, although it should be stressed that the accurate measurement of molecular weight by any technique is extremely difficult.

The secondary methods include gel permeation chromatography (GPC) and solution viscosity. Over the past decade GPC has made enormous strides both technically, with the development of the micro-particulate column packings allowing analyses to be performed in under 15 minutes, and in its general acceptance, being probably the most frequently encountered technique in laboratories carrying out routine molecular weight measurements. The undoubted advantage of GPC is its ability to 'fingerprint' the molecular weight distribution of a polymer in a single run and hence to allow all the molecular weight averages to be computed. Solution viscosity is also very widely used because its simplicity of technique and apparatus make it invaluable as a quality control method. Although capable of being transformed into molecular weight data, the solution viscosity measurements are more usually left in the form in which they are first obtained.

Because of its application to quality control and the inexpensiveness of the technique it is worth examining solution viscosity measurement in some detail.

For the other techniques, however, space does not permit a more detailed description to be given and the reader is referred to the standard works.[14–31]

6.3.2 Solution Viscosity—Definitions and Relationships with Molecular Weight

It has already been stated that this is a secondary, i.e. non-absolute, method of measuring molecular weight and, although considerable thought has been given to the theory of solution viscosity, as yet no fundamental relationship has been worked out completely to make it a primary one. For detailed considerations of the theory see Gordon[1] pages 220–237 and Tanford,[17] Allen[20] and Flory.[32] The following very brief treatment starts with the definition of essential terms.

It is regrettable that even now a multiplicity of names should still exist for the various terms used in dilute solution viscometry. The UK is in a particularly unfortunate situation since it strongly supports the adoption of ISO standards, which use the International Union of Pure and Applied Chemistry definitions and have done so since their inception, yet in 1975 a revision of BS 5168 *British Standard Glossary of Rheological Terms* was published which retained, as preferred names, the very names ISO is trying to make obsolete. This has resulted in the publication in 1979 of BS 2782, Method 730A 'Determination of Viscosity of Plastics in Dilute Solution,' which uses the BS 5168 definitions, while simultaneously the UK is actively involved in the revision of ISO R1628 (1970)[33] which employs the IUPAC definitions. It is to be hoped that the parties concerned with terminology at international level can liaise more closely in future so that this multiplicity of names can be buried once and for all. Table 6.1 gives the two systems of terminology.

The limiting viscosity number (LVN), $[\eta]$, is produced by plotting viscosity number or logarithmic viscosity number against concentration, which usually produces a straight line (at least at lower concentrations) and extrapolating to the $c = 0$ axis either by eye or, preferably, statistically by a 'least squares' method. Alternatively, if the particular polymer species, solvent and temperature have been reliably investigated, LVN may be calculated from a single solution viscosity determination at a known concentration by the Huggins equation:

$$\frac{\eta - \eta_o}{\eta_o c} = [\eta] + (k'[\eta]^2 c) \qquad (6.5)$$

where k' is the Huggins constant and is assumed to have been reliably determined previously.

The LVN having been so determined, to convert it to a value for average molecular weight involves the use of an appropriate mathematical relationship previously established between solution viscosity for the polymer, solvent and temperature in question, and molecular weight as measured by a primary method—most often that of osmotic pressure—using monodisperse samples as far as possible.

For many years the most favoured relationship has been the empirical Mark-Houwink equation:

$$[\eta] = KM^{\alpha} \qquad (6.6)$$

Table 6.1 Terminology Used in Dilute Solution Viscometry

Term	ISO (IUPAC)	BS 5168
η/ρ where η = dynamic or absolute viscosity; ρ = density	Viscosity/density ratio or Kinematic viscosity	Kinematic viscosity
η/η_0 where η = viscosity of the polymer solution; η_0 = viscosity of the pure solvent alone	Viscosity ratio	Relative viscosity
$(\eta - \eta_0)/\eta_0$	Viscosity relative increment	Specific viscosity
$(\eta - \eta_0)/\eta_0 c$ where c = polymer concentration (g/ml)	Viscosity number	Reduced viscosity
$\ln(\eta/\eta_0)/c$	Logarithmic viscosity number	Inherent viscosity
$[\eta]$	Limiting viscosity number (LVN)	Intrinsic viscosity

where K and α are constants and M = molecular weight; although more recent theoretical approaches, notably that of Fox and Flory,[32] have produced far more complicated equations, they can often be reduced by approximation to the above form.

The literature abounds with papers dealing with practical investigations establishing values of k', K and α for many polymeric species. Brandrup and Immergut[34] provide an excellent source of data in this respect. In using these values, it is imperative to satisfy oneself as to the reliability of the reported work in respect of:

(i) the efficiency of fractionation of the polymer studied—i.e. does it approach monodisperse?
(ii) the accuracy of the primary method of measuring molecular weight, e.g. in osmosis the efficiency of the semi-permeable membrane, the temperature control of the experimental set-up, the attainment of equilibrium, etc;
(iii) the general accuracy of the viscosity measurements with reference to such factors as temperature control, completeness of solution, cleanliness of solutions and of viscometer, and purity of solvent.

If all these factors seem in order, then the user of the reported data must be certain he is operating with precisely the same polymeric species, the same solvent and at the same temperature and is using the same concentration units (there is much confusion and indefinition in the literature about concentration units). Put very simply, solution viscosity assesses molecular weight because the larger the polymer molecule dissolved the greater the friction in flow, i.e. the greater the viscosity. Again, over-simplified, a 'good' solvent encourages the polymer chain to stretch out, whereas a 'poor' one causes it to coil up, so that a given polymer yields a lower viscosity solution in a 'poor' solvent than in a 'good' solvent, even when the viscosities of the solvents themselves are identical. Thus a given polymer could yield different solution viscosities, even though at the same concentration and temperature, in a pure solvent and in the same solvent contaminated with a small portion of a different one, if the polymer–solvent interactions were different. Again, in a given pure solvent, though two polymer species may have the same molecular weights, the solution viscosity of a homopolymer may be quite different from that of a copolymer even though the latter is based primarily on the same monomer unit. Both these effects will be seen to give different calculated (apparent) molecular weights from solution viscosity measurements on polymers of identical molecular weight.

When all is said and done, it is very rare that a knowledge of absolute molecular weight is needed; most often the trend, qualitative or quantitative, is of prime interest, be it to determine the effect of polymerisation conditions on molecular weight (decrease or increase and in what ratio) or the dependence of some physical property—say, softening point—on molecular weight and particularly where increase in the latter ceases to affect the former. Frequently, therefore, a guide to molecular magnitude, as from LVN, will be quite sufficient.

6.3.3 Measurement of Solution Viscosity

Standard methods in the UK generally follow the procedures laid down in BS 188 (1977)[35] which is a suitable manual for all such measurements on low viscosity liquids. ISO 3104 (1976)[36] and 3105 (1976)[37] combined are very similar, though not identical. The glass capillary viscometers described are almost invariably used; the other technique, falling sphere viscometry, is more suitable for 'dopes' because of the much higher viscosities involved.

Three types of capillary viscometer are described (Figures 6.1–6.3).

In the ultimate, every viscometer is calibrated using a liquid of known viscosity, though this calibration may be conveniently achieved using an unknown, homogeneous and stable liquid run through both the uncalibrated viscometer and a calibrated 'master' viscometer.

The kinematic viscosity of a liquid $\nu(=\eta/\rho)$ and its flow time, t, through a capillary viscometer are connected by a formula of the type

$$\nu = Ct \tag{6.7}$$

where ν is most often expressed in centistokes, t in seconds and C is the viscometer constant.

This equation is applicable in all cases except where the viscosity is so low, and the rate of efflux so great, that a correction must be made to allow for the effects of kinetic energy of efflux. In this case,

$$\nu = Ct - \frac{B}{t} \tag{6.8}$$

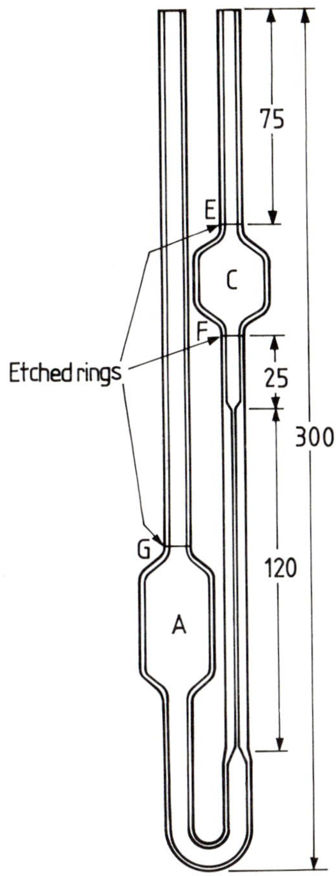

Figure 6.1 U-tube viscometer—dimensions in millimetres

B may be determined experimentally or calculated from the known dimensions of the viscometer. However, it is usually possible to use the simpler equation by choosing a viscometer to give a sufficiently large value of *t*. Under these conditions it will be noted that, for the determination of such values as viscosity ratio, viscosity number and logarithmic viscosity, the value of *C* need not be known (since it cancels out)—that is, the viscometer can be uncalibrated. However, it follows that one and the same viscometer must be suitable for both the solution and the (less viscous) solvent and this is often difficult to arrange. Table 3 of BS 188 describes a series of viscometers of different size in each type, and their usable ranges using the simple equation and the kinetic energy correction factor.

The use of the viscometers is fully described in BS 188; it suffices to confirm here that the efflux time *t* is the period of flow of the liquid between two etched marks (e.g. E and F in the U-tube viscometer) under the action of gravity. BS 188 lays down requirements for temperature measurement and control of the thermostat bath (±0·015°C in range 15°C to 100°C and ±0·03°C outside this), verticality of the viscometer and accuracy of the timing device (0·07 per cent over a period of not less than 15 min); it is essential to check stopwatches to

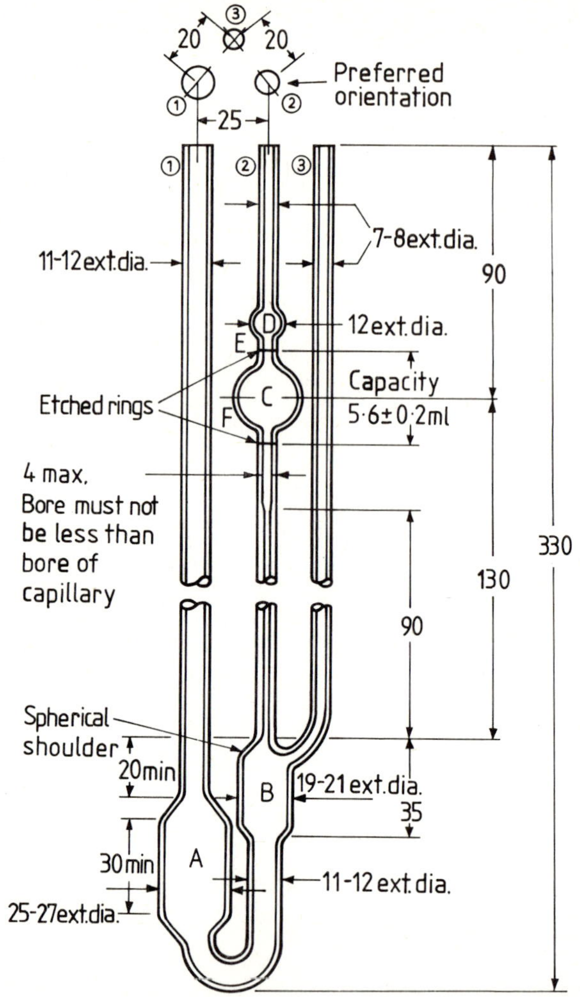

Figure 6.2 Suspended level viscometer—dimensions in millimetres

ensure that they meet this requirement. The British Standard also draws attention to the need for extreme cleanliness in undertaking such measurements.

American practice is laid down in ASTM D445 (1974),[38] which again covers all the important variables mentioned above and allows the use of some nineteen types of viscometer including those of BS 188. General guidance on German practice is given in DIN 51550 (1972).[39]

Viscous solutions may conveniently be examined by the 'falling sphere' method, which is fully described in BS 188. The kinematic viscosity ν (in centistokes) of the liquid is given by:

$$\nu = Mg \frac{(\delta - \rho)}{0{\cdot}03\pi v d\rho\delta} \times F \qquad (6.9a)$$

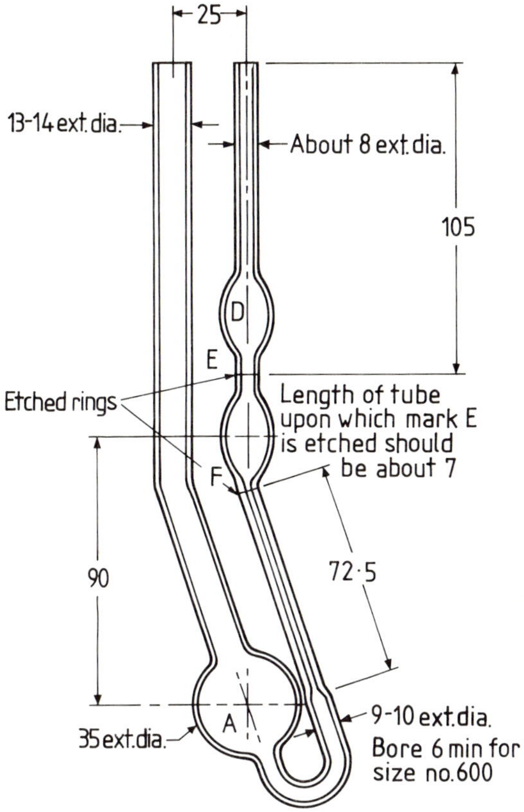

Figure 6.3 Cannon-Fenske viscometer—dimensions in millimetres

or

$$\nu = \frac{d^2 g(\delta - \rho)}{0{\cdot}18 v \rho} \times F \tag{6.9b}$$

where M is the mass of the sphere (g)
d is the diameter of the sphere (cm)
δ is the density of the sphere (g/cm^3)
ρ is the density of the liquid (g/cm^3)
v is the velocity of fall of the sphere (cm/s)
g is the local acceleration due to gravity (cm/s^2)
F is known as the Faxén term and is a correction for the effect of the wall of the fall-tube on the motion of the sphere: if D is the diameter (cm) of the fall tube, $F \simeq 1{\cdot}00 - 2{\cdot}104(d/D) + 2{\cdot}09(d^3/D^3)$.

Temperature control requirements are as before and an accurate timing device is again essential. Other than this, carefully machined steel balls and a 'fall tube', with parallel sides and horizontal timing marks, are all that is required.

6.3.4 Use of Dilution Viscometers

For the determinations of LVN $[\eta]$, the viscosities of solutions of various concentrations must be determined or, if the efflux times are long enough and the same viscometer is used throughout, just the flow times. Depending on the molecular weight of the polymer, the most usual concentrations to examine are in the range 0·05 to 5·0 g/100 ml, i.e. 0·0005–0·05 g/ml in the IUPAC units.

To obviate the tedium of making up these solutions, however, commercial viscometers are available which permit the determination of the efflux time of a precisely known volume of the most concentrated solution to be examined, followed by successive dilutions, with known volumes of solvent, and viscosity determinations at each stage; a final measurement with pure solvent is, of course, necessary. The feasibility of such a 'dilution viscometer' results from the fact that the suspended level viscometer (Figure 6.2 above) does not require the use of a carefully controlled volume of liquid, because of the 'free fall' of the liquid at the lower end of the capillary. Thus, by careful selection of the

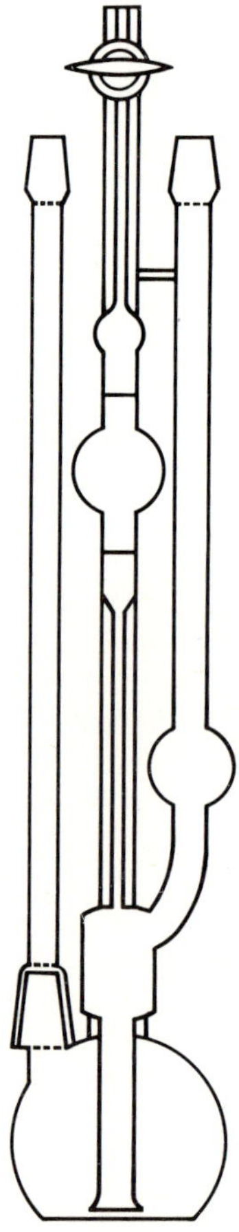

Figure 6.4 Suspended level dilution viscometer—filter type

dimensions of the bulbs, etc., dilutions of up to 10-fold can be made without upsetting the functioning of the viscometer. As a further refinement, a sintered glass filter has been built in (Figure 6.4).

These so-called 'Modified Ubbelohde Viscometers' are described in ASTM D1601 (1978).[40] Fully automated dilution viscometers are now commercially available which remove the tedium of making LVN measurements—at a price.

6.3.5 Solution Viscosity—Standard Test Methods

Table 6.2 (page 96) summarises some of the internationally and nationally standardised methods for measuring the viscosities of polymers in dilute solution.

REFERENCES

1. Gordon, M. (1963) *High Polymers, Structure and Physical Properties* 2nd edn, Iliffe, Chapter 6.
2. Frith, E. M. and Tuckett, R. F. (1951) *Linear Polymers*, Longmans, Green, Chapter 1.
3. Weissberger, A. (1949) *Physical Methods of Organic Chemistry* vol. 1 Part II, Interscience.
4. Kline, G. M. (Ed.) (1962) *Analytical Chemistry of Polymers* Part II, in *High Polymers* vol. XII, Interscience.
5. Ke, Bacon (1964) *Polymer Reviews*, vol. VI *Newer Methods of Polymer Characterisation*, Interscience/Wiley.
6. Miller, M. L. (1966) *The Structure of Polymers*, Reinhold.
7. Statton, W. O. (1959) in *International Symposium on Plastics Testing Standards*, STP No. 247, ASTM.
8. Bunn, C. W. (1959) in *The Physical Properties of Polymers*, SCI Monograph No. 5, Society of Chemical Industry, page 3.
9. Allen, G. (1967) *Review of Pure and Applied Chemistry*, **17**, 67.
10. Bovey, F. A. (1967) *Polymer Engineering and Science*, **7** (2), 128.
11. Slighter, W. P. (1968) *Journal of Chemical Education*, **1**, 10.
12. Kavesh, S. and Schultz, J. M. (1969) *Polymer Engineering and Science*, **9** (5), 331.
13. Kavesh, S. and Schultz, J. M. (1969) *Polymer Engineering and Science*, **9** (6), 452.
14. Potts, W. J. (1959) in *International Symposium on Plastics Testing Standards*, STP No. 247, ASTM.
15. Elliott, A. (1956) in *Symposium on Technical Polymer Science*, RIC Monograph No. 5, Royal Institute of Chemistry, page 26.
16. Weissberger, A. (Ed.) (1949/1954) *Physical Methods of Organic Chemistry* Vol. 1, Parts I and III, Interscience.
17. Tanford, C. (1961) *Physical Chemistry of Macromolecules*, Wiley.
18. Robb, J. C. and Peaker, F. W. (1961) *Progress in High Polymers*, Heywood.
19. Jonyhan, Ch'ien (1963) *Determination of Molecular Weights of High Polymers*, trans. Dr J. Schmorak, Israel Program for Scientific Translations, Jerusalem.
20. Allen, P. W. (1959) *Techniques of Polymer Characterisation*, Butterworth.
21. Smith, D. A. (Ed.) (1968) *Addition Polymers, Formation and Characterisation*, Butterworth, Chapter 5.
22. McIntyre, D. (1959) in *International Symposium on Plastics Testing Standards*, STP No. 247, ASTM.

Table 6.2 Standardised Procedures for Determining Solution Viscosity

Polymer type	Solvent	Concentration (g/100 ml)	Temperature (°C)	Viscosity result	Source
ISO					
PVC	Cyclohexanone	0·5	25	Viscosity number	ISO 174 (1974)
Polyamides	Formic acid or metacresol	0·5	25	Viscosity number	ISO 307 (1977)
Cellulose acetate	90% dichloromethane + 10% methanol	0·5	25	Viscosity number and ratio	ISO 1157 (1975)
Polyethylene and polypropylene	Decahydronaphthalene	0·02–0·5	135	Viscosity number and LVN	ISO 1191 (1975)
Poly(ethylene-terephthalate)	*o*-chlorophenol	1·0	25	Viscosity number	ISO 1228 (1975)
Methyl methacrylate polymers and copolymers	Chloroform	0·2–0·3	25	Viscosity number	ISO 1233 (1975)

See also: ISO R1628 (1970) *Directives for the Standardisation of Methods for the Determination of the Dilute Solution Viscosity of Polymers*
ISO 3104 (1976) *Petroleum Products: Transparent and Opaque Liquids—Determination of Kinematic Viscosity and Calculation of Dynamic Viscosity*
ISO 3105 (1976) *Glass Capillary Kinematic Viscometers—Specifications and Operating Instructions*

Polymer type	Solvent	Concentration (g/100 ml)	Temperature (°C)	Viscosity result	Source
BSI					
Polystyrene	Toluene	2·0	25	Viscosity (cP)	BS 2782, 404B (1970)
Unspecified*	Unspecified*	Unspecified*	25†	Viscosity number	BS 2782, 730A (1979)
Cellulose acetate	Acetone	6·0	25	Viscosity change	BS 1524 (1955)
Cellulose acetate	90% dichloromethane + 10% methanol	1·0	25	Viscosity ratio	BS 2880 (1957)

* BS 2782, 730A is a general test method for the measurement of dilute solution viscosity and is essentially equivalent to ISO R1628.

† 25°C is the preferred temperature, but where this is impractical one of the following temperatures should be selected: 40°C, 55°C, 70°C, 85°C, 100°C.

See also: BS 188 (1977) *Methods for the Determination of the Viscosity of Liquids*
BS 4708 (1971) *Methods for the Determination of the Viscosity of Transparent and Opaque Liquids (Kinematic and Dynamic)*

ASTM					
Cellulose nitrate	22% ethanol + 48·3% toluene + 17·5% ethyl acetate	12·2% wt/wt	25	cP by falling ball	ASTM D301 (1972)
Poly vinylidene chloride	*o*-dichlorobenzene	2·663	120	cP	ASTM D729 (1978)
Polyamide	90% formic acid or *m*-cresol	11 9·44	25	Viscosity ratio	ASTM D789 (1978)
Cellulose acetate, propionate or butyrate	not specified	0·26	25	LVN or cP by falling ball	ASTM D817 (1972)
Cellulose acetate	Conditions as for D817				ASTM D871 (1972)
Ethyl cellulose	Ethanol/toluene mixtures	5	25	cP	ASTM D914 (1972)
PVC	Cyclohexanone	0·2	30	Logarithmic viscosity number	ASTM D1243 (1966)
Polyethylene	Decahydronaphthalene	0·1–0·4	130	Viscosity ratio, logarithmic viscosity number or LVN	ASTM D1601 (1978)

See also: ASTM D445 (1974) *Kinematic Viscosity of Transparent and Opaque Liquids (and Calculation of Dynamic Viscosity)*
ASTM D1343 (1969) *Viscosity of Cellulose Derivatives by the Ball Drop Method*
ASTM D2857 (1970) *Dilute Solution Viscosity of Polymers*

Table 6.2 (*Continued*)

Polymer type	Solvent	Concentration (g/100 ml)	Temperature (°C)	Viscosity result	Source
DIN					
Polystyrene	Benzene	1	25	*k* number‡	DIN 7741 (1977)
Polycarbonate	Methylene chloride	0·5	25	*k* number‡	DIN 7744 (1976)
PMMA	Chloroform	0·25	25	Viscosity number	DIN 7745 (1976)
PVC	Cyclohexanone	0·5	25	*k* number‡	DIN 53726 (1978)
Polyamide	*m*-cresol	0·5	25	Viscosity number	DIN 53727 (1978)
Cellulose acetate	Methylene chloride/ methanol	0·5	25	Viscosity number	DIN 53728, Part 1 (1970)
Polyamide	Formic acid	10·1	25	Viscosity ratio	DIN 53728, Part 2 (1970)
Poly(ethylene terephthalate)	*o*-chlorophenol	0·5	25	Viscosity number	DIN 53728, Part 3 (1973)
Polyethylene and polypropylene	Decahydronaphthalene	0·5–0·03	135	Viscosity number or LVN	DIN 53728, Part 4 (1975)

‡ The *k* number is defined by the expression $\log(\eta/\eta_0) = [75k^2/(1+1{\cdot}5k.c)] + kc$, where c = polymer concentration (g/100 ml).

See also: DIN 51550 (1972) *Bestimmung der Viskosität*

23. Onyon, P. F. (1956) in *Symposium on Technical Polymer Science*, RIC Monograph No. 5, Royal Institute of Chemistry, page 4.
24. Zichy, E. L. (1963) in *Technical Polymer Science*, SCI Monograph No. 17, Society of Chemical Industry, page 122.
25. Quale, D. V. (1969) *British Polymer Journal*, **1** (1), 15.
26. Determann, H. (1968) *Gel Chromatography*, trans. E. Gross, Springer-Verlag.
27. Altgelt, K. H. and Segal, L. (1971) *Gel Permeation Chromatography*, Marcel Dekker.
28. Peebles, L. H. Jr (1971) *Molecular Weight Distributions in Polymers,* Interscience.
29. Green, J. H. S. and Dietz, R. (1973) *Industrial Polymers: Characterisation by Molecular Weight*, Transcripta Books.
30. Slade, P. E. Jr (1975) *Polymer Molecular Weights*, Marcel Dekker.
31. Billingham, N. C. (1977) *Molecular Mass Measurements in Polymer Science*, Kogan Page.
32. Flory, P. J. (1953) *Principles of Polymer Chemistry*, Cornell University Press.
33. ISO R1628 (1970) *Directives for the Standardization of Methods for the Determination of the Dilute Solution Viscosity of Polymers.*
34. Brandrup, J. and Immergut, E. H. (1975) *Polymer Handbook* 2nd edn, Part IV, Wiley.
35. BS 188 (1977) *Methods for Determination of the Viscosity of Liquids.*
36. ISO 3104 (1976) *Petroleum Products: Transparent and Opaque Liquids—Determination of Kinematic Viscosity and Calculation of Dynamic Viscosity.*
37. ISO 3105 (1976) *Glass Capillary Kinematic Viscometers—Specifications and Operating Instructions.*
38. ASTM D445 (1979) *Kinematic Viscosity of Transparent and Opaque Liquids* (*and the Calculation of Dynamic Viscosity*).
39. DIN 51550 (1978) *Viscometry: Determination of Viscosity—General Principles.*
40. ASTM D1601 (1978) *Dilute Solution Viscosity of Ethylene Polymers.*

Chapter Seven

Density and Dimensions

7.1 DENSITY

7.1.1 Definitions

Density	Mass per unit volume (at defined temperature) Unit: kg/m^3
Relative density (*formerly known as 'specific gravity'—a term no longer used in ISO and BSI standards*)	Mass (of substance) compared with the mass of an equal volume of a specific (reference) substance—most often water. In this instance the temperature of both the substance and the reference substance must be stated, e.g. relative density 23/23°C. Being a ratio, the property is dimensionless.

7.1.2 Standard Test Methods

The determination of the density of solid plastics, excluding cellular plastics, is detailed in ISO R1183 (1970),[1] which gives four methods according to the nature of the plastics material. This standard is undergoing revision in TC61, but at the time of writing was at a very early stage. The British equivalents are contained in BS 2782, Methods 620A to 620E (1980),[2] which closely follow the present ISO format. American standards are to be found in ASTM D792 (1966)[3] and D1505 (1968).[4]

The four basic methods are described below.

Displacement Method

This is Method A of ISO R1183. It is suitable for sheet, rod, tube and moulded articles, the density being determined at one of the standard laboratory temperatures of 20°C, 23°C and 27°C.

A specimen, with smooth surfaces free from crevices and dust and preferably of mass approximately 5 g, is weighed in air (W_1) and then in freshly boiled distilled water (W_2) at the requisite temperature, after sufficient time has been allowed for the specimen to reach this temperature.

If the density of the material under examination is less than that of water, a sinker is attached and the mass of this in water (W_3) and of the sinker plus the specimen also in water (W_4) are taken. Fine wire is used for the suspension of the specimen (and sinker)—the upthrust on part of the wire is sufficiently small to be ignored. All air bubbles must be removed, for example by use of a minute quantity of detergent.

$$\text{density} = \frac{W_1}{W_1 - W_2} \times \rho \tag{7.1a}$$

When a sinker is used,

$$\text{density} = \frac{W_1}{W_1 + W_3 - W_4} \times \rho \tag{7.1b}$$

where ρ is the density of the water, usually taken to be 1000 kg/m^3 but, for more accurate work, set to the value indicated in the table in the ISO standard.

Provision is made for using a liquid other than water if a polymer–water interaction occurs, in which case ρ is given by the density of that liquid at the test temperature.

The same basic procedure is followed in BS 2782 Method 620A and ASTM D792 Methods A1 to A3, with only slight changes in emphasis rather than principle.

Because density is frequently measured as a quality control parameter, relative density balances have been developed which enable the measurement to be made much more quickly. These devices are based on the displacement principle.

Pyknometer Method

This is Method B of ISO R1183. It is suitable for plastics in a preformed condition, such as powder, pellets, flake, etc. A pyknometer (Figure 7.1) is accurately weighed (W_1), a suitable quantity of the material (1–5 g) is added and the whole is reweighed (W_2). The material is covered with freshly boiled distilled water and the pyknometer is then placed in a vacuum dessicator and all air removed. The pyknometer is filled with water and placed in a constant temperature bath at 23°C ± 0·1°C (or other standard laboratory temperture). When the temperature has been achieved the capillary is filled or emptied as necessary to the appropriate level and the pyknometer is reweighed (W_3).

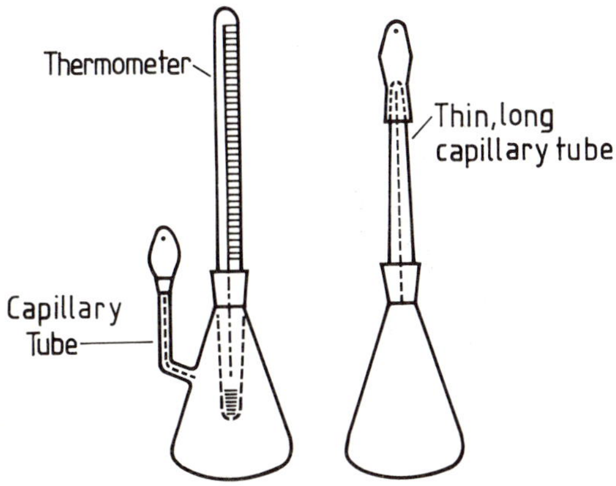

Figure 7.1 Typical pyknometers

The pyknometer is emptied, filled with freshly boiled distilled water and reweighed. The mass (W_4) is recorded.

$$\text{relative density } 23/23°\text{C} = \frac{(W_2 - W_1) \times d}{(W_3 - W_4) + (W_2 - W_1)} \tag{7.2}$$

For water d is exactly 1, but if another displacement liquid is used, for example when the material is not inert to water, d is the relative density of this liquid.

Density $\rho^{23°C}$ (kg/m^3) = relative density 23/23°C × 997·6 where 997·6 is the density of the water in kg/m^3 at 23°C.

The equivalent British Standard is BS 2782, Method 620B. This procedure is also contained in ASTM D792, Method B.

The pyknometer method is used for the determination of the density of liquid resins, as detailed in ISO 1675 (1975)[5] and BS 2782, 620E. BS 733 (1965)[6] gives details of pyknometer size, construction and design.

Sink–Float Method

This is Method C of ISO R1183. It is suitable for plastics in forms similar to those required for Method A and for pellets.

A series of liquids is prepared having a range of densities covering the anticipated range of interest and kept at 23°C ± 0·5°C (or other standard laboratory temperature). A small quantity of wetting agent is permitted if required. A test piece is placed in each liquid and the densities of the densest liquid in which the test piece sinks and the least dense liquid in which it floats are determined as quickly as possible by a standard method. The density of the plastic is, then, between these two densities.

Greater precision can be obtained by preparing a further set of liquids whose densities cover the range determined as above and the whole procedure repeated until the required precision is attained.

BS 2782 Method 620C uses the same procedure.

Density Gradient Method

This is Method D of ISO R1183. It is suitable for plastics in the same forms as Methods A and C.

Two solutions are made (A and B), of which A has a density lower than the lowest density of interest by about 20 per cent and B a density higher than the highest of interest by about 30 per cent. One litre of A is placed in Vessel 1 and one litre of B in Vessel 2, the taps being closed (Figure 7.2).

The magnetic stirrer in Vessel 1 is started (to give thorough mixing but not bubble formation) and the connecting tap is opened, when probably some of Solution B will flow into Solution A. When hydrostatic equilibrium has been established the right-hand tap is opened so that, in conjunction with the capillary bore, the column takes about 2 h to fill. While this happens, Solution B flows into, and is mixed with, Solution A in Vessel 1 which therefore increases in density; hydrostatic equilibrium between the two vessels must be maintained throughout and, for this, a wide-bore connecting tube is necessary with slow filling. Finally the filling tube, with the upper end closed, is removed carefully and slowly (see also Payne and Stephenson[7]).

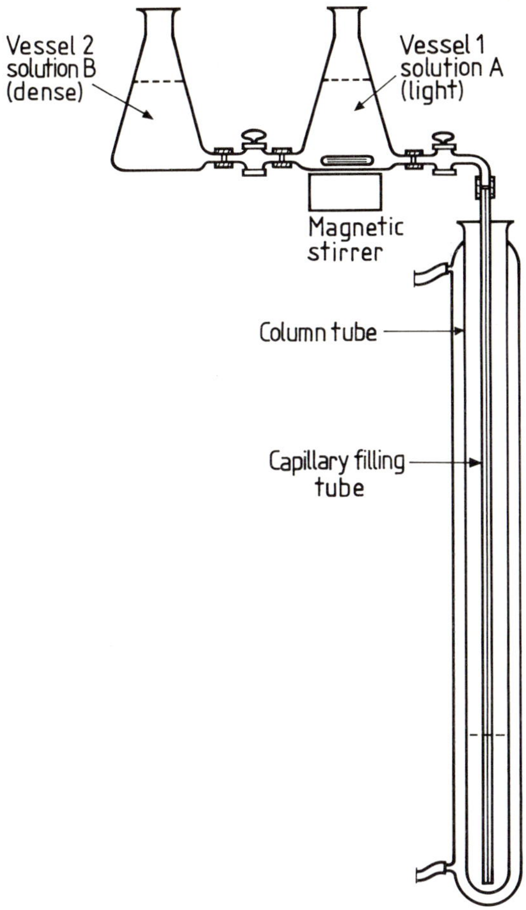

Figure 7.2 Method for filling density column

In use, careful control of temperature (±0·1°C) is of course essential. Total density gradients of more than 0·2 g/ml or less than 0·02 g/ml are not recommended. Needless to say, the mixed liquids must be without effect on the material under test, be transparent, of low volatility and low viscosity. The tube is calibrated with spherical glass floats, about 5 mm in diameter, the densities of which may, for instance, be obtained by a method such as that described in the previous section.

These floats are placed gently in the column and after at least 24 h the height of each above a reference level is determined by a cathetometer. The measured heights are plotted against density and a smooth curve, without discontinuities, should result. With careful use, a column should last several months.

Test specimens should be of easily identifiable shape and size, the latter being such that the centres of their volumes can be estimated to within 1 mm. At least two are placed gently in the column and their heights are recorded after not less than 10 min. The densities are read off the calibration graph and, in the case of polyethylene film for instance, should be within 0·0005 g/ml. Care must be taken to avoid air bubbles, dirt and interference between specimens and floats.

The column is cleared as and when necessary by a wire gauze basket, of the widest practicable mesh, drawn upwards through the column at a uniform rate not exceeding 3 cm/min, after which the basket is returned to the bottom of the column which is then recalibrated.

Table 7.1 lists typical liquid systems that are used, but is not meant to be exhaustive.

BS 2782 Method 620D (1980),[2] which makes reference to BS 3715 (1964),[8] and ASTM D1505 (1968)[4] also employ this technique.

Table 7.1 Liquid Systems for Density Gradient Columns

System	Density range (g/ml)
Methanol/benzyl alcohol	0·80–0·92
Isopropanol/water	0·79–1·00
Isopropanol/diethylene glycol	0·79–1·11
Ethanol/carbon tetrachloride	0·79–1·59
Ethanol/water	0·79–1.00
Toluene/carbon tetrachloride	0·87–1·59
Water/sodium bromide	1·00–1·41
Water/calcium nitrate	1·00–1·60
Zinc chloride–ethanol/water	0·80–1·70
Carbon tetrachloride/1,3-dibromopropane	1·60–1·99
1,3-dibromopropane/ethylene bromide	1·99–2·18
Ethylene bromide/bromoform	2·18–2·89
Carbon tetrachloride/bromoform	1·60–2·89

7.1.3 Special Methods

Apparent Powder Density

Two standard methods are available for determining the apparent density of moulding powders.

The first is a test developed for moulding material that can be poured from a funnel. This is detailed in ISO 60 (1977).[9] A funnel of the form shown in Figure 7.3 is used. It is mounted vertically with its lower orifice 20–30 mm above the top of a measuring cylinder (capacity 100 ml; internal diameter 40–50 mm) and coaxial with it. With the lower orifice closed, 110–120 ml of well mixed powder is poured into the funnel and then the powder is allowed to flow into the measuring cylinder, assisted if necessary by being loosened with a rod. When the cylinder is full, a straight-edged blade is drawn across the top of the cylinder to remove excess and then the contents are weighed. The mean of two determinations is taken and expressed in g/ml.

BS 2782, Method 621A (1978)[10] and DIN 53468 (1974)[11] are identical to this method.

ASTM D1895 (1969)[12] offers two similiar methods, but both with different

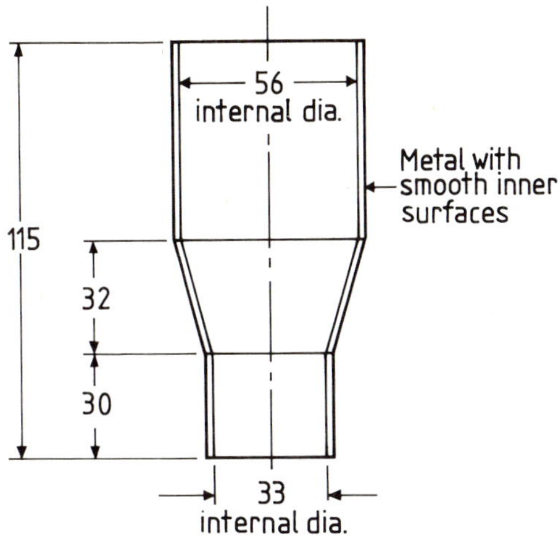

Figure 7.3 Standard funnel for apparent powder density—dimensions in millimetres

sized funnels from the above and some other, minor, variations. One of these funnels (of Method A) is used to measure pourability by timing the rate of flow of the powder out of the funnel.

The second method is for materials that cannot be poured from a funnel and is given in ISO 61 (1976).[13]

A cylinder is used which is 1000 ml in capacity and of internal diameter 90 ± 2 mm. Into this fits a plunger of slightly smaller diameter and with a total mass of 2300 g. A cylinder closed at its lower end and weighted with lead shot may be used.

60 g of the powder is dropped, little by little, into the cylinder so that it is evenly distributed and has a level surface. The plunger is then lowered on to the powder and allowed to remain there one minute before the height of the powder is measured with the plunger still in position. From the height of powder, the diameter of the cylinder and the weight of powder, the apparent density is calculated.

BS 2782, Method 621B (1978)[10] and DIN 53467 (1960)[14] are identical to this method, while Method C of ASTM D1895 (1969)[12] is essentially the same.

ISO 1068 (1975)[15] for PVC resins is a cylinder method for determining the compacted bulk density; a shaking machine is used to tamp down the material under a piston.

Bulk Factor

The bulk factor of a moulding is defined as the ratio of the volume of a given mass of moulding material to its volume in the moulded form. It is thus the ratio of the density of the moulded material to its apparent density before moulding.

ISO R171 (1961),[16] BS 2782, Method 501C (1970)[17] and the method of ASTM D1895 all require determination of apparent powder density by the corresponding techniques (as above) and moulded density by the appropriate method (§7.1.2). The German practice is similar and is to be found in DIN 53466 (1960).[18]

7.2 DIMENSIONS

7.2.1 Standard Test Methods

Fabricated Dimensions

At the time of writing there are within ISO three standards covering the measurement of film and sheet dimensions: ISO 4591 (1979)[19] dealing with gravimetric thickness, ISO 4592 (1979)[20] dealing with length and width measurement and ISO 4593 (1979)[21] dealing with thickness measurement by a mechanical scanning technique. It is anticipated that these will be adopted in British Standards, although they have not yet been included in BS 2782. The measurement of gravimetric thickness is already covered, however, in BS 2782, Method 512A (1970)[22] wherein an area of not less than 16 cm^2 (2.5 in^2) is cut with a punch, razor or sharp knife. The mass (W) of the specimen in grammes, its area (A) in square centimetres and its relative density (§7.1.2 above) are all determined to an accuracy of at least 1 per cent, at 23° ± 1°C:

$$\text{gravimetric thickness} = \frac{10{\cdot}0W}{A \times \text{relative density}} \text{ mm} \qquad (7.3a)$$

$$\text{or} \quad = \frac{0{\cdot}394W}{A \times \text{relative density}} \text{ in} \qquad (7.3b)$$

For rigid plastics such specialist methods are not required and the normal micrometer or dial gauge is most frequently employed, usually graduated in 0·02 mm (0·001 in). Micrometers are accurately standardised in BS 870 (1950)[23] and their use is described in ASTM D374 (1979),[24] Methods A and B. Likewise, dial gauges (or comparator gauges) are dealt with in BS 907 (1965)[25] and Method C of ASTM D374. Appropriate German standards are DIN 863 (1977)[26] and DIN 878 (1979).[27]

The measurement of soft materials which readily yield, particularly highly plasticised PVC and cellular plastics, presents difficulties in defining what the thickness (or width) really is. A figure under no compressive force is difficult to obtain, for optical means are hardly suitable for routine rapid working. The use of micrometers without ratchets clearly involves a largely subjective judgment of how hard the spindle is screwed down. Ratchet micrometers are little better, slipping at some arbitrary force. Many product specifications require the use of dial gauges, specifying the force on the plunger (which includes any spring loading) and the dimensions of the bearing faces ('anvils'), and there is much variation from one document to another. Each must be carefully studied and the instrument selected accordingly; a general method is given in Method 512B of BS 2782 (1970)[22] for flexible sheet, where the gauge must read 0·0025 mm (0·0001 in) for sheet up to 1·8 mm (0·070 in) thick, the foot of the anvil must be not less than 6·3 mm (0·25 in) in diameter and the pressure 10–20 kN/m^2 (0·1–0·2 kgf/cm^2; 1·5–3·0 lbf/in^2). For thicker sheet, a gauge reading to 0·025 mm (0·001 in) may be used.

Particle Size

The particle size of powders, granules or pellets may be of profound importance for it influences, for instance, the packing density of a moulding powder, and hence the charge in a fully positive mould, and the gelation characteristics of a polymer used in paste form (such as PVC). The simplest method, and the most widely used on a routine basis, is that of sieving. This technique is not normally recommended for particle sizes much below about 75 μm diameter (though ASTM D1921 (1963)[28] suggests 38 μm as minimum) where it is necessary to use sedimentation, optical or other methods.[29–33]

A word of warning is necessary in the use of test sieves, which are usually numbered to indicate directly or indirectly the number of apertures per unit dimension. Sieve screens may be made of woven silk or fine wire mesh and these two types differ quite considerably in their resistance to distortion under pressure. Wire mesh is most often specified, but again it is necessary to appreciate that a simple statement of the number of holes per unit dimension does not suffice since the diameter of the wire affects that all-important factor, the aperture size. Nor is specification of the latter alone adequate, since the diameter of the wire used to achieve that size controls the actual shape of the aperture, as the weave of the mesh must of necessity yield a non-planar hole. Obviously a coarse wire gives a more distorted (and larger) hole than one obtained from a fine wire, though in the main plane of the mesh both holes may be identical. For this reason it is absolutely essential to tie sieve analysis—as particle sizing by test sieves is termed—to the particular type of sieve used, for example as specified in BS 410 (1976)[34] which follows the recommendations of ISO Technical Committee TC24 and, in particular, ISO 565 (1972).[35]

There is a variety of ways of using test sieves, well illustrated by reference to the relevant ASTM standards (D1705 (1961)[36] and D 1921 (1963)[28]).

DIN 53477 (1965)[37] gives a method for sieve analysis of granular thermosetting compression moulding materials.

For normal dry sieving an appropriate range of sieves (usually 20 cm diameter)—appropriate, that is, in total span of aperture sizes covered and in the intervals of sizes between them—are 'nested' together, decreasing in aperture size from top to bottom; a tray or pan is fitted under the smallest mesh sieve.

A convenient quantity—say, 100 g—of sample is weighed out and transferred to the top sieve and the cover is placed on the top sieve. Although hand tapping of the assembled nest of sieves may be used it is both tedious and inefficient and therefore a mechanical vibrator is recommended (ASTM D1921 specifies one with a rotary motion and tapping at 150 taps per minute). After 10 min or some appropriate period to attain equilibrium the sieves and tray are separated and the contents weighed to the nearest 0·1 g either by direct weighing in the sieves (having already weighed the empty sieves) or by transferring the contents to weighing bottles, etc; in this latter case, extreme care must be taken to remove all particles from the walls and mesh of the sieve with a soft brush.

Normally the total recovered contents amount to 98 per cent or more of the original sample weight, in which case any lost is added to the figure for material passing through the finest sieve used, i.e. on to the tray.

An obvious variant of this method is to simplify it by employing just one, or a very limited number of sieves.

With vacuum dry sieving, smaller sieves, normally of 7 to 8 cm diameter, are used and, after being dried in a dessicator, are weighed individually before the test is started. They are then inserted in a conical adaptor of appropriate diameter to take the nested sieves and fitted with a lead at its smaller diameter to a vertical standpipe attached to a vacuum system. The suction of the latter is adjusted so that the finest sieve is not appreciably strained and then the vibrator is started. An appropriate quantity of sample is accurately weighed (1 g for very fine powders or 5–10 g for coarser ones) and carefully transferred to the top sieve; tapping of the sieve and light brushing of the sample assists. When no more powder is transferred through the top sieve, suction and vibration are ceased, the sieve is removed and any powder adhering underneath is brushed on to the next sieve below, when the operation is repeated and so on. The sieves with retained powders are weighed, whence the weights of the powder are obtained by difference.

A method is described in ASTM D1705 (1961)[36] for powdered polymers and copolymers of vinyl chloride. This uses the wet sieving technique, which eliminates static charges and troubles associated therewith such as agglomeration (it also reduces the tendency to 'flying' of powder and hence loss of sample).

Sieves are used as described above and are weighed prior to testing. An appropriate quantity of polymer is weighed out (25–100 g) into a beaker and approximately 300 ml of 0·5 per cent wetting agent solution (anionic sulphate or sulphonate type) is added with stirring. This mixture is poured into the top sieve and the powder is washed through this and the lower ones by more solution and then water. The sieves are dried in an oven and weighed.

In ASTM D1457 (1978)[38] for polytetrafluoroethylene materials the apparatus is illustrated in support of the description of the test method.

7.2.2 Non-standard Methods

Accurate measurement of large dimensions may be made by vernier caliper. Specialist methods for continuously measuring film or sheet thickness, such as beta ray absorption, capacitive and magnetic methods have been developed[39,40] but not so far standardised. There are also several methods of measuring pipe wall thickness given in the literature.[41–45]

7.2.3 Surface Roughness

It is not often that the surface roughness of a plastics material is required to be known so it is not surprising to find that no standard test methods exist.

For transparent plastics the surface roughness is of great importance since the light transmission and reflection characteristics are considerably dependent on it. In this case, however, the direct measurement of transmission, haze and gloss is carried out by methods of the type described in Chapter 14, rather than the measurement of roughness *per se*.

If measurements are attempted, then those developed for use with metals may be appropriate, although it must be borne in mind that the greater deformability of most plastics compared with metals may render the method unsuitable unless it can be modified to avoid deformation. The measurement of surface roughness is covered by ISO R468 (1966),[46] and also by the British equivalent, BS 1134 (1972),[47] in two parts, the first dealing with instrumentation and the second forming a general explanation. The British Standard is, therefore, a useful introduction to the subject.

7.2.4 Extensometry

The measurement of extension—or other forms of deformation—is an essential part of several tests, especially short-term mechanical tests of the type to be described in the next chapter. The method of measurement is, however, considerably dependent on the geometry of the test piece and the absolute deformation to be measured, and so the subject is best discussed in the context of the specific tests applied. Suffice it to say here that the precision required must be specified in the individual test method and is unlikely to be the same as that required for test piece dimensions.

7.2.5 Dimensional Stability

A number of standard test methods exist for the measurement of the dimensional stability of plastics but as these are intimately linked to the effect of temperature on plastics a description of them is deferred until Chapter 16.

7.2.6 Dispersion

The dispersion of pigments in plastics mouldings can be of profound importance in determining the suitability of the moulding for the job intended. In recognition of this, BS 2782 Methods 823A and B (1978)[48] describe two methods for assessing the uniformity of dispersion of carbon black in polyethylene, although the methods could be modified to be of more general application. The methods consist of examining a thin section of sample under a microscope using $\times 100$ linear magnification, the sections being obtained either by melting and pressing on a hot stage, as in Method A, or by microtomy, as in Method B. A visual assessment of the uniformity of dispersion is made based on the reference photographs presented in the standard.

It is appropriate to mention the valuable role the microscope can play in throwing light on the cause(s) of failure of a particular component. Examination of the failed component under a microscope may reveal the presence of weld or spider lines, poor pigment dispersion, incomplete blending of two polymers, insufficient rubber phase in a 'high impact' grade of polymer, etc. This use of the microscope as a powerful diagnostic tool is, sadly, often overlooked.

REFERENCES

1. ISO R1183 (1970) *Methods for Determining the Density and Relative Density of Plastics excluding Cellular Plastics.*
2. BS 2782 Methods 620A to E (1980) *Determination of Density of Solid Plastics excluding Cellular Plastics.*
3. ASTM D792 (1966) *Specific Gravity and Density of Plastics by Displacement.*
4. ASTM D1505 (1968) *Density of Plastics by the Density Gradient Technique.*
5. ISO 1675 (1975) *Plastics—Liquid Resins—Determination of Density by the Pyknometer Method.*
6. BS 733 (1965) *Density Bottles.*
7. Payne, N. and Stephenson, C. E. (1974) *Materials, Research and Standards*, **4** (1), 3.
8. BS 3715 (1964) *Concentration Gradient Density Columns.*

9. ISO 60 (1977) *Plastics—Determination of Apparent Density of Material that can be Poured from a Specified Funnel.*
10. BS 2782 (1978) Method 621A *Determination of Apparent Density of Moulding Material that can be Poured from a Funnel;* Method 621B *Determination of Apparent Density of Moulding Material that Cannot be Poured from a Funnel.*
11. DIN 53468 (1974) *Testing Plastics; Determination of Apparent Density of Moulding Material that can be Poured from a Specified Funnel.*
12. ASTM D1895 (1969) *Standard Test Methods for Apparent Density, Bulk Factor and Pourability of Plastic Materials.*
13. ISO 61 (1976) *Plastics—Determination of Apparent Density of Moulding Material that Cannot be Poured from a Specified Funnel.*
14. DIN 53467 (1960) *Testing Plastics; Determination of Apparent Density of Moulding Material that Cannot be Poured from a Specified Funnel.*
15. ISO 1068 (1975) *Plastics—PVC Resins—Determination of Compacted Apparent Bulk Density.*
16. ISO R171 (1961) *Plastics—Determination of Bulk Factor of Moulding Materials.*
17. BS 2782 (1970) Method 501C *Bulk Factor of Moulding Material.*
18. DIN 53466 (1960) *Testing Plastics; Determination of Bulk Factor of Moulding Materials.*
19. ISO 4591 (1979) *Plastics—Film and Sheeting—Determination of Average Thickness of a Sample and Average Thickness of a Roll, by Gravimetric Techniques (Gravimetric Thickness).*
20. ISO 4592 (1979) *Plastics—Film and Sheeting—Determination of Length and Width.*
21. ISO 4593 (1979) *Plastics—Film and Sheeting—Determination of Thickness by Mechanical Scanning.*
22. BS 2782 (1970) Method 512A *Gravimetric Thickness of Flexible Sheet*; Method 512B *Thickness by Direct Measurement of Flexible Sheet.*
23. BS 870 (1950) *External Micrometers.*
24. ASTM D374 (1979) *Standard Method of Test for Determining the Thickness of Solid Electrical Insulation.*
25. BS 907 (1965) *Dial Gauges for Linear Measurement.*
26. DIN 863 (1977) *Micrometers, External Micrometers, Terms and Definitions, Technical Requirements, Testing.*
27. DIN 878 (1979) *Dial Gauges.*
28. ASTM D1921 (1963) *Particle Size (Sieve Analysis) of Plastics Materials.*
29. BS 3406: Parts 1–4 (1961–3) *Methods for the Determination of Particle Size of Powders.*
30. Particle Size Analysis Subcommittee of the Analytical Methods Committee of the Society for Analytical Chemistry (1963) *Analyst,* **88** (144), 156.
31. Irani, R. R. and Callis, C. F. (1963) *Particle Size: Measurement, Interpretation and Application*, Wiley.
32. Dallavalla, J. M. (1948) *Micrometrics* 2nd edn, Pitman.
33. Orr, C., Jr and Dallavalla, J. M. (1959) *Fine Particle Measurement*, Macmillan.
34. BS 410 (1976) *Test Sieves.*
35. ISO 565 (1972) *Test Sieves—Woven Metal Wire Cloth and Perforated Plate—Nominal Sizes of Apertures.*
36. ASTM D1705 (1961) *Particle Size Analysis of Powdered Polymers and Copolymers of Vinyl Chloride.*
37. DIN 53477 (1965) *Testing Plastics; Determination of the Particle Size Distribution of Powdered Thermosetting Moulding Materials by Sieve Analysis.*
38. ASTM D1457 (1978) *Standard Specification for TFE–Fluorocarbon Resin Moulding and Extrusion Materials.*
39. Winzon Research (1978) *Plastics World*, **36** (5), 80.

40. Lake Wales Plastic Corporation (1977) *Plastics Technology*, **23** (5), 13.
41. Autech Corporation (1978) *Plastics Technology*, **24** (3), 9.
42. Brunner, M. (1977) *Plastics South Africa*, **6** (5), 39.
43. Barking, H. L. and Getachev, A. (1976) *Kunststoffe*, **66** (12), 807.
44. F. G. Industries (UK) Limited (1976) *Reinforced Plastics*, **20** (10), 279.
45. Lohrbaecher, V., Boes, D. and Schneiders, A. (1974) *Kunststoffe*, **64** (9), 438.
46. ISO R468 (1966) *Surface Roughness.*
47. BS 1134 (1972) *Method for the Assessment of Surface Texture.*
48. BS 2782 Methods 823A and 823B (1978) *Methods for the Assessment of Carbon Black Dispersion in Polyethylene using a Microscope.*

Chapter Eight

Short-term Stress–Strain Properties

8.1 HARDNESS

8.1.1 General Considerations

'Hardness' is not a fundamental property—indeed, the very interpretation of the word is decidedly subjective; the most widely accepted concept is probably that of resistance to identation, but others widely held include scratch resistance and rebound resilience. The last-mentioned is considered in Chapter 9 as a dynamic property, and scratch resistance in Chapter 10 under 'Wear'.

Resistance to indentation, which is invariably measured before fracture occurs, is readily envisaged as some function of rigidity or modulus and, like this property, measurement of hardness is subject to all the effects of temperature, time and other test variables mentioned in §8.2. Phillips and Ramakrishnan[1] have investigated the effect of time of indentation and of recovery on the Rockwell hardness of various thermoplastics. In common with the traditional methods used for metals, hardness measurement of plastics usually takes the form of forcing a standard indentor—often a hardened steel ball—under known load into a flat surface of the material under examination and measuring the resultant degree of penetration. The viscoelastic nature of plastics introduces two complications for, in addition to the dependence of depth of penetration on the time of application of the load, the results of the metals-type test are calculated from the diameter of the indentation (or some other characteristic parameter if the indentor is not sphere-ended) and so the values may well depend on how speedily the diameter can be measured after removal of the loaded indentor. A fairly successful way of overcoming this problem in one type of test is to interpose a piece of carbon paper between the ball-indentor and the test surface (carbon side to the latter) and measure the diameter of the carbon impression left after removal of the load.

There are a few attempts in the literature to relate indentation of plastics, but more often of elastomers, to modulus. Detailed consideration of the theories may be found, for instance, in Nielsen,[2] Ritchie,[3] Brown,[4] Gent,[5] Tangorra[6] and Soden,[7] the last being also a useful if now somewhat out-of-date review of methods for hardness testing of elastomers. Suffice it to state here that in addition to the effects of temperature and time and the nonlinear response of strain to stress, we have the fact that, at best, indentation is a complex function of properties including modulus, force, and indentor profile. It therefore follows that hardness values according to one method cannot generally be compared with those derived from another, though some *ad hoc* correlations have been published.

Work by Fett *et al.*[8] and, more recently, by Bowman and Bevis[9] has sought to relate hardness measurements to orientation conditions in a variety of thermoplastics.

General reviews of hardness testing are given by Davies *et al.*[10] and Fenner,[11] and for plastics specifically by Maxwell,[12] Lysaght,[13] Boor,[14] Haldenwanger,[15] Gouza[16] and Livingston.[17]

8.1.2 Standard Test Methods

ISO Standards

There are, at present, two standard methods in ISO: ISO 868 (1978)[18] for the measurement of Shore hardness and ISO 2039 (1973)[19] for a ball indentation method. The latter will be reissued along with a method for determining Rockwell hardness.[20]

ISO 868 describes two methods for the measurement of hardness using different shaped indentors. The Shore A method employs a truncated cone as the indentor and should be used for the softer range of plastics, a specimen thickness of at least 5 mm being required; the Shore D method, for harder plastics, employs a sharp cone with slightly rounded end and can be used on specimens down to 3 mm in thickness. Figure 8.1 gives the essential dimensions of these indentors.

It is recommended that measurements are not made less than 12 mm away from any edge.

As already mentioned and discussed in more detail in §8.2, the viscoelastic nature of the material necessitates that the reading be taken after a standard period of time. For these methods this is set at 15 ± 1 s, although an 'instantaneous' hardness may be estimated after a 1 s application of the durometer spring.

ISO 2039 employs a 5 mm diameter hardened steel ball which is pressed into the specimen under a specified load selected to give an indentation between 0·07 and 0·10 mm for Method A and between 0·15 and 0·35 mm for Method B. The

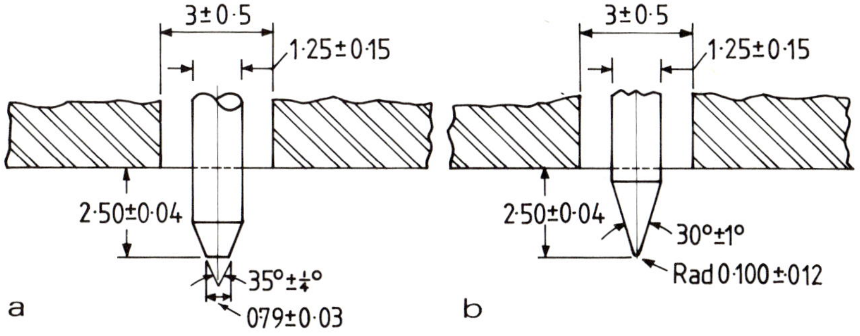

Figure 8.1 Shore durometer indentors: ISO 868—dimensions in millimetres (a) Type A (b) Type D

time of application of the load is 30 s before the depth reading is taken and a test piece thickness of 4 mm is recommended. The hardness value is defined as:

$$\text{ball indentation hardness} = \frac{\text{applied load}}{\text{surface area of impression}} \tag{8.1}$$

The revised version of ISO 2039, which at the time of writing was at Draft International Standard level, in addition to the ball indentation method will contain, as a second part, the measurement of Rockwell hardness. This follows the same procedure as laid down in ASTM D785, described more fully later. An annex to the ISO document describes the determination of Rockwell α, which, unlike the other Rockwell scales, correlates exactly with the hardness as determined by ball indentation. Fett[21] has shown that

$$H = \left(\frac{441{\cdot}4}{150 - R_\alpha}\right)^{1{\cdot}23} \tag{8.2}$$

where H = hardness by ball indentation and R_α = alpha Rockwell hardness number.

British Standards

At the time of writing there are three published methods for the measurement of the hardness of plastics: BS 2782, Method 365A (1976)[22] for 'Softness Number'; BS 2782, Method 365D (1978)[22] which is identical to the present ISO 2039 (1973) and so will not be considered further; and BS 2782, Method 1001 (1977)[23] which is for glass reinforced plastics and uses a Barcol impressor of Type 934-1. This last is identical to ASTM D2583 described below. There are also two standards in draft form, which are based on the corresponding ISO documents, so that it is intended to publish these as dual numbered ISO/BS standards, if possible. These are BS 2782, Method 365B, which uses the Durometer, and BS 2782, Part 3 (presumably to be 365C), on Rockwell hardness.

BS 2782, Method 365A has been brought into line with BS 903, Part A26 (1969),[24] rubber hardness, which is in technical agreement with ISO 48 (1979)[25] although the method of expressing the result is different. The method consists of measuring, at 23°C, the increase in depth of indentation of a steel ball 2·5 mm diameter into a flexible material when the force exerted on the ball increases from 0·3 N to 5·7 N. The softness number is defined as the depth of indentation of the ball in 0·01 mm units caused by this increase in force after it has been applied for 30 s. Figure 8.2 shows a typical instrument for carrying out this test.

The specimen should be between 8 and 10 mm thick for the standard test although other thicknesses are permitted provided the thickness is reported along with the softness number observed, an inverse relationship existing between the two. Edge effects are particularly important and no reading must be taken nearer than 10 mm to any edge.

A special note relating to PVC compounds is included as it is known that the hardness of plasticised compounds changes with time after moulding.[27] For such materials, conditioning should be carried out for 7 ± 0·2 days following moulding before the test is undertaken.

Figure 8.2 Dead load softness tester: BS 2782, Method 365A (*Courtesy H. W. Wallace & Co Ltd*[26])

USA Standards
Three standard methods are given: ASTM D2240 (1975)[28] which uses Shore A and D durometers; ASTM D785 (1965)[29] which uses the Rockwell scales; and ASTM D2583 (1975)[30] which uses the Barcol impressor.

D2240 is essentially the same as ISO 868 except that the 1 s reading is taken as standard, instead of 15 s.

D785 makes use of the well known 'Rockwell hardness tester'; a typical instrument is shown in Figure 8.3. Brinell hardness (see below) can also be measured on this instrument.

The scales (Table 8.1) overlap to a certain degree—the test increases in severity down the table—but correlation of scales is stated not to be desirable.

The test specimen (sheet 6 mm thick) rests on a flat anvil of at least 50 mm diameter. Two procedures are described. In Procedure A, essentially the minor load is applied for 10 s and then the major load for 15 s; the hardness reading is taken off the scale 15 s after the major load has been removed, but with the minor loading still operating. In Procedure B, which unlike A is limited to Scale

Figure 8.3 Rockwell and Brinell hardness tester
(*Courtesy Foundrax Inspection Equipment Ltd*)

Table 8.1 Rockwell Scales (ASTM D785)

Rockwell hardness scale	Minor load (kg)	Major load (kg)	Indentor diameter (in)	Indentor diameter (mm)
R	10	60	0·5000 ± 0·0001	12·700 ± 0·0025
L	10	60	0·2500 ± 0·0001	6·350 ± 0·0025
M	10	100	0·2500 ± 0·0001	6·350 ± 0·0025
E	10	100	0·1250 ± 0·0001	3·175 ± 0·0025
K	10	150	0·1250 ± 0·0001	3·175 ± 0·0025

R only, the essential principle is that indentation is recorded 15 s after application of the major load, but with the latter still applied. The test values are quoted as α (alpha) Rockwell Hardness Numbers and are obtained by subtracting the indentation reading from 150.

Further information on the Rockwell test as applied to plastics is given by Lysaght[13] and also by Boor,[14] who includes three scales not included in the above table but applicable to plastics.

D2583 describes the use of the Barcol impressor[31] of Type 934-1. A note refers to Type 935 for softer plastics but its use has not been standardised.

A diagram of the general style of the impressors is given in Figure 8.4. The indentor is a hardened steel truncated cone, having an angle of 26° with a flat top of 0·157 mm (0·0062 in) diameter. The indicating device has 100 divisions, each representing a depth of 0·0076 mm (0·0003 in) penetration—the higher the reading the harder the material. Test pieces must be at least 1·5 mm ($\frac{1}{16}$ in) thick and large enough in area to ensure a minimum distance of 3 mm ($\frac{1}{8}$ in) in any direction from the point of measurement to the edge of the specimen.

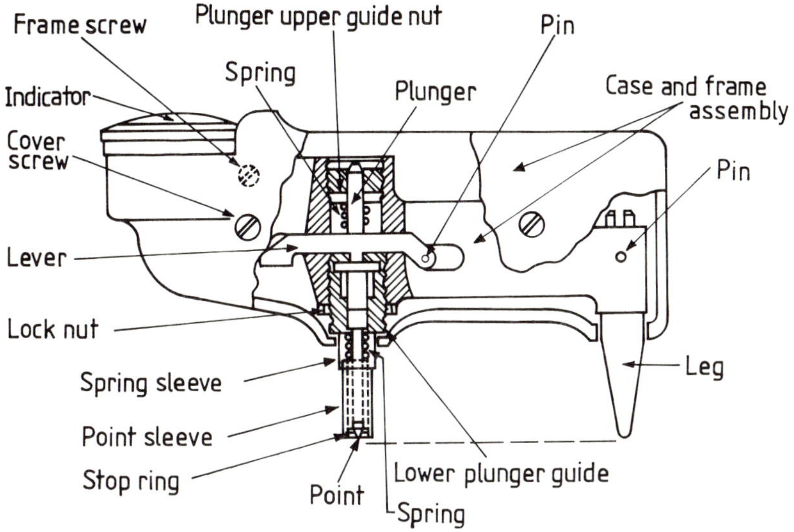

Figure 8.4 Barcol impressor:[31] ASTM D2583

When testing, the housing of the indentor is applied quickly by sufficient hand pressure to ensure firm contact with the test specimen and the highest dial reading is noted; the number of readings necessary ranges upwards from five for homogeneous materials.

German Standards
DIN 53456 (1973)[32] is essentially the same as ISO 2039 except that the major load must be selected from 49 N, 132 N, 358 N and 961 N, with a minor load of 9·81 N in all cases.

The formula gives

$$H = \frac{0{\cdot}0535F}{h - 0{\cdot}04} \qquad (8.3)$$

where H is the hardness
h is the penetration (mm)
F is the applied major load.

DIN 53505 covers the measurement of Shore hardness.

8.1.3 Other Methods

There are many other arbitrary hardness scales available for rubber and plastics and many which, fortunately, over the years have fallen into disuse. No attempt has been made to produce a comprehensive survey of all these methods and attention is directed below to only the more important.

The *Brinell method* already mentioned uses a spherical indentor and the hardness number is calculated from the formula:

$$\mathrm{HB} = \frac{2F}{\pi D^2 \{1 - [1 - (d/D)^2]^{1/2}\}} \tag{8.4}$$

where F = load (kg)
D = diameter of indentor (mm)
d = diameter of impression produced by indentor during period of 15 s (mm).

The indentors used are normally 1, 2, 5 or 10 mm in diameter and loads are selected to give an F/D^2 ratio of 30, 10, 5 or 1; the method was originally designed for metals and therefore an F/D^2 ratio of 5 or 1 is usually needed for plastics materials. The HB value is quoted with the F/D^2 ratio employed in the test.

The fundamental objection to the test for plastics materials is their recovery after removal of the load, and to some extent the indefinition of the indentation during test; the use of carbon paper (§8.1) affords some means of overcoming the former problem. Hounsfield[33] suggests the use of a *tensometer* to record the depth of penetration of the ball during test, but points out that the general principle is unsound as load is not proportional to depth of penetration; he prefers the use of a paraboloid indentor where the cross-section is proportional to the distance from the apex, i.e. the depth of penetration. He shows plots of load against depth of penetration for 'Perspex' and (low-density) polyethylene which are very nearly linear. The apparatus depicted in Figure 8.3 has conversion scales relating diameter of indentation to depth of penetration of some of its ball indentors, so that Brinell hardness can be obtained essentially while the load is still applied and the problem of recovery is thus overcome.

Brinell testing of metals is covered by BS 240: Part 1 (1962)[34] and ASTM E10 (1978),[35] the details of which differ in some degree from those given above.

The (*Vickers*) *diamond pyramid test* is another standard technique for metals which has been borrowed for hard plastics, from time to time. Its use for metals is described in BS 427: Part 1 (1961)[36] and ASTM E92 (1972).[37] The test employs a right diamond pyramid on a square base, with an apex angle of 136° between opposite facets, the object of which was to simulate one indentor in the Brinell range. The test is conducted in similar manner and the test values are computed as follows:

$$\mathrm{HV} = \frac{2F \sin(\theta/2)}{d^2} \tag{8.5a}$$

where F = applied load (kg)
d = mean diagonal width (mm)
θ = apex angle of the pyramid = 136°, i.e.

$$HV = \frac{1{\cdot}844F}{d^2} \tag{8.5b}$$

A 5 kg load is normally used for plastics.

Campbell[38] has conducted an investigation of the variables encountered in the Vickers test and included acrylics in his assessment of time effects, etc.

The Vickers diamond principle is used in an adaptation of the *Wallace micro-hardness tester*.[39,40] The apparatus is shown diagrammatically in Figure 8.5.

The loads employed are very small (a fraction of a gramme up to 3·5 kg) and the indentations correspondingly so; the test figures necessarily relate only to the surface of the test material, but the test is virtually non-destructive. The table on which the specimen is mounted is raised by the adjusting screw so that the indentor is returned to the exact position it occupied before the major load was applied. The indentor assembly is supported by leaf springs as shown, so that friction is eliminated, and as the principle is to return the indentor to the 'null' position, the leaf springs do not affect the result.

The 'null' position was formerly identified by means of headphones into which two oscillatory circuits were fed beating at nearly the same frequency. One of the circuits contained the capacitor shown and matters were so arranged that the frequencies were identical at the 'null' position, which could thus be readily identified by ear. An electronic 'null indicator' is now used. The wedge system and a suitable dial gauge permit indentation to be measured to 0·000 01 in (0·000 25 mm), clearly illustrating just how small the total indentation is in the test.

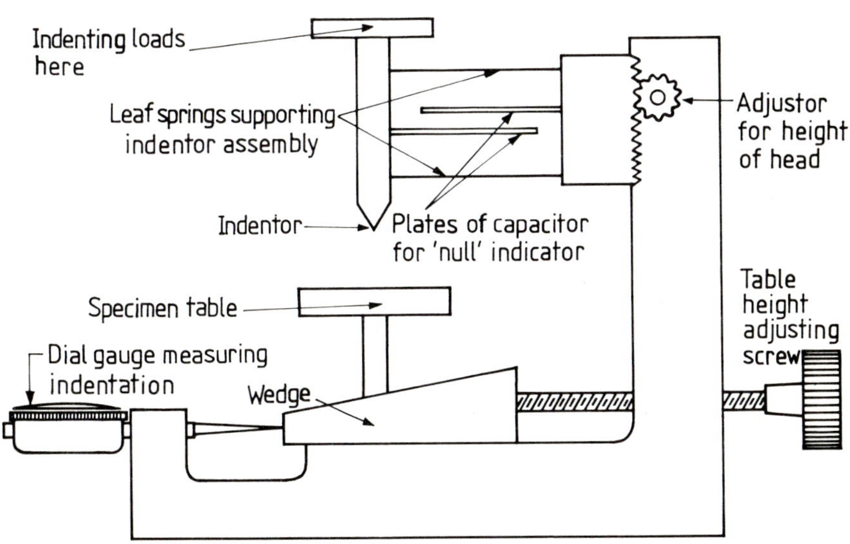

Figure 8.5 Diagrammatic view of Wallace micro-hardness tester

The same apparatus, but with a ball indentor 0·395 ± 0·005 mm in diameter, a contact force of 8·3 ± 0·5 mN and an indenting force increment of 145 ± 0·5 mN, is used for the micro-hardness testing of rubber (BS 903: Part A26 (1969),[24] Method M). The small size of the specimens required allows hardness measurements of small O-rings and irregularly shaped articles.

The results are read directly in 'International Rubber Hardness Degrees' but, depending on a number of variables, may not compare with those obtained from the already-mentioned Normal Method, N, of BS 903: Part A26. The microtest is best used as a method in its own right.

Likewise, use of a suitable dial gauge will give a 'microtest' suitable for softness numbers of PVC, but the agreement with BS 2782, Method 365A (see above) is not good and this variation is not standardised.

The microtest for rubber has now been standardised at international level and incorporated in ISO 48 (1979).[25]

Before leaving the closely allied subject of rubber hardness, mention must be made of BS 2719 (1975) *Pocket Type Rubber Hardness Meters* (*Methods of Use and Calibration*). One of the instruments described is the *Wallace pocket meter*, the mechanism of which (Figure 8.6) illustrates one method of using springs for this purpose. This instrument and the Shore A durometer give very similar results.

Knoop Hardness Number is obtained by using a Knoop indentor on the commercial instrument known as the 'Tukon' tester[41] which is also capable of carrying out the Vickers (136°) diamond test. The Knoop indentor is again a diamond, but with the ratio of long and short diagonals approximately 7:1 and, as a result, the depth of indentation is only about 1/30th of its length. Recovery problems are less[13] and only light loads need be applied, rendering the liability to shatter of brittle materials less than in the 136° test.

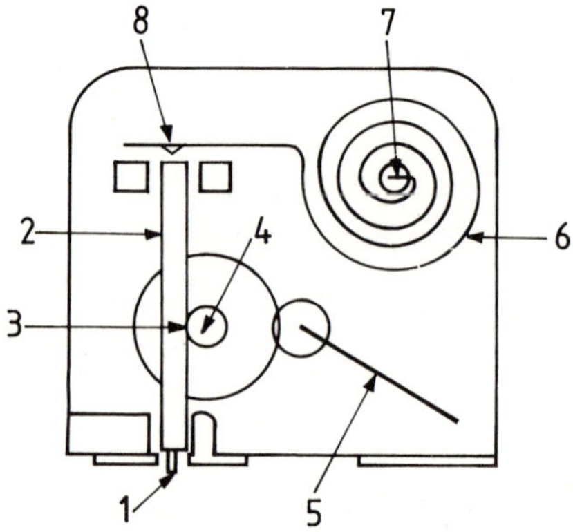

Figure 8.6 Mechanism of Wallace pocket hardness tester
The indentor (1) is fixed to a shaft (2), having a rack (3) rotating a pinion (4), which transmits movement through a train of gears with hairspring to the spindle carrying the pointer (5). Upward movement of the indentor and shaft is resisted by the clock-type spring (6), fixed at its inner end on the split spigot (7) and having a 'pip' on its free end (8) bearing on the top of the shaft. Spigot (7) is part of a circular plate on the back of the instrument, rotatable by means of a key to loosen or tighten the spring

The *TNO hardness tester*[42] employs an indentor of a sapphire polished into the form of a Vickers diamond pyramid and uses a capacitive technique for determining depth of penetration.

The *Sward hardness test* has been suggested for evaluating plastics materials non-destructively, its original development being for the examination of the hardness of paint films.[43]

The indentation testing of flooring materials is considered in detail by Gavan and Wein.[44]

The scleroscope is sometimes described as giving hardness values, but it is essentially a rebound (resilience) test. Likewise, Moh's test and pencil hardness are really for scratch hardness and are accordingly described in §10.2. Boor *et al.*[45] compare results from a number of the above hardness tests (several Rockwell scales, Knoop, Barcol, etc.) with those from scratch hardness (Bierbaum, pencil (Kohinoor) and Moh) and abrasion (ASTM Mar and Taber), examining a wide range of plastics materials. It is concluded, not surprisingly, that they mostly measure different parameters and therefore cannot be expected to give results which compare—even in placing materials in order of merit. Less comprehensive comparisons are given by Maxwell[12] and Lysaght.[13]

From time to time hot penetration tests have been suggested for checking the degree of cure of thermoset plastics;[46,47] the relationships, if any, are essentially empirical and subject to the influence of many compositional variables.

8.2 TENSILE STRESS–STRAIN

8.2.1 General Considerations

The most common type of stress–strain measurement is made in tension, that is by stretching the material. A tensile stress is thus applied, defined for a section of uniform cross-sectional area A_0 by the formula $\sigma_1 = F_1/A_0$ where σ_1 = tensile stress and F_1 = tensile force (Figure 8.7).

If this tensile stress induces a stretch to length l_1, the tensile strain ε_1 is defined as:

$$\varepsilon_1 = \frac{l_1 - l_0}{l_0} \qquad (8.6)$$

Taking the stressing operation to the ultimate, i.e. measuring the force until the material breaks, tensile strength (ultimate tensile stress)

$$\sigma = \frac{F}{A} \qquad (8.7)$$

where F = force at failure
A = area of cross-section at failure.

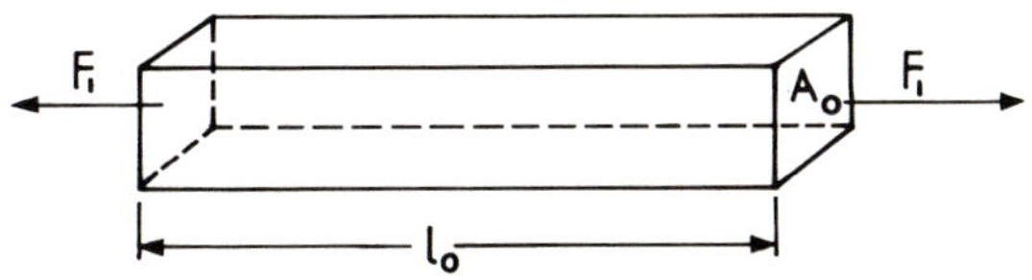

Figure 8.7 Material in tension

As the material stretches so its dimensions orthogonal to the axis of applied force decrease and thus the area of cross-section decreases. However, for experimental convenience most tensile strengths are based on the original cross-section (A_0) since this is easily measured before the test is started.

Ultimate elongation, or elongation at break, equals $l - l_0$ where l is the length at failure. This is usually expressed as a percentage of the original length:

$$\text{elongation at break} = \frac{l - l_0}{l_0} \times 100 \text{ per cent} \qquad (8.8)$$

According to Hooke's law, for an ideal elastic solid stress is proportional to strain (Figure 8.8). Even when we ignore the effects of temperature and most of the effects of time, no plastics material comes very close to this ideal. Plastics exhibit a whole spectrum of behaviour which, in qualitative terms, may be expected from such dissimilar materials as soft PVC, polystyrene, nylon and unplasticised PVC.

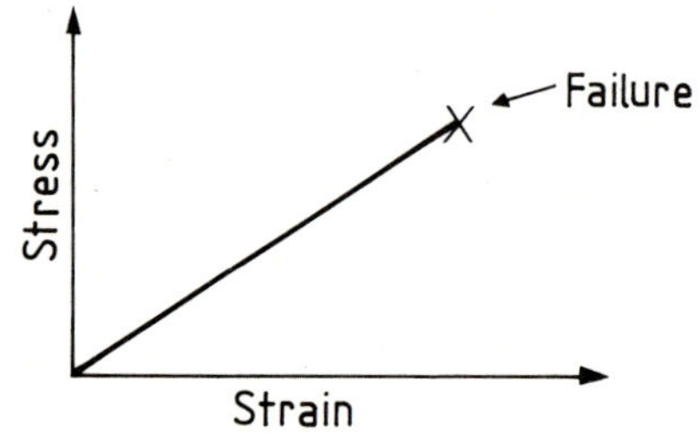

Figure 8.8 Hookean behaviour

Carswell and Nason[48] have envisaged five possibilities which they express in the form of stress–strain diagrams (Figure 8.9). It is immediately obvious that such terms as 'tensile strength' and 'elongation at break' can be misleading or gross oversimplifications even when used with the knowledge that such properties themselves have strictly limited application. For instance, in Figure 8.9a, is the correct value for tensile strength that deriving from the stress at break or that at the peak of the graph? Again, what significance has an elongation at break from a behaviour such as that depicted in Figure 8.9e, when we have a combination of steady extension with increase in stress, increase in extension with no increase in stress (even a drop in stress) and finally a rapid extension with relatively little increase in stress?

Nielsen[49] presents a general plot of stress–strain of the type shown in Figure 8.10.

A Hookean material has already been defined; the constant ratio σ/ε is called the Young's modulus, E. If a given plastics material has a stress–strain relationship which is initially linear, as in Figure 8.10 up to position L, $E = \sigma_L/\varepsilon_L$.

To this already complex situation must be added the further complication of the viscoelastic nature of the polymer, which causes the material's response to an applied force to vary according to the time-scale and temperature of the experiment. Thus, a tensile test carried out at 500 mm/min on a sample of polyethylene could result in a stress–strain curve of the type given in

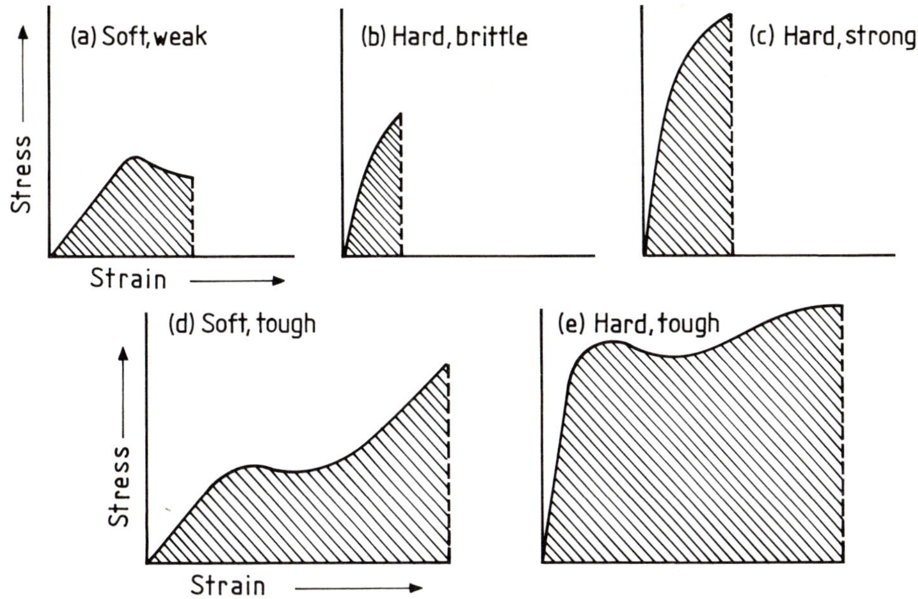

Figure 8.9 Stress–strain behaviour of various types of plastics[48]

Figure 8.9a, whereas at 5 mm/min the same material could exhibit the behaviour shown in Figure 8.9d. Similarly, conducting the test at two different temperatures could convert a hard, brittle plastic into a hard, tough one.

More detailed descriptions of stress–strain and viscoelastic behaviour are given by Nielsen,[49] Ritchie,[50] Alfrey,[51] Ogorkiewicz[52] and Ferry.[53] The fore-

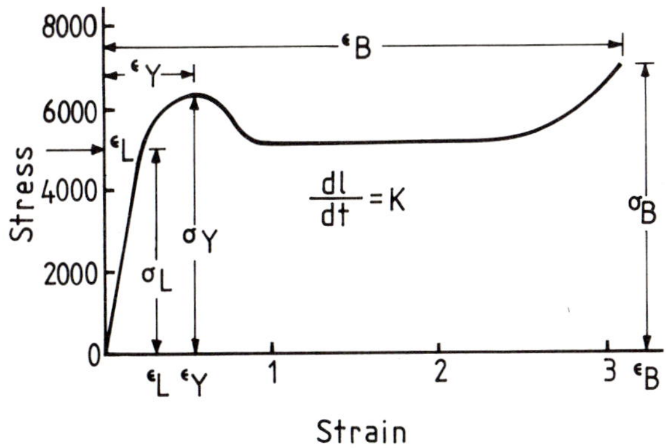

Figure 8.10 Schematic stress–strain curve[49]
σ_Y = yield stress
ε_Y = elongation at yield
σ_B = ultimate or tensile strength
ε_B = ultimate elongation or elongation at break
The precise location of the yield point, Y, is a matter of argument and, according to different authorities, may be anywhere between Y and L

going serves to illustrate that statements such as 'The tensile strength of poly *x* is *y* MPa' are meaningless unless the method of test is also given.

As with processibility tests discussed in Chapter 5, the standard methods for determining the short-term mechanical properties, outlined in this chapter, give only very limited information with regard to the behaviour of the material and are best thought of as instruments of quality control. Data to be used for engineering design purposes must be obtained under conditions which simulate as closely as possible the actual service conditions to be encountered, if grave errors are to be avoided. This will often involve considerable testing, including dynamic mechanical (Chapter 9), fatigue (Chapter 12) and long-term (Chapter 11) property measurements.

8.2.2 Testing Machines

The equipment required for carrying out tensile tests can vary from the very simple—adding weights to a freely suspended test piece until it breaks—to the very sophisticated—fully automatic machines where test pieces are placed in a loading carriage and tested one after the other with no further operator intervention. Built-in microprocessor units can effect initial area compensation and the digital output of strength and elongation can be fed to a printer for permanent record.

Of course such sophistication has to be paid for, but the past decade has seen a rapid growth in the use of electronic tensile test machines and there are available today quite simple and moderately priced machines which, nevertheless, offer a sufficiently wide range of test facilities to satisfy the average test laboratory. Such machines usually allow compression, bending and shear tests to be carried out also and are therefore generally referred to as 'universal' testing machines. The requirements for machines for testing plastics (and rubbers) are given in BS 5214(1975, 1978).[54]

Two types of testing machine may be defined: constant rate of loading (CRL) and constant rate of traverse (CRT), the latter being far and away the more common for polymer testing: all the standard test methods for tensile, compression and flexural tests specify this type of machine.

Possibly the most popular type of CRL machine is based on the steelyard principle (Figure 8.11). The poise weight is driven from left to right, from a position of equilibrium about the fulcrum such that there is no force on the specimen, along the arm at a steady rate, whereby the moment of the poise weight about the fulcrum is increased steadily and thus the force on the specimen is increased at a steady rate. The poise weight is driven by a separate engaging

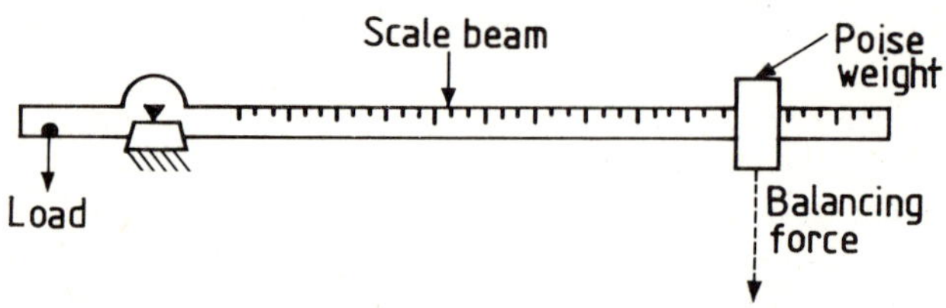

Figure 8.11 CRL steelyard-type testing machine

screw running parallel to the steelyard, or by a similar mechanism. Constant rate of loading conditions are most commonly called for in testing textiles and adhesives.

Constant rate of traverse machines are those where the test is controlled by the straining of the specimen and the resulting force is measured. This is a somewhat crude description because very few machines actually achieve the neat and tidy ideal of constant rate of extension of the specimen. In the first place the usual type of tensile specimen does not extend uniformly over its whole length, so that even if constant rate of separation of the jaws holding the ends of the specimen is achieved, extension will only be increased at a constant rate as an average over the whole length of the specimen. Secondly, the rate of separation of the driven jaws of the test machine is the sum of the deformation of the specimen plus movements in the jaws themselves and in the means of attaching the jaws to the machine, so that the rate of jaw separation, whether constant or not, is related to the specimen deformation (stretching, bending or compression) by a variable and probably unknown amount. Only by taking a gauge length of the specimen (see below), over which straining properties are uniform, and controlling the rate of separation of the jaws with reference to the extension of the gauge length, will true constant rate of extension be achieved. Thus constant rate of traverse, i.e. constant rate of separation of the crosshead holding one jaw from the other crosshead holding the other jaw, usually means no more than just that, i.e. 'CRT'. Moreover, many machines fondly thought to be 'CRT' are not this by any manner of means. The 'dynamometer' or force-measuring device may be found to 'yield' significantly under the load it is measuring and thus contribute, for instance, to the apparent extension of a tensile specimen: machines are termed 'soft' if this yield is appreciable relative to the deformation in the specimen and 'hard' if it is negligible. Finally, the mechanism driving or separating the jaws may alter in performance significantly according to the force it is acting against in the specimen.

Having drawn attention to all these shortcomings it is only fair to comment that with many machines the deviation from constancy of rate of transverse is not so serious as significantly to affect mechanical test results on plastics within practical limits, provided that the load is increased without 'surges' and the time to break is achieved within a defined period of time. Thus many specifications for mechanical tests speak in terms of 'extending the test specimen (or separating the jaws of the machine) at a rate which is essentially constant'.

Many machines on the market fall into the CRT category; Brown[55] gives details of current equipment.

8.2.3 Grip Systems

Of crucial importance to the successful completion of a tensile test is the grip system used to hold the test piece. This must ensure that the specimen is held firmly to minimise, or preferably prevent, slippage without crushing the tab ends; in addition, the design of the grips or chucks and their mode of attachment to the testing machine must be such as to ensure alignment of test piece in the direction of strain, without any bending or shearing component.

There are many designs of grips to achieve this aim, some complementary but mostly having special applications according to the nature of the material under test and its physical form, be it strong or weak, soft or hard, wide or narrow, thick or thin. Some designs are described below.

Probably the most widely used is the 'wedge action' type, suitable particularly for rigid flat sheeting (Figure 8.12). Some control over the friction between the test specimen and the gripping faces of the wedges can be obtained by having different surface patterns on the wedge faces, varying from serrated patterns to perfectly smooth; smooth faces may be further modified by insertion of emery paper between jaw face and test specimen, with the rough paper surface facing the test specimen. The final choice depends on trial and error.

Also very useful for general purpose work are pneumatically or hydraulically actuated grips: the grip pressure is easily adjusted prior to test and is then

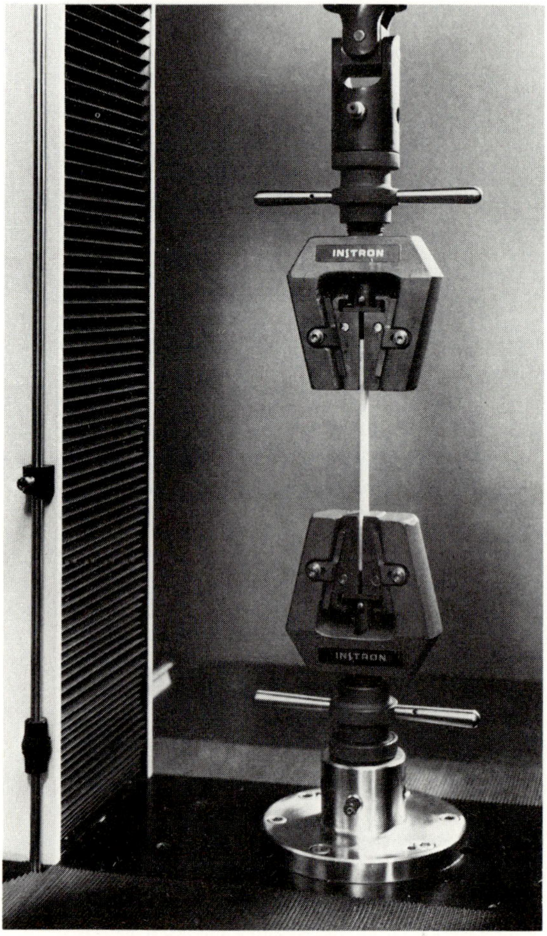

Figure 8.12 Wedge action grips
(*Courtesy Instron Ltd*)

maintained at a constant value throughout the test. Grip pressures from a few kilopascals to several tens of megapascals can readily be achieved.

Another type of grip, for the moulded thermoset type of tensile specimen, is illustrated in Figure 8.13.

For thin brittle films, vice chucks are often suitable (Figure 8.14), or windlass chucks (Figure 8.15). The film is gripped, vice fashion, between the 'flat' of a transverse rod of the chuck and the plate bolted thereto, and then the film and the rod–plate assembly are rotated (as shown by the arrow) so that the film has to wind round almost half a circumference of the rod before leaving the chuck to form a similar assembly at the other end with an identical chuck.

Very strong laminates and sheet materials are best gripped by strip chucks (Figure 8.16): the test piece is first drilled with a hole at each tab end to suit the bolts, which are screwed down hard on stout serrated washers which fit into chucks; the chucks are hollow, to fit round the free floating lugs of the test machine and be gripped by pins fitted in the holes shown, matching holes in the lugs. (This method of attaching the chucks to the test machine lugs is widely used.)

Finally, in this by-no-means exhaustive list, may be mentioned self-tightening jaw chucks used widely for soft rubber and plastics sheeting, such as highly plasticised PVC (Figure 8.17).

A self-aligning grip system for Instron machines has been described by Holland.[56]

8.2.4 Definition of Terms

Unfortunately at the moment there is no universally accepted set of definitions in relation to tensile tests, although the differences that do exist are of emphasis

Figure 8.13 Grips for moulded test pieces (*Courtesy Monsanto Ltd*)

Figure 8.14 Vice-type chucks
(*Courtesy Nene Instruments Ltd*)

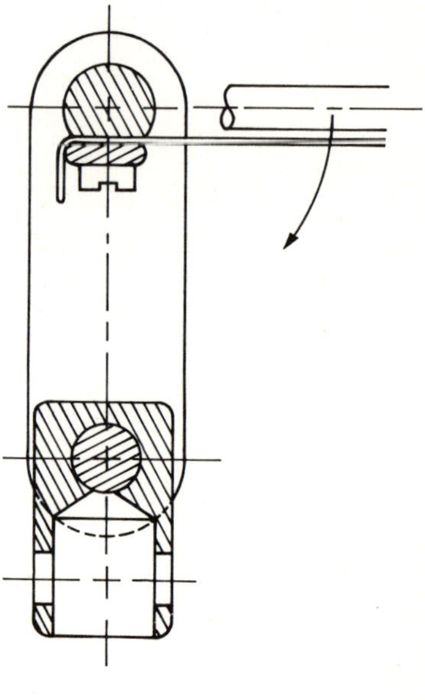

Figure 8.15 Windlass chuck action
(*Courtesy Tensometer Ltd*)

Figure 8.16 Strip chucks
(*Courtesy Monsanto Ltd*)

rather than of principle. The following have been taken from ISO/DIS 527,[57] which is a revision of ISO R527 (1966) currently being undertaken in TC61.

These definitions are virtually identical to those given in BS 2782, Methods 320A to F (1976):[58]

Tensile stress (nominal)	The tensile force per unit area of the original cross-section within the gauge length carried by the test piece at any given moment. The standard unit is megapascal (MPa) = MN/m^2 (meganewton/metre^2) $\equiv N/mm^2$
Tensile strength (nominal)	The maximum tensile stress (nominal) sustained by a test piece during a tension test.
Tensile stress at break	The tensile stress which occurs at break of the test specimen.
Yield stress	The tensile stress at which occurs the first marked inflection of the stress–strain curve. Where any increase in strain occurs without any increase in stress, this point is taken as the yield stress.
Offset yield stress	The tensile stress on the stress–strain curve where the curve departs from initial linearity by a specified strain.
Gauge length	The original length between two marks on the test piece over which the change in length is determined.
Strain	The change in length per unit original length of the measured gauge length of the test specimen. It is expressed as a dimensionless ratio.

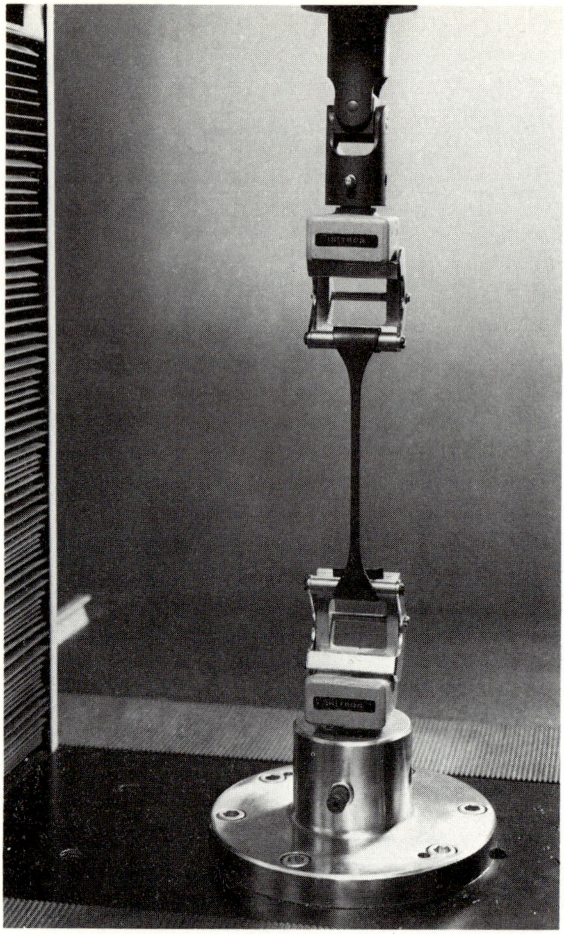

Figure 8.17 Self-tightening jaws
(*Courtesy Instron Ltd*)

Percentage elongation	The strain produced in the test piece by a tensile stress, expressed as a percentage of the gauge length.
Percentage elongation at yield	The percentage elongation produced in the gauge length of the test piece at the yield stress.
Percentage elongation at break, or at maximum load	The elongation at break, or at maximum load, produced in the gauge length of the test piece, expressed as a percentage of the gauge length.
Proportional limit	The greatest stress which a material is capable of supporting without any deviation from proportionality of stress to strain (Hooke's law).

Elastic modulus in tension (Young's modulus)	The ratio of tensile stress to corresponding strain below the proportional limit. The stress–strain relationship of many plastics does not conform to Hooke's law throughout the elastic range but deviates therefrom even at stresses well below the yield stress. For such materials the slope of the tangent to the stress–strain curve at a low strain is usually taken as the elastic modulus.
Secant modulus	In general, the ratio of stress to strain at any given point on the stress–strain curve.

8.2.5 Standard Types of Test Piece

In earlier chapters we have talked of preconditioning our test specimens and even how to mould or machine them, but there remains the question of the shape and size of pieces on which the measurements are to be performed. It is clear that different tests (for different properties) demand different test specimens—hence the reason for leaving the subject until discussion of each individual method—but it may be less obvious that, for instance, tensile tests on different materials employ specimens of different sizes; the effect of specimen size on tensile strength has been studied by Panov,[59] for instance. Again, this complication is due to the wide variety of types of material embraced in the term 'plastics', with quite different ultimate tensile strengths and percentage elongations at break.

To carry out a tensile test, i.e. a stretching test, some form of elongated specimen capable of being gripped at both ends is needed. The simple rectangular bar illustrated in Figure 8.7 is not suitable. Even if reduced in thickness to facilitate gripping, it is still often unsatisfactory in that fracture can occur anywhere along its length and most probably the point of fracture will be within one or other of the gripped portions, because of the weakening effect of the compression exerted by the gripping chucks (but see below). Thus a failure load may be obtained which is lower than the characteristic tensile value for the material in the specimen cross-section and, additionally, the failure may occur outside the portion of the test piece being examined for elongation (the gauge length). In order to select the portion in which failure takes place and obtain a breaking load on a cross-sectional area unbiassed by gripping mechanisms, a 'dumbbell' specimen is normally employed. For metals this is a fairly accurate description, since the test specimens are usually produced by lathe turning (Figure 8.18).

Many of the more rigid plastics may be machined on a lathe and so it is possible to produce similar specimens out of cast epoxide or poly(methyl methacrylate), for example; however, more often specimens have to be cut out

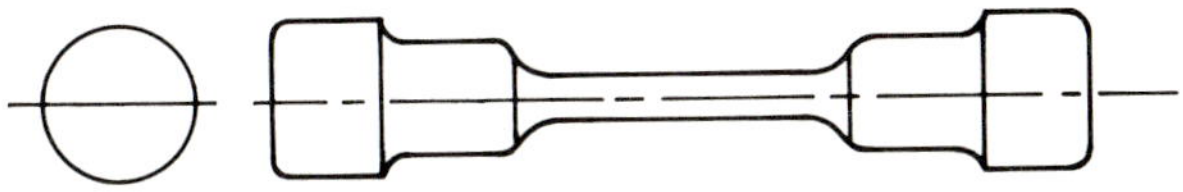

Figure 8.18 Metal dumbbell tensile test specimen

from (thin) flat sheet or moulded from powder or granules and the orthodox dumbbell specimen is not used as a standard test piece even when machining from castings. For sheet materials the specimens are 'two-dimensional dumbbells', retaining the flat profile of the sheet.

Four basic types of test piece are found in the standards literature: the narrow-waisted dumbbell, the broad-waisted dumbbell, the 'dog-bone' dumbbell and the parallel-side strip. These are illustrated in Figures 8.19–8.22, the dimensions being those contained in ISO/DIS 527.

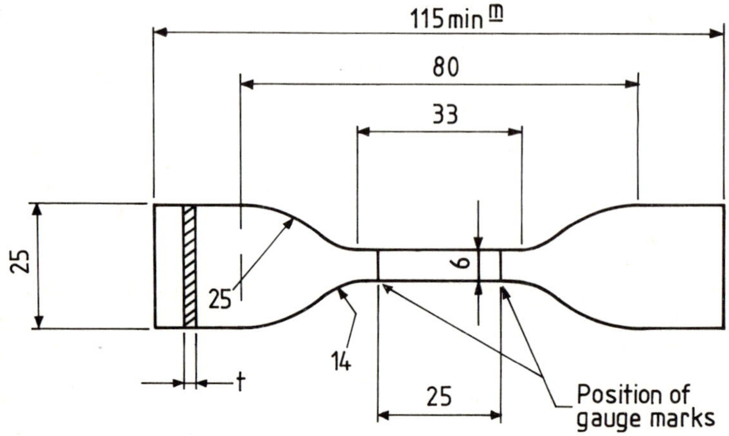

Figure 8.19 Tensile test piece: ISO/DIS 527 Type A (narrow-waisted dumbbell) t = 1 minimum, =3 maximum, =2 preferred—dimensions in millimetres

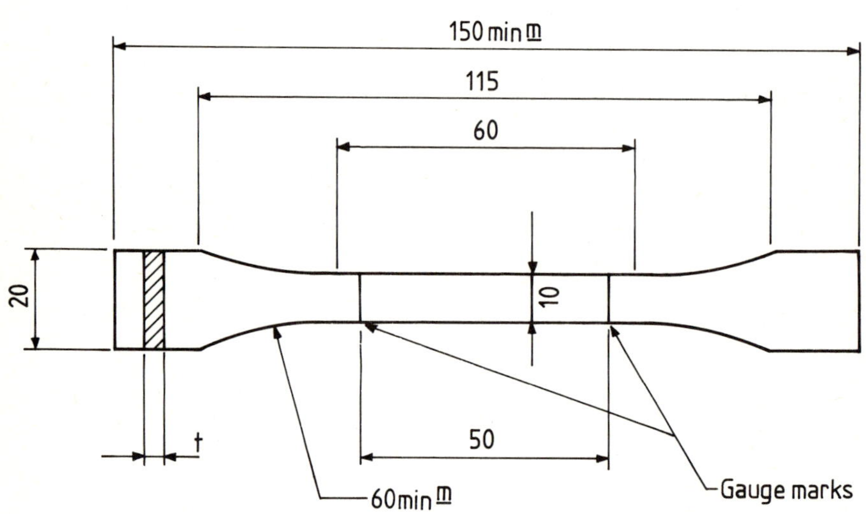

Figure 8.20 Tensile test piece: ISO/DIS 527 Type B (broad-waisted dumbbell) t = 1 minimum, =10 maximum, =4 preferred, =4 for moulded test pieces—dimensions in millimetres

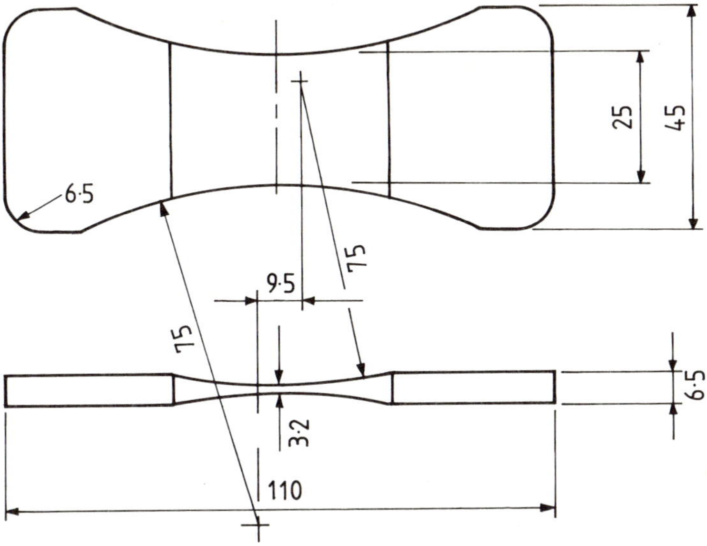

Figure 8.21 Tensile test piece: ISO/DIS 527 Type C ('dog-bone' dumbbell) —dimensions in millimetres

The narrow-waisted dumbbell is standardised in ISO R527 (1970) as the type 2 test piece, in ISO/DIS 527[57] as the Type A test piece, in BS 2782, Method 320A (1976),[58] in ASTM D638 (1977)[60] as the Type II or IV test piece and in DIN 53455 (1978)[61] as the Type 3 test piece. Apart from the American specimens, which differ in a number of dimensional details (being based essentially on Imperial units), these test pieces are identical. This type is particularly useful for testing plastics with a high elongation at break and is the same as is used for testing rubbers.

The same situation applies for the broad-waisted dumbbell, with the additional exception that the British Standard subdivides this class into two, according to thickness: Method 320B allows a maximum thickness of 4 mm and a minimum of 3·2 mm, while Method 320C has the same thickness range as ISO 527 but a preferred thickness of 3 mm instead of the ISO value of 4 mm. ASTM D638 has three test piece types in this class based on sample thickness, but other dimensions are also allowed to vary from test piece to test piece. This is the 'general purpose' test piece for use with materials having a low to moderate elongation at break.

The 'dog-bone' test pieces given in ISO R527 as Type 3, in ISO/DIS 527 as Type C and in BS 2782, Method 320D, are all identical. However, there appears to be little call for this test piece nowadays; it has been replaced to a large extent by the ISO Type B dumbbell, and there is a move within TC61 in ISO to delete this class from DIS 527 before it reaches the International Standard level. This type of specimen is used for thermosetting moulding powders only and is the only test piece which does not allow an elongation measurement to be made.

The final standard type of test piece is the parallel strip which, despite the disadvantage pointed out earlier, is found to be the most suitable for reinforced

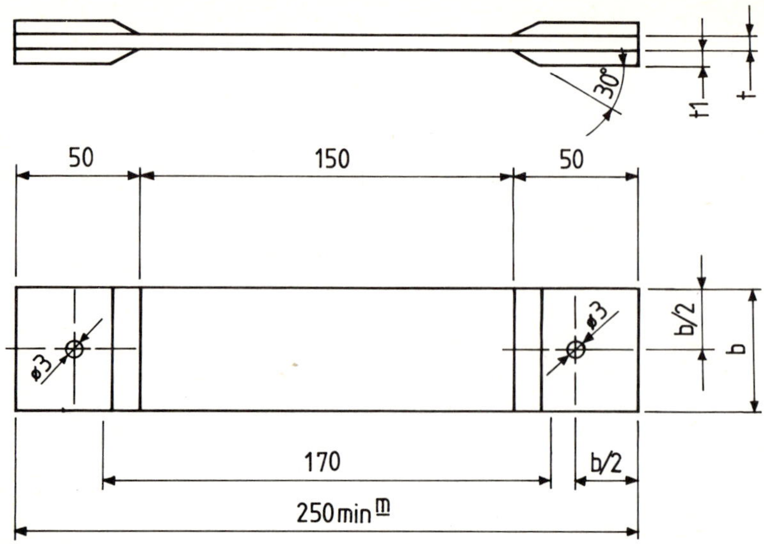

Figure 8.22 Tensile test piece: ISO/DIS 527 Type D (parallel-sided strip) t = 2 minimum to 10 maximum, t_1 = 3 minimum to 10 maximum, b = 25 or 50 —dimensions in millimetres

thermoplastic and thermosetting sheets. The problem of fracture in the grips is obviated by bonding end pieces onto the test piece (Figure 8.22). In addition, the tab ends so formed can be drilled so that the test piece can be bolted into the grips to prevent grip slippage.

Plastics films are also frequently tested using parallel strip specimens using the windlass-type chuck illustrated in Figure 8.15. There is still much lively debate, however, as to whether parallel strip or dumbbell shaped test pieces are to be preferred for testing plastics films.

ASTM D1708 (1979)[62] specifies the use of a microtensile dumbbell which can be used whenever a limited amount of material is available. The use of small test pieces, based on scaled-down versions of the current ISO Type 2 dumbbell, is also being pursued within TC61 and it is anticipated that an ISO standard covering such test pieces will be issued in the future.

8.2.6 Standard Test Methods

There are currently two internationally standardised test methods for the determination of the tensile properties of plastics: ISO R527 (1968) for testing plastics test pieces at least 1 mm in thickness and ISO R1184 (1970)[63] for test pieces less than 1 mm thick. The test pieces used in R527 and in its current revision DIS 527[57] have already been illustrated in §8.2.5; the test pieces for R1184 are of the parallel strip type with a length of at least 150 mm and a width between 10 and 25 mm.

The range of test speeds has been increased from five in R527 to nine in

DIS 527 and at present consists of

Speed A			1 mm/min ± 50%
Speed A1			2 mm/min ± 20%
Speed B			5 mm/min ± 20%
Speed C			10 mm/min ± 20%
Speed D*	20	or	25 mm/min ± 10%
Speed E			50 mm/min ± 10%
Speed F			100 mm/min ± 10%
Speed G*	200	or	250 mm/min ± 10%
Speed H			500 mm/min ± 10%

* The two D and G speeds are allowed because both are in very widespread use throughout the world.

ISO R1184 has a range of test speeds similar to the above, except that speeds A1, G and the 20 mm/min in Speed D are not allowed.

The British equivalent of R527 is BS 2782, Method 320 (1976)[58] and of R1184 is BS 2782, Method 326 (1977)[64] although these are not identical. The tensile testing of glass reinforced plastics is now specifically covered in BS 2782, Method 1003 (1977),[65] which follows the same basic procedure as Method 320 but with special emphasis on sample preparation and the problems of anisotropy. The measurement of modulus is also given rather more detailed description than is to be found in the more general test method.

Method 320A is for flexible thermoplastic sheet, extrusion and moulding compounds. The test speeds allowed are 100 mm/min and 500 mm/min and the test pieces are stamped from sheet.

Method 320B is for injection moulded thermoplastic materials including filled and reinforced compounds. The test speeds allowed are 1, 5, 25, 50 and 100 mm/min and the specimens must be injection moulded.

Method 320C is for rigid thermoplastic and thermosetting sheet; test speeds are the same as for Method B but test pieces are machined from sheet.

Method 320D, as previously mentioned, is for rigid thermosetting moulding materials including filled and reinforced compounds. The test speed is 5 mm/min and the test pieces must be either compression or injection moulded.

Method 320E is for fibre reinforced materials incorporating mat, cloth woven rovings. It can also be used for pre-pregs. Test pieces are machined from sheet and the allowed test speeds are 1, 5 and 10 mm/min.

Method 320F is for unidirectional fibre reinforced materials, including pre-pregs. The test piece as illustrated in Figure 8.23 is something of a hybrid between the 'dog-bone' and parallel strip in that it has bonded aluminium end pieces on a strip specimen but the centre section is reduced in thickness from 2 mm to 1 mm. The test speeds are 1 and 5 mm/min.

Method 301K[66] of the original 1970 version of BS 2782 has not yet been revised and is specific to PTFE. A narrow waisted dumbbell of 4 mm width is employed with a gauge length of 20 mm. The thickness of the test piece must not be greater than 2·5 mm and the test speed is 50 mm/min. The reason for having

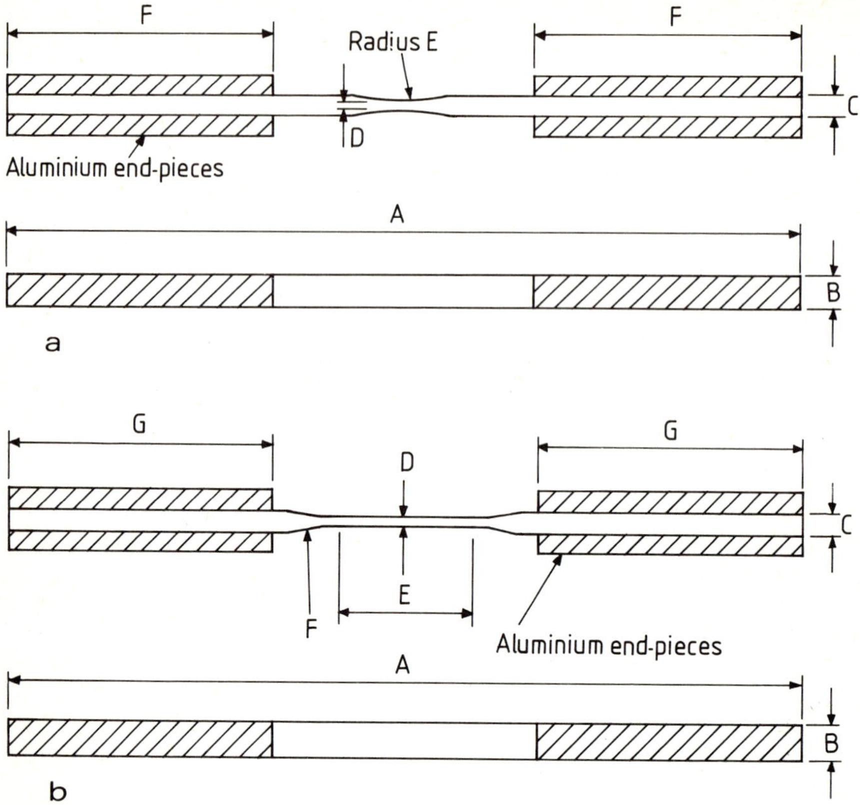

Figure 8.23 Tensile test piece for unidirectional fibre reinforced materials: BS 2782, Method 320F
(a) Type A test piece
(b) Type B test piece (for modulus measurement only)

a test specific to PTFE, however, is not the need to use a peculiar test piece but the need to use a non-standard test temperature, that being 35° ± 1°C. This is because PTFE has two molecular transitions in the range 19°C to 30°C and testing in the region of one of these would produce a large scatter of test results.

Methods 326A, B and C for films differ only in their test speeds, being respectively 5, 50 and 500 mm/min. The test pieces are strip specimens of at least 170 mm length and widths between 10 and 23 mm.

The two standard American tests are ASTM D638 (1977)[60] and D882 (1979)[67] and although, as previously mentioned, the test pieces differ dimensionally from European test pieces the test speeds and temperature are in agreement. For D638 the allowed speeds are 1, 5, 10, 50, 100, 500 mm/min according to the test piece selected; for D882 they are selected according to distance between the grips and the required initial strain rate, which can be 10 per cent, 50 per cent or 100 per cent per minute. A standard test temperature of 23° ± 2°C is used throughout.

ASTM D3039 (1976)[68] is specific to oriented fibre composites and uses the strip-type specimen of Figure 8.22 at a temperature of 23° ± 1°C and a strain rate of 1–2 per cent per minute.

ASTM D2289 (1969)[69] covers high-speed (not impact) tensile testing of plastics. The same type of test piece is used as in D638 and D882 but at test speeds of 2500, 25 000 or 250 000 mm/min. Tensile impact testing is standardised in ASTM D1822 (1979) and is dealt with in §8.7.

The standard German test is DIN 53455 (1978)[61] which includes the testing of films. The standard test speeds are 1, 5, 10, 25, 50, 100, 200 and 500 mm/min.

8.2.7 Measurement of Strain

As already mentioned in §8.2.2, it is generally inadequate to measure strain, or elongation, of the test piece simply by noting the rate of separation of the grips, because grip slippage, deformation of the grips and crosshead of the testing machine, as well as the necessary deformation of the force-measuring device all contribute in varying degrees to the apparent 'stretch' of the test piece. In addition, there is the extra complication that dumbbell-shaped test pieces do not stretch uniformly over their whole length.

There is, therefore, no substitute for using gauge marks to follow the elongation of a test piece, measuring the distance between the marks preferably continuously and in such a way that an electrical output, proportional to the elongation, can be generated and fed to the X axis of a chart recorder while the tensile force is simultaneously fed to the Y axis. Not surprisingly, in view of the wide range of elongations at break encountered with plastics materials, several types of device have been developed.

For materials such as reinforced plastics with very low elongations at break, strain gauges can be bonded directly to the test piece—as in ASTM D3039—while for strains in the range 1 per cent to 50 per cent the clip-on type of strain gauge extensometer is common. Strains in excess of 50 per cent can be measured with optical extensometers[26,70] of the type developed for rubber test pieces or with long-travel clip-on extensometers, or followed using a ruler or piece of string. In the last case, of course, only the ultimate elongation can be determined. H. W. Wallace & Co. Ltd[26] have recently introduced an optical follower system which is capable of measuring the strain to break of many thermoplastics materials with low-to-moderate elongations, thereby extending the range of the non-contacting follower systems. Such systems are to be preferred over the clip-on type which may induce premature failure.

Having stated that there is no substitute for using gauge marks, it must be admitted that the elongation of plastics films is often measured by grip separation, for four reasons:

(i) the tensile forces developed are relatively small, minimising errors due to the compliance of the machine;
(ii) a parallel strip specimen extends uniformly over its whole length, unlike a dumbbell;
(iii) the windlass type of chuck does not crush the end pieces of a specimen as do wedge grips;

(iv) a contacting type of extensometer cannot be used as the film is unable to support it without distortion.

This leaves only the optical follower system, which is relatively new, and the ruler or string technique which can only measure ultimate elongation, for direct measurement of gauge mark separation.

8.2.8 Measurement of Modulus

For a perfectly elastic solid, stress is proportional to strain and the constant of proportionality is called the modulus. For a given stress, therefore, the lower the modulus the greater the strain or deformation of the test piece. Thus, it can be seen that the modulus of a material is a measure of that material's ability to resist deformation and as such is an important engineering parameter.

Unfortunately, as seen in §8.2.1, plastics materials are far from ideal and many depart from Hookean behaviour at very low strains in a tensile test. Nevertheless the standard test methods include the measurement of tensile modulus (also known as elastic or Young's modulus) within their scope. The measurement is made by drawing a tangent to the initial—hopefully linear—part of the stress–strain curve and calculating the slope of the line. If no clearly defined linear portion exists, the decision as to where the line should be drawn is largely subjective and in such cases the 'secant modulus' can be evaluated with must greater precision (§8.2.4).

It is worth noting that the rubber technologist often refers to 100 per cent modulus, 200 per cent modulus, etc., when talking about tensile data on elastomers. These are not, however, modulus values but the values of the tensile stress at the stated elongation.

8.3 COMPRESSION STRESS–STRAIN

8.3.1 General Considerations

Compression stress–strain is the antithesis of tensile stress–strain, both being caused by the application of two equal and opposite colinear forces but acting in the reverse direction to each other.

This reversal of force application, tending to crush the material rather than to stretch it, solves some of the problems inherent in carrying out tensile tests but creates others unique to itself. Thus, while there is no difficulty in gripping the test piece and the deformation of the test piece can be related to the movement of the machine crosshead much more accurately than in a tensile test, there is a greater need to apply a truly axial force. The test piece has a tendency to bend or buckle and friction between the test piece and end plates tends to prevent lateral expansion of the specimen as it is compressed, leading to barrel distortion.

These considerations mean that the parallelism of the test piece—especially of the end faces—and its relative dimensions, as determined by the slenderness ratio (defined later), must be more rigorously controlled than for a tensile test piece.

8.3.2 Test Equipment

There are test machines available which are specifically designed for compression testing but, as mentioned in §8.2.2, the trend today is towards 'universal machines' which allow several types of mechanical test to be applied. These can be used either directly in compression, using a compression load cell to measure the force, or in the tensile mode using a compression cage (Figure 8.24) to reverse the test direction on the specimen while maintaining a tensile force on the load cell.

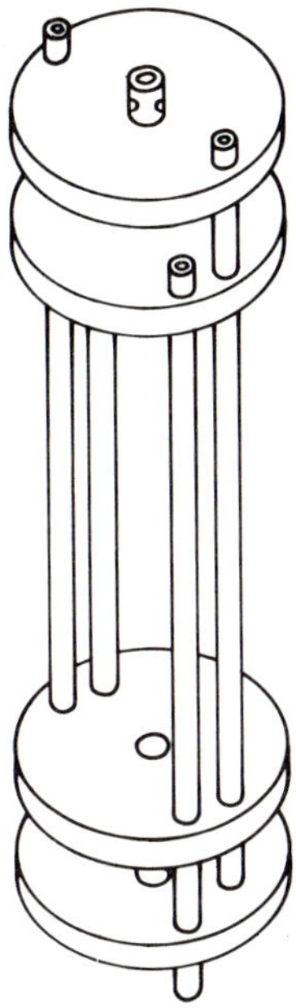

Figure 8.24 Compression cage for tensile machine

Since such cages may generate frictional forces during test, the compressive forces to be measured must be much greater than the frictional forces if serious errors are to be avoided.

It is worth noting that the compressive forces generated in a compression test are often considerably greater than the tensile forces generated in a tensile test because of the larger dimensions of the test pieces. If a single testing machine is to be used for both types of test, therefore, it may have to be of greater capacity than would be required for tensile testing alone, and, as the forces are high and deformations low, a particularly 'hard' or stiff machine is required.

8.3.3 Definitions

ISO 604 (1973),[71] Section 3 defines the various terms used in compression stress–strain tests which for the most part are analogous to the tensile definitions given above in §8.2.4. One term, however, has no counterpart in the tensile test and that is the already briefly mentioned 'slenderness ratio'. This is defined as 'the ratio of the length of a solid of uniform cross-section (column) to its least radius of gyration'. The least radius of gyration, i, is quoted for the commonly used test piece shapes.

Thus, for a right square prism,

$$i = \frac{a}{3{\cdot}46} \qquad (8.9a)$$

For a right rectangular prism,

$$i = \frac{b}{3{\cdot}46} \qquad (8.9b)$$

For a right circular rod,

$$i = \frac{d}{4} \qquad (8.9c)$$

For a right circular tube,

$$i = \frac{\sqrt{(D^2 + d_1^2)}}{4} \qquad (8.9d)$$

where a = length of the side of the square prism
b = length of the shorter side of the rectangular prism
d = diameter of the cylinder
D = outer diameter of the tube
d_1 = inner diameter of the tube.

8.3.4 Standard Test Methods

ISO Standards

ISO 604 (1973)[71] defines four types of test piece that may be used for carrying out compressive tests: the right square prism, the right rectangular prism, the right cylinder and the right circular crown tube. For each of these the test requires that the ends of the test piece, i.e. the load-bearing faces, be parallel to within 0·1 per cent of the height of the test piece.

The preferred test piece height is 30 mm but the allowed range is from 10 mm to 40 mm, the standard slenderness ratio being set to 10. For materials which buckle under test this is reduced to 6. The rate of deformation of the test piece (mm/min) is 0·3 times the test piece height (mm).

British Standards

BS 2782, Method 345A[72] is technically similar to the above ISO standard but allows a greater number of test pieces to be selected: only the Type 4 test pieces conform to ISO.

The Type 1 test piece is a cube of side 12 ± 0·1 mm and is suitable for thermoplastic or thermosetting sheet and for flat compression or injection mouldings, including filled or reinforced compounds, of thickness not less than 12 mm. For thicknesses greater than 12 mm, the test piece is machined on one face only to reduce the thickness to 12 mm.

The Type 2 test piece is similar to the above but used when the sheet or moulding thickness is less than 12 mm. Provided the sheet or moulding is at least

8 mm thick, the height of the test piece is the same as the sheet thickness. For materials less than 8 mm thick, however, test pieces are plied up and machined as appropriate to produce an overall test piece height of 12 mm.

The Type 3 test piece is suitable for casting and laminating resin systems containing no fibrous reinforcement. The test piece is either a cylinder of diameter 10 ± 0·2 mm and height 10 ± 0·2 mm or a cube of side 10 ± 0·2 mm.

The Type 4 test pieces are identical to those of ISO 604.

The Type 5 test piece is for thin sheet materials, glass fibre reinforced laminates and flat injection or compression moulded test pieces. Figure 8.25 illustrates a typical test piece of this type, although parallel-sided test pieces have been found to be satisfactory.

Unlike ISO 604, the test speed is selected according to the type of test piece to be tested.

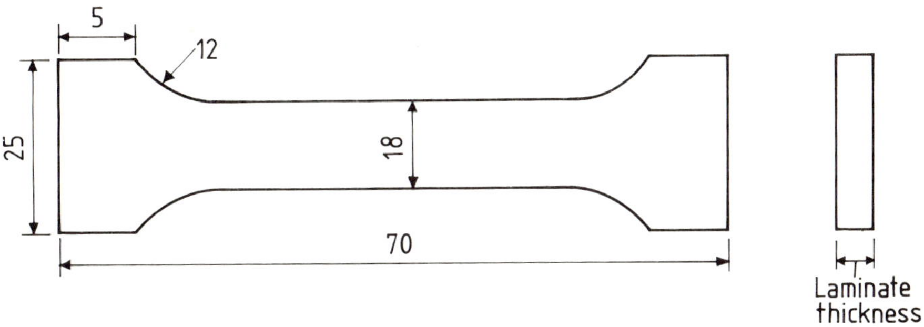

Figure 8.25 Type 5 test piece

USA Standards

ASTM D695 (1977)[73] employs the same technique as above but again the test piece dimensions differ, the preferred dimensions being 12·7 mm square by 25·4 mm high for a right prism, and 12·7 mm diameter by 25·4 mm high for a right cylinder.

For rod material the diameter is that of the rod to be tested and the length is such as to produce a slenderness ratio in the range 11 to 15:1.

ASTM D621 (1964)[74] also employs a compression device but is primarily a heat resistance test and is briefly described in Chapter 16.

German Standards

German practice is covered in DIN 53454 (1971)[75] and is very similar to the BS standard in the form and dimensions of the test pieces that may be selected. Test speeds and other conditions are also comparable.

8.3.5 Plain Strain Compression Test

Williams and Ford,[76] using the fact that strain is much easier to measure in compression tests than over gauge lengths, have evaluated the 'plain strain compression test', first developed for metals, as a means of determining the permanent and total deformation curves for plastics up to high levels of strain such as may be encountered in engineering applications. The outline of the experimental set-up is shown in Figure 8.26.

By this technique, the area under load remains constant. The specimen faces can be lubricated with a number of materials: molybdenum disulphide grease was used by Williams and Ford.

The application of the technique to the examination of polymers at large strains has been further described by Williams[77] and the reader is referred to these two papers in particular for the full experimental details, mathematics and some results.

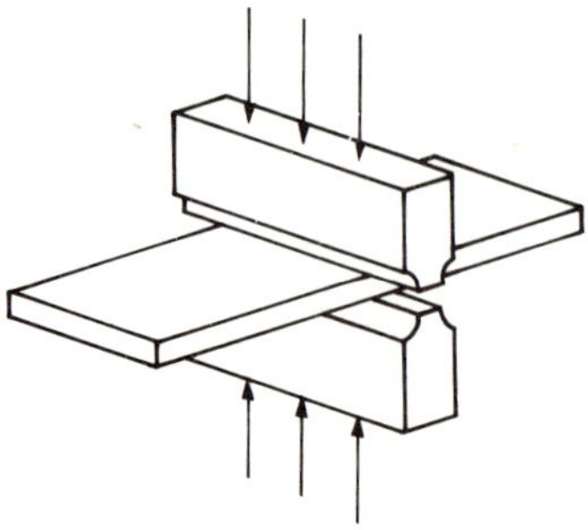

Figure 8.26 Plain strain compression test[76]

8.4 SHEAR STRESS–STRAIN

Shear, unlike tensile and compressive forces which are normal to the plane on which they act, acts parallel to the plane (Figure 8.27):

$$\text{shear stress, } \tau = \frac{F}{a^2} \tag{8.10}$$

$$\text{shear strain, } \gamma = \frac{\delta a}{a} \tag{8.11}$$

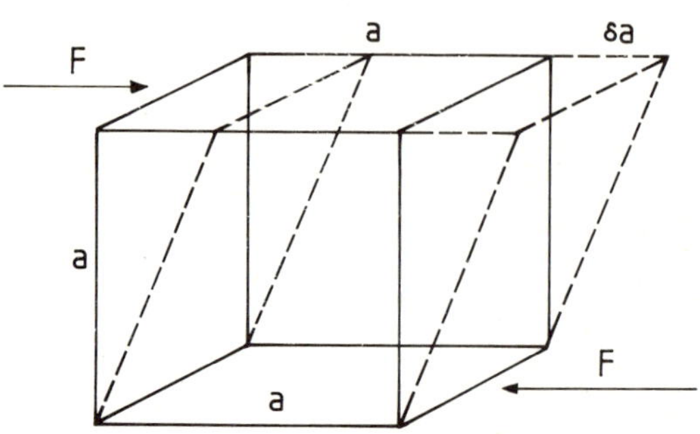

Figure 8.27 Material in shear

If Hooke's law is obeyed,

$$\text{shear modulus, } G = \frac{\tau}{\gamma a} = \text{modulus of rigidity} \tag{8.12}$$

Amongst a number of loading systems which give rise to shear forces is the instance where parallel but opposite forces act through the centroids of sections that are spaced an infinitesimal distance apart. This would lead to pure shear but cannot be realised in practice; somewhere near an approximation comes from the punch shear test where a punch bears on a flat sheet of test material resting on a die; the nearer the internal diameter of the die and the external diameter of the punch, the closer the approximation to pure shear. This type has found wide popularity as a standard test for plastics.

BS 2782, Method 340A,[78] for moulding materials, and 340B,[78] for sheet materials, both use the same die and punch assembly (Figure 8.28).

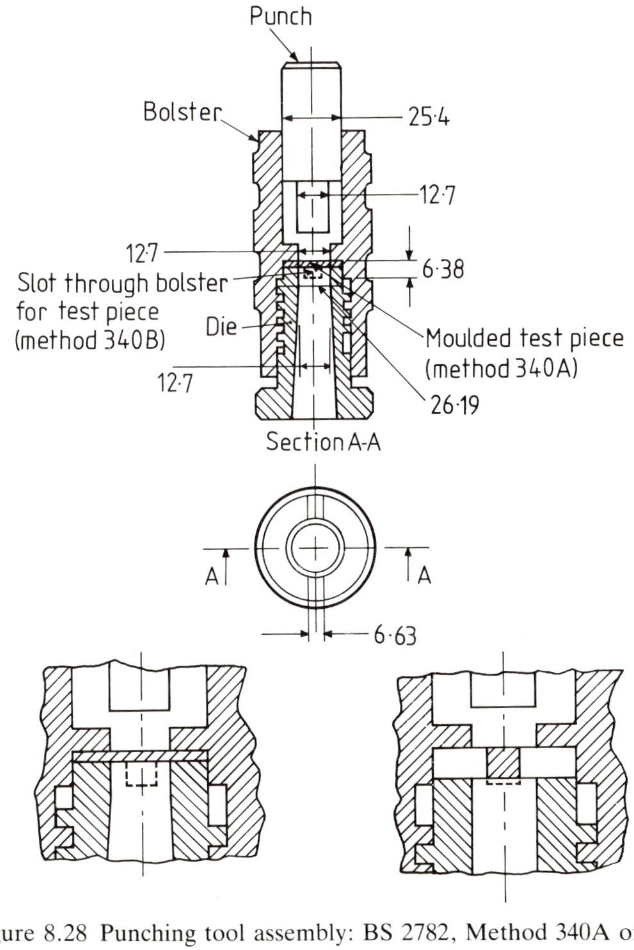

Figure 8.28 Punching tool assembly: BS 2782, Method 340A or B

In Method 340A the test pieces are moulded discs 25·3 ± 0·1 mm in diameter and 1·6 ± 0·1 mm thick, while in Method 340B the test pieces are rectangular bars 6·4 ± 0·2 mm wide, at least 32 mm long and of thickness equal to that of the sheet under test, except where this exceeds 6·35 mm, when the test piece is machined on one surface only to reduce the thickness to 6·10 ± 0·25 mm.

The test is carried out with the tool mounted in a testing machine operated as for compression and must be completed between 15 and 45 s from first application of load. The shear strength for Method A is given by

$$\text{shear strength} = \frac{F}{\pi D T} \tag{8.13a}$$

and for Method B by

$$\text{shear strength} = \frac{F}{2{\cdot}096 B T} \tag{8.13b}$$

where F = force at fracture
D = diameter of the punch
B = mean width of the test piece
T = mean thickness of the test piece.

The factor 2·096 is required to correct for the curvature of the sheared surfaces in Method B.

ASTM D732 (1978)[79] employs a similar technique but allows any thickness between 0·125 and 12·5 mm to be used and the test piece is drilled centrally to locate a guide pin.

At low temperatures torsional tests giving an apparent shear modulus are most frequently used (Chapter 16).

8.5 FLEXURAL (BENDING) STRESS–STRAIN

8.5.1 General Considerations

When a rectangular beam is bent a continuous change occurs from maximum tensile stress on one surface through the thickness to maximum compressive stress on the other.

In Figure 8.29 (for a homogeneous isotropic material in pure bending) the top surface is in tension and the under surface in an equal compression; if the tensile and compressive moduli are equal, at the midpoint of thickness the stress is zero, where tension diminishes to zero before compression starts building up, and the dotted line represents this neutral axis.

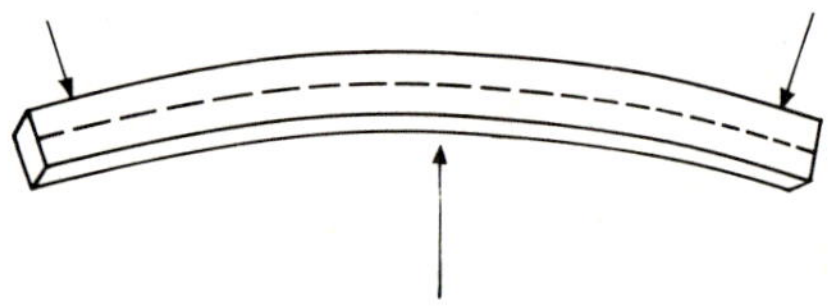

Figure 8.29 Material in bending

It is beyond the scope of this book to develop the formulae for bending; suffice it to state that pure bending, i.e. the combination of only tensile and compressive forces, is never achieved and there is always some transverse shear component.

The bending formulae for small deformations given below have been derived assuming pure bending, which is assisted by having a span (length of loaded section) of beam large by comparison with the thickness so that bending approaches a true arc of a circle. Heap and Norman[80] describe flexural testing of plastics in detail and develop equations of more general applicability. For three-point bending of plastics, as in the diagram above, the span–depth ratio should be about 16:1, but this cannot be taken as invariably satisfactory; the appropriate figure for span–depth ratio depends on the characteristics of each individual material.

Three-point Bending

Three-point bending is the type of bending most commonly used in standard tests: for a rectangular beam supported at the mid-point, the flexural stress (or 'maximum fibre stress') is given by

$$\sigma_F = \frac{3FL}{2bh^2} \tag{8.14}$$

where F is the force at the mid-point and the other symbols are as shown in Figure 8.30. At fracture, therefore,

$$\text{flexural strength} = \frac{3F_mL}{2bh^2} \tag{8.15}$$

where F_m is the maximum force recorded. Flexural strength is also known as cross-breaking strength.

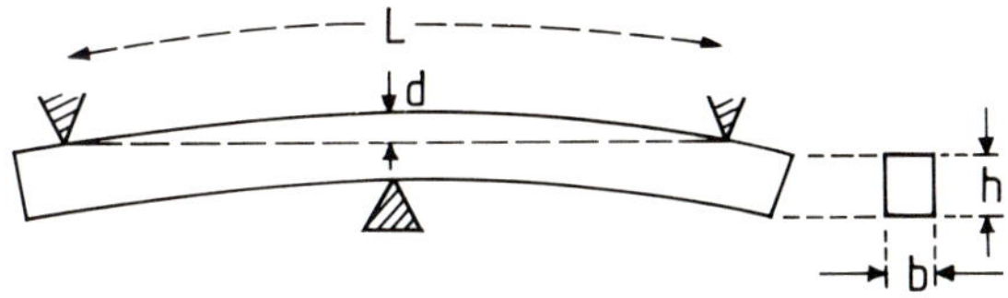

Figure 8.30 Three-point bending

A more accurate expression, which takes into account the horizontal component of the flexural moment, is given by

$$\sigma_F = \frac{3FL}{2bh^2}\left(1 + \frac{4d^2}{L^2}\right) \tag{8.16}$$

The apparent modulus of elasticity in flexure (equal to Young's modulus only as a first approximation) is given by

$$E_b = \frac{L^3}{4bh^3} \times \frac{F}{Y} \tag{8.17}$$

where F/Y is the slope of the initial linear force–deflection curve.

For a circular rod,

$$\text{flexural strength} = \frac{8F_m L}{\pi D^3} \tag{8.18}$$

$$\text{modulus of elasticity} = \frac{4L^3}{3\pi D^4}\left(\frac{F}{Y}\right) \tag{8.19}$$

where F_m, L, F and Y have the same significance as before and D = diameter of rod.

Four-point Bending
For a rectangular beam (Figure 8.31),

$$\text{flexural strength} = \frac{6F_x L_1}{bh^2} \tag{8.20}$$

For a circular rod,

$$\text{flexural strength} = \frac{32F_x L_1}{\pi D^3} \tag{8.21}$$

where F_x = force on each bearing point, i.e. half total force.

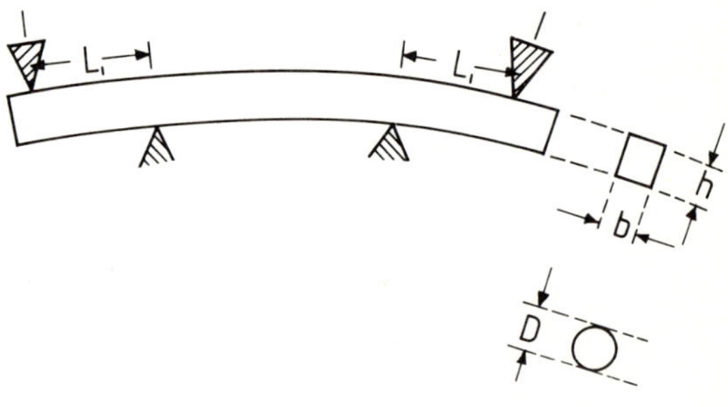

Figure 8.31 Four-point bending

Cantilever Bending
For a rectangular beam (Figure 8.32),

$$\text{flexural strength} = \frac{6FL}{bh^2} \tag{8.22}$$

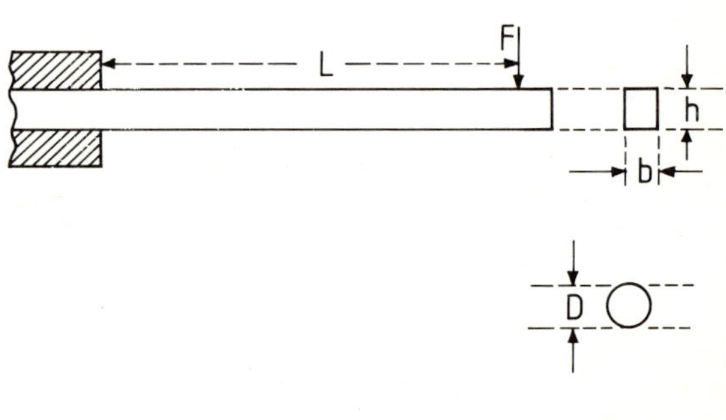

Figure 8.32 Cantilever bending

For a circular rod,

$$\text{flexural strength} = \frac{32FL}{\pi D^3} \tag{8.23}$$

8.5.2 Definitions

We have seen in the previous section most of the terms used in flexural tests and these are defined in a manner analogous to their counterparts in tensile and compression tests. There is, however, a term which has no counterpart in the other two modes of test. This is the 'flexural stress at the conventional deflection', which is defined in ISO 178 (1975)[81] as the flexural stress at a deflection of 1·5 times the thickness of the test piece. This parameter is particularly useful for materials which do not fracture under test.

8.5.3 Standard Test Methods

Bending tests may be carried out in tensile or compression test machines (Figure 8.33).

In addition to the usual variables of specimen size and shape, test speed and temperature, there must be added the radius of curvature of the bearing rods, which must not be too sharp to cause fracture. Cross-breaking strength tests are popular because the test specimens are simple and therefore easy to prepare and the loads to be measured are relatively low; also gripping problems are eliminated and deflection data more easily obtained. Theoretically, in pure bending, cross-breaking strength and bending modulus are equal to tensile strength and Young's modulus respectively, but because bending is not pure and because many materials tested are not homogeneous and isotropic, these equalities are often very approximate at best.

Again, three-point bending is generally preferred in standard tests, but there are many advocates of four-point tests because in the latter the stress is equal, under practical conditions, over the whole of the span between the inner two supports, in contrast to the local maximum stress which occurs opposite the centre support in three-point bending which is essentially non-pure; clearly errors may result if the material in the region of the centre support is not representative of the whole.

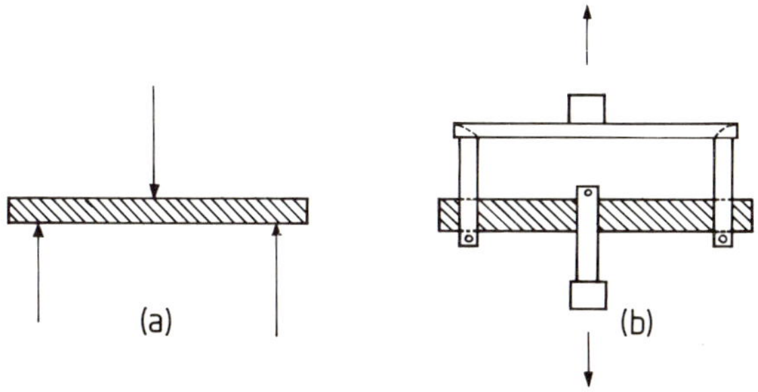

Figure 8.33 Bending test (a) in compression machine, (b) in tensile machine

ISO Standards

ISO 178 (1975)[81] specifies the standard test piece as being at least 80 mm long, 10 ± 0·5 mm wide and 4 ± 0·2 mm thick. It does, however, acknowledge that such a test piece may be unsuitable for some materials, in which case the dimensions of the test piece must be selected so that the length is at least 20 times the thickness and the width should be between 10 and 25 mm for normal materials or between 20 and 50 mm for materials containing coarse fillers.

The span length must be set to between 15 times and 17 times the test piece thickness for normal testing, although for very thick and unidirectional fibre-reinforced test pieces this can be exceeded to prevent delamination due to shear forces and for very thin materials the span-to-thickness ratio can be reduced.

Unless otherwise given in the product specification the test speed in mm/min is selected either to be a half the test piece thickness (mm) for qualification testing, or 10 mm/min for routine and screening tests.

British Standards

BS 2782, Method 335A (1978)[82] is identical to ISO 178 and need not be considered further, while Method 1005 (1977)[83] is technically equivalent insofar as it deals with glass reinforced plastics only.

Method 336B of BS 2782 (1978)[84] employs the cantilever bending mode; the apparatus is illustrated in Figure 8.34. Test pieces 70·0 mm long, 25·4 mm wide and 1·50 mm thick are moulded under the conditions appropriate to the material and a hole between 2·00 and 2·02 mm in diameter is drilled centrally through each, 50·8 mm from one end. A test piece is clamped as illustrated and the loading arm, without the 50 g weight, is attached. The cursor reading against the mirror scale is recorded after 5 min, the 50 g weight is attached to the cursor and the new scale reading recorded, again after 5 min. The deflection in bend is calculated from the difference in these two scale readings.

There are two other methods within BS 2782 which, while not being flexural tests in the same sense as the above two methods, nevertheless employ the three-point loading principle.

The first measures the stiffness of plastics film and is detailed in Method 332A (1976).[85] The test measures the force required to push a sample of film into the rectangular gap formed between two metal plates (Figure 8.35).

The second, detailed in Method 341A (1977),[86] gives a measure of the interlaminar shear strength of composite materials containing fibrous reinforcement. The main difference between this method and method 335A above is that the span-to-thickness ratio is reduced from 16:1 to only 6:1, thereby increasing the shear forces so as to induce shear failure—a tensile type failure (illustrated in the standard) is considered unsatisfactory. The shear strength is calculated according to the expression

$$S = \frac{0{\cdot}75F}{bh} \tag{8.24}$$

where the symbols are as before.

USA Standards

The principal test method is given by ASTM D790 (1971),[87] which lays down two methods and two procedures. Method I uses a three-point loading system

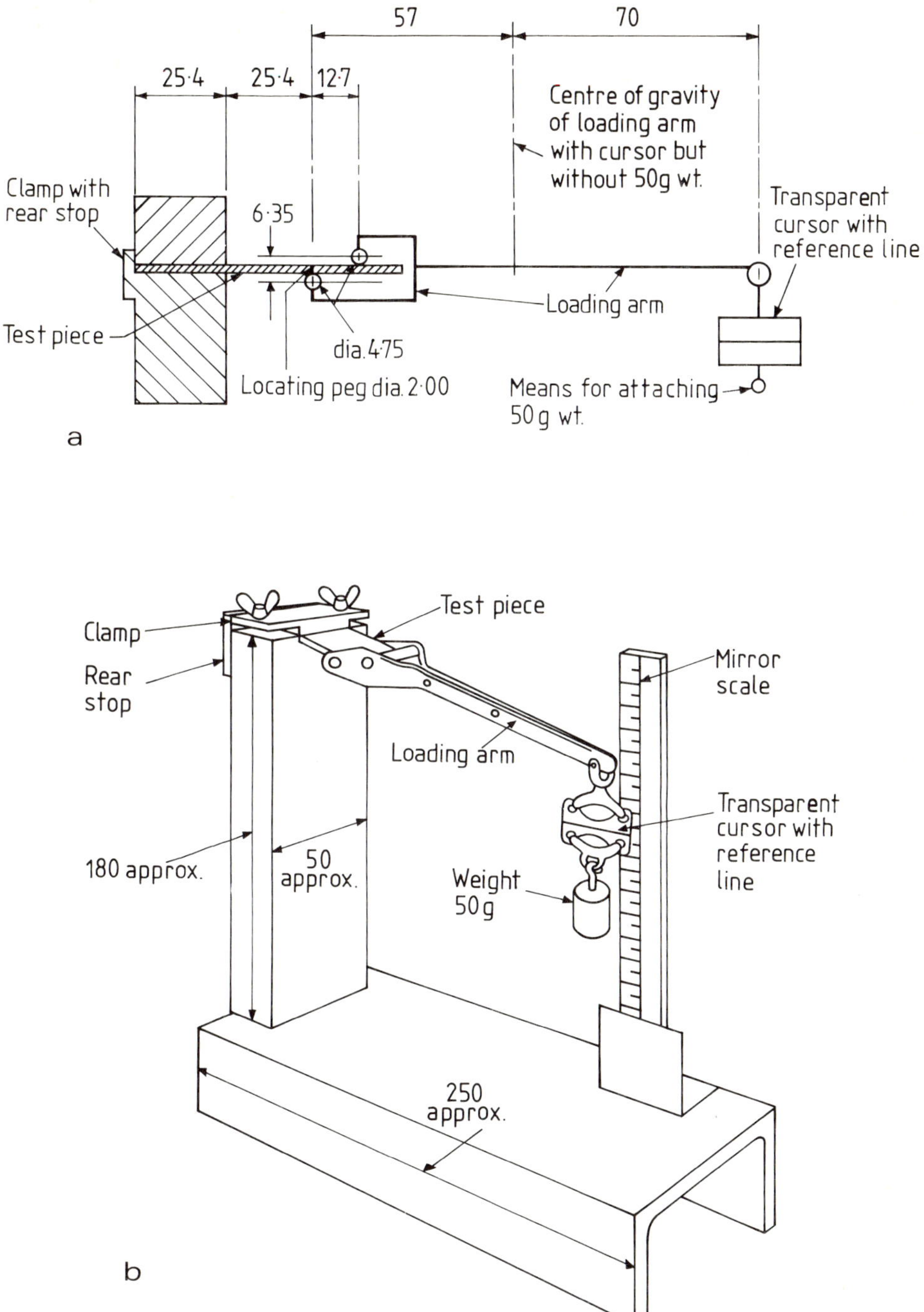

Figure 8.34 Apparatus suitable for use with BS 2782, Method 336A—dimensions in millimetres

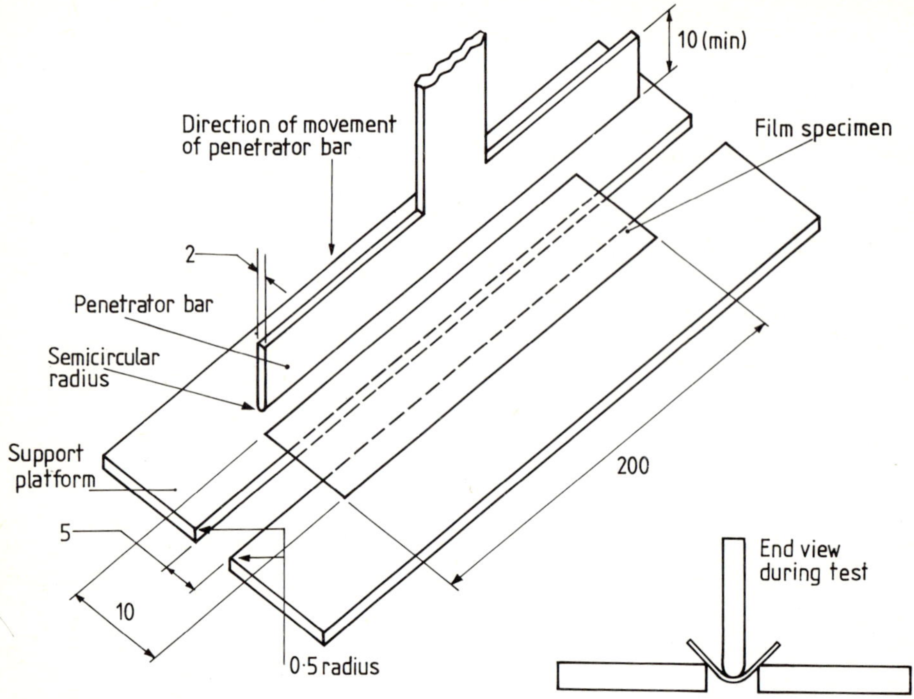

Figure 8.35 Apparatus for determining stiffness of film: BS 2782, Method 332

and Method II a four-point loading system, wherein the two load points are equally spaced from their adjacent support points and the distance between the load points is one-third of the support span. Procedure A is followed when the material under test fractures at small deflections and Procedure B is for materials undergoing large deformations during test. A range of support span-to-depth ratios is allowed according to the nature of the material.

ASTM D747 (1970)[88] is somewhat similar to BS 2782, 336A although the apparatus and technique are more complex.

German Standards
The standard test method is detailed in DIN 53452 (1977)[89] and is very similar to the ISO Standard, using three-point loading and the same, preferred, test piece with other dimensions being permitted if required. A four-point bending test is also given, however, in DIN 53435 (1977),[90] which uses the Dynstat apparatus.

8.6 TEAR TESTS

8.6.1 General Considerations

Tear tests on plastics materials are invariably applied to plastics films, unlike their counterparts for rubbers which are more often applied to standard test sheets of 2 mm nominal thickness. It is easy to see why this should be so, since plastics film finds such widespread use in the packaging industry; obviously puncture resistance (§8.7) and tear resistance are significant parameters to be determined in assessing the suitability of a film material for a particular application.

Although the tests themselves are generally straightforward enough the reproducibility is often rather poor and usually more test pieces are tested than is normal for, say, a flexural test. Films are also frequently more anisotropic than is the case with, for example, injection moulded test pieces, and mono- or biaxially oriented film is common. The properties along and across the machine direction can differ significantly, therefore, and both should be determined if incorrect conclusions are to be avoided.

A common difficulty encountered with tests of the 'trouser tear' and Elmendorf types is that the tear does not propagate in the direction of the test piece length but wanders off at some angle to it. Standards usually specify the amount of wander that is acceptable before the result be discarded and another test piece substituted. The trouser tear test suffers the additional disadvantage that a steady tearing force is not always observed and the precise manner in which the force–extension curve should be treated to extract the tearing force in such cases is often glossed over, leaving considerable room for subjective interpretation.

These points are under discussion in the various standards committees dealing with the preparation of tear tests and it is likely that with the passage of time standard test methods will become more precise in their interpretation—a method especially to deal with the interpretation of tear (or adhesion) force traces is being developed by TC45 for rubbers, for example.

A point worth mentioning here is that the force required to tear a given test piece is not a simple function of thickness. For this reason tear strength values obtained on films of widely differing thicknesses are not, strictly, comparable and should be used with caution. Not all standards make this point clear.

8.6.2 Standard Test Methods

There are at present no standard test methods within ISO although two documents are currently being considered by TC61. These are DIS 6383, which employs the 'trouser tear' test piece, and a document which specifies the Elmendorf tester. Both these methods of test will shortly be published by BSI within BS 2782 but these may not be identical to the final ISO format.

At the time of writing, the current tear tests in BS 2782 are Methods 360B (1980),[91] which uses the 'trouser tear' test piece and replaces the 1970 version of Method 308A (which used the 90° 'angle' test piece of ASTM D1004), and 308B (1970),[92] which uses the Elmendorf apparatus. The latter method is under review and will be replaced shortly by Method 360A[91] which will retain the Elmendorf test.

The Elmendorf apparatus (Figure 8.36) consists of a stationary jaw and a movable jaw carried on a pendulum, preferably formed by a sector of a wheel or circle, which is free to swing about a substantially frictionless bearing. The test piece is of the 'trouser tear' type and is held in the two jaws. When the sector release mechanism is operated the potential energy of the pendulum causes the cut in the film to propagate along the film's length and the energy absorbed by

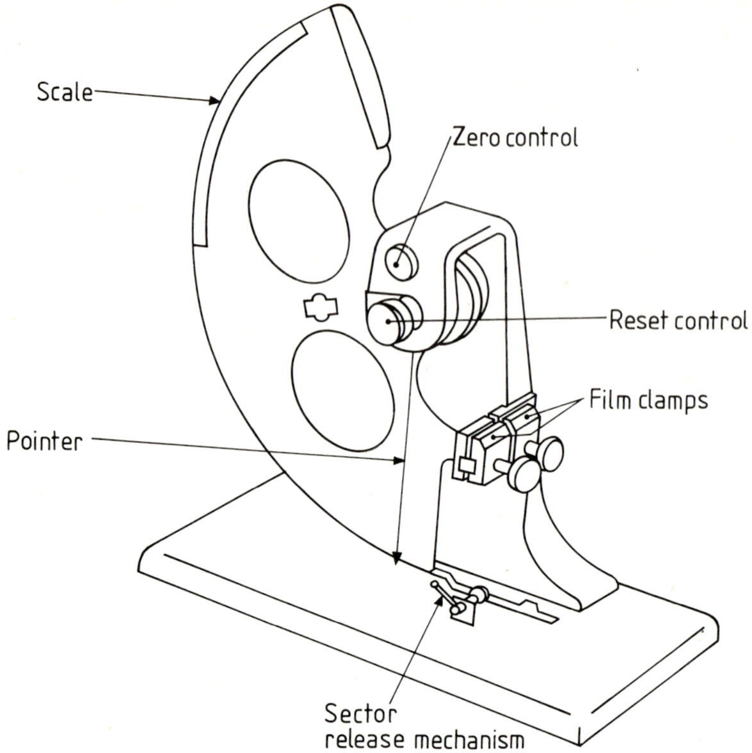

Figure 8.36 Tear testing apparatus of the Elmendorf type

this process is read directly off the calibrated scale on the pendulum; the scale is so positioned that, with no film sample present, a reading of 0 is obtained when the pendulum is released. The scale reading is converted to tear strength via the known calibration factor for the apparatus, the length of the tear produced and the sample thickness.

The 'trouser tear' method makes use of a normal tensile testing machine, the test piece being a strip of film 200 mm long by 50 mm wide with a clean cut made in one end to a distance of 75 mm, the cut being central to the width of the test piece. The two 'trouser legs' are held in the two jaws of the testing machine and a grip separation rate of 250 mm/min is employed. After an initial rise in the force–extension curve, a plateau region is generally observed as the tear propagates. The force in this plateau area divided by the film thickness is defined as the tear strength. The trouser tear test piece is now preferred to the earlier angle tear test piece which also made use of a tensile testing machine.

ASTM has no fewer than four test methods covering the tearing properties of sheet and film. ASTM D1004 (1966)[93] uses the angle test piece now being withdrawn in the United Kingdom. This test piece measures the resistance of the film to tear initiation and is tested at 51 mm/min, the maximum recorded force being divided by the film thickness to give the tear resistance. This is essentially the same test piece as is used for elastomers in ASTM D624 (1973).[94]

ASTM D1922 (1967)[95] is similar to BS 2782, 360A, being of the Elmendorf type, although the dimensions of the apparatus are somewhat different. D1938 (1967)[96] is a 'trouser tear' test but the dimensions of the test piece differ from those in the United Kingdom. The test speed is the same, however, at 250 mm/min, as is the treatment of the data.

ASTM D2582 (1967)[67] attempts to combine the effect of tear initiation and propagation in one test. A weighted pointer is released from a given height and it falls down a carriage onto a film, clamped in a curved holder. The holder and pointer are positioned in such a way that the pointer first punctures and then tears the film, its kinetic energy being absorbed by doing so. After the pointer has come to rest, the length of the cut produced is measured from a suitably positioned scale and the tear resistance is calculated from this, the weight of the pointer assembly and the drop height.

8.7 IMPACT TESTS

8.7.1 General Considerations

Mention has already been made (§8.2.1) of the effect of speed of test on the property value being determined and reference has been given to standard works on the viscoelastic nature of polymers. Very few high-speed—as opposed to impact—test methods have been standardised, although much work has been undertaken to assess the properties of certain materials as a function of test speed.[98–122]

It is ironical that, in contrast to the scarcely standardised subject of (scientific) high-speed mechanical testing in general, and for tensile, flexural, compressive and shear properties in particular, the decidedly *ad hoc* field of impact strength and testing has received considerable attention in official standards, materials data sheets and the literature at large; yet the result of an impact test is basically no more than one point on the general curve of studying strength properties as a function of speed of testing. The one advantage they offer is a ready measure of the actual energy required to break an (arbitrary) test piece, which information can only be calculated from stress–strain diagrams in, say, tensile or flexural tests with some effort.

'Toughness' is a concept most people can appreciate and a broadly accepted mathematical definition is the work done in breaking a specimen, moulding, article, etc. As such it may obviously be derived from the load–extension graph by computing the area under the curve (e.g. see Figure 8.10) since this will integrate all the units of force × distance (i.e. extension over which force operated) to give the total work, i.e. the toughness. As calculated from high-speed tests, such data may be quite useful for prophesying the behaviour of plane face articles, but the area under a conventional tensile or bending test curve is of very limited value because in practice we are mainly interested in toughness under conditions of rapid deformation, i.e. where an article is dropped or has something dropped on it—hence the concept of impact resistance or 'strength' and the introduction of impact tests, or shock resistance and its assessment. The subject is vast and at least one whole book has been given over to it[123] with a whole chapter (II) devoted to plastics and rubber. Davis *et al.*[124] and Fenner[125] give general descriptions of conventional impact testing.

Standard impact tests for plastics divide themselves into (a) those which use instruments where a pendulum of known energy strikes a specimen of defined size and shape and (b) those where weights or other impactors are allowed to fall freely through known heights on to specimens and the impact strength is computed from the minimum combination of height and weight required to cause fracture. Tests of type (a) are further subdivided into cantilever (Izod) and supported beam (Charpy) variants, using specimens of flat plane face or more often containing a moulded or machined notch in order to assess sensitivity to weakening by notches, and tensile impact using dumbbell-shaped test pieces. Data from tests of type (a), and particularly the Izod version, are open to much criticism; they and the other important standard tests will be described first, after which the practical significance or otherwise of the results is discussed briefly.

8.7.2 Pendulum Impact Tests

Izod Test

The basic principle of the Izod test is to allow a pendulum of known mass to fall through a known height and strike a standard specimen at the lowest point of its swing, and to record the height to which the pendulum continues its swing. If the striking edge of the pendulum is sited to coincide with the centre of percussion of the pendulum, the bearings of the pendulum are frictionless and there is no loss of energy to windage, then the product of the mass of the pendulum and the difference between the fall distance and the height it reaches after impacting the test specimen is the impact 'strength' of the latter. (It will be seen that in fact, whatever property is determined by this test, it is certainly not a 'strength' as so often it is called—even in official standards.) The result is usually expressed in joules and, since there is by and large no simple or reliable method of reducing the values to a property figure independent of the specimen cross-section and of the distance between the specimen support(s) and the point of striking (see below), the test values must be referred to the original specimen only; sometimes they are reduced to unit width across the specimen, although the area behind the notch is becoming the preferred reference variable. The test may be carried out on plain rectangular bars, but most often a carefully defined notch is moulded or machined into the face to be struck. The reason for this is that impact tests are often regarded as a means of assessing the resistance of a material to shock where notches or 'stress raisers' generally are present; the ratio of the impact strengths unnotched/notched is sometimes regarded as a measure of the notch sensitivity of a material.

The ISO test method, ISO R180 (1961)[126] is under revision in TC61 and at the time of writing had advanced to draft international standard level. There will undoubtedly be changes before this achieves standard status, but the main principles are already well established. Two alterations to the original R180 method which are expected to be retained are the abandonment of the so-called 'toss factor' and the expression of the impact resistance as energy per unit area behind the notch rather than energy per unit length of notch. As mentioned above, there is no satisfactory way of isolating the impact strength from the dimensions of the test piece, but normalising with respect to area behind the notch gives a better correlation than normalising with respect to notch length.

The velocity of the striker on impact has been standardised at 3·5 ± 10 per cent m/s with impact energies of 1·0, 2·75, 5·5, 11·0 and 22·0 J. Four test pieces are permitted, as detailed in Table 8.2, with Type 4 as the preferred size.

Table 8.2 Standard Izod Test Bars (ISO/DIS 180)

Specimen type	Length (mm)	Dimension *y* (mm)	Dimension *x* (mm)
1	63·5 ± 2	12·7 ± 0·2	12·7 ± 0·5
2	63·5 ± 2	12·7 ± 0·2	6·4 ± 0·3
3	63·5 ± 2	12·7 ± 0·2	3·2 ± 0·2
4	80·0 ± 2	10·0 ± 0·2	4·0 ± 0·2

Two types of notch may be cut in the test piece (moulded notches are not permitted); both form an angle of 45° but the radius at the tip of the notch is 0·25 mm for Type A (preferred) and 1·0 mm for Type B. A Type C test piece is formed by reversing the Type A or B test piece to produce, essentially, an 'unnotched' specimen. The notch is always cut into the *y* dimension and is of depth *y*/5 so that the *x* dimension becomes the notch length.

The designation of *x* and *y* dimensions is especially helpful when referring to tests on laminated test pieces, which may be tested in edgewise and/or flatwise directions (Figure 8.37). Reference to thickness and width, as hitherto, in relation to testing laminates has always led to confusion over which dimension is which.

The height to which the pendulum rises after impact is recorded by an idler pointer on a scale which reads directly in energy absorbed.

Regrettably, to cover the whole range of impact energies several machines are required as interchangeable pendulum assemblies are not generally supplied for fitting to one chassis, making this an expensive test when a wide range of materials needs to be examined.

BS 2782, Method 306A (1970)[127] is broadly similar to the above standard although it does differ in one or two important areas. This method is due for revision and it is anticipated that the ISO standard when it becomes available will be adopted to replace it.

Perhaps the most obvious discrepancy between the two documents is in the impact velocity, which for the British Standard is 2·44 m/s, and in the different impact energies of 1·36, 4·07 and 13·6 J. The test pieces have dimensions conforming to Types 1 and 2 above but the British Type A test piece must have a moulded notch, cut notches being confined to the Type B and C test pieces. Only the Type B notch, of 1 mm radius, may be used in the British Standard. Laminated products may be tested but in this case the notch is produced across the laminations by drilling and then opening out the drill hole with a saw cut (Figure 8.38).

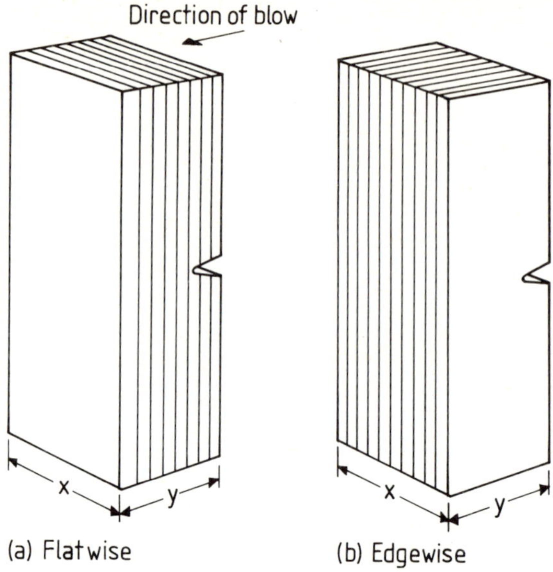

Figure 8.37 Izod impact testing laminated sheet

ASTM D256 (1978)[128] also follows the same basic principles as the above methods but is more closely aligned to the ISO method. The ASTM method has retained the units, energy per unit length of notch, however, and in Method C, which is used for materials with impact strengths below 27 J/m of notch, the toss factor has also been retained. This purports to allow for the energy required to toss the broken half of the test piece after fracture has occurred, although there has been considerable criticism of the technique. The toss factor is estimated by replacing the broken half of the test piece on the clamped portion and striking it a second time with the pendulum released from a height corresponding to that to which it rose following the breakage of the test piece. The 'energy of toss', which is subtracted from the impact energy, is then the difference between this reading and the free swing reading obtained from this height.

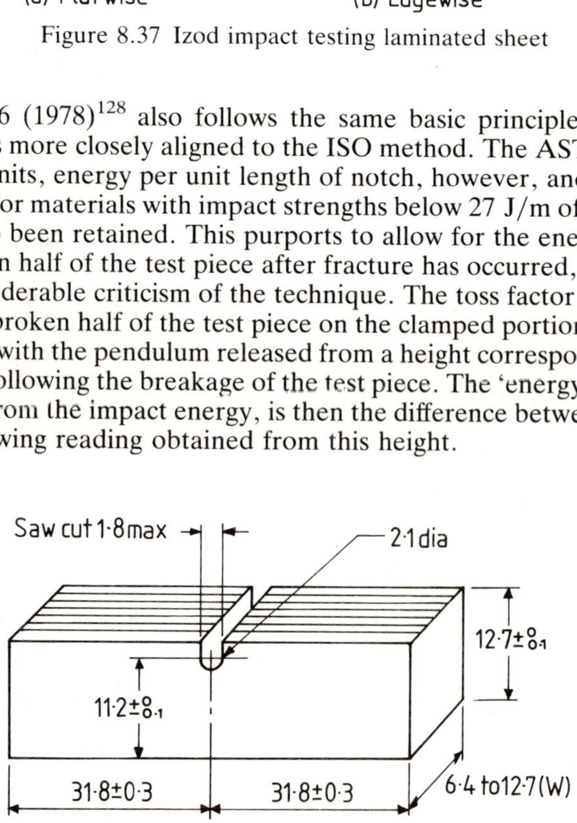

Figure 8.38 Izod impact resistance test piece—machined laminate: BS 2782, Method 306

Charpy Test

Although the Charpy test is similar to the Izod test, in that both are flexural impact tests performed by a pendulum striking a bar-shaped test piece, there are a number of significant differences between them and no general correlations relating the data obtained from each have been developed. The Charpy method has long been preferred in Europe.

As with the Izod standard, the original ISO Charpy recommendation, ISO R179 (1961),[129] is currently under review in TC61 and at the time of writing has just been advanced to DIS level. The British Standard is described first, therefore, and then the main differences between this and the present ISO text.

BS 2782, Method 351A (1977)[130] details the UK procedure. A rectangular bar, which may be notched or unnotched, is supported at both ends in such a position that the pendulum, when it is released, strikes the test piece in the centre and directly behind the notch, if present. The energy absorbed in the impact is then read directly by means of an idler pointer from a scale calibrated to allow for frictional and windage errors. Figure 8.39 illustrates a suitable machine.

Three standard test pieces are defined and for each there are three notches that may be used. The preferred test piece is 120 × 15 × 10 mm with a span of

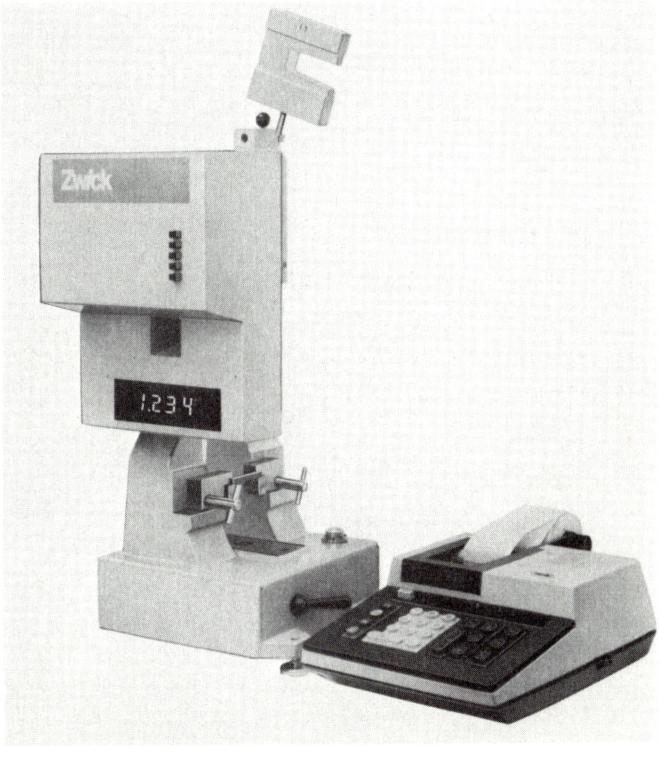

Figure 8.39 Charpy impact tester
(*Courtesy Zwick Instruments*)

70 mm; the Type A notch, which is of square section, 2 mm wide and one-third the specimen thickness in depth, has been adopted as the standard for quality control purposes. Notches B and C are V-shaped with base radii of 0·25 mm and 1·00 mm respectively. The depth of these notches is set to one-fifth thickness. The standard impact energies are 0·5, 1, 4, 15 and 50 J.

Because of the popularity of the Charpy test it has proved impossible to confine the ISO document to so small a number of variables and still retain an acceptable measure of international agreement. The number of test pieces is four, the first three of which are as in the British Standard and the fourth 125 mm long by 13 mm square. The same notch types are used but the characteristics of the pendulum have a very much broader range: two striking velocities are permitted, 2·9 m/s and 3·8 m/s with respective energy levels of 0·5, 1·0, 2·0, 4·0, 5·0 J and 2·7, 7·5, 15·0, 25·0 and 50·0 J. Moulded notches are permitted but a note explains that these do not give the same results as machined notches, which are preferred.

DIN 53453 (1975)[131] details the German practice of measuring Charpy impact strength. It is very similar to the UK standard, using the same impact energies, pendulum velocities, standard test piece dimensions and notch Type A, although the tolerances are not always identical. Notches B and C are not specified.

(An impact test can also be carried out on the Dynstat apparatus using an unnotched test piece but the clamping geometry is unique, making the test neither a Charpy nor Izod test. This test is detailed in DIN 53435, which also deals with the normal flexural test).

ASTM D256 Method B gives the current USA practice, which differs from the European standards in test piece dimensions and impact energy, being essentially an 'Izod' test with a Charpy mode of support.

8.7.3 Falling Weight Tests

At the time of writing no ISO standard exists dealing with this type of test, although a document is currently being prepared. This is at such an early stage of development, however, that it would be inappropriate to give any details here.

The standard UK test is given by BS 2782 Methods 306B and C (1970).[127] These describe two variants of an essentially simple impact-strength test which assesses the resistance to shock of a thin flat disc of material by finding how resistant is such a specimen to fracture when impacted by allowing a 'ball' to fall centrally on its surface. The height of fall of the 'ball', and the mass thereof, may be altered and the minimum value of the product of fall height and mass which causes fracture is the impact strength of the material. The concept is simple, but the standardisation is not, and a little thought will soon indicate an important reason why. Thus, if the same disc is continually impacted by increasing impact blows, how is one to know how much 'damage' (not actually causing fracture) has been sustained before the impact value causing fracture is finally reached? If such a progressive damage is to be tolerated, how may it be standardised?

In practice, the BS method tests each specimen only once and the impact resistance is computed from the general spectrum of behaviour of 'go/no-go' fracture against regularly varied impact values; the method is therefore somewhat profligate in its use of specimens.

The specimen is a 57–64 mm diameter disc or a 57–64 mm square. For moulded or extruded material the specimen thickness is 1·52 ± 0·05 mm, for sheet it is the sheet thickness. At least twenty such specimens are required and no specimen is used more than once.

The apparatus may be of the form shown in Figure 8.40 (suitable for Method 306B). The specimen support is a hollow cylinder of internal diameter 50·80 ± 0·05 mm, external diameter not less than 57·2 mm and length not less than 25·4 mm. The axis of the cylinder coincides with the line of fall of the striker and a soft shock-absorbing disc is placed inside the cylinder to rest on its base. The specimen may be clamped on to the support.

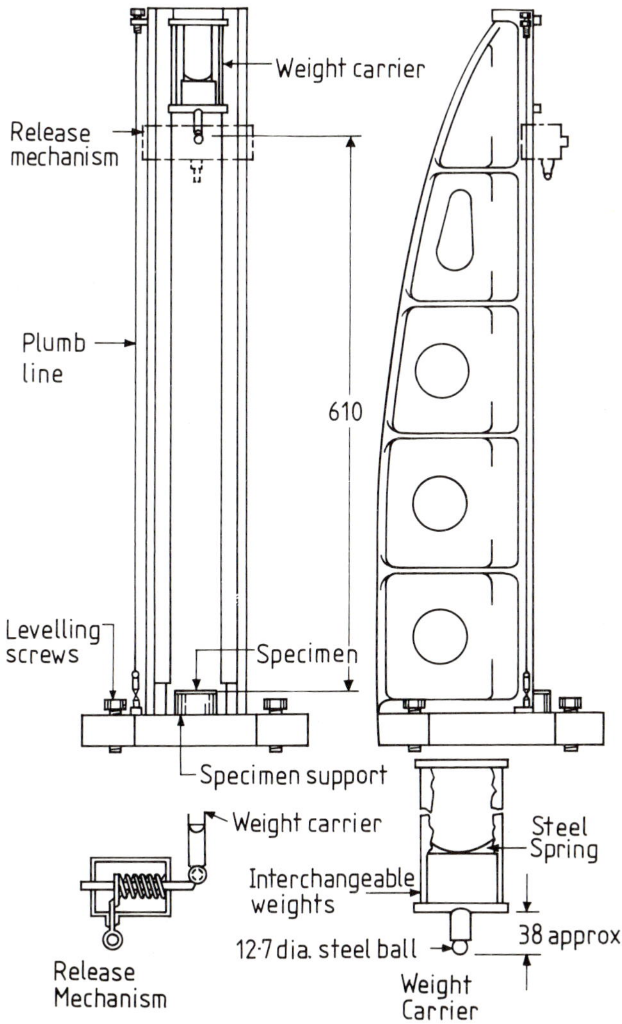

Figure 8.40 Falling weight impact apparatus: BS 2782, Method 352D—dimensions in millimetres

The striker has a hardened hemispherical striking surface, 12·7 ± 0·05 mm diameter (the 'ball'), and is fitted with a carriage to take weights so that a specified series of increments of energy may be obtained if the striker is released from a height of 610 ± 2 mm (Method 306B) or 305 ± 1 mm (Method 306C) above the upper surface of the specimen. The striker is allowed to fall with or without guides; in the former instance the fall must be substantially without friction. The striker can conveniently be supported electromagnetically and released by switching off the current.

To carry out the test a 'trial run' is undertaken first. The striker is loaded in such a way that the product of weight and fall height is equal to the expected impact strength. If, on release of the striker, the specimen does not fracture, or cracks on one surface only, the result is recorded as 'unbroken'; if the specimen breaks or cracks or tears through from one surface to the other, the result is recorded as 'broken'. Thereafter:

(i) If the result was 'broken', a second specimen is tested with an impact energy less than the first by a specified amount *S*. If this second specimen is 'broken', a third is tested with an impact energy *S** less than that applied to the second—and so on until a specimen does not break.
(ii) If the first result was 'unbroken', a second specimen is tested with an impact energy greater by amount *S* obtained as before. As in (i), this is repeated, this time until a specimen does break.

The 'test run' is now carried out, using the remaining test specimens, the energy of the blow applied to any specimen being Y more or less than that on the previous specimen, respectively, according to whether the latter remained unbroken or was broken, with Y being specified in the BS method. The test run thus uses a maximum of 18 specimens (since at least two of the twenty must be required for the 'trial run'), but must not use less than twelve.

The impact strength is calculated as follows:

$$\text{impact strength} = \frac{1}{21 - m}(\gamma_{m+1} + \gamma_{m+2} + \cdots + \gamma_{21}) \qquad (8.25)$$

where m = number of blows in trial run
γ_{m+1} = impact energy of first blow of the testing run
γ_{m+2} = impact energy of second blow of the testing run
γ_{21} = impact energy of twentieth blow, decreased or increased by Y according to whether the specimen broke or did not break.

An alternative method of calculating the impact strength of a material is by the probit method, as used in BS 2782, Method 352D (1979)[132] for film materials, and as discussed in BS 4618: Section 1.2 (1972).[133] In this method, ten (usually) test pieces are tested at each impact energy, the impact energy just to give 0 per cent failure and the energy just to give 100 per cent failure being determined roughly by trial and error. After this the energy range between these levels is divided into approximately 5 equal increments and then 10 pieces are tested at an energy level equal to the 0 per cent failure level plus one increment, the number of failures being recorded. The energy level is raised by a further increment and another 10 test pieces are tested, and so on up to the maximum energy. The percentage failure at each energy is plotted against the energy (on probability × linear graph paper) and the best fit straight line to the data obtained is drawn in. The energy required to produce a 50 per cent chance of failure is taken as the impact resistance, with a standard deviation given by the energy difference between the 50 per cent and 16 per cent failure probabilities.

Similar techniques, albeit with rather different dimensions and impact energies, are used in USA and German standard methods.[134–136]

8.7.4 Consideration of Impact Strength Data

Of the two classes of standard test for impact strength just described, the pendulum method is the traditional and more widely used method. From the brief discussion earlier of the effect of rate of strain, and of other test variables, on mechanical properties of plastics, it may well be imagined that this type of test does not yield what might be termed 'fundamental data'; furthermore, accepting the arbitrary selection of variables which constitutes the standard ASTM Izod test, for instance, the values obtained are still of doubtful significance, a point well illustrated by the correction made for 'toss factor' in Method C of ASTM D256 (1978) (see above), which refinement is not included in any other of the standard tests of a similar nature.

Although, as has already been seen, the various plastics materials may show an infinite range of 'strength' values under a given set of test conditions, very broadly the mode of failure is generally tough or brittle. Which of these is exhibited depends as stated before on whether the material yields prior to crack initiation (tough failure) or a crack forms at a stress lower than that of the yield stress (brittle failure). Since brittle fracture is if possible to be avoided in service, the ostensible aim of impact testing is to investigate those conditions under which a material will fail in a brittle manner; the purpose of introducing the notch in the pendulum tests is to concentrate the stress and increase the straining rate at the root of the notch, thus inducing brittle behaviour in a wide range of materials at normal testing temperatures and simply obtained speeds of testing. Vincent,[137] however, takes notched impact testing to task and makes, *inter alia*, the following points:

(i) The test may be too severe—i.e. causes brittle behaviour unrealistically and penalises a satisfactory material.
(ii) It may give a result dependent more on crack propagation resistance than ability to resist crack initiation.
(iii) It chooses an arbitrary combination of speed, specimen dimensions, notch profile and temperature which may give quite misleading results even on a comparative basis.
(iv) It is an inaccurate test, very dependent on precisely how the test is carried out (see below).
(v) Not surprisingly, in view of (i)–(iv), the test conditions are probably unrelated to service conditions.

Telfair and Nason[138] have analysed the factors comprising 'impact strength' as measured by the Izod test as follows:

(a) Energy to initiate fracture of specimen
(b) Energy to propagate fracture across specimen
(c) Energy to deform the specimen plastically
(d) Energy to throw the broken end of the test specimen
(e) Energy lost through vibration of the apparatus and its base, and through friction

(a) is probably the value of real interest, though the addition of (b) and (c) has obvious practical significance. (d) is the 'toss factor' referred to in Method C of ASTM D256 (1973) and the significance of the correction in certain cases is well illustrated by Ritchie[139] in a general discussion on the Izod impact test: with low impact strength materials the 'toss factor' can be the major proportion of the measured value. (e) requires a rigid design of pendulum arm in particular and a massive foundation.

Ritchie also mentions further factors which may influence results and lead to errors, illustrating each with results from practical investigations:

(i) Variation of clamping pressure in vice
(ii) Failure to strike the specimen squarely
(iii) When notches are machined, the state of the cutter and actual cutting technique used

The effect of such variables as temperature and notch radius on the impact strength of polyolefins in particular has been reported by Horsley and Morris,[140] who have also examined the influence of processing variables on falling weight impact strength.[141] The ASTM D256 technique of reducing impact strength to unit width of specimen is open to criticism as a result, for example, of the work of Wolstenholme *et al.*[142] who showed that impact strength per unit width is not invariably independent of specimen width tested. A wide range of thermoplastics was examined at specimen notch widths between $\frac{1}{8}$ in and $\frac{1}{2}$ in (3 and 13 mm) and many showed marked differences in 'reduced' impact strength even from widths between $\frac{1}{4}$ and $\frac{1}{2}$ in (6 and 13 mm). (It has been stated that, with polycarbonate, as little as 0·01 in difference in the width of specimens can alter the mode of failure from tough to brittle—see also Shoulberg and Gouza.[143]

In 1972 the BSI published BS 4618: Section 1.2 (1972)[133] which gives the recommended practice for deriving impact data to be used for design criteria. The standard deals with the philosophy and methods of obtaining multi-point data derived from unnotched specimens and notched specimens with various notch radii, at various temperatures. Although the time–temperature interrelationship suggests that the effect of test speed can be obtained from temperature data, the standard points out that direct multi-speed testing may still be necessary in certain cases. Impact energy—normalised with respect to area behind the notch—is plotted as a function of temperature and, for a given temperature, against notch depth and/or notch tip radius to build up a composite picture of the impact behaviour of the plastics material. The standard also makes specific reference to the effect of processing conditions and describes experiments whereby these effects may be evaluated. Drop weight impact testing of finished products is also strongly recommended and the design of suitable experiments discussed.

In recent years the science of fracture mechanics, through the use of instrumented impact testers, has been enjoying increasing popularity, with many papers appearing in the literature, covering both the theoretical aspects and equipment design.[144–154] As yet, however, there are no standard test methods, although it is likely that the next decade will see the emergence of methods yielding rather more fundamental impact data than are at present to be found.

8.7.5 Tensile Impact Strength—Standard Test Methods

All the tests so far described have used a flexural mode of deformation. However, in service a component may be subject to shock loading in a tensile configuration and two standard test methods have been developed to enable data to be gathered from this type of loading. Both ASTM D1822 (1979)[155] and DIN 53448 (1977)[156] employ pendulum impact machines, although the test piece dimensions and equipment design are different in the two methods. In the ASTM method one end of the dumbbell-shaped test piece is held in the pendulum head with the other in a crosshead clamp. On release of the pendulum, the head is able to pass between two halves of a rigidly mounted anvil while the (following) crosshead is not (Figure 8.41). The test piece is thus fractured in tension and the absorbed energy of impact is read directly from the subsequent swing of the pendulum head as in a Charpy or Izod test.

At least five, and preferably ten, specimens are tested and the results calculated on a unit cross-sectional area basis after applying correction factors for friction and windage and the 'bounce' of the crosshead. The method is described in rather greater detail by Westover and Warner[157] who give typical test results obtained in the evolution of the method. Richter[158] has examined the problems of tensile impact testing with different forms of test piece on a range of thermoplastic materials. Therberge and Hall[159] found a high degreee of correlation between tensile impact values and unnotched 'impact strengths' for glass-filled SAN, polystyrene, polycarbonate, polyethylene, polypropylene, PVC, polyurethane, polysulphone, polyester and nylons −6, −6·6 and −6·10; not surprisingly, the correlation did not exist for notched Izod impact strengths.

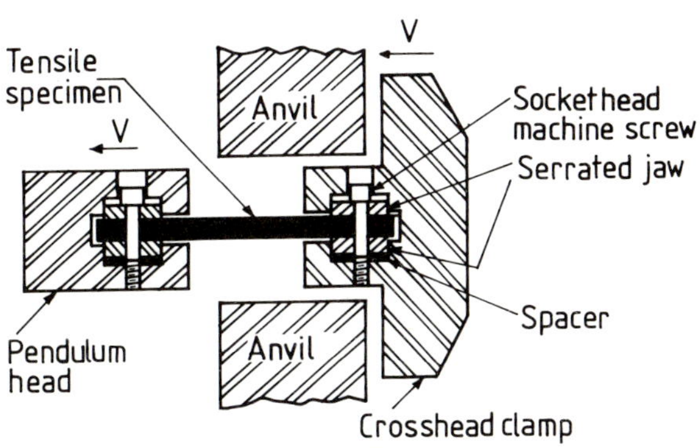

Figure 8.41 Schematic diagram of tensile impact testing machine: ASTM D1822

In the German variant the components in Figure 8.41 are reversed, in that the anvil becomes the pendulum, the pendulum head becomes the rigidly mounted base and the crosshead clamp becomes the movable head.

Because a relatively thin test piece is used, the test may be particularly suitable for measuring the effects of weathering since most of the attack is in the surface layers.

REFERENCES

1. Phillips, P. J. and Ramakrishnan, N. R. (1978) *Polymer Engineering and Science*, **18** (11), 869.
2. Nielsen, L. E. (1962) *Mechanical Properties of Polymers*, Reinhold, page 220.
3. Ritchie, P. D. (1965) *Physics of Plastics*, Iliffe, page 185.
4. Brown, R. P. (1979) *Physical Testing of Rubbers*, Applied Science, Chapter 8.
5. Gent, A. N. (1958) *IRI Transactions*, **34** (2), 46.
6. Tangorra, G. (1966) *Rubber Chemistry and Technology*, **39** (5), 1520.
7. Soden, A. L. (1951) *A Practical Manual of Rubber Hardness Testing*, MacLaren.
8. Fett, T., Nothdurft, W. and Racke, H. H. (1973) *Kunststoffe*, **63** (2), 107.
9. Bowman, J. and Bevis, M. (1977) *Colloid and Polymer Science*, **255** (10), 954.
10. Davies, H. E., Troxall, G. E. and Wiskoch, C. T. (1964) *The Testing and Inspection of Engineering Materials* 3rd edn, McGraw-Hill, Chapter 7.
11. Fenner, A. J. (1965) *Mechanical Testing of Materials*, Philosophical Library, Chapter 8.
12. Maxwell, B. (1955) *Modern Plastics*, **32** (9), 125.
13. Lysaght, V. E. (1948) *Materials and Methods*, 27, 84.
14. Boor, L. (1960) *ASTM Bulletin*, 244, 43.
15. Haldenwanger, H. (1961) *Kunststoffe*, **51** (2), 82.
16. Gouza, J. J. (1966) in *Testing of Polymers* vol. 2, J. V. Schmitz, Ed., Interscience, Chapter 7.
17. Livingston, D. I. (1967) in *Testing of Polymers* vol. 3, J. V. Schmitz and W. E. Brown, Eds., Interscience.
18. ISO 868 (1978) *Plastics—Determination of Indentation Hardness by means of a Durometer (Shore Hardness)*.
19. ISO 2039 (1973) *Plastics and Ebonite—Determination of Hardness by Ball Indentation Method.*
20. ISO/DIS 2039/II *Plastics—Determination of Rockwell Hardness.*
21. Fett, T. (1972) *Material Prüfung*, **14** (5), 151.
22. BS 2782 Method 365A (1976) *Determination of Softness Number of Flexible Plastics Materials*; Method 365D (1978) *Determination of Hardness by Ball Indentation Method.*
23. BS 2782 Method 1001 (1977) *Measurement of Hardness by means of a Barcol Impressor.*
24. BS 903; Part A26 (1969) *Methods of Testing Vulcanized Rubbers—Determination of Hardness.*
25. ISO 48 (1979) *Vulcanized Rubbers—Determination of Hardness (Hardness between 30 and 85 IRHD)*.
26. H. W. Wallace & Co. Ltd, St James's Road, Croydon CR9 2HR.
27. Bessant, K. H. C., Dilke, M. G., Hollis, C. E. and Millane, J. J. (1952) *Journal of Applied Chemistry*, **2** (9), 501.
28. ASTM D2240 (1975) *Rubber Property—Durometer Hardness.*
29. ASTM D785 (1965) *Rockwell Hardness of Plastics and Electrical Insulating Materials.*

30. ASTM D2583 (1975) *Indentation Hardness of Rigid Plastics by means of a Barcol Impressor*.
31. Barber Colman Co., Rockford, Illinois.
32. DIN 53456 (1973) *Testing of Plastics—Indentation Hardness Test*.
33. Hounsfield, L. H., *Commercial Testing*, Part II, *Plastics*, Tensometer Ltd, 81 Morland Road, Croydon CR9 6HG.
34. BS 240: Part 1 (1962) *Method for Brinell Hardness Test—Testing of Metals*.
35. ASTM E10 (1978) *Brinell Hardness of Metallic Materials*.
36. BS 427: Part I (1961) *Method for Vickers Hardness Test—Testing of Metals*.
37. ASTM E92 (1972) *Vickers Hardness of Metallic Materials*.
38. Campbell, R. F. (1967) *Materials, Research and Standards*, **7** (10), 443.
39. Anon. (1955) *British Plastics*, **28** (3), 110.
40. Bennett, F. N. B and Hayes, R. (1955) *Plastics, London*, **20** (217), 282.
41. Lysaght, V. E. (1946) *ASTM Bulletin*, **138**, 39.
42. Grodzinski, P. (1953) *Plastics, London*, **18** (194), 312.
43. Roberts, J. and Steel, M. A. (1966) *Journal of Applied Polymer Science*, **10** (9), 1343.
44. Gavan, F. M. and Wein, J. T. Jr (1965) in *Testing of Polymers* vol. I, J. V. Schmitz (Ed.), Interscience, Chapter 12.
45. Boor, L., Ryan, J. D., Marks, M. W. and Bartoe, W. F. (1947) *ASTM Proceedings*, 17, 1017.
46. Bennitt, J. H. and Avenall. C. E. (1952) *Chemistry and Industry*, 39, 936.
47. McSheehy, W. H. (1964) *Plastics Technology*, **10** (11), 44.
48. Carswell, T. S. and Nason, H. K. (1944) *Modern Plastics*, **21** (10), 121.
49. Nielsen, L. E. (1962) *Mechanical Properties of Polymers*, Reinhold, Chapter 5.
50. Ritchie, P. D. (1965), *op. cit.*, Chapter 2.
51. Alfrey, T. Jr (1948) *Mechanical Behaviour of High Polymers*, Interscience, Chapter F.
52. Ogorkiewicz, R. M. *et al.* (1969) *SPE Journal*, **25** (3), 43.
53. Ferry, J. D. (1970) *Viscoelastic Properties of Polymers* 2nd edn, Wiley.
54. BS 5214 Part 1 (1975) *Tensile, Flexural and Compression Testing Machines for Rubbers and Plastics*; Part 2 (1978) *Constant Rate of Force Application Testing Machines for Rubbers and Plastics*.
55. Brown, R. P. (1979) *RAPRA Guide to Rubber and Plastics Test Equipment*, RAPRA.
56. Holland, J. R. (1967) *Journal of Scientific Instruments*, **44** (5), 389.
57. ISO/DIS 527 *Plastics—Determination of Tensile Properties*.
58. BS 2782, Methods 320A to F (1976) *Plastics—Tensile Strength, Elongation and Elastic Modulus*.
59. Panov, P. (1969) in *Testing of Polymers* vol. 4, W. E. Brown (Ed.), Interscience.
60. ASTM D638 (1977) *Tensile Properties of Plastics*.
61. DIN 53455 (1978) *Testing of Plastics: Tensile Test*.
62. ASTM D1708 (1979) *Tensile Properties of Plastics by Use of Microtensile Specimens*.
63. ISO R1184 (1970) *Plastics—Determination of Tensile Properties of Films*.
64. BS 2782, Methods 326A to C (1977) *Determination of Tensile Strength and Elongation of Plastics Films*.
65. BS 2782, Method 1003 (1977) *Glass Reinforced Plastics—Determination of Tensile Properties*.
66. BS 2782, Method 301K (1970) *Tensile Strength and Elongation at Break of PTFE (Excluding Rod)*.
67. ASTM D882 (1979) *Tensile Properties of Thin Plastic Sheeting*.
68. ASTM D3039 (1976) *Tensile Properties of Oriented Fibre Composites*.

69. ASTM D2289 (1969) *Tensile Properties of Plastics at High Speed.*
70. J. J. Lloyd Instruments Ltd, Brook Avenue, Warsash, Southampton SO3 6HP.
71. ISO 604 (1973) *Plastics—Determination of Compressive Properties.*
72. BS 2782, Method 345A (1979) *Determination of Compressive Properties by Deformation at Constant Rate.*
73. ASTM D695 (1977) *Compressive Properties of Rigid Plastics.*
74. ASTM D621 (1964) *Deformation of Plastics under Load.*
75. DIN 53454 (1971) *Testing of Plastics—Compression Test.*
76. Williams, J. G. and Ford, H. (1964) *Journal of Mechanical Engineering Science*, **6** (4), 405.
77. Williams, J. G. (1967) *Transactions of the Plastics Institute*, **35** (117), 505.
78. BS 2782 (1978) Method 340A *Determination of Shear Strength of Moulding Materials*; Method 340B *Determination of Shear Strength of Sheet Material.*
79. ASTM D732 (1978) *Shear Strength of Plastics by Punch Test.*
80. Heap, R. D. and Norman, R. H. (1969) *Flexural Testing of Plastics*, Plastics Institute.
81. ISO 178 (1975) *Plastics—Determination of Flexural Properties of Rigid Plastics.*
82. BS 2782, Method 335A (1978) *Determination of Flexural Properties of Rigid Plastics.*
83. BS 2782, Method 1005 (1977) *Glass Reinforced Plastics—Determination of Flexural Properties. Three Point Method.*
84. BS 2782, Method 336B (1978) *Determination of Deflection in Bend under an Applied Force.*
85. BS 2782, Method 332A (1976) *Stiffness of Plastics Film.*
86. BS 2782, Method 341A (1977) *Determination of Apparent Interlaminar Shear Strength of Reinforced Plastics.*
87. ASTM D790 (1971) *Flexural Properties of Plastics and Electrical Insulating Materials.*
88. ASTM D747 (1970) *Stiffness of Plastics by Means of a Cantilever Beam.*
89. DIN 53452 (1977) *Testing of Plastics—Bending Test.*
90. DIN 53435 (1977) *Testing of Plastics—Determination of Flexural Properties and Impact Resistance with Dynstat Test Specimens.*
91. BS 2782 Method 360A (to be published) *Tear Strength of Film* (*Elmendorf Method*); Method 360B (1980) *Tear Strength of Sheeting* (*Trouser Tear Method*).
92. BS 2782 (1970) Method 308B *Resistance to Tear Propagation of Thin Flexible Sheet.*
93. ASTM D1004 (1966) *Initial Tear Resistance of Plastic Film and Sheeting.*
94. ASTM D624 (1973) *Rubber Property—Tear Resistance.*
95. ASTM D1922 (1967) *Propagation Tear Resistance of Plastic Film and Thin Sheeting by Pendulum Method.*
96. ASTM D1938 (1967) *Tear Propagation Resistance of Plastic Film and Thin Sheeting by a Single-Tear Method.*
97. ASTM D2582 (1967) *Puncture-propagation Tear Resistance of Plastic Film and Thin Sheeting.*
98. Dietz, A. G. H., Gailus, W. J. and Yurenka, S. (1948) *Symposium on Speed of Testing*, ASTM, page 32.
99. Dietz, A. G. H. and McGarry, F. J. (1956) in *Symposium on Speed of Testing of Non-metallic Materials,* STP 185, ASTM, page 30.
100. Hall, Warburton H. (1959) in *International Symposium on Plastics Testing Standards*, STP No. 247, ASTM, page 137.
101. Supnik, R. H. (1962) *Materials, Research and Standards*, **2** (6), 498.
102. Goldfein, S. (1964) *Modern Plastics*, **41** (12), 149.
103. Strella, S., Sigler, H., Chmura, M. and Holman, B. (1958) *ASTM Bulletin*, 228, 50.
104. Dorsey, J., McGarry, F. J. and Dietz, A. G. H. (1956) *ASTM Bulletin*, 211, 34.

105. Strella, S. (1959) *ASTM Bulletin*, 236, 59.
106. Ely, R. E. (1957) *Plastics Technology*, **3** (11), 900.
107. Ely, R. E. (1957) Report PB. 151573, US Dept of Commerce, Office of Technical Services.
108. Anon, (1959) *Rubber Age, London*, **85** (1), 140.
109. Sandek, L. (1962) *Plastics Technology*, **8** (2), 26.
110. Silberberg, M. and Supnik, R. H. (1962) *SPE Transactions*, **2** (2), 140.
111. Heimerl, G. L. and Manning, C. R. Jr (1962) *Materials, Research and Standards*, **2** (4), 270.
112. McAbee, E. and Chmura, M. (1963) *SPE Journal*, **19** (1), 83.
113. Strella, S. and Gilman, L. (1957) *Modern Plastics*, **34** (8), 158.
114. Ely, R. E. (1959) *Plastics Technology*, **5** (6), 36.
115. Holt, D. L. (1968) *Journal of Applied Polymer Science*, **12** (7), 1653.
116. Barnes, C. B., Hawkins, E. L. and Davis, M. V. (1966) *Materials, Research and Standards*, **6** (11), 560.
117. Jones, J. W. (1960) *Journal of Applied Polymer Science*, **4** (12), 284.
118. D'Amato, D. A. (1964) *Journal of Applied Polymer Science*, **8** (1), 197.
119. Patterson, G. D. Jr and Miller, W. H. Jr (1960) *Journal of Applied Polymer Science*, **4** (12), 291.
120. Amborski, L. E. and Mecca, T. D. (1960) *Journal of Applied Polymer Science*, **4** (12), 332.
121. Grimminger, H. and Jacobshagen, E. (1962) *Kunststoffe*, **52** (5), 254.
122. Dietz, A. G. H. and Eirich, F. R. (1960/61/62/64/65) *High Speed Testing* vols I, II, III, IV, V, Wiley/Interscience.
123. Späth, W. (1961) *Impact Testing of Materials* (rev. M. Rosner), Thames & Hudson.
124. Davis, H. E., Troxell, G. E. and Wiskocil, G. T. (1964) *The Testing and Inspection of Engineering Materials,* 3rd edn, McGraw-Hill, Chapter 8.
125. Fenner, A. J. (1965) *Mechanical Testing of Materials*, Philosophical Library, Chapter 7.
126. ISO R180 (1961) *Determination of the Izod Impact Resistance of Rigid Plastics.*
127. BS 2782 (1970) Method 306A *Impact Strength* (*Pendulum Method*); Methods 306B and C *Impact Strength* (*Falling Weight Method with Sheet Specimen*).
128. ASTM D256 (1978) *Impact Resistance of Plastics and Electrical Insulating Materials.*
129. ISO R179 (1961) *Determination of the Charpy Impact Resistance of Rigid Plastics.*
130. BS 2782, Method 351A (1977) *Determination of Charpy Impact Resistance of Rigid Plastics and Ebonite.*
131. DIN 53453 (1975) *Testing of Plastics*; *Determination of Impact Resistance.*
132. BS 2782 (1979) Method 352D *Falling Weight Impact Resistance of Films.*
133. BS 4618: Section 1.2 (1972) *Recommendations for the Presentation of Plastics Design Data—Impact Behaviour.*
134. ASTM D1709 (1975) *Impact Resistance of Polyethylene Film by the Free Falling Dart Method.*
135. ASTM D3029 (1978) *Impact Resistance of Rigid Plastic Sheeting or Parts by Means of a Tup* (*Falling Weight*).
136. DIN 53443 (1975) *Testing of Plastics*: *Multiaxial Impact Behaviour; Falling Weight Test.*
137. Vincent, P. I. (1962) *Plastics, London*, **27** (294), 116.
138. Telfair, D. and Nason, H. K. (1943) *Modern Plastics*, **20** (11), 85.
139. Ritchie, P. D. (1965) *op. cit.*, Chapter 3.
140. Horsley, R. A. and Morris, A. C. (1966) *Shell Polyolefins Engineering Design Data*, Shell International Chemical Co. Ltd/Shell Chemicals UK Ltd.
141. Horsley, R. A. and Morris, A. C. (1966) *Plastics, London*, **31** (350), 1551.

142. Wolstenholme, W. E., Pregun, S. E. and Stark, C. F. (1964) in *High Speed Testing* vol. IV, Wiley/Interscience, page 19.
143. Shoulberg, R. H. and Gouza, J. J. (1967) *SPE Journal*, **23** (12), 32.
144. Marshall, G. P., Williams, J. G. and Turner, C. E. (1973) *Journal of Materials Science*, **8** (7), 949.
145. Blumenauer, H. and Schroeder, M. (1973) *Plaste und Kautschuk*, **20** (11), 832.
146. Racke, H. H. and Fett, T. (1974) *Kunststoffe*, **64** (9), 481.
147. Matonis, V. A. (1974) in *Plastics in Surface Transportation, National Technical Conference, Detroit, Mich., November*, Confer 63TR, Society of Plastics Engineers, page 238.
148. Bader, M. G. and Ellis, R. M. (1974) *Composites*, **5** (6), 253.
149. Plati, E. and Williams, J. G. (1975) *Polymer Engineering and Science*, **15** (6), 470.
150. Butcher, B. R. (1976) *Composites*, **7** (1), 12.
151. Gonzalez, R. and Stowell, W. J. (1976) *Journal of Applied Polymer Science*, **20** (5), 1389.
152. Ireland, D. R. (1976) *Materials Engineering* **83** (5), 64.
153. Tak, A. G. M. (1977) *Polymer Engineering and Science*, **17** (10), 733.
154. Radon, J. C. (1978) *Journal of Applied Polymer Science*, **22** (6), 1569.
155. ASTM D1822 (1979) *Tensile Impact Energy to Break Plastics Materials*.
156. DIN 53448 (1977) *Testing of Plastics: Tensile Impact Test*.
157. Westover, R. F. and Warner, W. C. ((1961) *Materials, Research and Standards*, **1** (11), 867.
158. Richter, K. (1967) *Plaste und Kautschuk*, **14** (1), 23.
159. Therberge, J. E. and Hall, N. T. (1969) *Modern Plastics*, **46** (7), 114.

Chapter Nine

Dynamic Stress–Strain Properties

9.1 INTRODUCTION

A dynamic test is one in which the plastics material is subjected to a cyclic, usually sinusoidal, deformation with the stress and strain being recorded continuously in most cases. Traditionally, dynamic tests do not include those tests where the prime objective is to assess the life of a component which will be subject to cyclic deformation in service. Such tests are more properly considered as fatigue tests and are discussed in Chapter 12.

Dynamic tests yield data concerning the appropriate modulus of the material, according to the mode of deformation being applied, and, perhaps more significantly, the mechanical damping exhibited by the material under that particular deformation pattern. This information is not obtainable from the 'static' tests detailed in the previous chapter. Such information, while being of immediate value to the design engineer, has also proved to be of immense value in the study of the molecular transitions that occur in high polymers. In order to exploit this, however, the equipment used in such tests must be capable of operating over as wide a frequency and/or temperature range as possible. This has resulted in the widespread application of these techniques to fundamental studies, but an almost total absence of use for routine and quality control purposes because of the complexity of the apparatus. The only method in the plastics field to have been standardised at national and international level is the torsion pendulum. In the field of rubbers rather more standards exist, dealing with general testing requirements as well as particular test methods. Brown[1] discusses these in detail.

There are basically two types of dynamic test: the free vibration method in which the test piece is initially deformed, then released and allowed to oscillate without any further input of external energy; and the forced vibration method in which the oscillation of the test piece is maintained by external means. The latter method may also be subdivided into those methods operating at resonance and those operating away from resonance. Before dealing with such methods in more detail, however, it would be useful to outline, briefly, the principles and terms encountered. Ferry[2] and Nielsen[3] provide excellent detailed texts on this subject.

As mentioned in the previous chapter, the behaviour of a plastic during deformation is not perfectly elastic but depends on the time-scale of the experiment. It behaves, to a first approximation, as a series combination of Voigt and Maxwell elements. The former consists of a perfectly elastic spring in parallel with a perfectly viscous dash-pot and the latter consists of a series combination of spring and dash-pot. The Maxwell element can be ignored in dynamic tests (provided there is no static strain component of deformation) and the behaviour explained by the Voigt element alone.

If a sinusoidally varying stress is applied to the element, the stress at any time is given by

$$\sigma = \sigma_0 \sin \omega t \tag{9.1}$$

where σ_0 is the maximum applied stress
ω is the angular frequency
t is time.

Because the dash-pot cannot respond instantaneously to the applied stress level, but rather is rate-dependent, the resultant strain lags behind the stress by an amount dependent on the viscous and elastic constants of the element. Thus,

$$\gamma = \gamma_0 (\sin \omega t - \delta) \tag{9.2}$$

where γ_0 is the maximum strain attained and δ is the phase angle (Figure 9.1).

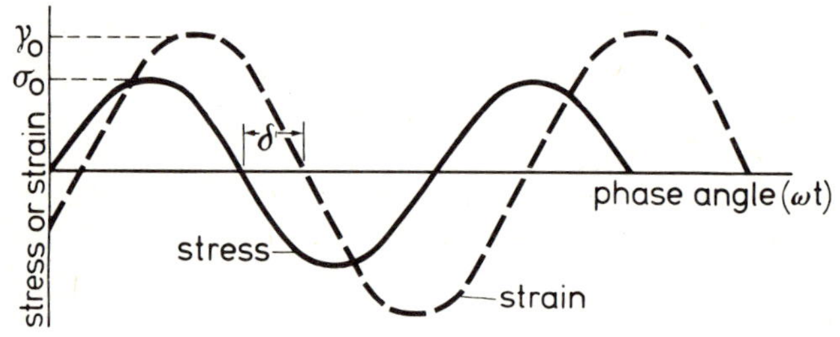

Figure 9.1 Stress–strain relationship for sinusoidally varying applied stress

This use of the simple model above leads to the very useful, if oversimplified, concept of polymer behaviour consisting of two simultaneously acting parts: a perfectly elastic part in which no energy is lost and the applied stress and resultant strain are in phase with each other; and a purely viscous part in which energy is not stored but is dissipated as heat and where the strain vector is 90° behind the stress. This in turn leads to the definition of two moduli, the in-phase or storage modulus given by the ratio of the peak in-phase stress to peak strain and the out-of-phase or loss modulus given by the ratio of the peak out-of-phase stress to peak strain. Reference to Figure 9.2 shows that

$$M' = \sigma_0'/\gamma_0 \tag{9.3}$$

$$M'' = \sigma_0''/\gamma_0 \tag{9.4}$$

where M' = storage modulus and M'' = loss modulus (the actual modulus obtained depends upon the mode of deformation: for shear modulus, the symbol usually employed is G and for extensional or Young's modulus, E).

The complex modulus, M^*, is defined as

$$M^* = \sigma_0/\gamma_0 \tag{9.5}$$

so that

$$M^* = \sqrt{(M'^2 + M''^2)} \tag{9.6a}$$

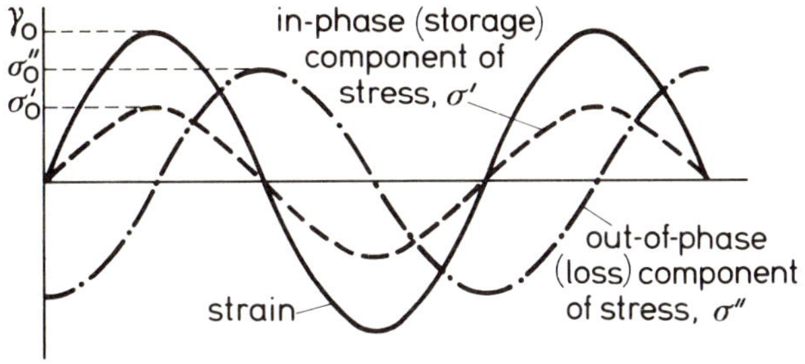

Figure 9.2 Stress–strain relationship of Figure 9.1 resolved into storage and loss components

although this is more usually written as

$$M^* = M' + iM'' \tag{9.6b}$$

where $i = \sqrt{-1}$.

The loss tangent or loss factor is given by tan δ and it can be shown that

$$\tan \delta = \frac{M''}{M'} \tag{9.7}$$

In practice, the amount of deformation that can be applied to a test piece before non-linear viscoelastic effects become important is very small, although shear tests have a larger linear range than tensile or compression tests. For this and other reasons shear tests are usually preferred.

9.2 FREE VIBRATION METHODS

One of the simplest pieces of apparatus and perhaps the most widely used in the study of the dynamic mechanical properties of polymers is the torsion pendulum. This has been standardised for plastics in ISO R537 (1967),[4] ASTM D2236 (1970)[5] and DIN 53445 (1979),[6] but at present no such standard exists in BS 2782. The ISO recommendation is under revision but the present draft international standard is a somewhat unsatisfactory document in that it is attempting to satisfy both quality control purposes and fundamental polymer studies. A more irreconcilable combination is hard to imagine. Many amendments will be required if this is to be brought, as it is hoped, within the scope of BS 2782.

The basic requirements of a torsion pendulum are illustrated in Figure 9.3. The test piece is rigidly clamped at one end, while the other end is attached to an inertia member which is free to oscillate about the axis of the test piece. The temperature of the test piece is controlled by a thermostated oven and some means of measuring the amplitude of the oscillations without interfering with the damping characteristics of the system is required—an optical system, as illustrated, is frequently used. A variant on this is to counterbalance the weight of the inertia member to minimise the tensile force on the test piece.

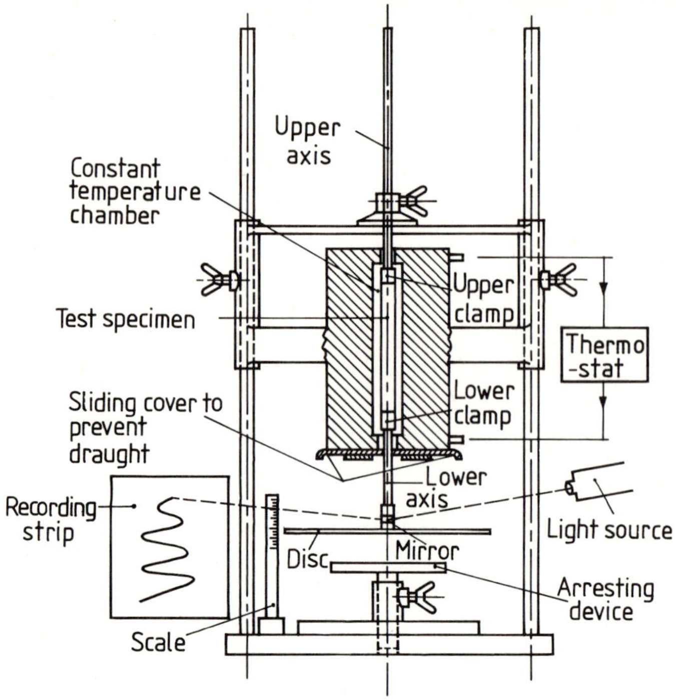

Figure 9.3 Torsion pendulum apparatus

In ISO R537 the recommended test piece has the dimensions 60 mm × 10 mm × 1 mm and an inertia member having a moment of inertia of 310 g cm^2 is suggested as being the most suitable. The thermostat should be capable of controlling the temperature to at least ±1°C and have a range from −60°C to +250°C. The maximum oscillation should be set to no more than ±3° from the rest position and the frequency range covered is 0·1 to 10 Hz.

According to this standard, the shear modulus (which one is not stated!) is given by

$$G = \frac{12\mu^2 \times I \times f^2 \times l}{b \times h^3 \times C} - \frac{m \times g \times b}{4 \times h^3 \times C} \tag{9.8}$$

where I = moment of inertia of the inertia member
f = frequency of oscillation
l = free length of the test piece
b = test piece width
h = test piece thickness
m = tensile force on the test piece
g = acceleration due to gravity
C = correction factor (or shape factor) defined by the ratio of b to h

and the loss tangent by

$$\tan \delta = \frac{\Lambda}{\pi[1 + (\Lambda/2\pi)^2]} \tag{9.9}$$

where Λ = logarithmic decrement, defined as the natural logarithm of the ratio of the amplitude of successive oscillations in the same direction, i.e. $\Lambda = \ln(A_n/A_{n+1})$.

The ASTM method allows the use of cylindrical test pieces and the calculation of the modulus is somewhat different, but otherwise it follows the ISO pattern very closely. The German standard appears to have been the model for the ISO recommendation and is almost identical to it.

Both Fritzsche *et al.*[7] and Bell *et al.*[8] describe the development of automated torsion penduli, the latter with dedicated data-reduction facility to reduce significantly the number of man-hours required in carrying out a full temperature scan, while Aghili-Kermani[9] and co-workers describe a high-sensitivity instrument designed for films and fibres covering the temperature range −269°C to 180°C. Other papers deal with the use of a torsion pendulum to determine dynamic mechanical properties of polymers.[10–14]

A method not strictly falling within the scope of free vibration methods, but one which it is convenient to discuss briefly at this point, is rebound resilience. The Shore Schleroscope is perhaps the best known example.

The technique is to allow a spherical (or hemispherical) striker to fall vertically from a given height on to a temperature controlled test piece, the rebound height being recorded. The resilience, which is defined as the ratio of the rebound height to the drop height, is related to the loss tangent of the material by the expression:

$$R \simeq \exp(-\pi \tan \delta) \tag{9.10}$$

Thus, by carrying out the experiment over a wide temperature range, the loss tangent of the material can be evaluated and the transition temperatures located. One property it certainly does not measure, despite occasional claims to the contrary, is that of hardness!

In view of its extreme simplicity and ruggedness it is perhaps surprising that it has not been used more, especially in the context of quality control.

Instead of a falling ball, the rebound of a pendulum can be used to measure resilience and such methods are widely employed and standardised for rubbers.[1]

9.3 FORCED VIBRATION METHODS

There are a great many designs of machine for carrying out forced vibration testing of plastics and it would be inappropriate to deal with the subject in any depth here, both because of the amount of information already published in the literature[1–3,15–25] and because, ordinarily, such testing is only rarely used for routine material characterisation and quality control.

Two instruments worthy of direct mention here, however, both as illustrations of the two modes of testing and by reason of their popularity, are (a) the Rheovibron, which operates away from resonant frequency, and (b) the vibrating reed, which operates at the resonant frequency.

9.3.1 Forced Vibrations Away from Resonance

The Rheovibron has found increasing popularity since its development in the early 1960s[26] because of its versatility. It requires only a small test piece, can operate in tension, compression (although not usually for plastics in this mode), or shear and is capable of operating over a wide temperature range. Its main disadvantages are: system resonance; end-effect errors which need correction; and (for the non-automated versions at least) labour intensiveness.

One end of the test piece is gripped in a clamp which is driven at the selected frequency (between 3·5 and 110 Hz) while the other is gripped in a rigid clamp attached to a piezoelectric crystal which detects the force generated. The instrument gives a direct reading of tan δ. A recent publication by Kenyon *et al.*[27] describes the automation of the Rheovibron to effect continuous measurements of stress, strain and phase angle as a function of temperature. Massa *et al.*[28] review the capabilities of the instrument, discuss sources of error and equipment modifications. Yee and Takemore[29] and Locati[30] also give details of the use of the Rheovibron.

The use of ultrasonics to measure the moduli of plastics is an example of the forced vibration method away from resonance but this subject is deferred to Chapter 19 on Non-Destructive Testing.

9.3.2 Forced Vibrations at Resonance

In the vibrating reed apparatus one end of the test piece, which may be of circular or rectangular cross-section, is held in a relatively massive vibrating head while the other end is free to oscillate. The applied frequency of the head is altered via a sine-wave generator and the amplitude of the free end of the test piece is measured by some suitable means. This amplitude varies with the applied frequency and reaches a maximum at the resonant frequency of the reed. Young's modulus for the material can be determined from the expression

$$E = \frac{38{\cdot}24\, dL^4 f_r^2}{D^2} \tag{9.11}$$

where d = density of the plastic
L = free length of the reed
D = thickness of the reed
f_r = resonant frequency.

The mechanical damping term can be calculated from the width of the resonance peak by the formula

$$\tan \delta = \frac{\Delta f}{f_r} \tag{9.12}$$

where Δf is the width of the resonance curve at the half power point (70·7 per cent of the peak height).

Vibrating reed tests are described by Newman,[31] Atkinson and Eagling,[32] Strella,[33] Scherr,[34] Fielding-Russell and Wetton[35] and more recently by Washburn,[36] who used the technique for studying discontinuities in compression moulded composites and fatigue properties of carbon fibre reinforced phenolics.

More general information concerning dynamic mechanical testing is given by Dean and Read[37] and Bert and Clary,[38] while Koo[39] and Murayama[40] provide recent reviews of techniques and the type of information obtainable.

REFERENCES

1. Brown, R. P. (1979) *Physical Testing of Rubbers*, Applied Science, Chapter 9.
2. Ferry, J. D. (1970) *Viscoelastic Properties of Polymers*, Wiley,
3. Nielsen, L. E. (1962) *Mechanical Properties of Polymers*, Reinhold.
4. ISO R537 (1967) *Testing of Plastics with the Torsion Pendulum.*
5. ASTM D2236 (1970) *Dynamic Mechanical Properties of Plastics by means of a Torisonal Pendulum.*
6. DIN 53445 (1979) *Testing of Plastics: Torsion Pendulum Test.*
7. Fritzsche, C., Hoechli, B. and Moser, K. (1975) *Kunststoffe*, **64** (11), 675.
8. Bell, C. L. M., Gillham, J. K. and Benci, J. A. (1974) in *SPE 32nd Annual Technical Conference, San Francisco*, page 598.
9. Aghili-Kermani, H., O'Brien, T, Armeniades, C. D. and Roberts, J. M. (1976) *Journal of Physics, E*, **9** (10), 887.
10. Nielsen, L. E. (1960) *SPE Journal*, **16** (5), 525.
11. James, D. I., Norman, R. H. and Payne, A. R. (1959) in *SCI Monograph No. 5*, Society of Chemical Industry, page 233.
12. Lord, P., Pithey, E. R. and Wetton, R. E. (1961) *Laboratory Practice*, 10, 884.
13. Lepie, A. H. and Adicoff, A. (1967) *Review of Scientific Instruments*, **38** (11), 1615.
14. Roller, M. B. and Gillham, J. K. (1978) *Journal of Coatings Technology*, **50** (636), 57.
15. Crissman, J. M. and McCammon, R. D. (1962) *Journal of the Acoustical Society of America*, **34** (11), 1703.
16. Lee Tung-Ming (1963) *Journal of Applied Physics*, **34** (5), 1524.
17. Learmonth, G. S. and Pritchard, G. (1968) *SPE Journal*, **24** (11), 47.
18. Hanson, O. C., Fabry, T. L., Marker, L. and Sweeting, O. J. (1963) *Journal of Polymer Science*, **1** (5), Part A, 1585.
19. Kline, D. E. (1956) *Journal of Polymer Science*, **22** (102), 449.
20. Payne, A. R. (1961) *Materials, Research and Standards*, **1** (12), 942.
21. Maxwell, B. (1956) *ASTM Bulletin*, 215, 76.
22. Sharma, M. G. and Hung, C. S. (1975) *Polymer Engineering and Science*, **15** (11), 805.
23. Koga, K., Kajiyama, T. and Takayanagi, M. (1975) *Journal of Physics, E*, **8** (4), 299.
24. Davis, W. M. and Macosko, C. W. (1977) *Polymer Engineering and Science*, **17** (1), 32.
25. Braginskii, A. I. and Zelenev, Yu. V. (1975) *Polymer Science USSR*, **17** (5), 1356.
26. Takayanagi, M. (1963) *Memoirs of the Faculty of Engineering, Kyushu University*, 23, 41.
27. Kenyon, A. S., Grote, W. A., Wallace, D. A. and Rayford, M. (1977) *Journal of Macromolecular Science, B*, **13** (4), 553.
28. Massa, D. J., Flick, J. R. and Petrie, S. E. B. (1975) *Coated Plastics Preprints*, **35** (1), 971.
29. Yee, A. F. and Takemore, M. T. (1977) *Journal of Applied Polymer Science*, **21** (10), 2597.
30. Locati, G. (1978) *Polymer Engineering and Science*, **10** (10), 793.
31. Newman, S. (1959) *Journal of Applied Polymer Science*, **2** (6), 333.
32. Atkinson, E. B. and Eagling, R. F. (1959) *SCI Monograph No. 5*, Society of Chemical Industry, 197.
33. Strella, S. (1956) *ASTM Bulletin*, 124, 47.
34. Scherr, H. J. (1966) *Materials, Research and Standards*, **6** (12), 614.
35. Fielding-Russell, G. S. and Wetton, R. E. (1970) *Plastics and Polymers*, **38** (135), 179.
36. Washburn, R. M., (1974) *Sampe Quarterly*, **6** (1), 22.

37. Dean, G. D. and Read, B. E. (1976) *Plastics and Rubber: Materials and Applications*, **1** (3/4), 145.
38. Bert, C. W. and Clary, R. R. (1973) in *3rd Conference on Composite Materials, Testing and Design, Williamsburg*, Conf. 627, ASTM, page 250.
39. Koo, G. P. (1974) *Plastics Engineering*, **30** (5), 33.
40. Murayama, T. (1978) *Materials Science Monographs* vol. 1, Elsevier Scientific.

Chapter Ten

Friction and Wear

10.1 INTRODUCTION

Traditionally, laboratory measurements of friction and wear have been made as independent, almost unrelated, topics despite the fact that in the 'real world' there is a very close link between the two. However, this is not so surprising when one considers that generally it is either the one or the other property, rather than both simultaneously, that is of real significance in a given situation. A clear exception to this is the use of plastics in bearings applications where both parameters are of crucial importance. Also, of course, when high resistance to wear is required, friction is but one of the many parameters which determine that resistance.

The measurement of abrasion resistance is encountered rather more frequently than friction measurement and this is reflected in the number of standards documents dealing with these subjects.

10.2 FRICTION

10.2.1 General Considerations

The basic 'laws' of friction have been known for several centuries and are usually stated as follows:

1. The frictional force, F, opposing motion is proportional to the normal force, N, between the contacting surfaces, i.e.

$$F = \mu N \tag{10.1}$$

where the constant of proportionality, μ, is called the coefficient of friction.
2. The coefficient of friction is independent of the apparent area of contact.
3. The coefficient of friction is independent of the velocity between the contacting surfaces provided this is not zero.

These three 'laws', although rarely obeyed by polymers, still form the basis on which the frictional properties of these materials can be studied. James[1] details the effect of various parameters on the frictional behaviour of plastics and rubbers and how these are related to the three classical laws and to modern theories of friction. The most important parameters, as regards methods of test, are described briefly below.

No matter how smooth a surface may appear to be, it is, at the microscopic level, rough. This means that the actual area of contact between two faces is considerably less than the apparent area and it is the real contact area which determines the frictional force, experiments having shown a proportional relationship between frictional force and real area of contact. For materials exhibiting a definite yield stress, the real contact area between two faces is proportional to the normal force, from which the first friction law naturally follows. Metals are generally in this class. For substances which deform elastically the situation is more complex and, in general, frictional force is proportional to normal force raised to a power varying between $\frac{2}{3}$ and 1. Only where it is unity is the coefficient of friction constant and for all plastics materials the coefficient decreases as normal force increases. This implies that if a friction measurement is to be made on a plastics material for other than routine inspection purposes then the test should be undertaken using a variety of normal loads.

Standard textbooks on physics often refer to 'static' and 'dynamic' friction, although truly static friction is an impossibility since relative motion between contacting surfaces is unavoidable if frictional force is to be measured. The normal meaning of static friction, therefore, is the force required just to induce motion, i.e. the frictional force at very low speed. Dynamic friction on the other hand usually means the force required to maintain a constant, perceptible, relative motion between the surfaces. Clearly, therefore, velocity is a factor in determining the frictional force, although for many materials the variation in the coefficient of friction over many decades of velocity is so small as to be negligible—hence the third law of friction. For polymers, however, this is not so and variations in the coefficient of friction may be as much as three- or four-fold over a few decades of velocity. Unfortunately the effect of velocity cannot be isolated from that of temperature, but clearly an adequate characterisation of the frictional behaviour of a plastic cannot be carried out without examining the effect of velocity.

A phenomenon which sometimes occurs is that of 'stick–slip', in which the two surfaces, or small parts thereof, momentarily cease their relative motion causing the frictional force to rise until a critical value is reached at which point the surfaces slip past each other with the frictional force decreasing to some lower value before the cycle is repeated. For stick–slip to occur the coefficient of friction must fall as velocity increases. The frequency of this stick–slip cycle is often in the audible frequency band giving rise to the familiar squealing sound sometimes heard as surfaces move past each other. The occurrence of stick–slip in service, therefore, can be of great nuisance value.

Obviously atmospheric conditions and the presence of lubricants and dust influence the friction, so test conditions should be adjusted to simulate service conditions as far as is practically possible.

10.2.2 Methods of Measuring Friction

At present there are relatively few standard test methods for measuring friction and none at international level, although a document is being prepared by TC61. At the time of writing this was only at the level of a study item and so it will be many years before it achieves the status of an International Standard.

BS 2782, Method 311A (1970)[2] describes the measurement of the static and dynamic coefficients of friction for polyethylene film. A piece of film is held on a horizontal bed by means of a vacuum-operated clamp and another, similar, piece of film is attached to the underside of a sled of sufficient mass to give a contact pressure of 490 N/m^2 when placed on the bed. The sled is driven at 800 mm/min over the bed and the forces required to initiate relative motion and to maintain constant motion are recorded.

ASTM D1894 (1978),[3] for plastic film and sheeting in general, is very similar although a wider variety of apparatus is allowed (Figure 10.1).

A method for measuring static friction only is the inclined plane method given in BS 3424 (1973).[4] In this the test piece is fastened as before to a sled which rests on an initially horizontal bed. The bed is tilted at a given rate until the sled just starts to slide, at which point the drive mechanism is switched off, usually via a microswitch, and the angle of inclination is read off a suitable scale. The static coefficient of friction is then equal to the tangent of this angle.

Despite the paucity of standard test methods a large number of apparatuses have been constructed to carry out friction measurements. Gough[5] describes the construction of an apparatus consisting of four levers supporting a test platform at its corners in such a way that each lever is pivoted at a point that maintains neutral equilibrium in the mechanism. The test surface is drawn over the platform and the coefficient of friction is given by the tangent of the angle of displacement of the levers. Bowers *et al.*[6] describe the use of a surface grinder modified to allow frictional measurements to be made and James[7] describes a similar technique using a lathe bed. Frictional 'drag' fed to the secondary winding of a differential transformer via a leaf spring has been used by Lauer and Friel[8] in a machine they claim has sensitivity over a wide range of values. The effect of velocity on friction has been investigated by Westover and Vroom[9] who describe their apparatus as a 'variable speed frictionometer'. The use of this apparatus has now been standardised in ASTM D3028 (1972).[10]

A recently designed piece of apparatus developed at the Rubber and Plastics Research Association and described by James and Newell[11] allows frictional measurements to be made over a wide range of velocities and normal loads. The apparatus (Figure 10.2) is fixed to a normal tensile testing machine, giving a very stiff system which enables small forces to be measured accurately. It can be used inside an environmental chamber so that the effect of temperature can be investigated. The test piece geometry is readily changed, allowing tests to be performed on finished products or parts of products. James and Newell, in addition to describing the RAPRA/Daventest friction tester in detail, also provide an excellent source of references for friction test methods and theory.

An apparatus which is of value in measuring the friction of large surfaces in service—flooring or polymeric sports surfaces, for example—is the skid tester developed by the Road Research Laboratory.[12] This is a pendulum device which, when released, brushes against the surface to be measured; the energy absorbed is read from the resultant angle of swing of the pendulum in a manner analogous to the pendulum impact testers described in Chapter 8.

In 1975 the British Standards Institution published Part 5.6 of BS 4618[13] dealing with the presentation of plastics design data. Part 5.6 is a guide to sliding friction and is a useful introduction to the subject in the context of plastics materials.

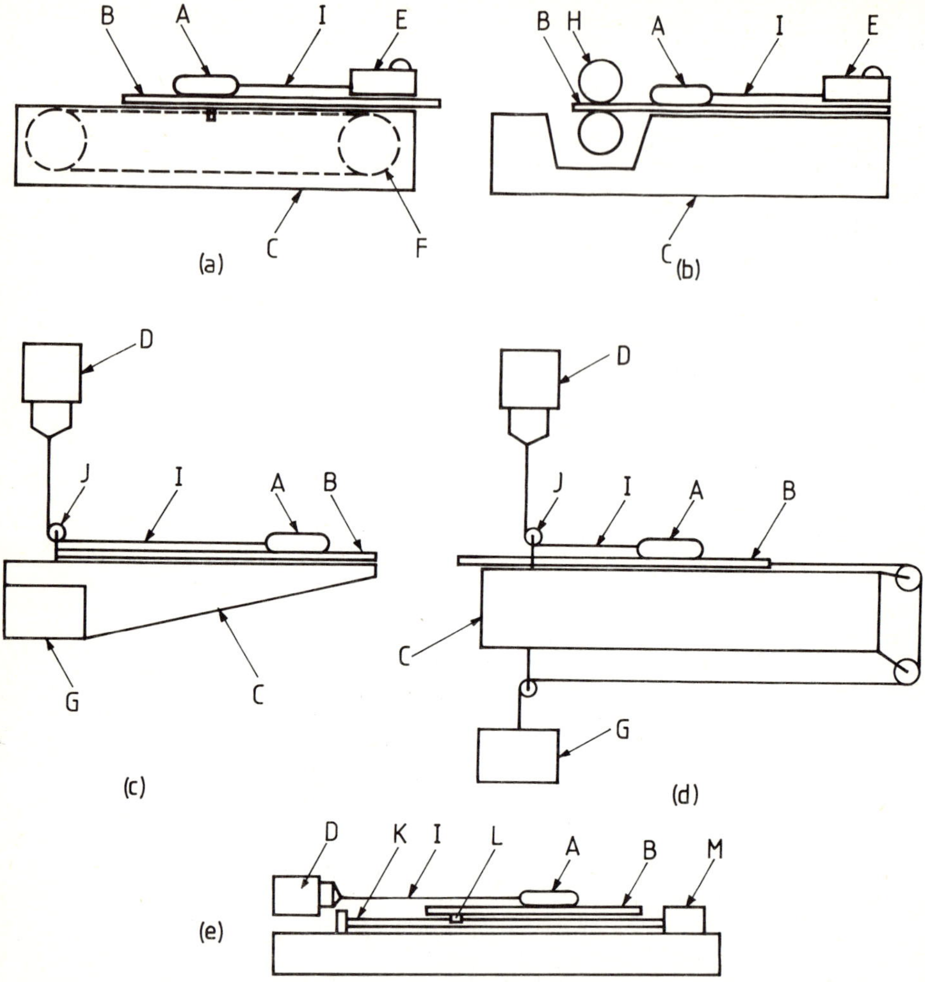

Figure 10.1 Apparatus for measuring coefficient of friction of film: ASTM D1894
A Sled
B Plane
C Supporting base
D Strain gauge
E Spring gauge
F Constant-speed chain drive
G Constant-speed tensile tester crosshead
H Constant-speed-drive rolls
I Nylon monofilament
J Low-friction pulley
K Worm screw
L Half nut
M Hysteresis, synchronous motor

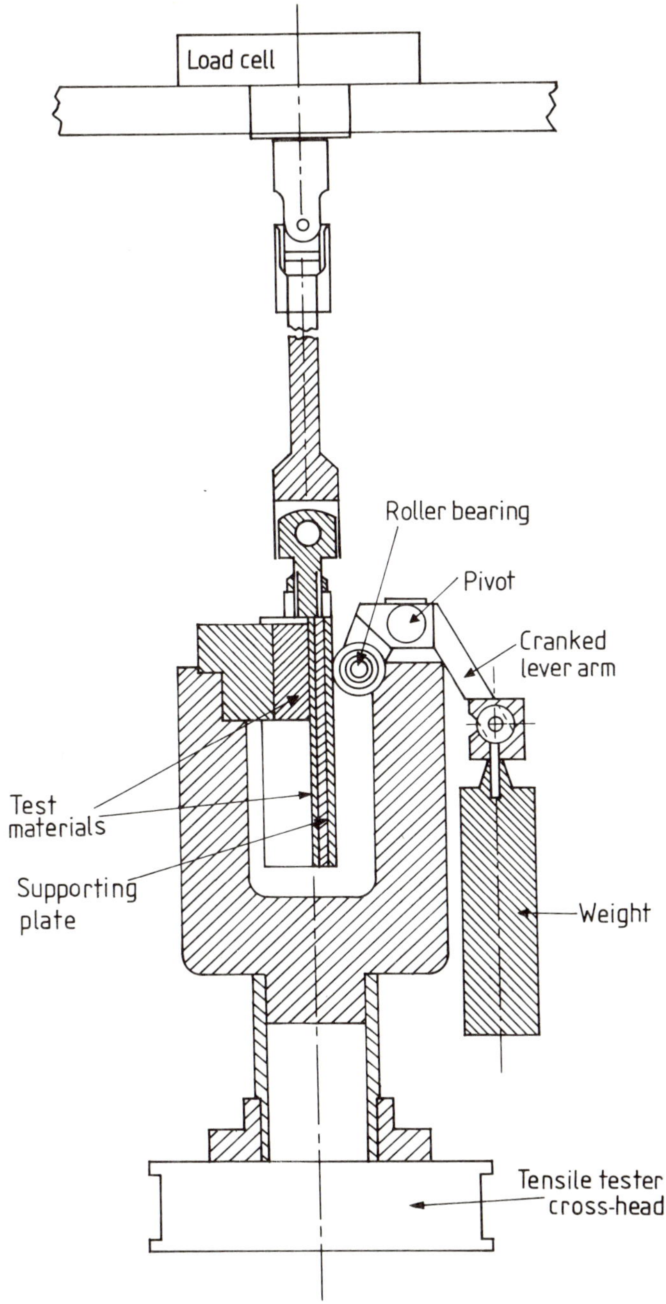

Figure 10.2 RAPRA/Daventest friction apparatus

10.3 WEAR

10.3.1 General Considerations

The practical requirement of knowing how well a material will stand up to abrasive wear has been felt for centuries, and probably the oldest 'standard' method is the use of Moh's scale, whereby the ability of the material to resist surface damage is assessed by progressively scratching the surface with numbered minerals of increasing hardness. The material's resistance is expressed as the number of the mineral which just causes surface damage. A similar test is the pencil hardness test which uses the range of draughtsmen's pencils as the abradants. Bierbaum's hardness test, which employed a diamond scribe, was once proposed as a standard method in ASTM but, like the previously mentioned methods, has since been discontinued, probably due to lack of reproducibility. It is clear that the above methods simultaneously measure mar resistance, which is a wear mechanism, and surface hardness (see §8.1), which is not, so that they perform neither function satisfactorily—this may be the more telling reason for their disappearance.

The mechanisms of wear are complex and highly interactive; Lancaster[14] in his review concentrates on the three major processes: abrasion, adhesion and fatigue. More detailed descriptions are given in Volumes 5A and B of the Polymer Science and Technology series[15] and a new review of friction and wear mechanisms in polymers has recently been produced by Briscoe and Tabor.[16]

The very complexity of wear processes has given rise to a profusion of test apparatus—and standard methods—while at the same time rendering the quantitative correlation between methods non-existent in general,[17–19] although for a limited range of materials it may be possible to develop approximate interrelationships between test methods. Even more regrettable is the fact that different methods do not always rank materials in the same order, although it must be stressed that in this case the agreement between methods is considerably improved.[20]

The foregoing points clearly indicate that in order for meaningful data to be generated in the laboratory, the tests that are applied must simulate the real service conditions as closely as possible. Even then, the experimenter must usually settle for merely ranking the materials of interest rather than deriving quantitative figures for them.

10.3.2 Standard Test Methods

ISO Standards

There is at present no standard test method for abrasion within ISO although TC61 does have, as a study item, a document dealing with friction and abrasion of materials in linear contact.

British Standards

The only abrasion test within BS 2782 at the time of writing is Method 310B (1970),[21] the apparatus for which is illustrated in Figure 10.3. The test measures print adhesion on PVC sheet. Test pieces 229 mm × 51 mm are used and first

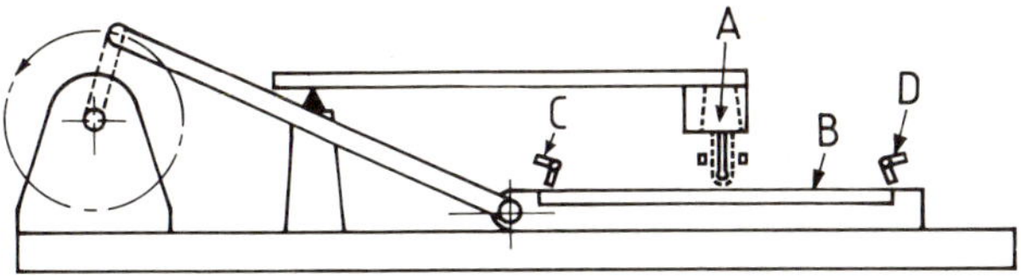

Figure 10.3 Print adhesion test: BS 2782, Method 310B
The abrading member (A) is a brass peg the centre line of which is at right angles to glass plate (B) over which the test surface is clamped by grips (C, D). The brass peg exerts a force of 9 N on the test piece, through a piece of carefully specified bleached cotton fabric secured round the peg. The rate of reciprocation is 15 ± 2 cycles (each of two strokes) per minute

washed with a soap solution, rinsed, dried and conditioned. The number of cycles to cause visible damage of the print is recorded.

BS 3424, Method 16 (1973)[22] for coated fabrics specifies an abrasion tester called the Rubfastness Tester, which employs a reciprocating table upon which the test piece is placed and a loaded metal knife which rubs against the test piece.

BS 3794, Appendix B (1973)[23] for decorative laminates employs the well known Taber Abraser, described below, and at present British Standards Committee PLC/17 is considering a draft standard for inclusion in BS 2782 which uses a rotary platform double-head abrader of the Taber type (Figure 10.4). In the decorative laminates specification, use is made of reference sheets of laminate of specified wear resistance obtainable from BSI. The laminate under test

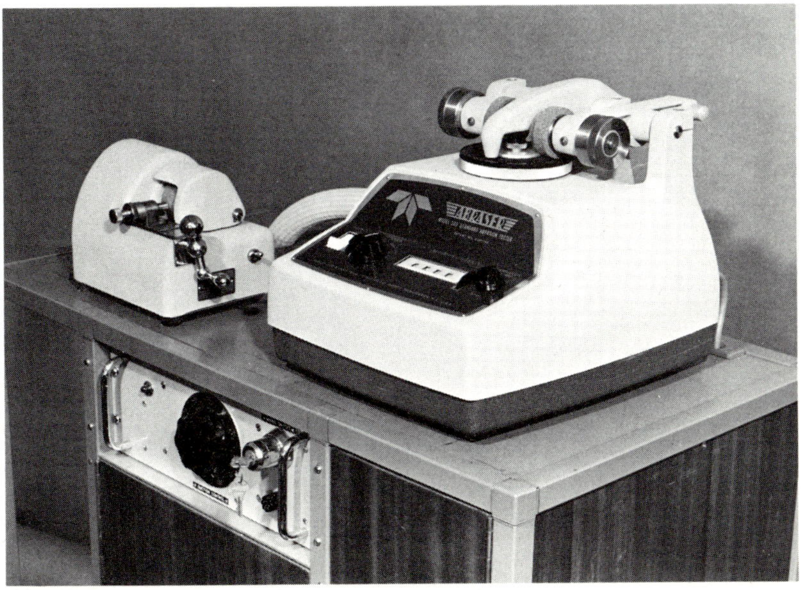

Figure 10.4 Taber Abraser
(*Courtesy Funditor Ltd*)

and a sheet of the reference laminate are clamped side by side on the table of the abrader in such a way that the abrasive wheels mark out equal lengths of track on each laminate on each revolution of the table. The number of revolutions to cause the same degree of damage to the surface layers of the reference and test laminates are recorded, from which the wear resistance, $(WR)_1$, of the test laminate is calculated by the formula:

$$(WR)_1 = \frac{(WR)_0 \times (REV)_1}{(REV)_0} \tag{10.2}$$

where $(WR)_0$ = wear resistance of reference piece
$(REV)_0$ = number of revolutions at end-point for reference piece
$(REV)_1$ = number of revolutions at end-point for test piece.

In this way the variability of the wear resistance due to variations between different abrasive wheels and between different abraders can be overcome to a large extent, though not entirely. The use of standard reference materials with abrasion tests is rather more common for rubbers than plastics, as explained by Brown.[24]

USA Standards
ASTM D1044 (1978)[25] covers measurements of abrasion of transparent plastics using the Taber Abraser with 'Calibrase' CS-10F abrading wheels. The basic principle is the rotation, at a prescribed speed, of test discs 100 mm diameter, or 100 mm squares, under the abrading wheels freely resting thereon with loads acting on them of 250, 500 or 1000 g. To obtain abrasion the line joining the centres of the wheels is offset from the axis of rotation of the table. Abrasion resistance is measured by the amount of transmitted light which is diffused by the abraded track, using the ASTM D1003 *Test for Haze and Luminous Transmittance of Transparent Plastics* (Chapter 14).

The more conventional use of the Taber Abraser is by determining the weight loss of the test disc or square induced by the abrading wheel under a specified load and as a result of a specified number of revolutions—with a particular type of the four variants of resilient wheels ('Calibrase') or four variants of hard wheels ('Calibrade'). This form of test is included in Test LPS-106 of the National Electrical Manufacturers' Association standard for decorative laminated plastics. Great care must be exercised in the use of the wheels: for example, they must be kept clean at all times—an air jet is normally used to blow off abraded material during the test—and the flexible wheels have a very limited life. A study of the reliability of the Taber methods has been made by Hill and Nick[26] who conclude that it is best to take the mean of a considerable number of test runs; they also correlate results with the hardness of the (flexible) test wheels.

A great advantage of the Taber Abraser is its versatility, in that the severity of abrasion can be altered by using different types of wheel under different loads. The test can also be conducted in the presence of a liquid medium since the normal table can be replaced by a lipped table to hold the fluid. Its major disadvantage, particularly significant for testing finished products, is the large area of flat specimen required.

Two methods of testing the abrasion resistance of plastics are included in ASTM D1242 (1956).[27] Method A uses loose abrasive applied under defined conditions to the test surface and results are expressed as volume loss (from weight loss and relative density). Method B uses a 'bonded abrasive abrading machine' (Figure 10.5) which also gives results in volume loss.

The resistance of glossy plastics to abrasion may be measured by loss of gloss caused by the abrasive action of impacting carborundum grit (Figure 10.6); such is the principle of ASTM D673 (1970).[28] The gloss before and after abrasion is measured with a gloss meter (Figure 10.7).

Wiinikainen[29] has described a 'round robin' of scratch resistance and abrasion tests on transparent plastics materials.

German Standards
DIN 53754 (1977)[30] describes a test with the Taber type of geometry, but perhaps the best known abrader is the so-called 'DIN Abrader' which is

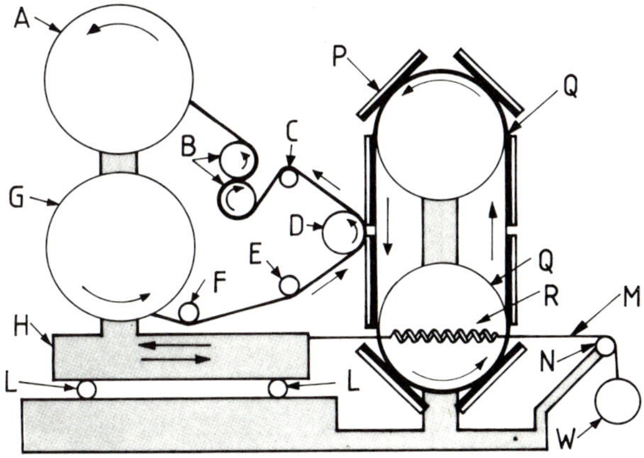

Figure 10.5 Bonded abrasive abrading machine: ASTM D1242, Method B
A Take-up drum
B Constant-speed-driven rolls
C Abrasive tape
D Steel contact roll
E, F Slotted guide roll
G Abrasive tape supply drum
H Carriage
L Roller bearings
M Spring cable
N Pulley
P Specimen mounting plates
Q Constant-speed-driven pulleys with continuous link belt
R Spring
W Dead weight
Essentially the machine comprises two independent units. On the right-hand side is a vertical unit which carries the test specimens to meet the abrasive tape (C), which rotates in the opposite direction and is kept in contact by being mounted on carriage (H) on roller bearings (L), and tensioned by weight (W) through spring (R)

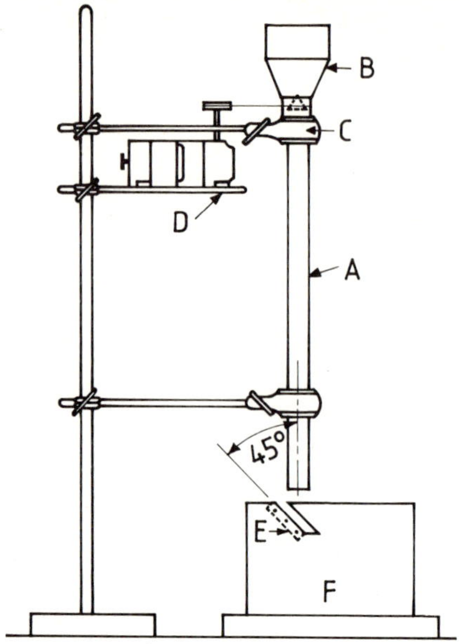

Figure 10.6 Mar-resistance abrader: ASTM D673
Abrasive grit is allowed to fall on to the test specimen resting in groove (E) from a hopper (B) rotated to produce a controlled rate of feed. The grit falls down tube (A) on to the specimen and then into receptacle (F)

described in DIN 53516 (1977);[31] it is designed primarily for rubbers, but it can also be used successfully for plastics. With this abrader, the test piece is held in a chuck and traversed over a rotating drum covered with a sheet of the abradant. In this way there is a large area of abradant exposed to the test piece so that the wear of the abradant is slow and uniform. The main advantages of the DIN Abrader are the small test piece required—only 16 mm diameter by, at least, 6 mm thick, and the ease of operation which makes it ideal for quality control

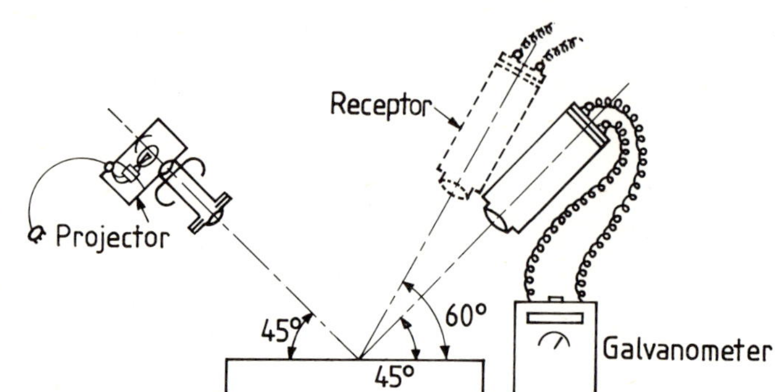

Figure 10.7 Mar-resistance gloss meter: ASTM D673
The photoelectric receptor is set at 45° and the galvanometer reading is taken (l_1); the reading at 60° is then taken (l_2). Gloss (per cent) = 100 $(l_1 - l_2)/l_2$

purposes. It does, however, lack the versatility of the Taber even if the abradant and load are changed from those specified in the standard.

REFERENCES

1. James, D. I. (1973) *RAPRA Members Journal*, July, 170.
2. BS 2782, Method 311A (1970) *Coefficients of Friction of Polythene Film.*
3. ASTM D1894 (1978) *Static and Kinetic Coefficients of Friction of Plastic Film and Sheeting.*
4. BS 3424, Method 12 (1973) *Methods of Test for Coated Fabrics—Determination of Surface Drag.*
5. Gough, V. E. (1953) *Journal of Scientific Instruments*, 30, 345.
6. Bowers, R. C., Clinton, W. C. and Zisman, W. A. (1954) *Modern Plastics*, **31** (6), 131.
7. James, D. I. (1961) *Journal of Scientific Instruments*, **38** (7), 294.
8. Lauer, J. L. and Friel, P. J. (1957) *Review of Scientific Instruments*, **28** (4), 294.
9. Westover, R. F. and Vroom, W. I. (1963) *SPE Journal*, **19** (10), 1093.
10. ASTM D3028 (1972) *Kinetic Coefficient of Friction of Plastics Solids and Sheeting.*
11. James, D. I. and Newell, W. G. (1978) *RAPRA Members Report*, No. 20.
12. Giles, C. G., Sabey, B. E. and Cardew, K. H. F. (1964) *Road Research Technical Paper No. 66*, HMSO.
13. BS 4618: Section 5.6 (1975) *Recommendations for the Presentation of Plastics Design Data: Guide to Sliding Friction.*
14. Lancaster, J. K. (1973) *Plastics and Polymers*, **41** (October), 297.
15. Lieng-Huang Lee (Ed.) (1974) *Advances in Polymer Friction and Wear, Polymer Science and Technology* vols 5A and B, Plenum.
16. Briscoe, B. J. and Tabor, D. (1978) in *Polymer Surfaces*, D. T. Clark and W. J. Feast (Eds.), Wiley, page 1.
17. Harper, F. C. (1961) *Wear*, **4** (6), 461.
18. Anon. (1961) *Wear*, **4** (6), 479.
19. Anon. (1964) *Wear*, **1** (3), 302.
20. Frick, O. F. V. (1969) *Wear*, **14**, 119.
21. BS 2782, Method 310B (1970) *Adhesion of Print on Flexible PVC Sheet.*
22. BS 3424, Method 16 (1973) *Methods of Test for Coated Fabrics—Determination of Fastness of Colour to Wet and Dry Rubbing.*
23. BS 3794, Appendix B (1973) *Decorative Laminated Plastics Sheet—Determination of Resistance to Surface Wear.*
24. Brown, R. P. (1979) *Physical Testing of Rubbers*, Applied Science,
25. ASTM D1044 (1978) *Resistance of Transparent Plastics to Surface Abrasion.*
26. Hill, H. E. and Nick, D. P. (1966) *Journal of Paint Technology*, **38** (494), 123.
27. ASTM D1242 (1956) *Resistance of Plastic Materials to Abrasion.*
28. ASTM D673 (1970) *Mar Resistance of Plastics.*
29. Wiinikainen, R. A. (1969) *Materials, Research and Standards*, **9** (12), 17.
30. DIN 53754 (1977) *Testing of Plastics; Determination of Abrasion by the Abrasive Disc Method.*
31. DIN 53516 (1977) *Determination of Abrasion.*

Chapter Eleven

Creep, Relaxation and Set

11.1 INTRODUCTION

We have seen in earlier chapters how the viscoelastic nature of a plastics material renders its short-term and dynamic mechanical stress–strain behaviour considerably more complex than that of more perfectly elastic materials such as metals. Here and in the following chapter the long-term effects of applied stress and strains on plastics materials are examined. For the design engineer the long-term behaviour of the material is often of considerably greater importance than the short-term behaviour, since distortion of a component under load (creep), loss of applied force (relaxation) and component failure under loads less than the 'static' load to failure (fatigue) can all seriously reduce the useful life of the component.

Consideration of fatigue testing, which almost invariably results in the catastrophic failure of the item under test, is deferred to Chapter 12, while creep and stress relaxation are discussed below. Set, which is seldom encountered in the literature on plastics, is mentioned briefly at the end of the chapter.

11.2 CREEP

11.2.1 General Considerations

The principle of creep testing is very simple: a fixed load is applied to a test piece and the deformation of the test piece as a function of time is recorded. Parameters such as temperature and humidity, where the latter is known to affect the creep rate, such as for polyamides, are kept constant. In practice, of course, the situation is much more involved and great care must be exercised if reproducible results are to be obtained. The deformation-measuring device must be capable of high resolution, in order to measure the (usually) small strains involved, while at the same time having a high stability to long-term thermal or electrical drift. It must not interact significantly with the test piece, thereby modifying the creep behaviour. The test piece should also be well characterised chemically and physically since the time-scale of a creep test allows the individual viscoelastic nature of each test piece to exert itself to the full and slight differences between test pieces, which might not be detectable in short-term tests, may become significant in a long-term test. Effects such as an increase in crystallinity for a crystallisable polymer, post-curing in a thermoset, or different moulding stresses in a thermoplastic, can alter the creep behaviour of the material significantly. All this means that the test environment is normally controlled rather more rigorously than is usual for short-term tests and temperature control is often required to be within ±1°C or better. Humidity control at ±2 per cent may also be required for materials which are sensitive to water content.

11.2.2 Definitions

The following definitions are taken from BS 4618 Section 1.1 (1970):[1]

Creep strain	The total strain, which is time-dependent, resulting from an applied stress or system of stresses.
Creep modulus	The ratio of applied stress to creep strain.
Creep lateral contraction ratio	The ratio of lateral strain to longitudinal strain measured simultaneously in a creep experiment (also known as Poisson's ratio).

The above definition of creep strain is the one currently favoured in plastics engineering practice. There is, however, a 'scientific' definition of creep strain which subtracts the instantaneous, or elastic, strain from the total strain to give the creep strain. Such a definition follows naturally from the phenomenological models of the Kelvin and Voigt type mentioned in Chapter 9 and used to approximate the viscoelastic behaviour of polymers, but in practice the truly elastic, i.e. instantaneous, deformation cannot be separated from the time-dependent deformation. It is worth noting that when ASTM D674 was replaced by ASTM D2990[2] in 1977, the definition of creep strain was changed from the 'scientific' to the 'engineering', thus bringing it into line with current ISO and BS practice.

11.2.3 Standard Test Methods

Creep tests may be carried out in tension, compression, flexure, shear or torsion, although to date only the first three have been standardised; tests in tension are the most popular.

ISO Standards
At the time of writing only one creep test exists as an International Standard, this being ISO R899 (1968).[3] The standard is under revision but as this is only at the draft proposal stage the document is not referred to further since many changes are likely to be made before it reaches standard status. A similar situation relates to the draft proposal on flexural creep which is being prepared in committee TC61. This document has its origins in ASTM D2990 (1977),[2] which is described more fully below.

ISO R899 describes the apparatus requirements (which are rather more exacting in terms of axiality of load application than for the corresponding short-term test method) and type of grip system that may be used. The loading system must be designed to apply a load within ±1 per cent of the desired load and be of such a construction that if a test piece breaks the energy released is not transmitted to any of the other test pieces. The extensometer must be capable of measuring to ±1 per cent of the total strain anticipated. Contactless, optical systems are recommended especially for creep rupture tests and the comment is made that electrical resistance gauges should only be used for short-duration tests and then only when perfect adhesion to the test material can be guaranteed.

The test pieces are as specified for the determination of tensile properties but the variation in cross-sectional area within the gauge length should not exceed 2 per cent and dimensional measurements should be made with equipment having an accuracy of at least 0·01 mm. It is recommended that an allowance be made for any electrical and thermal drift of the strain-measuring device by means of the use of a control test piece which is prepared and tested under identical conditions to the real test piece, except that no load is applied to it.

The loading of the test piece is very important and this should be carried out smoothly and continuously within a period of from 1 to 5 seconds from the start of loading, zero time for the experiment being taken at the point of full loading. At no time during loading must the stress exceed the level at zero time. The strain is recorded either automatically (preferably) or manually at approximately equally spaced logarithmic time intervals.

The experiment is usually carried out over a range of applied stresses since the creep modulus is not independent of stress level except at low stress levels (in the linear viscoelastic region) and it is frequently convenient to derive isochronous stress–strain curves from the strain–time curves obtained under different stresses. Figures 11.1 and 11.2 illustrate this operation. Temperature, too, may be considered as a parameter to be examined especially if the data are to be used for design purposes, in which case the whole experiment is repeated at several temperatures.

Creep recovery tests may also be carried out by removing the load, in the same manner in which it was applied, after some suitable time and then following the decrease in strain as a function of time from the instant of unloading.

British Standards

There is, at present, no test method dealing with creep in BS 2782, although it is anticipated that once the tensile and flexural creep tests in ISO are standardised these could be brought within the scope of BS 2782.

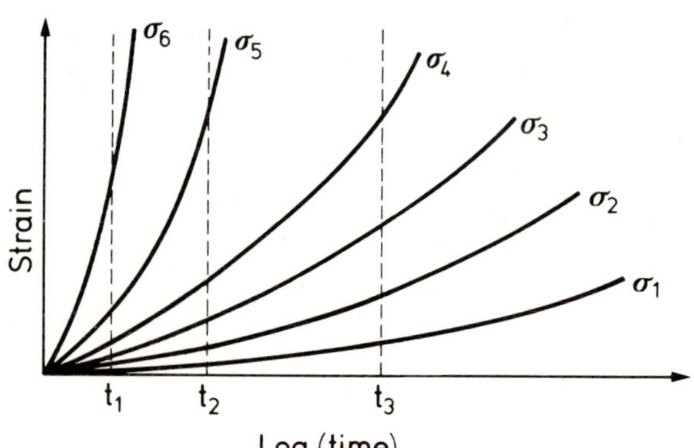

Figure 11.1 Typical creep curves at different stress levels

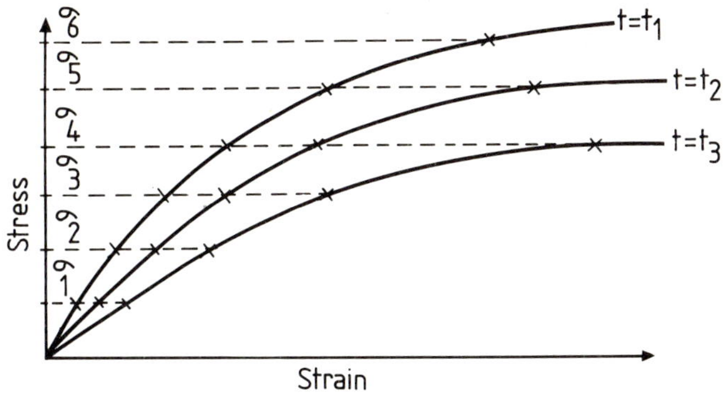

Figure 11.2 Isochronous stress–strain curves

BS 4618 Section 1.1 (1970)[1] is the the relevant UK document in this context. It is considerably more detailed than ISO R899 and is an excellent source of information on the philosophy and procedures for creep testing of plastics. Creep rupture, however, is not included within its scope.

This document makes the point not made in ISO R899 that the strain response immediately after the application of the load depends upon the precise loading path taken and recommends that strain measurements made within a period equal to ten times the loading period be ignored.

Section 1.1 describes an experiment whereby the isochronous stress–strain curve (for a short time) may be derived from a single test piece, but warns that for data suitable for engineering design purposes 50 to 100 test pieces may need to be tested. The minimum programme for characterising a non-linear viscoelastic material is given as:

(i) the determination of the isochronous stress–strain curve at a particular temperature in the range 19° to 24°C at some specified short time such as 60 s or 100 s;
(ii) for the same temperature and material state as in (i), the determination of at least four creep curves at different stress levels, preferably of duration at least one year;
(iii) at the end of each creep test, the determination of the creep recovery for at least 10 per cent of the period under stress and preferably longer;
(iv) for the same material state as in (i), determination of the effect of temperature by means of isochronous stress–strain curves at various temperatures and creep curves at selected stress levels and temperatures;
(v) determination of the effects of factors such as fabrication variables, environment and humidity by means of isochronous stress–strain tests.

It is not surprising, in view of such a list, that creep data are rather less readily available than tensile, flexural, etc., data!

The final part of Section 1.1 deals with the presentation of data and gives typical creep, isochronous stress–strain, stress–temperature, creep recovery, etc., curves.

Subsection 1.1.1 deals specifically with uniaxial creep in both tension and compression, giving criteria of suitability for test pieces and apparatus as well as describing test conditions and experimental programmes. A point worthy of note is the classification of the strain-measuring device according to precision: Category A has a minimum detectable increment of 0·002 per cent strain, Category B 0·008 per cent and Category C 0·02 per cent. The requirements of axiality of the test machine are discussed in relation to these three categories.

Subsection 1.1.2 deals with flexural creep in three-point loading. The advantages of using flexural creep tests are:

(a) in the linear viscoelastic region, i.e. at low strains, data of high accuracy are more readily obtained than in tension and compression;
(b) the apparatus can be simpler;
(c) data can be obtained that are directly relevant to flexural deformation in structures.

These three considerations lead to three categories of testing, the requirements of the first two of which are described in detail in the standard. Category 1 is for the measurement of flexural creep modulus to better than ±3 per cent, Category 2 for measurement poorer than ±3 per cent and Category 3 where the degree of accuracy depends on the purpose for which the data are required (and therefore must be stated in the presentation of the data) but where the general principles of the previous categories are still applied.

Subsection 1.1.3 deals with the measurement of Poisson's ratio, which is one of the four fundamental parameters required to describe uniquely the deformation behaviour of an isotropic, linear, viscoelastic material (the other three being the bulk, Young's and shear moduli). As only two of these four parameters are independent for a given material, the measurement of Poisson's ratio and tensile creep modulus, which can be carried out simultaneously on the same apparatus, enables an isotropic linear viscoelastic material to be fully characterised. Two strain gauges making measurements in the longitudinal and transverse directions are required for this test and the same classification of strain gauge precision as described in Subsection 1.1.1 also applies.

USA Standards

The standard test method is ASTM D2290 (1977),[2] which covers the measurement of creep in tension, compression and flexure. Creep rupture of plastics is also included.

The experimental procedures follow much the same pattern as the ISO and BS Standards previously mentioned, although the test temperature control is only to within ±2°C which is rather surprising. For creep rupture tests it is recommended that a minimum of seven stress levels be selected in order to give failure in approximately 1, 10, 30, 100, 300, 1000 and 3000 hours. The standard points out that for simple material comparisons, such as for data sheets, a useful parameter to determine is the stress to give 1 per cent strain at 1000 hours. This is obtained from the 1000 hour isochronous stress–strain curve, usually by interpolation. The effect of temperature on the material is readily shown via this parameter by carrying out the test over a suitable temperature range.

A creep rupture test specifically designed for plastics pipe is ASTM D1598 (1976)[4] in which a length of pipe is pressurised by means of liquid or gas, as appropriate, and the time to failure is recorded.

German Standards
There is only one creep test for plastics in Germany, that being DIN 53444 (1968)[5] which covers tensile creep only and which follows the same general procedures as described above.

11.2.4 Practical Systems

Having looked at the standard test methods available, it is appropriate to examine briefly some practical systems that have been used to produce creep data. More detailed information on creep, its measurement and the effects of molecular parameters on it is given by Nielsen[6] and Ferry[7] and also by Gittus,[8] who provides a particularly detailed account of creep in materials generally, rather than plastics alone.

The simplest possible experimental technique is described by Findley[9] who studied creep in cellulose acetate specimens at $25\pm\frac{1}{2}$°C and 50 ± 2 per cent r.h. at periods up to 7000 h (about ten months). The stresses involved ranged between 3·45 MPa and 18·6 MPa (based on original cross-sectional area) and these were achieved by the application of simple dead loads of from 140 to 700 N to the specimens. Under these conditions extensions in the range 1–45 per cent were produced, and measurements of adequate accuracy were obtained with a travelling microscope, which was used to determine the distances between pairs of marks clamped to the specimens. As would be expected the specimens decreased in thickness significantly during the progress of the tests, and it was necessary to spring load the clamps to accommodate this reduction (see Figure 11.3).

Results of this experimental work are shown in Figure 11.4. As can be seen, this technique is capable of demonstrating the creep characteristics of this material over a range of stresses and deformations which might be relevant to consumer goods and non-critical applications, though the low creep region, of interest for serious engineering projects, is less adequately covered.

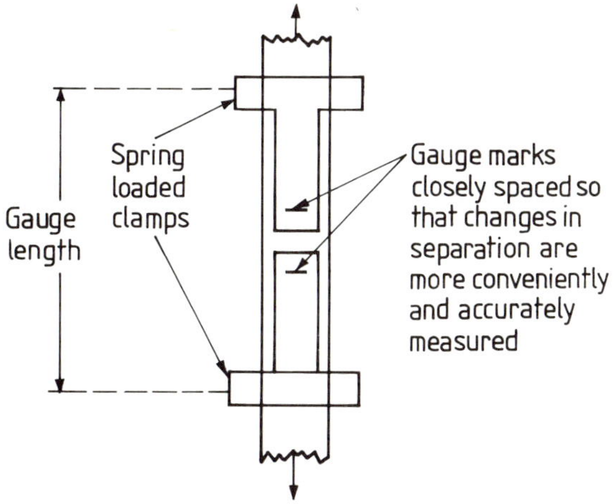

Figure 11.3 Arrangement of gauge marks[9]

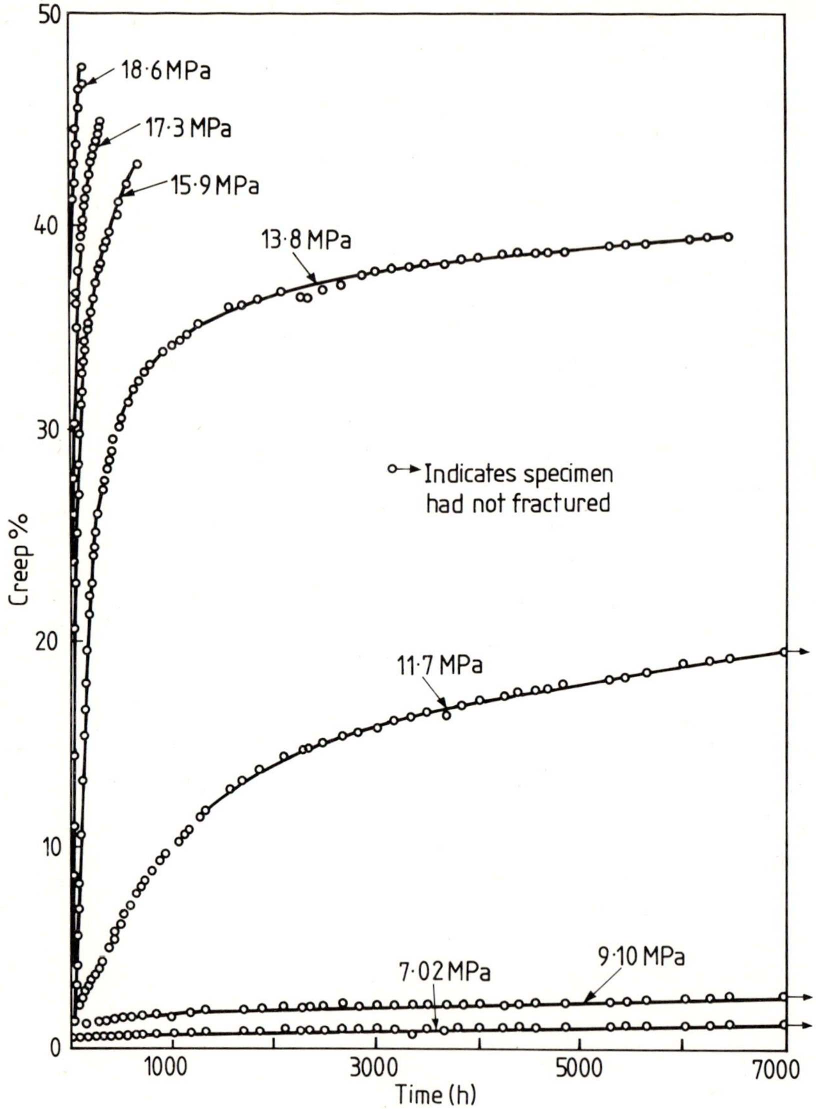

Figure 11.4 Creep–time diagram of cellulose acetate at 25°C[9] (*Reproduced by permission of McGraw-Hill Inc.*)

Thermosets generally creep less than thermoplastics and Telfair *et al.*[10] used a similar loading technique for their work on various filled phenolics, but improved the sensitivity of strain measurement by the use of resistance strain gauges. Two such gauges were bonded to each test specimen, one on each of the two wider faces (presumably to counteract the effect of fortuitous bending as well as increase the sensitivity of the measurement) and two similarly connected gauges, bonded to an unstressed specimen, were included in the measuring circuit to minimise the effect of minor temperature changes. With this system, strains as small as 0·005 per cent (0·000 05 mm/mm) were 'easily detected'. Various test temperatures and stresses were investigated, resulting in creep extensions of less than 1 per cent at times up to 1000 h. The work also paid attention to the recovery of the specimens after the removal of the stressing loads (just after completion of 1000 h); a typical creep and recovery curve is shown in Figure 11.5.

Subsequently Staff *et al.*[11] reported their examination of filled phenolics, using techniques similar to those described above, but extending the time scale to 5000 h at least, and 14 500 h in some cases. The description of their work is distinguished by their practical comments on the use of strain gauges and the precautions they took to avoid erroneous results from the gauges. They also comment on the prediction of performance of the materials they examined for service of up to five years.

More elegant techniques of creep determination involve the use of extensometers based on the optical lever principle and lever loading systems. The two basic optical levers, both developed for conventional modulus measurements, are those of Lamb and Martens and they are adequately described by Fenner[12] who indicates that they will detect extensions of a few millionths of an inch. Dunn *et al.*[13] modified the conventional Lamb extensometer so that it was more suitable for use with thermoplastics (which suffer extensions rather greater than those of metals, but are less able to support the weight of an extensometer or the stresses which have to be applied to locate the knife edges). A schematic diagram of this modified extensometer is shown in Figure 11.6.

As can be seen, the instrument consists of two pairs of knife edges affixed one to each of four light rigid members. As the pairs of knife edges move apart one

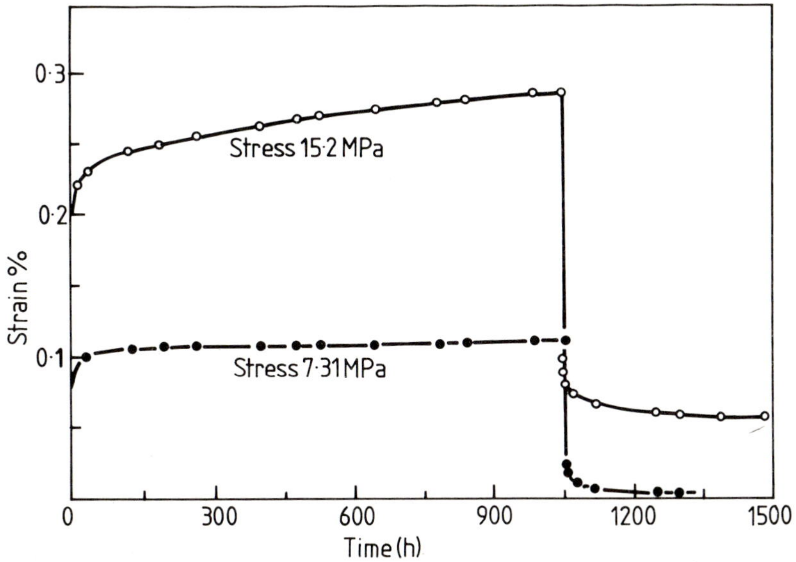

Figure 11.5 Creep and recovery of wood-filled phenolic at 25°C[10] (*Reproduced by permission of McGraw-Hill Inc.*)

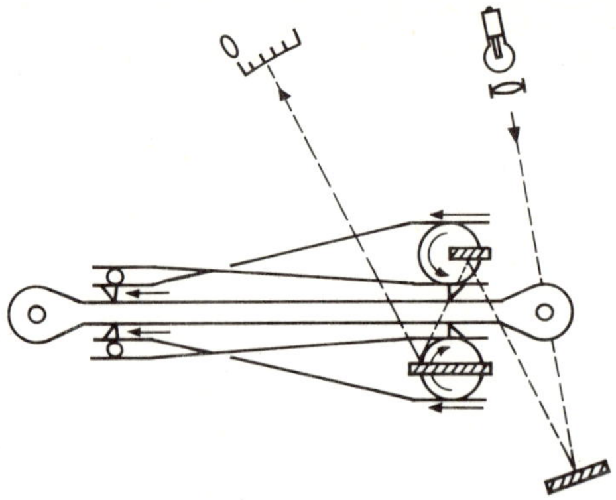

Figure 11.6 Schematic diagram of modified Lamb extensometer[13]
(*Reproduced by permission of 'British Plastics'*)

pair from the other, due to extension of the gauge length of the specimen, two rollers are caused to rotate by relative motion of the rigid members. The rollers carry small plane mirrors, their rotation resulting in deflection of a light beam reflected by both mirrors in turn on to a circular scale which is centred at the second mirror. The components of the extensometer are held together, and the knife edges are held in contact with the specimen, by sprung members approximately in line with the pairs of knife edges. It is said that when the scale is 500 mm from the final mirror and the gauge length is 80 mm a strain of 0·01 mm/mm results in a 128 mm displacement of the reflected light beam.

In the same paper there are descriptions of two other extensometers for use with materials which could not support the weight of the type described above: one for polyethylene which, while being optical in principle, requires the insertion of locating pins (which actuate levers) into the test specimen; the other for films or monofilaments, which is completely optical in principle, and involves the minimum of disturbance of the specimen. All three extensometers are used with single lever loading machines of the basic type shown in Figure 11.7.

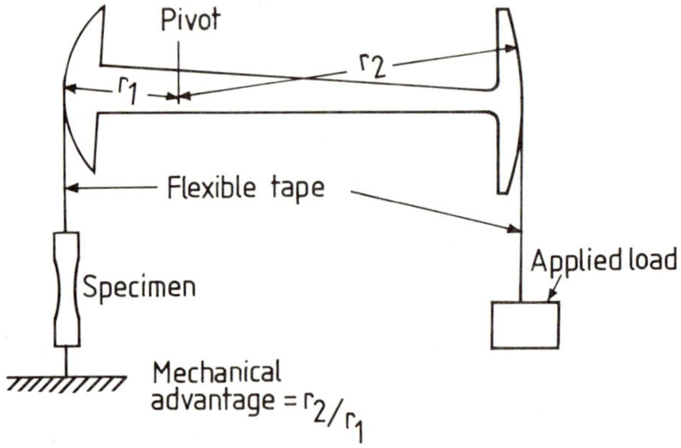

Figure 11.7 Typical single lever loading system

Considerable ingenuity has been applied to the problems of measuring creep extension and three further examples are cited. Mullen and Dolch[14] describe apparatus devised for use with Portland cement, but which could well be modified for use with low creep plastics. It is suitable for use with specimens immersed in liquids, or which are inaccessible for other reasons, as the indication is remote from the specimen itself. In essence the extensometer consists of two concentric lengths of fine hypodermic needle tubing, arranged and coupled to the specimen as shown in Figure 11.8.

The fiducial marks, shown inset, are cut with a single-edge razor blade while the tube is rotated in a watch-maker's lathe; it is claimed that if they are illuminated in a standard fashion their separation can be determined to an accuracy of 0·001 mm (0·000 040 in, i.e. 0·000 010 in/in on a 4 in gauge length) with an appropriate linear traverse microscope.

Creep apparatus designed and constructed by BTR Industries Ltd[15] was devised to enable materials to be investigated while they were immersed in various liquids at both above and below ambient temperatures, using horizontal specimens. Light Invar structures are mounted at the gauge marks and the separation of these is measured by a portable extensometer, consisting of two micrometers reading in units of 0·000 01 in (0·000 25 mm) mounted end to end, as shown in Figure 11.9.

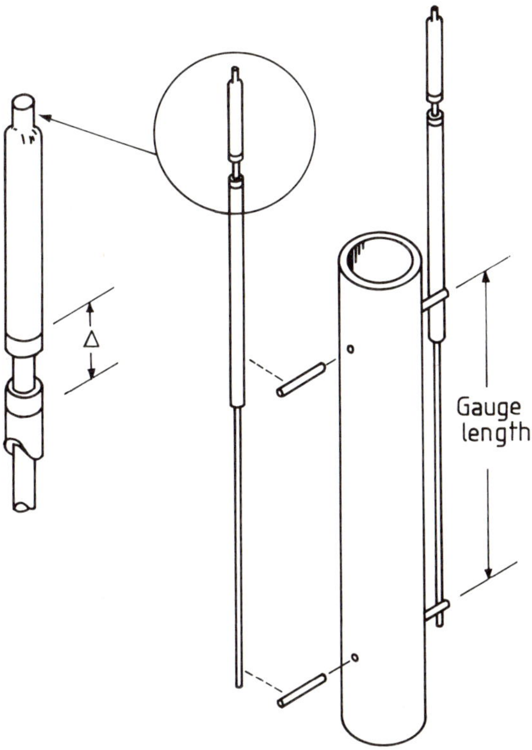

Figure 11.8 Periscope strain gauge[14]

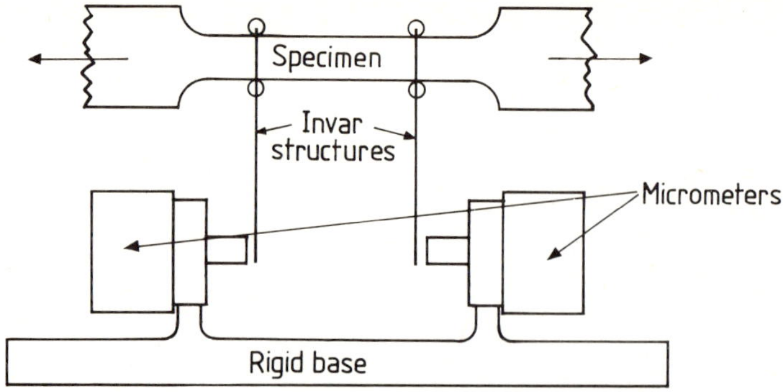

Figure 11.9 Micrometer extensometer

The micrometer drums are rotated until fine foil contacts on the ends of the spindles just touch vanes mounted on the Invar structures, this end point being indicated electrically by a suitable simple circuit. Because the extensometer is portable one of these relatively expensive units can serve several specimens, making a useful reduction in the capital costs of a creep test facility.

The third apparatus, which is also based on an optical system, is one developed at RAPRA and described in detail by Wright.[16] This uses a Moiré fringe extensometer and so, unlike the previous methods, is essentially digital in operation. This results in a number of advantages over analogue extensometers, in that it has long-term stability, no calibration is required, a 20 per cent strain limit with no loss in resolution is achievable and strain is the independent variable, i.e. the strain is automatically recorded at equal strain increments rather than equal (or logarithmic) time increments. The Moiré fringes are generated via two different gratings inclined at a small angle to each other. One grating is attached via a suitable carriage to one gauge mark of the test piece and the other grating to the other gauge mark. As the test piece extends, the Moiré fringe pattern moves over the surfaces of suitably positioned photocells which are connected to a counting circuit; as each fringe is detected, the time is recorded by means of a data logger. The apparatus, of Class A sensitivity (BS 4618), is illustrated in Figure 11.10.

Special difficulties arise if it is necessary to determine creep at constant stress—that is, if it is necessary to reduce the applied load in step with the reduction in area of the specimen as it extends. Andrade[17] stressed specimens by means of an appropriately shaped weight (hyperboloid of revolution) which was partially immersed in a liquid. As the specimen extended and decreased in cross-section the load was decreased by buoyancy of the liquid. Ward and Marriott[18] used a single lever machine, but the circular arc at the 'dead-weight' end of the lever in the conventional machine (as shown in Figure 11.7) was replaced by a specially profiled cam. In computing the shape of the cam it was assumed that Poisson's ratio was 0·5, which is approximately true for many plastics. No allowance appears to have been made for the fact that elongation and section reduction of the parts of the specimen immediately adjacent to the grips will be less than occurs in the central part of the specimen.

The application of the creep load has been given considerable attention. The

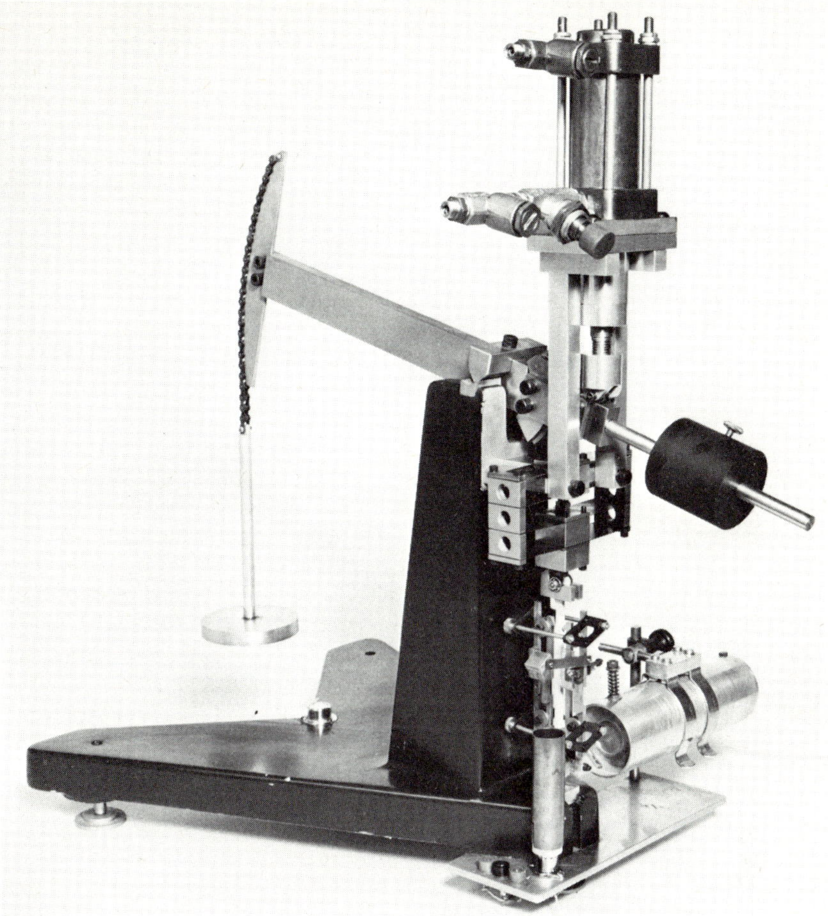

Figure 11.10 RAPRA–Hampden creep apparatus

investigator is required to apply the load rapidly (to avoid the theoretical embarrassment of a stress which increases over a significant period of time) and reproducibly but without shock or vibration. Findley[9] supported his dead weights 'on planks, blocked up in such a way that they could be used as levers to lower the weights quickly but gently until they were supported by the specimen . He then 'immediately' recorded extension and time. Dunn *et al.*[13] describe a system for automatic shock-free loading which can apply loads up to 30 kg without vibration, in times considerably below $\frac{1}{2}$ s. Telfair[10] increased the difficulties of interpretation by adding weights in 'convenient increments', taking strain readings at each increment so that modulus data could be obtained. The advantages of knowing the modulus for the actual creep test specimen may be considerable, but this technique must considerably increase the difficulties of mathematical interpretation of the results. The RAPRA apparatus uses a hydraulically actuated piston which applies the load in approximately 1 s.

The examination of specimens in compression is subject to the difficulties which apply to compression tests in general, described in Chapter 8. Of these, the tendency of the specimens to buckle is the most serious, as this would obviously invalidate the measurements of small creep deformations. Dunn *et al.*[13] and Thomas[19] suggest the use of a hollow cylindrical specimen to prevent or reduce buckling, while Jones *et al.*[20] suggest the use of a square-section specimen which is almost totally enclosed in a close-fitting tube (the method is unsuitable for large deformations). In all three cases the intention is to use a specimen which is sufficiently slender to be stressed without the use of excessively large loads. Decrease in specimen length is assumed to be directly related to the relative movement of the compression platens (which is recorded by dial gauge) in the work reported by Jones *et al.* and Dunn *et al.*, although the latter do indicate that this assumption is of doubtful validity and recommend the use of extensometers; Thomas shows that, for some materials at least, the error associated with the use of platen-mounted dial gauges is unacceptably large.

Little published work appears to have been carried out in flexure, although Hammant and Roberts[21] describe a novel technique in which the centre-loading nose has the same radius as the radius of curvature of the test piece, the load being rapidly applied via electromagnets. Electromagnetic actuators were also employed by Bergfjord and Penwell[22] in their development of a tensile creep instrument for thin films for which normal extensometers cannot be used. Direct movement of the lower grip, measured with a differential transformer, was used to indicate the strain of a parallel-sided test piece.

Finally, it is perhaps worth mentioning that a number of more commonplace tests, such as hardness tests (load applied for perhaps 30 s), indentation tests on floor tiles, etc., are really elementary creep tests, insofar as they require the application of a constant load to a specimen and the measurement of the resulting deformation. Indeed, Gonzalez[23] has taken this to its logical conclusion and modified a Rockwell hardness tester to set up a screening test for the torsional creep of plastics. Correlation between the degree of penetration and the (measured) torsional creep appears to be good in 11 out of the 13 materials examined, the two exceptions being reinforced plastics.

11.3 STRESS RELAXATION

In stress relaxation tests, instead of holding the stress (or, more correctly, the load) constant and measuring strain as a function of time, the strain is fixed and the stress is measured. By contrast to creep tests, stress relaxation is seldom measured in plastics materials and this is reflected in the paucity of standard test methods. This situation is reversed for elastomers, as detailed by Brown.[24]

Nielsen[9] describes a simple, basic test equipment for the measurement of stress relaxation, and indicates that simple instruments are suitable for use with low modulus (and, by inference, highly extensible) polymers (Figure 11.11). Rigid polymers present much greater difficulties, and for accurate work the apparatus must be very stiff because otherwise its deformation may be comparable to that of the polymer. Nielsen cites a relaxation experiment in which a rigid specimen, 1 in long, stretched by 0·001 in. The stress-measuring device

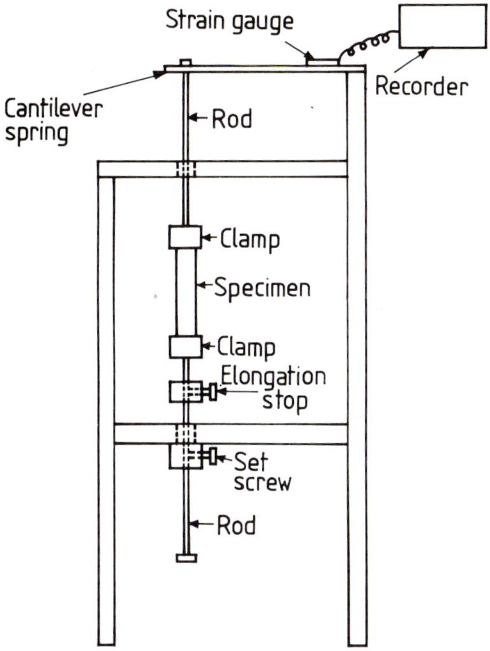

Figure 11.11 Simple apparatus for measuring stress relaxation[6]

(and the apparatus) must not deform by more than 1 in $\times 10^{-5}$ (2·5 mm $\times 10^{-4}$) if stress measurement accurate to 1 per cent is to be achieved.

A more detailed description of a specific instrument is given by Curran, Andrews and McGarry.[25] This consists essentially of a rectangular aluminium frame, which is several orders of magnitude more rigid than the test specimens, with the specimen stretched along its longer axis. The requisite stress is applied to the specimen via one grip which is positioned by a keyed micrometer screw passing through one short frame member, and the (initial) strain in the specimen is indicated during the setting up process by a clip-on extensometer, which is subsequently removed. The stress is recorded by a group of four resistance strain gauges which are bonded to a thin strip steel tensile member, which restrain the lower grip. Typical results are shown in Figure 11.12. This method is suitable for tests involving liquid environments, and a suitable environmental cell is described by the authors.

For the most accurate work this technique would not be acceptable because there are two mechanisms whereby the strain in the specimen could change during the course of the test. The strip steel load weighing member will decrease in length, if only marginally, as the stresses in it (and the specimen) decrease. More significantly, no attention is paid to the changes in strain distribution within the specimen as the test progresses. If one considers a typical dumbbell specimen it is obvious that the stress will be greater in the narrow portion of the length. Under certain conditions of stress, this narrow portion will creep, permitting the larger sectioned parts of the specimen to retract, and thus invalidating the concept of constant strain. This type of mechanism is also possible in

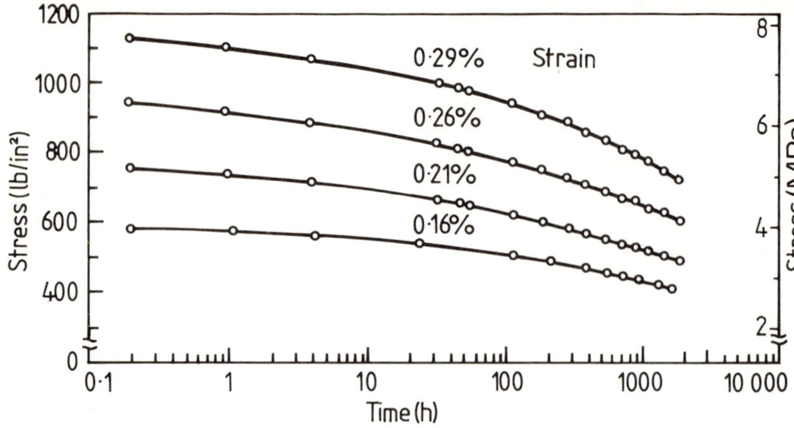

Figure 11.12 Typical stress relaxation data for 'multiphase' thermoplastics[25] (*Reproduced by permission of 'Modern Plastics', McGraw-Hill Inc.*)

parallel-sided specimens, as their stress distribution is far from uniform because of the disturbance caused by the gripping members, although in this case a simple commonsense analysis is not possible.

More recently an apparatus suitable for testing plastics has been described by Kaufmann and Hofmann.[26]

The only standard method for the measurement of stress relaxation in plastics is given in ASTM D2991 (1971).[27] This recommended practice avoids giving detailed test procedures but does indicate environmental and test piece tolerances as well as giving guidance on the measurement of load and strain, and the presentation of data.

ASTM D2552 (1969),[28] which deals with stress rupture, is more properly regarded as a fatigue test and is discussed in Chapters 12 and 17.

11.4 SET

Set is the percentage residual deformation remaining in a test piece after the application of a given reference deformation for a given time followed by a given period of recovery after removal of the load required to maintain that deformation. It is, therefore, essentially a stress relaxation followed by a creep recovery test.

This is not a measurement which is ordinarily carried out on plastics materials although it is commonly used in the rubber industry, partly because the apparatus is very simple and partly because set appears at first sight to be the important parameter for judging sealing efficiency. Stress relaxation is, in fact, the more significant in this context.

Set is dealt with in detail by Brown;[24] briefly, a 'normal' compression set test consists of compressing three discs of the material by 25 per cent in a rigid clamp, keeping this at perhaps 70°C for 22 hours, releasing the clamp and then measuring the thickness of the discs after a 30 minute recovery period. Compression set is then given by:

$$\text{compression set} = \frac{(\text{initial thickness}) - (\text{recovered thickness})}{(\text{initial thickness}) - (\text{compressed thickness})} \times 100 \text{ per cent}$$

Clearly, this is not a suitable test for rigid materials.

REFERENCES

1. BS 4618: Section 1.1 (1970) *Recommendations for the Presentation of Plastics Design Date: Creep*; Subsection 1.1.1 (1970) *Uniaxial Creep*; Subsection 1.1.2 (1976) *Creep in Flexure at Low Strains*; Subsection 1.1.3 (1974) *Creep Lateral Contraction Ratio (Poisson's Ratio)*.
2. ASTM D2990 (1977) *Tensile, Compressive and Flexural Creep and Creep-rupture of Plastics*.
3. ISO R899 (1968) *Determination of Tensile Creep of Thermoplastics*.
4. ASTM D1598 (1976) *Time to Failure of Plastic Pipe under Constant Internal Pressure*.
5. DIN 53444 (1968) *Testing of Plastics: Tensile Creep Test*.
6. Nielsen, L. E. (1965) *Mechanical Properties of Polymers*, Reinhold.
7. Ferry, J. D. (1970) *Viscoelastic Properties of Polymers*, Wiley.
8. Gittus, J. (1975) *Creep, Viscoelasticity and Creep Fracture in Solids*, Applied Science.
9. Findley, W. N. (1942) *Modern Plastics*, **19** (12), 71.
10. Telfair, D., Carswell, T. S. and Nason, H. K. (1944) *Modern Plastics*, **21** (6), 137.
11. Staff, C. E., Quackenbos, H. M. and Hill, J. M. (1950) *Modern Plastics*, **27** (6), 93.
12. Fenner, A. J. (1965) *Mechanical Testing of Materials*, Philosophical Library.
13. Dunn, C. M. R., Mills, W. H. and Turner, S. (1964) *British Plastics*, **37** (7), 386.
14. Mullen, W. G. and Dolch, W. L. (1966) *Materials, Research and Standards*, **6** (4), 191.
15. Cameron, J. B. (1967) *Transactions of the Plastics Institute, London*, **35** (119), 681.
16. Wright, D. C. (1971) *RAPRA Bulletin*, **25** (6), 133.
17. Andrade, E. N. da C. (1910) *Proceedings of the Royal Society, A*, LXXXIV, 1.
18. Ward, A. G. and Marriott, R. R. (1948) *Journal of Scientific Instruments*, **25** (5), 147.
19. Thomas, D. A. (1969) *Plastics and Polymers*, **37** (131), 485.
20. Jones, E. D., Koo, G. P. and O'Toole, J. L. (1966) *Materials, Research and Standards*, **6**(5), 241.
21. Hammant, B. L. and Roberts, J. (1967) *Engineering Materials and Design*, December, 1922.
22. Bergfjord, J. A. and Penwell, R. C. (1977) *Modern Plastics*, **54** (3), 84.
23. Gonzalez, H. (1976) *Journal of Testing and Evaluation*, **4** (1), 21.
24. Brown, R. P. (1979) *Physical Testing of Rubbers*, Applied Science.
25. Curran, R. J., Andrews, R. D. and McGarry, F. J. (1960) *Modern Plastics*, **38** (3), 142.
26. Kaufmann, W. and Hofmann, D. (1969) *Kunststoffe*, **59,** 173.
27. ASTM D2991 (1971) *Standard Recommended Practice for Testing Stress Relaxation of Plastics*.
28. ASTM D2552 (1969) *Environmental Stress Rupture of Type III Polyethylenes under Constant Tensile Load*.

Chapter Twelve

Fatigue

12.1 INTRODUCTION

Fatigue is the name given to the process whereby a component fails after some period of time throughout part, or the whole, of which it has been exposed to loads of lesser magnitude than the static ultimate load value. As such it should include creep and stress relaxation failures but these are normally thought of as separate from fatigue which ordinarily, though not exclusively, is associated with the action of fluctuating stresses. Creep and stress relaxation have been dealt with in the previous chapter.

The phenomenon of fatigue bedevils probably the whole field of materials science: virtually all materials—natural and synthetic—suffer from its effects. Apart from metal fatigue,[1] which must be known to almost everyone, wood,[2] concrete,[3] bituminous materials[4] and, of course, plastics[5–8] all come under the umbrella of this pervasive 'disease'. It has been estimated[9] that some 80–90 per cent of all service failures of equipment are due to fatigue.

Even for the traditional engineering materials such as metals, for which fatigue has been a recognised problem for over a century, the fundamental causes of failure are still not fully understood, with all too familiar and often tragic consequences. It should come as no surprise to learn, therefore, that for the newer plastics materials the fatigue processes are only just beginning to be understood and that fatigue data are very sparse. A considerable amount of work is being done to fill this gap but it is unlikely that plastics design engineers will be provided for many years to come with fatigue data comparable in quality or quantity to those available to engineers using more traditional materials.

Despite the volume of work being undertaken to map the fatigue properties of plastics, very little is carried out to any nationally, much less internationally, recognised test method. Since only two such methods, one American and one German, exist at present specifically for plastics materials, this is understandable. This is in marked contrast to fatigue testing of elastomers for which a number of standard test methods exist, as described in detail by Brown.[10] Fatigue testing of metals is also well covered in the standards literature; in particular, ISO R373 (1964),[11] BS 3518: Part 1 (1962)[12] and ASTM E206 (1972)[13] contain much valuable information relating to terminology, data handling and test methods. Test methods and procedures developed for metals can be, and have been, adapted for use with plastics but care must be exercised since metals are low damping, high thermal conductivity materials. Deformation frequencies for plastics testing must, therefore, be much less than for metals testing, as a rule.

The fatigue properties of plastics are often considerably influenced by the medium in which the plastics material is situated. A classic example of this is the well known phenomenon of environmental stress cracking. This effect and the test methods for studying it are considered in Chapter 17.

12.2 CYCLIC TESTS

12.2.1 General Considerations

A large number of cyclic deformation patterns is possible but the data are almost invariably presented as the number of cycles to failure at a given maximum stress level—the so-called *S–N* curve (Figure 12.1). From plots such as this the fatigue life and fatigue limit may be derived.

The fatigue life, or endurance, is the number of cycles to failure at a specified stress level. The greater the applied stress, the fewer cycles needed to break the test piece. Generally, the fatigue life of a polymer is reduced by increasing the temperature.

The fatigue limit, or endurance limit, is the stress condition below which the material may endure an infinite number of stress cycles without failure. For most real materials the fatigue limit is not as clear-cut as depicted in Figure 12.1 and estimates based on statistical procedures are all that can be obtained. For many polymers the fatigue limit is between 25 and 30 per cent of the static tensile strength.[14]

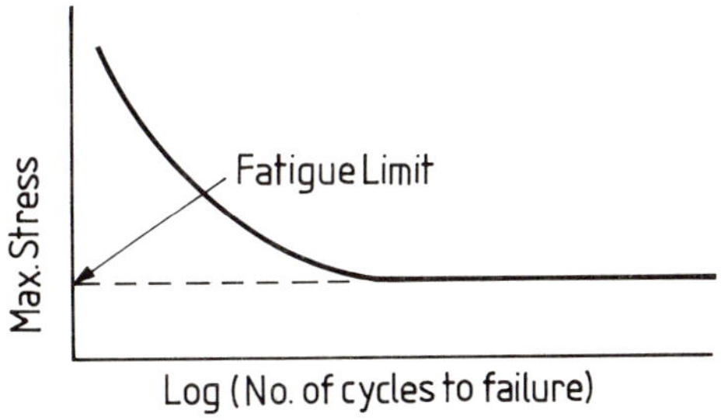

Figure 12.1 Typical *S–N* fatigue curve

The deformation imposed may be tensile, compressive, flexural or shear or any combination of these, generally with sinusoidally varying stress, since this is the easiest to generate, although saw-tooth and square-wave functions may be obtained without undue difficulty. Tests may be carried out with constant amplitude of cyclic stress or constant amplitude of cyclic strain. The latter method has the advantage of ease of setting up but also has disadvantages: stiff materials are subjected to greater stresses than soft (i.e. low modulus) materials—it is, therefore, a more severe test for such materials; furthermore, once cracks develop the imposed stresses lessen, so that often a considerable number of cycles is required to break the test piece. In constant stress tests, on the other hand, once cracks start to develop failure usually follows very quickly.

Further variants on the above methods are to superimpose the oscillating stress (or strain) on to a static stress (or strain) or to increase the amplitude of the cyclic stress (or strain) with time, as employed by Lazar.[15]

One popular test method, borrowed from the field of metals testing, is to rotate a cylindrical test piece about its axis of symmetry while the ends are constrained by bearings which are symmetrically misaligned. Figure 12.2 shows a schematic drawing of the apparatus. Each elementary volume of the test piece is subjected to a stress which varies sinusoidally between similar maxima of tension and compression. The frequency of test is readily altered, as is the overall applied stress. Lazar[15] describes tests made on PMMA, polystyrene and nylon using this technique while Romualdi *et al.*[16] describe a similar technique, but with the additional refinement that the test pieces can be subjected to a superimposed tensile load, so that alternating stresses varying from complete reversal to pure tension can be developed by varying the magnitudes of the axial and bending stresses. Findley *et al.*[17] developed another machine suitable for use down to temperatures of −196°C.

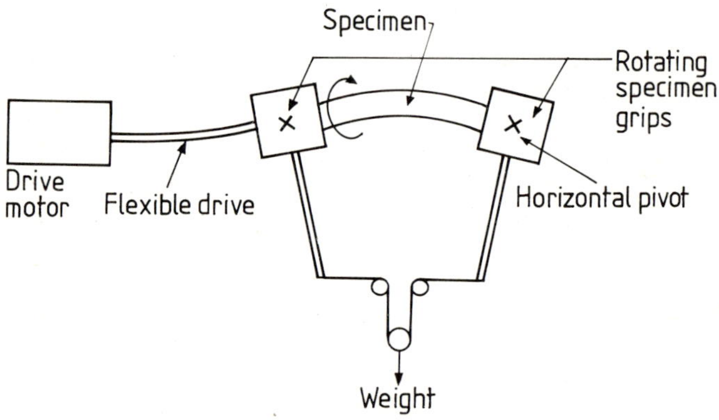

Figure 12.2 Basic design of rotational fatigue apparatus

Axially applied tensile and compressive stresses have been considered by Lazar[15] and Boller[18] and also by Owen,[19] who describes a tension fatigue testing machine in some detail.

Flexural tests (without rotation) can be performed on rectangular bars and may be carried out in any of the modes available to static flexural tests (§8.5). Thus, Gotham[5] and Cessna *et al.*[20] employed cantilever bending, while Lavengood and Gulbransen[21] used three-point loading and Hutchinson and Benham[22] used four-point loading. Flexural tests are somewhat easier to perform than tensile–compression tests since the problems of axiality are less severe, but the data obtainable are more limited. Reinforced thermosetting resins are almost invariably tested in tension–compression because of the technical superiority of this mode, while for thermoplastics (reinforced and unreinforced) flexural tests are often adequate.[6]

Another approach to fatigue testing, which is particularly applicable to flexible materials, involves the repeated severe bending of thin sheet specimens. An appropriate technique is described by Carey[23] and the De Mattia fatigue tester is readily used for plastics coated fabrics, although its use is more common in the rubber industry.[10]

12.2.2 Standard Test Methods

USA Standards
ASTM D671 (1971)[24] utilises the constant amplitude of force approach mentioned above. It describes terms relevant to the subject and indicates that the fluctuating stresses and strains need not be of constant amplitude or frequency as they can result from random variations. Two test pieces are described, to be selected according to the test piece thickness and stress range over which the measurements are to be made. Both specimens are of triangular form, with a rectangular cross-section, thus providing a uniform stress distribution over their respective test spans. Details of machining these rather awkwardly shaped test pieces are given, as sample preparation techniques can significantly influence the fatigue life. Having pointed out the importance of frequency of testing, the standard goes on to describe the apparatus which, in fact, only operates at a fixed frequency of 30 Hz. The setting up procedure also seems somewhat involved. Brief descriptions of producing *S–N* curves and of deriving fatigue life and fatigue limit data are given.

One of the major disadvantages of this apparatus is its inflexibility in terms of test frequency, which is considered by some workers to be too high for the majority of plastics materials and may explain why it is seldom used—in the UK at least. As already mentioned, plastics are relatively high damping, low thermal conductivity materials, which means that repeated straining of an article leads to a temperature rise generated within and throughout its body. If the stress–strain cycle is sufficiently rapid, significant heating of the article can occur, possibly leading to the glass transition boundary being crossed (see §16.4) and inducing thermal failure.[25,26] While the existence of this phenomenon is important to note, its quantitative effect is highly frequency-dependent and test frequency becomes an important parameter if the phenomenon is to be adequately evaluated. Where the thermal effect is to be minimised, and cooling of the test piece is inadequate since the heat is generated within the body and not on the surface of the test piece, much lower frequencies—of the order of a few Hz—must be employed. This contrasts very sharply with metals testing, where frequencies in excess of 30 Hz are common, although it has been observed[27] that carbon fibre reinforced plastics can be tested at over 100 Hz without significant heating.

German Standards
The German test method, DIN 53442 (1975),[28] by contrast, employs the constant amplitude of deformation mode in flexure and is more of a general test document than the ASTM standard—the test frequency is not specified and the comment is made that fatigue effects are frequency-dependent and that frequency should, therefore, be one of the parameters investigated. The apparatus briefly described and illustrated schematically in the standard is capable of operation over a range of deformation amplitudes and at non-zero mean strain if required. If this latter mode is selected, the standard explains, relaxation (see Chapter 11) of the test piece will mean that a non-constant mean stress level will

be experienced by the test piece during the test. Dumbbell-shaped test pieces are used, similar to tensile dumbbells but with the parallel-sided gauge section omitted.

One interesting difference from the ASTM standard is the requirement that surface temperature be recorded as a function of number of cycles as well as the stress levels (measured by means of a suitable load cell attached to the non-driven end of the test piece). Thus, in addition to the normal S–N curve, a plot of surface temperature against log N is also constructed.

The notes at the end of the standard explain that, while fatigue testing of plastics follows the same lines as fatigue testing of metals, seldom can a machine designed for use with metals be used on plastics without suitable modifications.

12.3 STATIC STRESS RUPTURE

Stress rupture tests are of a form similar to the long-term continuous stress tests described in Chapter 11. They are, in fact, creep tests divested of the onerous task of determining the associated (usually) small elongations. Specimens are subjected to a stress, often defined as some fraction of their ultimate short-term failing stress, and the time which elapses between application of load and ultimate failure is recorded. For short-time tests it is usually sufficient for the specimens to be observed by an operator with a stop watch or process timer; in the other extreme, a calendar may be sufficient! Tests of intermediate duration—days, weeks or months—may involve the use of more elaborate devices to ensure that the time of failure is recorded with sufficient accuracy. Stress rupture test results are influenced by temperature, vibration, etc., as are creep and stress relaxation tests.

Examples of this type of test occur only infrequently in standard specifications. It is perhaps appropriate that two 'stress rupture' tests described in British Standards are relevant to plastics pipes, as it was necessary to justify the use of plastics in this field by extensive tests of this nature. BS 3505 (1968)[29] and BS 3506 (1969)[30] both describe a 'long-term hydraulic test'. A specimen of pipe is maintained at a constant (internal) pressure so selected that the specimen will burst after between one and ten hours. Throughout the test the internal pressure is kept constant within ±2 per cent and the specimen is immersed in a water bath maintained at 20° ± 1°C. The time to burst is recorded. A second specimen is tested in a similar fashion, except that the applied pressure is selected to give a burst after between 100 and 1000 h. The minimum wall thickness and mean outside diameter of each specimen are determined before the test, and the circumferential stress in the specimen is derived from

$$P = \frac{2\delta t}{D - t} \qquad (12.1)$$

where P = pressure to be applied (bar)
δ = circumferential stress (bar)
t = minimum wall thickness (mm)
D = mean outside diameter (mm).

As a quality control procedure on a pipe production unit, the pipe manufacturer is required to record data for all pipe produced to the above-mentioned specifications, grouped according to size and class (pressure rating) of pipe.

Subsequently circumferential stress is plotted against time on a log–log scale, and a regression line drawn through all the production data is extrapolated to yield a 50-year (438 000 h) value for circumferential stress. Even on a logarithmic scale this is a large extrapolation, but in this case data exist[31] that support its validity for unplasticised PVC. Both BS 3505 and BS 3506 require that the 50-year value of burst (circumferential) stress should exceed 230 bar (23 MPa) for 7 in (178 mm) diameter and smaller pipes, and 260 bar (26 MPa) for larger pipes.

Subsection 1.3.1 of BS 4618 (1975)[32] deals with aspects of static fatigue failure, describing experimental procedures for deriving data suitable for use as design criteria.

A stress rupture test method similar to those in BS 3505 and BS 3506 is described in ASTM D2239 (1974).[33] ASTM D1598 (1976)[34] also applies. Another ASTM method, D2552 (1969)[35] is also worthy of mention. This describes a standardised method for submitting dumbbell specimens to continuously applied forces, while the specimens are immersed in a surface active agent which is maintained at a constant temperature. The time taken for specimens to fracture is recorded, and specimens showing ductile failures involving 'a noticeable degree of uniform cold drawing' are not considered in the assessment of results. The end-point of a given test (at one test–temperature combination) is defined as the 'probable time required for 50 per cent of the specimens to fail in a brittle manner' (see also Chapter 17). Finally, when considering specifications it should be remembered that the guidance given in ASTM D2990 (1977)[36] and DIN 53444 (1968),[37] directed specifically towards creep tests (§11.2), is almost all equally applicable to 'stress rupture' tests; ASTM D2991 (1971)[38] is of general use in listing certain essential test requirements.

For most situations apparatus and techniques used for creep tests are equally applicable to stress rupture measurements. Carey[39] describes a simple laboratory apparatus in which horizontally mounted tensile specimens (of polyethylene) are stressed by a simple multiplying lever. Elongations of up to about 20 per cent are possible in the apparatus, and the specimen is immersed, if required, in a temperature-controlled liquid. Numerous results are shown graphically.

Millane[40] has examined a series of glass reinforced plastics laminates, based individually on melamine, epoxide, phenolic, silicone and polyester resins. As an essential preliminary, the flexural strengths of groups of three specimens, 6 in long and $\frac{1}{8}$ in thick (152 × 3 mm) were determined at a 5 in (127 mm) span. Subsequently single specimens of similar size were stressed in the same configuration, under various applied loads, in air or water at various temperatures. Stress levels were selected to give flexural stress–rupture time curves extending to 1000 h; examples are shown in Figure 12.3.

Millane's paper describes some practical aspects of the requisite apparatus, including an electro-mechanical timing device. However, simple elapsed time counters are probably the most suitable equipment for this duty, and it is relatively simple to arrange for a counter to be switched off at time of fracture of the specimen.

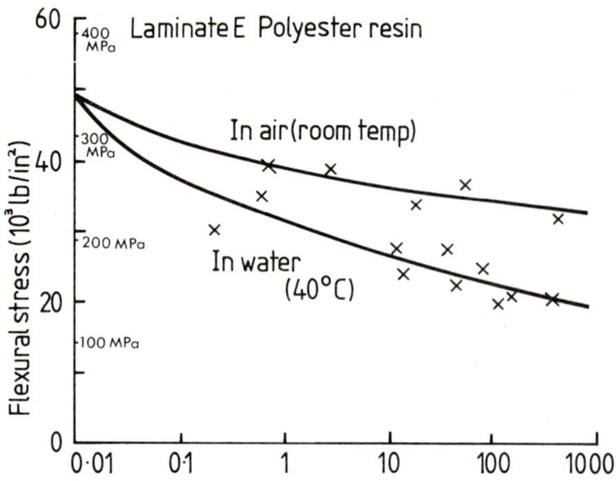

Figure 12.3 Typical stress rupture data for glass cloth reinforced laminates[40] (*Reproduced by permission of 'British Plastics'*)

Static stress–rupture measurements are very often used as a measure of resistance to environmental stress cracking (§17.3.3).

REFERENCES

1. Forrest, P. G. (1962) *Fatigue of Metals*, Pergamon.
2. Lewis, W. C. (1946) *ASTM Proceedings*, 46, 814
3. Ople, F. S. and Hulsbos, C. L. (1966) *Proceedings of the American Concrete Institute*, 63, 59.
4. Pell, P. S. and Taylor, I. F. (1969) in *International Conference on Civil Engineering Materials, Southampton University*.
5. Gotham, K. V. (1969) *Plastics and Polymers*, 37, 309.
6. Turner, S. (1972) *Mechanical Testing of Plastics*, Iliffe.
7. Nielsen, L. E. (1965) *Mechanical Properties of Plastics*, Reinhold.
8. Schultz, J. M. (1977) *Treatise of Materials Science and Technology* vol. 10, Part B, Academic Press, page 599.
9. Bennet, J. A. and Quick, G. W. (1954) *NBS Circular 550*, September.
10. Brown, R. P. (1979) *Physical Testing of Rubbers*, Applied Science, Chapter 12.
11. ISO R373 (1964) *General Principles for Fatigue Testing of Metals*.
12. BS 3518: Part 1 (1962) *General Principles (of Fatigue Testing of Metals)*; Part 5 (1966) *Guide to the Application of Statistics*.
13. ASTM E206 (1972) *Definition of Terms Relating to Fatigue Testing and the Statistical Analysis of Fatigue Data*.
14. Lazan, B. J. and Yorgiadis, A. (1944) *STP No. 59*, ASTM, page 66.
15. Lazar, L. S. (1957) *ASTM Bulletin*, 220, 67.
16. Romualdi, J. P., Chang, Chiao-Lio and Peck, C. F. Jr (1954) *ASTM Bulletin*, 200, 39.
17. Findley, W. N., Jones, P. G., Mitchell, W. I. and Sutherland, R. L. (1952) *ASTM Bulletin*, 184, 53.
18. Boller, K. (1964) *Modern Plastics*, **41** (10), 145.
19. Owen, M. J. (1967) *Journal of the Plastics Institute*, 35, 353.

20. Cessna, L. C., Levens, J. A. and Thomson, J. B. (1969) *Polymer Engineering and Science*, **9** (5), 339.
21. Lavengood, R. E. and Gulbransen, L. B. (1969) *Polymer Engineering and Science*, **9** (5), 365.
22. Hutchinson, S. J. and Benham, P. P. (1970) *Plastics and Polymers*, **38** (134), 102.
23. Carey, R. H. (1955) *ASTM Bulletin*, 206, 52.
24. ASTM D671 (1971) *Flexural Fatigue of Plastics by Constant Amplitude of Force.*
25. Riddell, M. N., Koo, G. P. and O'Toole, J. L. (1966) *Polymer Engineering and Science*, 6, 363.
26. Constable, L., Williams, J. G. and Burns, D. J. (1970) *Journal of Mechanical Engineering and Science*, 12, 20.
27. Owen, M. J. and Morris, S. (1970) *SPE Conference* Section 8E.
28. DIN 53442 (1975) *Testing of Plastics: Fatigue Test in the Field of Bending Strain of Flat Specimens.*
29. BS 3505 (1968) *Unplasticised PVC Pipe for Cold Water Services.*
30. BS 3506 (1969) *Unplasticised PVC Pipe for Industrial Uses.*
31. Richard, K. and Diedrich, G. (1955) *Kunststoffe*, **45** (10), 429.
32. BS 4618; Section 1.3.1 (1975) *Recommendations for the Presentation of Plastics Design Data: Static Fatigue Failure.*
33. ASTM D2239 (1974) *Standard Specification of Polyethylene (PE) Plastic Pipe (SDR-PR).*
34. ASTM D1598 (1976) *Time to Failure of Plastic Pipe under Constant Internal Pressure.*
35. ASTM D2552 (1969) *Environmental Stress Rupture of Type III Polyethylenes under Constant Tensile Load.*
36. ASTM D2990 (1977) *Tensile, Compressive and Flexural Creep and Creep Rupture of Plastics.*
37. DIN 53444 (1968) *Testing of Plastics: Tensile Creep Test.*
38. ASTM D2991 (1971) *Stress Relaxation of Plastics.*
39. Carey, R. H. (1958) *Industrial Engineering and Chemistry*, **50** (7), 1045.
40. Millane, J. J. (1960) *British Plastics*, **33** (5), 199.

Chapter Thirteen

Electrical Properties

13.1 INTRODUCTION

The tests described in this chapter are principally material tests, although some tests for semi-finished products such as sheet, film and tape are covered. Tests on cables and finished products are not specifically included as these become rather specialised, although the principles of the tests mentioned in this chapter are often applicable to products.

Reference is made to tests drawn up by the International Electrotechnical Commission (IEC) since the ISO does not draw up electrical standards when adequate IEC standards exist.

No attempt is made to describe the detailed phenomena underlying the electrical tests. The theory of electrical behaviour is described by Baird[1] and Blythe[2] (general), and by Daniel[3] (dielectric properties).

Fairly elementary descriptions of dielectrics and dielectric processes, mostly with special reference to polymeric materials, are given by Moullin,[4] Swiss and Dakin,[5] Hoffman,[6] Vail,[7] Sharbaugh[8] and Devins and Sharbaugh.[9] These references are interesting, not least for the variety of approaches to the subject adopted by the authors.

Broader descriptions of electrical insulation properties and test methods considered mainly from the application viewpoint are given by Mason,[10] Stark[11] and Baker.[12]

A number of useful chapters[13–22] dealing with electrical properties are to be found in more general books on polymers.

13.2 WHY MEASURE ELECTRICAL PROPERTIES?

The purpose of electrical insulation may generally be described as the prevention of unwanted contacts between electrical conductors at different potentials.

A simple example such as two mounted terminals can usefully serve as an illustration. The insulation must maintain the separation under a wide variety of environmental hazards which include humidity, temperature, vibration, radiation, the presence of gases, moisture and other contamination. It must be sufficiently strong physically to withstand the mechanical forces which may be exerted by the conductors and at the same time must prevent any significant flow of current between the conductors.

No insulator is perfect, but what constitutes a significant flow of current depends upon the application. A few microamps in an electrical supply circuit would be negligible, but in an electrometer input terminal could ruin any measurement. In every case the current must be sufficiently small to eliminate secondary effects such as temperature rise, mechanical stress and electrochemical action. A significant current may arise in various ways.

When a small or moderate d.c. voltage is applied across an insulator, a current flows. To minimise this the material must have an adequately high resistivity (§13.3). The current may be predominantly through the volume of the material, when the property involved is the *volume resistivity*. On the other hand, the current may be predominantly through a surface layer of different composition (e.g. due to adsorbed materials) and the property involved is then the *surface resistivity*. Since surface resistivity is often related to contamination, it is not generally a material property, although material properties may influence contamination and hence surface properties.

It is sometimes useful, especially in testing products, to measure the *insulation resistance*, which is the resistance between two defined electrodes, no differentiation being made between the passage of current along the surface and that through the bulk of the material (§13.4).

There are uses of plastics where the passage of a d.c. current is desirable, e.g. for the dissipation of static electrical charges, for heating, or in plastics potentiometers, and special tests are required for *antistatic* and *conductive* plastics (§13.5).

The ability of an insulator to 'resist' the passage of direct current is not necessarily synonymous with its ability to resist alternating currents; whereas in the former case the resistivity of the material is of prime importance, in the latter the *power factor* and *permittivity* (§13.6) are more relevant, since they determine the power loss occurring in alternating fields, which loss may vary enormously with applied frequency. These two properties are collectively known as *dielectric properties*. The power factor is due to the orientation of dipoles and the movement of charged ions in an alternating field. These motions, together with the movements of electrons with respect to such dipoles and ions and also to atoms, and of adjacent atoms with respect to each other when the field is applied, contribute to the permittivity.[1] In many insulation applications power losses are important (and undesirable) and thus low values of both power factor and permittivity are desirable. Since capacitance is proportional to permittivity, and in most cases the capacitance of a given conductor–insulator system is an unwanted effect which must be tolerated, the capacitance must be kept low by using low permittivity material. In screened cables, for example, of small diameter and hence low thickness of dielectric, it is fairly common practice to use dielectrics of cellular plastics materials to reduce the capacitance, and such materials may have effective permittivities approaching unity, the value for air. The problem of unwanted capacitance becomes more acute with increasing frequency.

The major field of application for dieletric materials where a high permittivity is the chief factor is in capacitors. Here the emphasis is continually towards reduction in physical size and, since the capacitance of a given electrode–dielectric system is proportional to the permittivity of the material (capacitance being defined as the ratio of stored charge to applied voltage), the importance of a high permittivity need hardly be stressed. As far as plastics materials as capacitor dielectrics are concerned, the available range of permittivity is approximately 2–20, so that the possibilities of size reduction by choice of permittivity alone are limited; these possibilities are reduced even further when power factor is considered, since virtually all the low power factor materials have permittivities at the lower end of the range. Capacitance is also inversely proportional to the dielectric thickness and so the only real direction of advance at present is in the use of thinner films; plastics foils of thickness 1 μm or less have been developed and used to a limited extent—limited because the problems of handling and manipulating such material are severe; some work is in progress on the vapour deposition of polymeric dielectric films *in vacuo*.

Power factor and permittivity are seldom significant in 250 volt power frequency cables, but become important when the frequency and/or voltage are raised.

It is sometimes desirable that the product of power factor and permittivity be high, so that the energy supplied by a high-frequency dielectric field may be converted into heat. In dielectric or microwave heating, it is possible to produce more rapid heating throughout a body than would be possible by heat conducted from the surface. In a related application, synthetic resin adhesives used in furniture may be cured by selective heating of the glue lines. The use of water-soluble adhesives such as those based on urea formaldehyde eases the problem, since the permittivity of water is approximately 80 and it also has a high power factor. High frequency fields are also used to weld, for example, two sheets of PVC together, where very rapid heating–cooling cycles can be attained. Another rather more specialised application is in energy-absorbent materials for use at microwave frequencies.

So far, we have discussed only weak field phenomena, and those properties which are fairly well defined and constant for given environmental conditions, and the tests themselves do not cause any irreversible alteration to the insulation. Under the influence of relatively strong fields, however, plastics materials may be permanently altered, either in form or character, and, under continual stress, degradation may occur and ultimately lead to failure.

The *electric strength* of a material (§13.7) is a measure of the ability of a material to withstand a high voltage, usually in the presence of discharges controlled by the rounded edges of electrodes, and is important for high voltage applications.

The tests for *resistance to surface discharges, tracking and arcs* (§13.8) relate to surface rather than bulk breakdown and emphasise the effects of discharges of various types on the surface of the test piece.

In the vicinity of high voltage conductors ionisation of the air occurs due to the high local electric stresses, and insulation subjected to prolonged exposure to the consequent discharges may fail through degradation or erosion. *Discharge resistance* is therefore of practical importance in high voltage bushings and insulators, which often utilise epoxide resins and other thermoset materials.

Tracking resistance is important even at low voltages since the presence of an aqueous contamination can cause localised stresses and result in the formation of a conductive carbonised track across the surface.

Arcs may be produced in many electrical devices, even at low voltages, and we need therefore to know how materials will withstand contact with an arc. In *arc resistance* tests, an arc produced between two surface-mounted electrodes is allowed to play along the surface of the insulation so that the predominant mode of failure is thermal degradation. However, the failure is accelerated by the surface carbonisation that occurs, resulting in current actually flowing within the surface layers of the material. This property is important in high current switchgear and circuit breakers, where arcs occur on making and breaking contacts.

Two other electrical tests, one for the *corrosion of copper* by adhesive tapes and one for the determination of *conductivity of water extract* are mentioned briefly in §13.9.

13.3 RESISTIVITY OF INSULATING PLASTICS

The methods used are different from those used for conductive and antistatic plastics which are described in §13.5.

13.3.1 Definitions and Qualifications

Volume resistance is the quotient of a direct voltage applied between two electrodes in contact with opposite faces of a specimen and the current between the electrodes after a specified time, excluding current along the surface, and neglecting possible polarisation phenomena at the electrodes.

Volume resistivity is the volume resistance reduced to a theoretical cube with unit side. It is the quotient of the nominal voltage gradient (applied voltage divided by thickness) and current density, although the true voltage gradient may differ from the nominal value because of contact effects.

Surface resistance is the quotient of a direct voltage applied between two electrodes on a surface of a specimen and the resulting current between the electrodes at a given time of electrification.

Surface resistivity is the surface resistance reduced to a theoretical square area; the size of the square is immaterial. It is the quotient of the nominal voltage gradient and the current per unit width of current path.

Both volume and surface resistivities depend on a large number of factors including the magnitude of the applied voltage, the time of application of the voltage, the nature and geometry of the electrodes, the temperature and humidity of the atmosphere and the moisture content of the test piece. The changes of surface resistivity may follow very rapidly (in seconds) upon any humidity changes. In addition there may be slow irreversible changes (especially of surface resistivity) due to chemical changes.

In explaining various facts it is often useful to use the terms *conductance* and *conductivity*, the reciprocals of resistance and resistivity respectively.

It is important to note that in any surface resistivity test, even though a guard plate is used, some of the current passes (between the two electrodes on one surface) through the volume of the material. Thus if the surface layer is not of lower 'resistivity' than the bulk or is sufficiently thin, the measured value of 'surface resistivity' is substantially a function of the volume resistance, and the 'surface resistivity' has no significance. This was discussed by Geppe.[23] With most standard tests a useful guide is to ignore surface resistivity results when the volume resistance is not greater than 10 per cent of the measured surface resistance or, what amounts to a similar criterion in most cases, where the volume resistivity in ohm cm is not numerically greater than the calculated surface resistivity in ohms. Samples which are thin (relative to the distance between the two electrodes on the same surface), are anisotropic, or have high contact resistances may give meaningless surface resistivity results.

13.3.2 Units, Ranges and Precision of Values

The basic SI unit of volume resistivity is the ohm metre (Ω m), derived from $\Omega\ m^2/m$. In practice a sub-multiple of the SI unit, the Ω cm, is more commonly used.

The volume resistivity of common insulating plastics covers a span of some ten decades between, very approximately, 10^8 and 10^{18} Ω cm; with the conventional size of specimen in sheet form, the corresponding resistances to be measured lie between 10^6 Ω and 10^{16} Ω. At the upper extremity of this range, above approximately 10^{17} Ω cm, lie the non-polar plastics, PTFE, polyethylene, polystyrene, etc., and measurements on these materials are considered sufficiently difficult (and the results perhaps of rather academic interest only) to be excluded from our present terms of reference. Dorcas and Scott[24] and Sazhin and Skurikhina[25] give further details.

The SI unit of surface resistivity is the ohm, Ω, derived from Ω m/m. It is sometimes referred to as 'Ω per square'.

Because of the variation of resistance of a given test piece with test conditions and because of non-uniformity of the same material from test piece to test piece, determinations of surface and volume resistivity are rarely reproducible to closer than ± 10 per cent and are often much more widely divergent (the ratio of the maximum and minimum values may be as high as 10). This has led some people to quote the logarithm of the resistivity, which is very useful in tabulation and graphical presentation, but clumsy in textual references. As a general guide to the accuracy expected from industrial materials, two materials would not be considered very significantly different unless their resistivities differed by at least an order of magnitude.[26] Measurements on at least two test pieces are considered essential.

13.3.3 Measurement Techniques

Only the measurements of resistance up to approximately 10^{15} Ω, which is the upper limit of commercially available resistance meters, will be considered, the normal lower limit of such instruments being of the order of 10^6 Ω. As surface resistance values for plastics are in general of the same order as, or lower than, the volume resistance for standard specimens, what follows applies to both properties. The techniques given below are suitable for materials with volume resistivities above about 10^8 Ω cm. For more conductive materials, see §13.5.

The usual measurement technique is best illustrated by considering the simple case of a two-terminal resistance, for example, a resistor, before turning to the additional complications involved with measurements on insulation.

In general the unknown resistance is connected in series with a current-measuring instrument across a high potential supply. The potential is high to provide the necessary sensitivity and is commonly of the order of 100–500 V d.c. The current-measuring instrument has historically been a sensitive galvanometer, but is now more commonly a valve or solid state circuit which measures the small voltage drop across its input resistance. This resistance is related to (may actually be) a known fixed value resistor, which may be switched to provide a wide series of ranges. The current is the voltage developed divided by the input resistance. The resistance of the 'unknown' is usually calculated by dividing the applied voltage by the current, since the voltage drop across the measuring instrument is generally less than 1 per cent and may be as low as 0·001 per cent of the applied voltage.

Multi-range megohmmeters are available which have built-in voltage sources and indicate the resistance directly. Factors which must be considered in purchasing a megohmmeter are considered in §13.3.4.

When measurements on insulating materials are considered it is usual in specification tests to use a sheet specimen in the form of a disc or a square. A two-terminal specimen consisting of such a disc with a suitable electrode applied to each surface would be simple to measure, but the current passing through the body of the material would be augmented by current passing over the surface and round the edge of the specimen, so that it would be impossible to separate the volume conductance from the total measured conductance. To obviate this difficulty a third electrode, called a guard ring, is placed concentrically around one of the existing electrodes (which is then known as the guarded electrode and which is reduced in size and separated from the guard electrode). By modified circuitry the desired volume conductance can now be measured direct, and further, by a simple interchange of connections, the surface resistance can be measured also, using the same specimen. The electrodes, current flows and theoretical circuits are shown in Figures 13.1–13.3, the symbols R_v, R_s and R_g representing the volume, surface and guard resistances, respectively, and R the input resistance of the measuring instrument.

In the circuits of Figures 13.2–13.3 the input resistance of the current measuring instrument is shunted respectively by the surface resistance in one case and the volume resistance in the other.

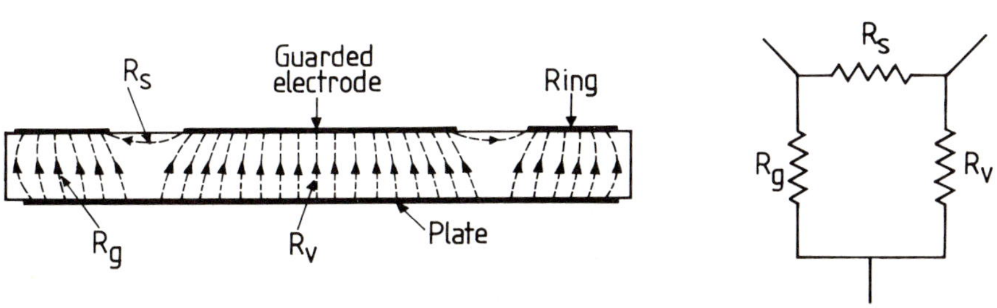

Figure 13.1 Electrode arrangement and equivalent circuit

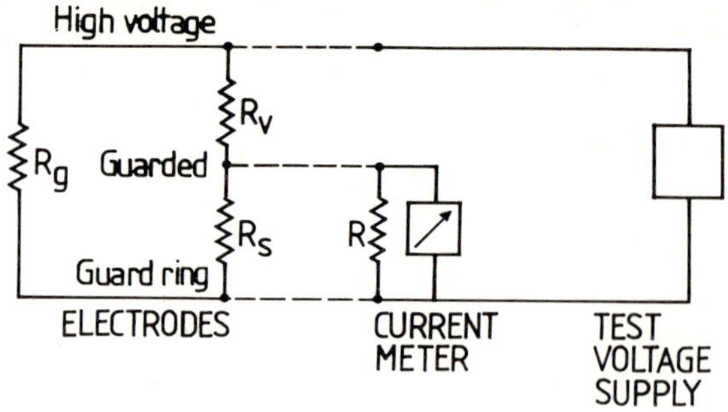

Figure 13.2 Volume resistivity—theoretical circuit

Suppose volume resistance is being measured (Figure 13.2). In many cases R_s is of the same order as R_v and if the voltage developed across the measuring instrument is small compared with the test voltage, R is much less than R_s and the current through R_s is negligible. Very occasionally R_s may be of the same order as R or even much less than R, so that much of the current which should go through the measuring instrument is diverted. It is therefore useful to be able, at the end of the test, to carry out a 'leakage check'.

For this, the guard is connected to a positive potential slightly greater than that developed across the current-measuring instrument during the resistance measurement (Figure 13.4 shows a suitable circuit). Any current through R_s will then augment the measured current, so that a significant change in meter reading will indicate a significant error. This can generally be overcome by cleaning the gap between the guard and guard ring. The corresponding change of potential of the guard plate is useful in a surface resistivity test (Figure 13.5), especially on thin test pieces, as it may indicate a conductive flaw through the test piece. If the screen of the lead from the guarded electrode is connected to the guard as shown, this procedure also checks the adequacy of the insulation on this lead. (It will be necessary to disconnect the screen on this lead at the connection to the current-measuring instrument if a facility for this check is to be incorporated.)

The resistance between the high voltage electrode and the guard electrode in both cases is a shunt across the supply voltage and therefore has no effect on the measurement, unless it lowers the applied voltage.

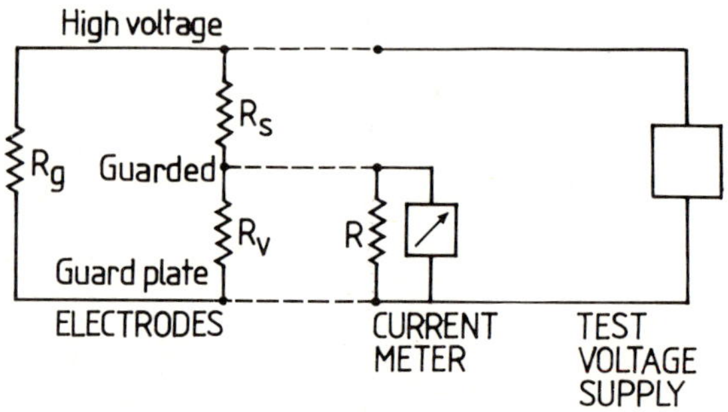

Figure 13.3 Surface resistivity—theoretical circuit

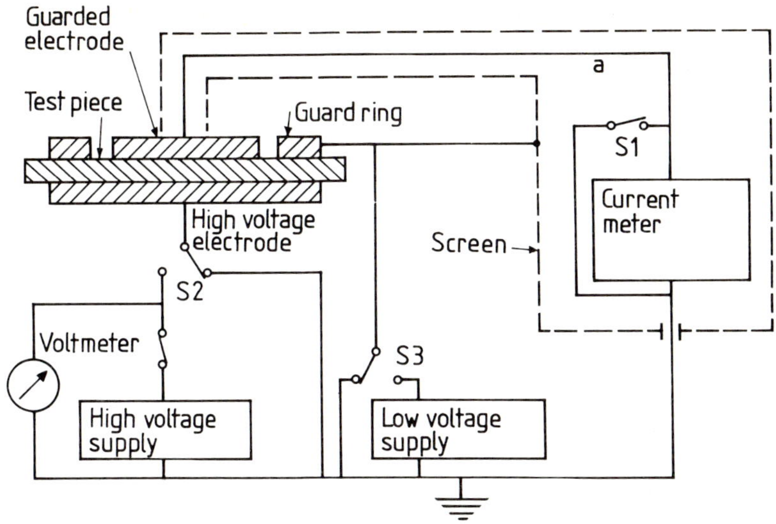

Figure 13.4 Volume resistivity—practical circuit

The circuits of Figures 13.4 and 13.5 have been shown separately for clarity but it will be obvious that the addition of a changeover switch in the leads to the ring and the bottom plate provides one circuit for both surface and volume resistivity tests.

A sample which has been charged, e.g. by accidental rubbing or by direct application of a voltage, may show a delayed discharge current or residual current and, to avoid error, this must be negligible compared with the current to be measured for the latter to be unaffected. A 'residual charge check' should be

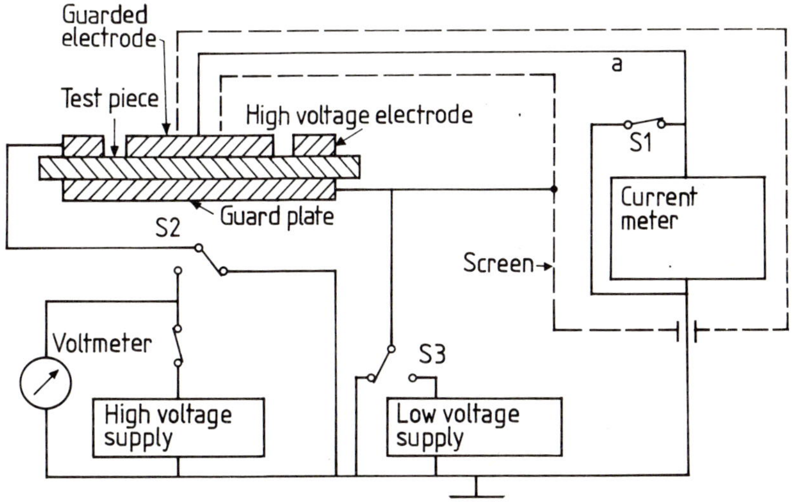

Figure 13.5 Surface resistivity—practical circuit

carried out before the test is started by connecting the sample to the measuring instrument, all other electrodes being earthed (in Figures 13.4 and 13.5 by leaving S2 and S3 in the positions shown and opening S1), and measuring the discharge current. If it later proves that the discharge current was not negligible, then all electrodes should be earthed for some time and the sample retested. Munick[27] has investigated transient currents in plastics due to residual charge effects.

13.3.4 Choice/Design of Test Equipment

It is not possible to buy a complete megohmmeter which includes all the facilities described above. However, the check-charge and check-leakage facilities can often be appended as small simple add-on systems.

The alternative is to buy a power supply unit and a picoammeter and to assemble the control circuitry. If this is done, care must be taken that all the circuit (leads, plugs, sockets, switches) connected with the guarded electrode (a, Figure 13.5) has an insulation resistance much above the input resistance of the current-measuring instrument and that this situation is not nullified by dirt or careless assembly. It is also essential that the above-mentioned parts of the circuit are adequately screened in a rather special sense. Not only must they be free of any electrostatically induced external influences, but any leakage path over or through any piece of insulation on this part of the circuit must terminate on part of the circuit connected to earth or preferably to the guard electrode.

In buying complete instruments or components, the following points should be checked.

(i) The applied voltage must be within the range required by the specification (commonly 100 V ± 10 V or 500 V ± 50 V) over the complete range of resistances to be measured.

(ii) The applied voltage must not vary during the test. The required stability is indicated in §13.3.6.

(iii) If surface or volume resistivity and not merely insulation resistance tests are to be carried out, the instrument must be suitable for use with a guard plate or ring.

(iv) It is necessary that the guard electrode and one side of the high voltage supply can be at earth potential.

(v) The voltage developed between the guarded and guard electrodes (i.e. across the current-measuring equipment) must be small. One volt should be regarded as a maximum although a few millivolts is preferable.

(vi) It should be possible to fit the 'residual-charge check' and 'leakage check' facilities.

(vii) The meter should be adequately protected from damage if the electrodes should short-circuit or a conductive test piece is accidentally introduced. This can be done by ensuring that the current-measuring instrument will withstand the full test voltage. This may not be possible on the higher current ranges, but a rapidly acting current-limiting system on the voltage supply can deal with this.

(viii) In addition to using the above check list, the purchaser should ascertain from the supplier that equipment is suitable in other ways for tests to the required specifications.

A suitable screened cable with high insulation resistance must be used for the connection to the guarded electrode. Ordinary screened cable can produce quite large amounts of electrical noise when flexed, but non-microphonic types are available in which the noise level is considerably reduced by the insertion of a semiconducting layer between the dielectric and the screening braid. It is essential that this layer be removed at the cable ends, preferably for at least 10 mm.

In the author's experience, the measurement of high resistivity values is beset with difficulties. To avoid uncertainty about the functioning of the measuring apparatus it is useful to have a set of high resistances to hand, ranging from 10^6 to 10^{13} Ω. Such resistances are now available with accuracies of 2 per cent or better, even for the higher values, but care should be taken with the selection, since the voltage coefficient is large. This is normally negative, so that a nominally 10^{13} Ω component can have a value much lower at 500 V than the manufacturer's reported value, usually measured at a few volts only. The manufacturer should be consulted for details of the voltage coefficient of his resistors. France[28] gives details of measurement of high value resistors.

Hitchcox[29] has written extensively on high resistance measurements and the use of electronic techniques for extending the upper limits of measurement.

13.3.5 Electrodes

Many types of electrode are available. Among the most widely used are the following: graphite, conducting paints, metallic foils applied with a thin layer of viscous liquid, vacuum-deposited metallic coatings and conducting rubber sheets. Most of these require solid metal backing plates to contact the electrodes proper, but whichever type is used, intimate contact with the specimen is essential and the choice will be dictated by environmental conditions.

Some specifications include mercury electrodes but the use of these is now deprecated because they present a toxic hazard. They should certainly not be used for tests at elevated temperatures. Colloidal graphite suspensions painted on to the test specimen are often successful, but with some materials adhesion is poor and (especially with flexible plastics) they tend to part from the surface. Colloidal graphite in alcohol wets most plastics better than the water suspension which is often specified and may be useful for plastics not affected by alcohol.

Conductive paints, often based on a suspension of metallic silver in a suitable resin or resin–solvent system, have the virtue of a high conductivity, but care should be taken that the solvent does not adversely affect the specimen; residual solvent can produce odd results. All the brushed-on coatings require care in ensuring that the guard ring gap is not encroached upon.

Thin metallic foils (about 25 μm thick) may be applied with a thin layer of petroleum jelly, silicone grease or a mixture of poly(ethylene glycol), water, a wetting agent and an electrolyte. The foils, rolled on to the wetted surface very forcibly with a metal roller, are easily removed and therefore useful where other tests may be required to be made on the same specimen. Their effectiveness depends on the thickness of 'adhesive' being extremely small so that, despite its fairly high resistivity, its actual volume resistance is relatively low. This limitation makes foils generally unsuitable for low resistivity materials and also, since it is difficult to avoid contamination of the guard gap, they are not recommended for surface resistivity measurement. Gold leaf could be used, since it may be applied without grease.

Conducting rubber sheets have been used for rapid comparative measurements on materials, but fairly high contact pressure is required and conducting rubbers tend to become less conducting with age. Too high a pressure may change the conductivity of the rubber before the contact resistance is adequately reduced. The contact resistance may be variable and much higher than the resistance of the rubber electrode.

The dimensions of the electrodes for volume resistivity are not critical, provided that the guard ring gap is not too large. A popular size, calling for a 100 mm test piece approximately 3 mm thick, consists of a guarded disc electrode 50 mm in diameter, a guard ring of 60 mm internal diameter and 80 mm external diameter, and a bottom (applied voltage) disc electrode at least as large as the outer diameter of the guard ring. A larger electrode area and a thinner test piece are advantageous for material of very high resistivity. Some volume resistivity standards regard the area through which the current flows as being that of the guarded electrode, while others use the area of the circle having a diameter equal to the mean of the inner guard ring diameter and the guarded electrode diameter. The difference between these is small (21 per cent) for the electrodes described compared with errors generally outside the control of the operator, and in the UK it is disregarded. In the USA (see ASTM D257 (1978)[30]) not only is the mid-diameter used, but an additional fringing correction is also applied, based on the dimensions of the electrodes and the thickness of the test piece, where a high precision is required! The resistivity is given by AR_v/t Ω cm where A is the area (cm^2), t the thickness (cm), and R_v the measured volume resistance (Ω) between centre and lower electrodes.

If D and d are the diameters corresponding to the outside and inside, respectively, of the electrode gap, and provided that $(D - d)$ is small compared with D,

$$\text{surface resistivity} = \pi \frac{D + d}{D - d} R_s \qquad (13.1)$$

where R_s is the surface resistance measured between the two upper electrodes.

Note the absence of length dimensions in the surface resistivity expression, which is thus stated in ohms. The units of volume resistivity are often misquoted as Ω/cm^3 or Ω/cc, which are dimensionally incorrect and misleading.

13.3.6 Effect of Time of Application of Voltage

All resistances have an associated capacitance. Thus, when a voltage is applied, a relatively large current will flow to charge this capacitance and it will be necessary to short-circuit the current-measuring instrument for a short period after applying the voltage to prevent overloading it. After this time, the current will still fall with increasing time of application of voltage (a delayed charging current or alternatively one due to the sweeping of mobile ions to the electrodes) over a period which may be weeks. To standardise in reporting values of resistivity, it is customary to adopt the value attained one minute after the initial application of voltage, and strictly such values should be reported as *apparent resistivity* (*one minute value*) though this is seldom done. These values for polar materials will generally be within an order or two of the true infinite time resistivity. Reddish[31] has shown the errors that can occur with plasticised PVC compounds, where the one-minute value is predominantly due to polarisation current and may be useless as a criterion even for comparison of such compounds.

Another effect of the capacitance of the test piece is that any variation of the applied voltage will give rise to a current equal to the product of the capacitance and the rate of change of voltage. For example, if the test piece has a capacitance of 100 pF (10^{-10} farads) and the voltage changes at $-0{\cdot}001$ V/s, a current of -10^{-13} A will flow. If the resistive current being measured were only 10^{-13} A, this would give rise to a zero current and an apparently infinite resistance.

The voltage supply must therefore have extremely good short-term stability. A bank of lead–acid cells is probably ideal but generally impracticable. A well stabilised mains-driven power supply is normally used, but an adequate time should be allowed for warming up of the power supply unit. Dry cell batteries should not be used because they can drift at a relatively constant rate giving rise to a steady capacity current. If there are occasional large fluctuations of the mains supply so that some significant changes in output of a stabilised supply occur, these generally give rise to a fluctuating capacity current which is easily distinguished from the current to be measured. When fluctuations occur, the test must be repeated at a more opportune time.

13.3.7 Effect of Temperature

Resistivity is a temperature-dependent property and for non-metallic materials it invariably decreases with increasing temperature, unlike the behaviour of many metals. The decrease is always relatively large and of the order of several decades when the increase is from room temperature to, say, 100°C. Sensitivity to temperature change is possibly more marked for volume resistivity than surface resistivity, but it is advisable always to consider the possible effects of humidity and temperature variations on both properties. It is important in measurements at any temperature to ensure that temperature is maintained very constant during the test. Fluctuations in temperature produce changes in measured current which cannot be distinguished from the normal time-dependent charging current, and may cause errors greatly in excess of those due to the temperature dependence itself.

It is common to plot log resistivity against the reciprocal of absolute temperature, $1/T$, in common with many other phenomena, and frequently the result is a straight line for the long-time steady values. A precaution in determinations at elevated temperatures is to measure with increasing temperature and to ensure that the specimen is fully discharged and short-circuited before changing the temperature. It is possible to produce a residual charge on a specimen by cooling with the voltage applied, which charge may only disappear after a very long time interval.

It is always advisable finally to cool and retest, to ensure that initial and final measurements at room temperature agree, any difference suggesting a change in cure, for example, or some other non-reversible temperature effect.

At sub-ambient temperatures an additional hazard is caused by the possible condensation of moisture on insulators, and this can be avoided by using an evacuated system or one which can be mounted in a vacuum chamber placed in a low-temperature bath. Alternatively the electrodes can be swept continuously with very dry nitrogen.

13.3.8 Effect of Humidity

Resistivity is often very dependent on humidity, differences in which may cause orders of magnitude changes, dependent upon the type of plastic. Not only is surface resistivity particularly dependent upon ambient humidity, but in some cases it responds in fractions of a second, as may be observed by breathing on the sample during test. On the other hand, the time taken for equilibration of the volume resistivity of some materials (especially thick test pieces) with the atmosphere may be so long (e.g. months) that an acceptable period of conditioning before test can at best only minimise the effects of test piece history. The importance of ambient conditions upon the measurement of resistivity is underlined by the requirements of most specifications, which generally call for tolerances of ±5 per cent in relative humidity and ±2°C for the preconditioning period, usually of a minimum 18 h duration. In view of what has been said about the speed of response to humidity changes, some official specifications quite inadequately specify the conditions during measurement, which should be as rigorous as those of the preconditioning cycle where surface resistivity is concerned.

13.3.9 Standard Test Methods

The resistivity tests given in BS 2782 (1970)[32] and BS 903 (1956)[33] are at present being revised and brought into line with each other. These methods are similar to some of those given in IEC 93 (1958)[26] which itself is being revised. Circular electrodes are used. Several electrode materials are accepted but it is likely that the mercury electrodes will be omitted in the revised documents. The standard test conditions in both British Standards will be 23°C and 50 per cent r.h. The IEC document contains much information, especially about some sources of error and precautions to be taken.

Surface resistivity tests are sometimes required to be made after immersion in distilled water, but in that case the precise conditions between immersion and testing may influence the result.

Volume resistivity measurements are seldom required to be made at elevated temperatures, but BS 3815 (1964)[34] is an exception and a test at 90°C is involved. Here colloidal silver or graphite electrodes are stipulated.

IEC 645 (1971)[35] describes methods for making high-temperature surface and volume resistivity tests relying on IEC 93 for the electrical measurement techniques.

The blanket ASTM specification covering resistivity measurements is ASTM D257 (1978).[30] Electrode sizes are not stipulated for sheet material (nor even the shape, since round, square or rectangular types are permitted). In the case of surface resistivity the gap width between guard ring and centre electrode is made approximately equal to twice the specimen thickness. The test voltage is usually 500 V applied for one minute as in the British test. The ASTM specification gives a great deal of additional information about the techniques and hazards of measurement and also covers methods for tubing.

DIN 53482 (1967)[36] uses methods similar to some of those in IEC 93 (using silver or graphite painted electrodes) for volume resistivity, but requires a different electrode system for surface and volume resistivity. A narrow guard gap (1 mm) makes it difficult to avoid short-circuiting the electrodes. As an alternative to expressing the result as a volume resistivity, it may also be expressed as a volume resistance referred to unit area. Of the three electrode systems specified for surface resistance, only one uses IEC 93 type electrodes, the other two being essentially insulation resistance tests.

BS 4618 (1975)[37] prescribes the presentation of volume (Section 2.3) and surface (Section 2.4) resistivity data. The recommendation to precondition the test piece at not more than 1 per cent r.h. for the study of effect of temperature is necessary to minimise changes in moisture content as the temperature is raised, but it may be queried whether the data obtained are then relevant to design data. The test methods quoted are those in IEC 93, but the limit below which these become of doubtful use (given in BS 4618 as 10^3 Ω cm) should be about 10^8 Ω cm (§13.5).

ISO 1325 (1973)[38] for thin sheet and film uses the methods of IEC 93 for volume resistivity, but with a test voltage of 100, 10 or 1 V according to the thickness of the material, and the electrodes may be of evaporated metal, silver paint or metal foil. A document covering the same subject is under consideration by IEC, and ISO 1325 may be withdrawn.

ASTM D2305 (1972)[39] gives a volume resistivity test for thin films. Conductive silver electrodes of 25·4 mm diameter are painted on each side of a test piece and no guard ring is required for films less than 0·13 mm thick. The test voltage is 100 V d.c. applied for one minute.

13.4 INSULATION RESISTANCE

13.4.1 General Considerations

Insulation resistance is the resistance between defined electrodes applied to a defined piece of sheet, rod or tube or to a product. It, or rather its reciprocal, insulation conductance, is a summation of a surface conductance and a volume conductance, although frequently predominantly the former. Insulation resistance is a function of the shape of the sample and the electrode system and its value is useless without reference to the test method. In practice, the utility of the test is confined to comparative measurements on a few materials of similar type, e.g. synthetic resin bonded paper or other laminates. In such cases the test is commonly used to assess the effect of moisture penetration.

The unit of insulation resistance is the ohm (Ω) and values may range from a few megohms or less to more than 10^{15} Ω. Most of what has been said in §13.3 is true of insulation resistance, although different electrodes, generally of solid metal, are used. In addition, in the case of insulation resistance there is no guard electrode and the resistance is measured using only two electrodes. There is therefore no guarded electrode–guard leakage problem, although the procedure of raising the potential of the screen of the lead connected to the current-measuring instrument may be used to check the adequacy of the insulation of the lead.

13.4.2 Standard Test Methods

The methods given in BS 2782 (1970),[32] based on IEC 167 (1964),[40] are being revised and combined with those of ISO 2951 (1974)[41] to give a common document suitable for insertion in BS 903 (1956)[33] and BS 2782. The first three methods in the new draft document are primarily for rigid materials and involve immersion of the test piece in distilled water for 24 h before the test. It is probable that the new document will place both upper and lower limits on the time between immersion and testing and prescribe a maximum humidity during the test in an attempt to avoid errors due to variable degrees of drying. An earlier method using three brass screws as electrodes will probably be dropped, leaving two methods for rigid rod and tube, one using a 25 mm length with plane parallel ends clamped between flat metal plate electrodes under a pressure of 35 kPa and another using taper pin electrodes. The remaining method for rigid sheet also uses taper pin electrodes.

Two further methods which do not involve water immersion are prescribed primarily for flexible materials. The first uses stripes of conductive paint as electrodes and is applicable to sheet, rod or tube, while the second, for sheet and tape materials, specifies metal bar clamps as electrodes.

The ASTM 257 (1978)[30] insulation resistance tests are somewhat similar to the BS ones, but still include nut and bolt (or binding post) electrodes spaced 32 mm apart in a line (ASTM D700 (1975)[42] uses a circular arrangement). DIN 53482 (1967)[36] includes a method using either taper pins or cylindrical plugs pushed into holes in the test piece. It also includes a so-called 'surface resistance' test in which two forms of the electrode actually measure an insulation resistance, although the closeness of the conductive paint electrodes probably makes surface conduction predominant. The other electrode system consists of segmented springy brass contacts.

The method using bar clamp electrodes may be used for testing the insulation resistance of thin films as in ASTM D2305 (1972),[39] where it may additionally be used to detect the tendency of the films to corrode metals.

For tests on adhesive tapes BS 3924 (1978)[43] prescribes bar clamp electrodes (see Figures 13.6 and 13.7). In ASTM D1000 (1978)[44] the test piece, clamped in the electrodes, is preconditioned at 23°C, 96 per cent r.h. for 18 h and the resistance between pairs of electrodes is measured at 100–130 V after 15 s as an indication of the tendency of the tape to corrode metals. The specially-built humidity chamber described in BS 3924 seems inadequate and the method of guarding is inferior to that in the new draft of BS 2782.

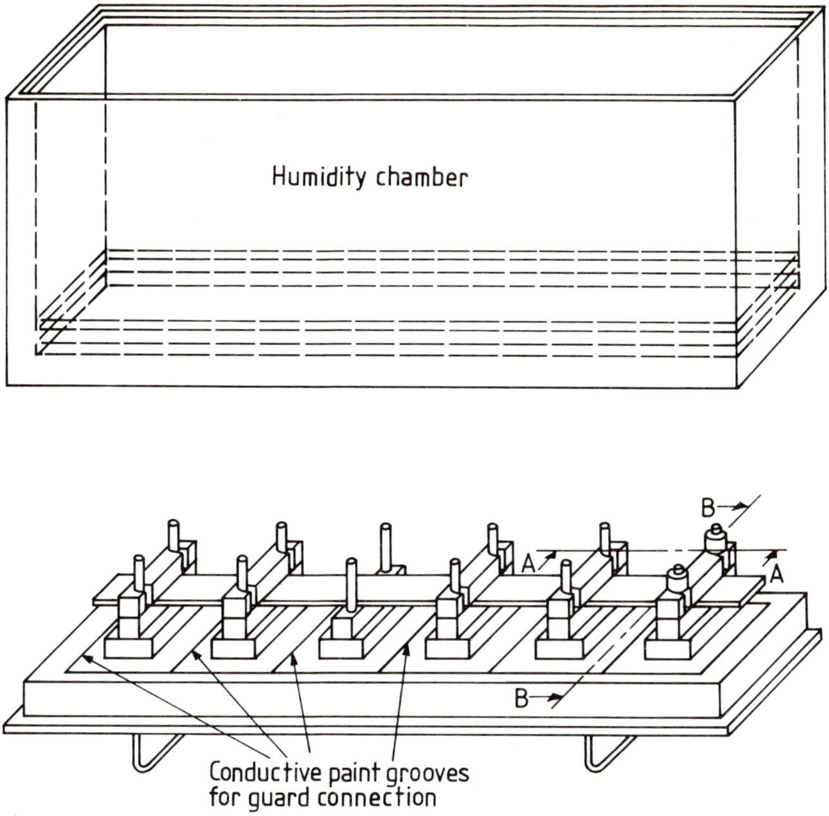

Figure 13.6 Bar clamp electrode: BS 3924

Insulation resistance tests on PVC tubing are covered by ASTM D876 (1971),[45] which requires that a U-shaped specimen filled with a dilute salt solution be immersed in a salt water bath and the resistance measured between a wire inserted through the length of the specimen and the external water.

Tests on insulating sleeving are usually restricted to electric strength and insulation resistance. In BS 2848 (1973)[46] insulation resistance is determined on specimens with a copper wire or tube insert, three foil electrodes being wound externally and connected so that end leakage is avoided. The test is conducted either at 50 per cent r.h. 23°C, or after three cycles of accelerated damp heat conditioning.

13.5 TESTS FOR ANTISTATIC AND CONDUCTIVE PLASTICS

13.5.1 General Considerations

Electrostatic charges may be unbalanced whenever an insulating material is separated from another (or from a conductor) leaving an excess or deficit of electrons on the insulating material. The problem of electrostatic charging is a

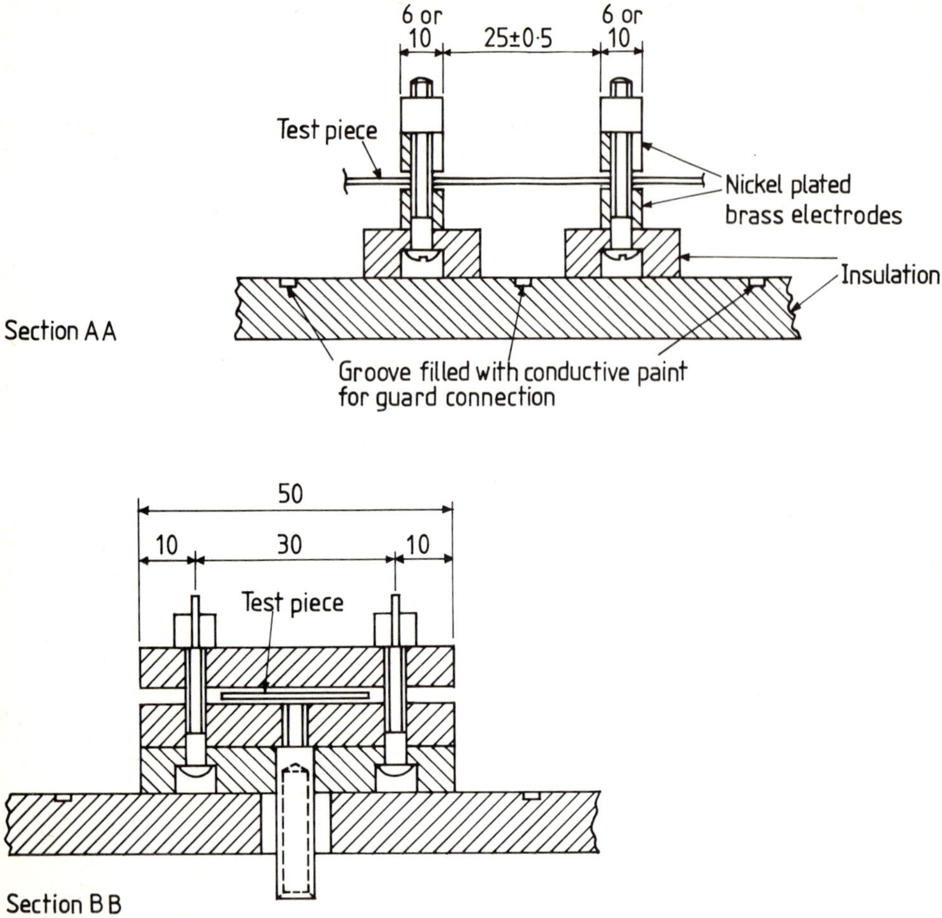

Figure 13.7 Detail of bar clamp electrode: BS 3924— dimensions in millimetres

serious one in several industries.[47] Many situations involving rubbers and plastics, where 'static' can be built up to such an extent that a spark occurs, have an inherent fire hazard because of the presence of flammable solvents or explodable gas mixtures. Mining, the petroleum industry and hospital operating theatres are examples. The packaging industry has an interest in the problem of dust and dirt pick-up on plastics which is aggravated by electrostatic attraction due to surface charges. It also has sometimes a problem in handling of charged plastics films.

A review by Morris[48] covers the theory of electrostatic charging in some detail, with an extensive bibliography.

The use of the terms 'conductive', 'conducting' and 'antistatic' is very confusing. The first two are interchangeable, although under most circumstances the first is more accurate. 'Antistatic' originally referred to articles which would dissipate static charges without being sufficiently conductive to cause electric shock or a fire hazard if they contacted live electric mains. This distinction is no longer valid in relation to specifications; for the purposes of this book the three terms may be regarded as synonymous.

It is theoretically possible for two materials to be 'electrically matched',[49,50] and thus cause no charge separation; however, each of these may charge if separated from a third (or if separated from each other when surface conditions have changed as a result of a change in atmospheric conditions). Thus the only sure way to prevent the build-up of static charge is to make all materials at least somewhat conductive, so that separated charges leak away sufficiently fast. In the author's opinion, this makes measurements of charge developed by (say) rubbing, which are notoriously variable, invalid as specification tests. It is also questionable whether charge decay measurements, which are limited in the useful range of times (compared with the wide range of resistivities which can be measured), give any useful information which cannot be gained from a suitable resistivity test. Beesley and Norman[51] showed that a good correlation could be obtained between decay time and surface resistivity for a range of commercial PVC films (Figure 13.8). However, all three types of measurement have been used and will therefore be outlined.

Conductive coatings may be used on plastics for electromagnetic screening. Klouda[52] describes screening measurement techniques using a signal generator and a receiver. He indicates that, for initial selection of coatings for screening, a resistivity test can be used.

In applications other than antistatic ones, e.g. plastics potentiometers, a resistivity test on the material or a resistance test on the product (using its own terminals as electrodes) is again the obvious choice.

Further details on types of test for conductive and antistatic materials are given by Norman,[47] Chapter 8.

13.5.2 Effects of Temperature, Humidity and Strain

There are two general types of conductive plastics. The first is one which is filled with conductive filler particles, such as carbon black or metal powders. Such

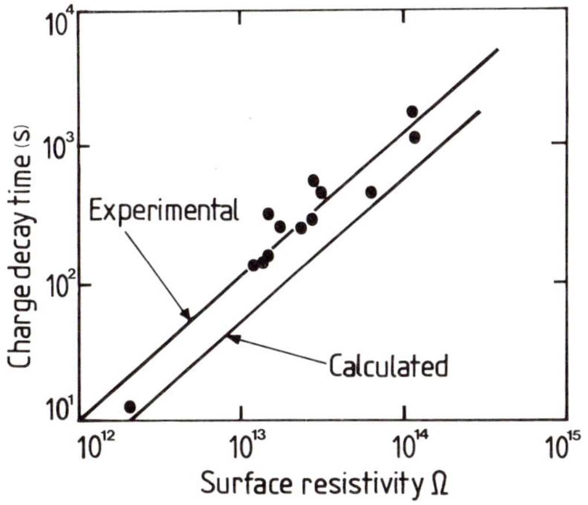

Figure 13.8 Relationship between charge decay time and resistivity[51]

plastics generally have resistivities from about 10^7 ohm cm down to about 10^{-2} ohm cm. In this range they are relatively insensitive to temperature, humidity and voltage, but if they are flexible they are sensitive to strain history.

The second type contains additives known as antistatic agents, which are organic molecules that diffuse to the surface and either dissociate to give mobile ions or are hydrophilic and collect a film of moisture on the surface. These are sometimes highly sensitive to humidity and temperature but little is known about their sensitivity to voltage, and they are insensitive to strain.

Antistatic agents are capable of giving resistivities down to 10^7 ohms in plasticised PVC, and can then be used for some of the antistatic purposes where a hazard is involved. Antistatic agents in other plastics tend to give resistivities from about 10^{10} ohm cm upwards and are suitable then only to combat 'nuisance static', i.e. to prevent dust pick-up and, with thin films or fibres, handling problems.

13.5.3 Charge Measurements

In charge measurements a high impedance instrument such as an electrostatic field meter or an electrometer is used to measure the *net charge* developed (or a figure proportional to the net charge developed) on a material or article under service or production conditions or as a result of some arbitrary treatment such as rubbing in a (hopefully) consistent manner with a specified material.

Electrostatic field meters are described by Cross[53] and Shashoua;[54] Langdon[55] gives details of suitable enclosures and techniques for comparing fabricated articles and films. The actual net charge cannot generally be ascertained without enclosing the test piece or article in a 'Faraday ice pail' or conductive container

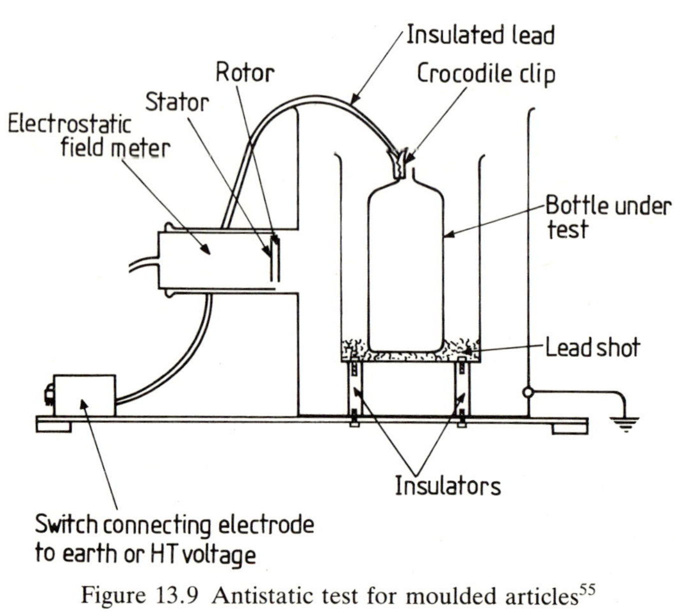

Figure 13.9 Antistatic test for moulded articles[55]

of known capitance. Figure 13.9 shows an apparatus used by Langdon, based on this principle.

The words 'net charge' have been used because the surface, especially of a newly moulded object, may have charges of each sign at different parts of its surface.[56]

Methods other than those using the 'Faraday ice pail' are used to measure a figure proportional to the charge (although it is not generally possible to state the constant of proportionality), but must employ a fixed geometry of apparatus inside an earthed screen.

Dust or smoke pick-up tests are sometimes used to provide qualitative information.

There appears to be no specification test for the direct measurement of surface charges on plastics materials, probably because of the difficulty (or even impossibility) of providing a reproducible method of charging which bears any resemblance to common charging mechanisms.

One logical application for measurement of charge developed was described by Moreno and Gross[57] in connection with the corona charging of electrets (solid dielectrics possessing a persistent dielectric polarisation) for use in electret microphones. Their method permits a measurement while the electret is being charged or discharged.

13.5.4 Tests for Rate of Charge (or Discharge)

Three methods of test are described in BS 2782 (1976)[58] for use on film materials.

Method 250A (See Figure 13.10) is based on a method described by Langdon[55] and uses a long strip of film threaded over a series of rollers. The rollers A and B are initially connected to a source of 5000 V and, after a given period, are connected to earth; the time of fall of the reading of the field meter to 50 per cent of its maximum value is measured.

Method 250B is suitable only for very flexible films. The equipment is rather like a gold leaf electroscope, where the test piece replaces the gold leaf. A voltage (10 kV) is applied to the test piece support and the time taken for the sample ends to diverge to an angle of 90° is recorded.

Method 250C uses a test piece which is clamped between 'brims' on two coaxial 'top hat' cavities. In one cavity is an electrode which may be connected to 300 V or to earth. In the other is an electrode connected to an electrometer. When the first electrode is switched from 0 to 300 V, the electrometer reading rises sharply to a maximum and then decays. The time for it to decay to half the steady reading which would be obtained in the absence of the film is recorded.

The useful range of the above three methods is generally 1–1000 s, the lower value being often limited by instrumental time constants in Methods A and C and by the mechanical time constant of the sample in Method B, although the standard implies that 0·1 s is attainable. Readings of greater than 1000 s may be related to other effects than the sample properties.[54]

The above tests are suitable for ensuring there will be no 'nuisance static'. Where 'hazardous static' is a problem, much shorter discharging times must be attained and resistance or resistivity measurements are required (§13.5.5).

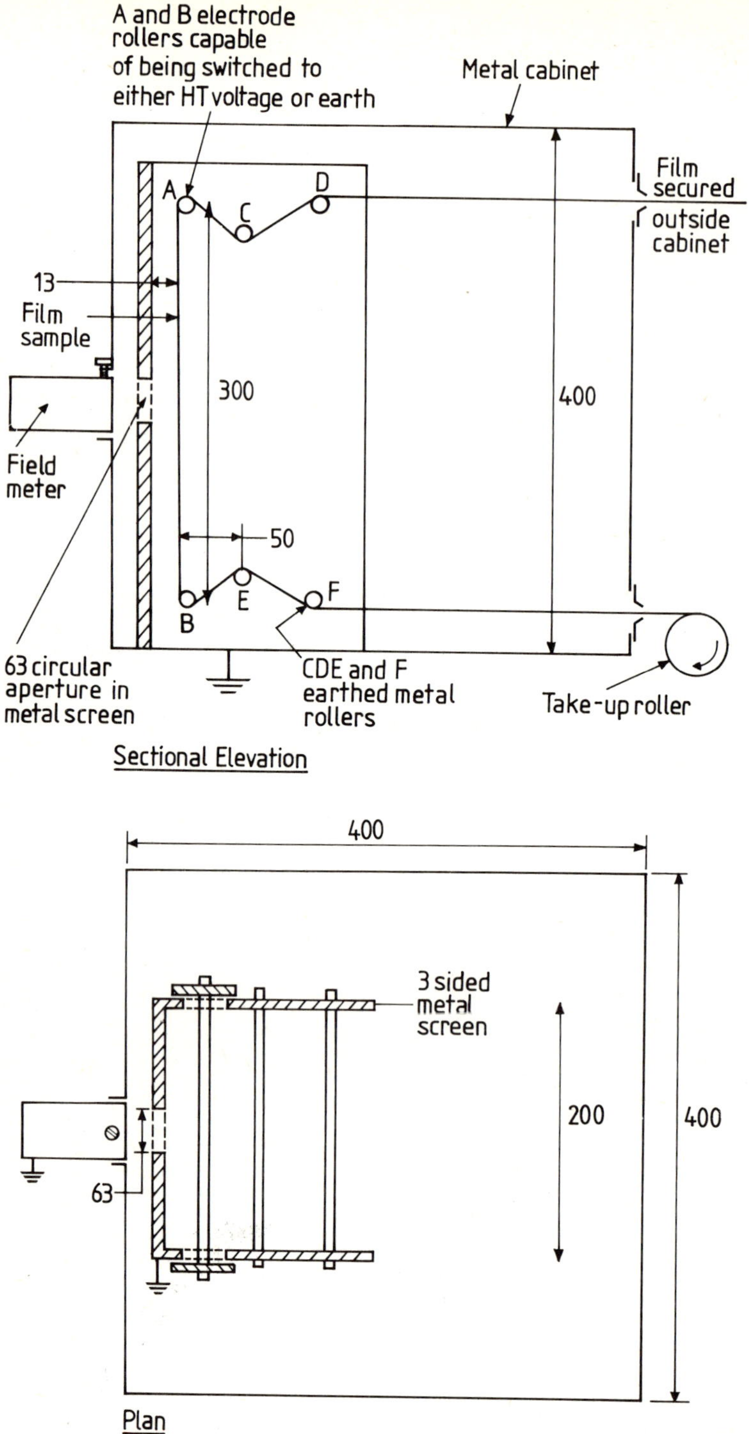

Figure 13.10 Antistatic test for film: BS 2782, Method 250A—dimensions in mm

13.5.5 Resistance and Resistivity Tests

The resistivity tests for *materials* having resistivities below about 10^8 Ω cm differ from those described in §13.3 for the following reasons:

(a) the high voltage gradients used for insulating rubbers would cause more conductive materials to heat up or even ignite;
(b) contact resistance effects, while possibly less than for insulating materials, are practically of greater significance;
(c) with materials of resistivity below about 10^7 ohm cm, the effects of humidity are usually negligible;
(d) the resistivity of some conductive materials is highly sensitive to strain.

Materials having resistivities above about 10^8 Ω cm should be tested by the methods described in §13.3. For materials containing antistatic agents a surface resistivity test is more appropriate.

Resistance tests for antistatic and conductive *products* are in essence insulation resistance tests (§13.4), although with such products most of the conductance is usually a volume conductance. A series of test methods (which involve applying electrodes except where metal parts of the product are used as electrodes) using a conventional ohm-meter is described in ISO 2878 (1978)[59] for rubber and almost identically in BS 2050 (1978)[60] for flexible polymeric materials generally. In most cases the electrodes consist of a mixture of poly(ethylene glycol), water, wetting agent and electrolyte, backed by brass electrodes. The test is made with an instrument which has an open circuit voltage of 500 V, but whose output resistance limits the maximum power dissipation in the product. DIN 51953 (1975)[61] uses a somewhat similar test for flooring.

A number of British Standard product specifications use methods similar to those in BS 2050, sometimes with the omission of the electrolyte. Some refer to BS 2050 for the test method; others, e.g. BS 3289 (1960),[62] describe in detail a test method.

BS 2044 (1978)[63] describes three methods of measuring the *resistivity* of conductive and antistatic rubbers. One of these methods, a potentiometric test, is described in ISO 1853 (1975)[64] for rubber, and with slight modification (because of the greater difficulty of making a stiffer test piece conform with the electrodes) is under consideration in ISO as a plastics test. In this method a current is passed between two electrodes at the ends of the test piece and the voltage drop is measured between two electrodes at intermediate positions. This test method eliminates the contact resistances at the two ends of the test piece. The contact resistances of the potential electrodes may be very high and it is therefore essential that the current supply and measuring circuit be isolated (to better than 10^{12} Ω) and that the input impedance of the voltage-measuring instrument be greater than 10^{11} Ω.

13.6 POWER FACTOR AND PERMITTIVITY

13.6.1 General Considerations

The measurement of power factor and permittivity and the related parameters outlined in §13.6.2 (often collectively known as *dielectric properties*) may need to be carried out over a wide range of frequencies from a few Hz to several tens of GHz, but most measurements are made between 50 Hz and 100 MHz.

The range of permittivities which may be encountered is from just over 1 for foams to 20 or more. Power factors may range from 0·0002 for polyethylene to 0·1 or even higher for a poor dielectric. (Note that a poor dielectric is a good material for dielectric heating!)

Materials with anisotropic structures may have anisotropic dielectric properties.

13.6.2 Definitions and Equivalent Circuits

The *relative permittivity* of an insulating material is the ratio of the capacitance of a capacitor, in which the space between (and around) the electrodes is entirely and exclusively filled with the insulating material in question, to the capacitance of the same configuration of electrodes in a vacuum (or to an accuracy of 0·05 per cent in air).

The test methods used are such that the words 'and around' can be ignored and we need only consider that part of the test piece between the electrodes, with a correction for the fringe field at the edge if necessary.

The word 'relative' in the context of tests on plastics is usually dropped for convenience, although the unqualified term 'permittivity' has another meaning. There is no cause for confusion since the magnitudes expressing relative permittivity are in the range 1 to about 80 (dimensionless) whereas absolute permittivities are of the order of $10^{-11}-10^{-9}$ F/m.

The *dielectric constant* is the same as relative permittivity and the terms capacivity and specific inductive capacity, found in American usage, also have the same meaning.

The *loss angle* (δ) of an insulating material is the angle by which the phase difference between applied voltage and the resulting current deviates from $\pi/2$ radians when the dielectric of the capacitor consists exclusively of the dielectric material. It is sometimes expressed in microradians.

The *dissipation factor*, often known as tan δ or loss tangent, is the tangent of the loss angle.

The *power factor* is the ratio of the power loss in watts dissipated in the material to the product of the r.m.s. voltage applied and the r.m.s. current passing through the material. It is equal to the cosine of the phase angle, i.e. to sin δ.

It is common practice in dielectric work to use the tangent of the loss angle rather than the sine, since many of the theoretical equations for dielectrics are then simpler. Additionally, the equations relating to measuring circuits invariably give simpler relations to the tangent. With good dielectrics (tan δ less than 0·1) the difference between power factor and tan δ is less than 0·5 per cent. Thus the habit which has grown up among people dealing with dielectrics (and even in British Standards) of using the word 'power factor' when 'tan δ' is implied does not generally lead to significant error.

The *loss index*, formerly called *loss factor*, is the product of the relative permittivity and tan δ. For a fixed geometry of capacitor subjected to a fixed frequency and voltage, the power dissipated is proportional to the loss index.

The power factor has been defined in terms of energy dissipation, and the dissipation factor in terms of the loss angle, but in considering measurements at frequencies up to 10^8 Hz it is often convenient to consider equivalent circuits, i.e. series or parallel arrangements of a resistor and a capacitor (Figure 13.11) which *at a single frequency* have the same electrical characteristics as the capacitor formed by the electrodes and the dielectric material in question.

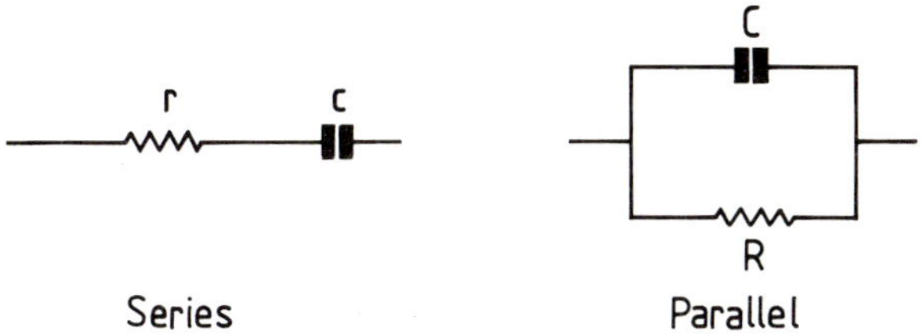

Figure 13.11 Equivalent circuits

It must be emphasised that for any dielectric the equivalent values of c and C change slowly with frequency while the values or r and R change rapidly, often approximately inversely proportional to frequency. For good dielectrics, the values of c and C are nearly equal while the values of R and r are very different. The choice of the equivalent circuit is usually made on the basis of the measuring circuit employed, since in general the measuring circuit components are so arranged as to give the appropriate capacitance and power factor in terms of either the series or the parallel equivalent.

The condition that the series and parallel equivalents are identical at a given frequency is that their impedance (ratio of current to voltage) is the same with respect to both magnitude and phase. Using complex notation ($j^2 = -1$), the impedances of the series and parallel equivalents are, respectively,

$$Z = r - \frac{j}{\omega c} \quad \text{and} \quad Z = \frac{R}{1 + \omega^2 C^2 R^2} - \frac{j\omega C R^2}{1 + \omega^2 C^2 R^2} \qquad (13.2\text{a,b})$$

where $\omega = (2\pi \times \text{frequency})$ and the units of capacitance and resistance are respectively farads and ohms. Equating real and imaginary parts leads to the following relations:

$$r = R\frac{\tan^2 \delta}{1 + \tan^2 \delta} \quad \text{and} \quad c = C\,(1 + \tan^2 \delta) \qquad (13.3\text{a,b})$$

The equations for the series and parallel circuits above lead respectively to

$$\tan \delta = \omega c r \quad \text{and} \quad \tan \delta = \frac{1}{\omega C R} \qquad (13.4\text{a,b})$$

For values of $\tan \delta = 0{\cdot}1$ or less, equations (13.3a,b) reduce within 1 per cent to

$$r = R \tan^2 \delta \quad \text{and} \quad c = C \qquad (13.5\text{a,b})$$

To illustrate the significance of these equations, consider the example of a low-loss test piece of polyethylene with a capacitance of 10 pF and loss tangent of 0·0001. At $\omega = 10^4$ rads/s (frequency = 1600 Hz approximately), the series equivalents would be $r = 1000\ \Omega$ and $c = 10$ pF and the parallel equivalents $R = 10^{11}\ \Omega$ and $C = 10$ pF. At $\omega = 10^7$ rads/s ($f = 1{\cdot}6$ MHz approximately), the equivalents would be $r = 1\ \Omega$, $R = 10^8\ \Omega$, both capacitances again being 10 pF. The small value of r at the higher frequency also illustrates the importance of keeping any lead resistances in the measuring circuit small, since if they are included in the measurement the loss tangent will be augmented accordingly by an amount $\omega c r'$, r' being the extraneous resistance and c the measured capacitance. At the lower frequencies the reverse situation applies and it is important to ensure that no extraneous resistance in parallel with the specimen is included in the measurement unless it is extremely high compared with the specimen equivalent parallel resistance. This latter resistance should not be confused with the d.c. resistance of the specimen or component and it cannot be measured by d.c. methods.

13.6.3 Effects of Temperature, Frequency and Humidity

The effects of temperature and frequency are interrelated. With non-polar materials the changes of properties with frequency and temperature are small. With polar polymers very large changes may occur. For example, with plasticised PVC the power factor may change by a factor of two over a 10°C range in temperature or a tenfold change in frequency, and the permittivity may change by 20 per cent over similar ranges.[65]

With polar materials (or even impure non-polar materials), especially at low frequencies, the effect of humidity can be large, e.g. a 30-fold increase in loss index and a 20 per cent increase in permittivity may be produced by the absorption of 0.5 per cent of water between a dry condition and 75 per cent r.h.[66]

13.6.4 Electrodes

As in resistivity measurements, a sheet dielectric material with an electrode applied to each surface would be simple to measure, but the capacitance between the electrodes due to the test piece would be augmented by extraneous capacitance round the edges of the test piece either wholly or partly in air, which would be difficult to assess and allow for. A third electrode, placed concentrically around one of the existing electrodes (which is diminished in area) and as close to it as possible without touching, serves as a guard ring. When this is used with appropriate circuitry it reduces the uncertainty of the measured capacitance to an extremely small amount, since the effective area of the guarded electrode very nearly corresponds to that based on the mid-gap radius. Usually the requirement of the measuring circuit is that the guarded and guard electrodes are maintained at exactly the same potential and phase during measurement to ensure that the capacitance and associated loss between them do not contribute to the measured values; alternatively, the capacitances of the guard/guarded electrode and the guard/non-guarded electrode may, with some circuits, be connected so that they do not affect the balance or accuracy.

Three-terminal measurements are more accurate than two-terminal ones, but are mainly limited to bridge measurements and are thus rather impracticable above about 1 MHz, although there is an increasing tendency for radio frequency bridges to be designed to accommodate them. Above 10 MHz two-terminal systems are standard, and corrections may need to be made for edge effects.

Consideration will now be given to the means by which contact is established between dielectric test pieces and the measuring instrument. The importance of understanding the principles and techniques involved cannot be overstated, since probably more difficulties and errors are associated with electrodes than with the measurement itself. Much of what has been said in §13.3.5 about electrodes in connection with resistivity measurements is applicable here also, with the recognition that the requirements are probably more stringent with dielectric measurements, since greater accuracy can be obtained and is usually desired.

In what follows, essentially only sheet test pieces are considered, but the techniques are commonly applicable to tubes with appropriate changes in the permittivity formulae.[67]

All electrodes are designed either to ensure as intimate a contact as possible between the test piece and the measuring apparatus, or alternatively and rather paradoxically, to ensure a precisely defined air or liquid gap between the test piece and the electrodes, for which accurate allowance can be made. Amongst the *contacting electrodes* appear conducting paints, vacuum deposited metallic films, mercury, conducting rubbers, metal foils applied with grease, and colloidal graphite. Although each of these types may have some utility under particular circumstances, the most widely used is probably that involving metal foils, and possibly the best is that utilising metal films deposited *in vacuo*. Conducting paints and graphite suspensions tend to give relatively high-resistance films, which cause spuriously high loss values on low-loss materials especially at high frequencies, and the solvent base may attack the material. Conducting rubbers have relatively high resistance and are thus only suitable for 'lossy' materials, but have the virtue of being speedy and 'clean' to apply. Mercury presents a toxic hazard and requires carefully designed clamps and considerable care in avoiding air bubbles. Vacuum deposition of metallic films such as gold or aluminium ensures intimate contact; appropriate masks enable the electrode diameter to be defined accurately.

Tin or aluminium foils applied with silicone grease or petroleum jelly and rolled on very hard with a roller are convenient and simple. The foils are usually 0·03–0·05 mm thick and the rolling process produces an extremely thin film of grease, which must have a capacitance at least two orders greater than the test piece to reduce the measurement error to 1 per cent. This is not difficult to achieve with low permittivity test pieces 2 mm thick or so, but errors may become very large with high permittivity or thin test pieces ($<0{\cdot}5$ mm, say) and these foils are not generally suitable for measurements on films. Gold foil does not require grease but it is difficult to handle; on the other hand, it does provide very intimate contact and can be used even on thin films. It should be emphasised that no one electrode material or technique is superior to all others and necessarily gives the 'correct' answer. The only procedure with a doubtful result is to substantiate it by at least one other completely independent method.

All the above electrodes need metal backing plates of the same size as the electrodes. The use of flat plates alone is permitted both in ASTM D150 (1978)[68] and in BS 2782 (1970)[32] (for 50 Hz only), but this is qualified by the requirement to use them only under very high pressure (17·3 MPa) or if they can be shown to give the same results as accepted methods. DIN 53483 (1969)[69] includes ground flat plates using silicone oil to ensure a good contact.

If d_o is the diameter midway in the gap between a circular guarded electrode and its guard ring, the permittivity is given by

$$\varepsilon = \frac{144tC}{d_o^2} \tag{13.6}$$

where d_o = mean diameter (mm)
t = test piece thickness (mm)
C = capacitance (pF).

For two-terminal electrode systems, when the test piece is at least equal in diameter to the electrodes, the permittivity is given by

$$\varepsilon = \frac{144t(C - C_e)}{d^2} \tag{13.7}$$

where d = electrode diameter (mm)
t = test piece thickness (mm)
C = capacitance of test piece/electrodes (pF)
C_e = edge capacitance of electrodes (pF).

The edge correction may be obtained either from the Kirchhoff relation[70,71] or by calibration with test pieces of known permittivity.

The power factor is not related to the electrode dimensions.

A difficulty with most contacting electrodes is that any preconditioning of the test piece must be carried out before their application, but this difficulty does not arise with *non-contacting electrodes*. With *air gap* systems the electrodes are spaced apart and the test piece is introduced leaving a deliberate gap. Two measuring techniques are then available: *fixed gap* and *fixed capacitance*. In the first, the capacitance and loss tangent of the system are measured with the test piece in and then with the test piece removed, the electrode spacing being left untouched throughout. The characteristics of the material are given by

$$\varepsilon = \frac{t_s}{t_s - t_o[1 - (C_o/C_i)]} \tag{13.8}$$

$$\tan\delta = \tan\delta_i\left(1 + \varepsilon\frac{t_o - t_s}{t_s}\right) - \varepsilon\frac{t_o}{t_s}\tan\delta_o \tag{13.9}$$

where t_s = test piece thickness
t_o = electrode separation
C_i = capacitance, test piece in
C_o = capacitance, test piece out
$\tan\delta_i$ = loss tangent, test piece in
$\tan\delta_o$ = loss tangent, test piece out

ε = test piece permittivity
tan δ = test piece loss tangent.

For a properly designed electrode system, where tan δ_o would normally be negligibly small compared with tan δ_i, the second term would vanish.

The units of capacitance and distance are arbitrary (but consistent) since only ratios are involved. The equations are rather cumbersome, but their virtue is that the calibration of the electrodes need not be known, since their area is not involved, and no electrode adjustment is required, although this limits the permissible test piece thickness as it is desirable for the sake of sensitivity to keep the gap small. As a guide, the gap should be of the order of 10–20 per cent of the test piece thickness. Smaller gaps tend to cause difficulty due to dust particles and fibres, and with larger gaps the electric field tends to distort and cause errors.

With a fixed capacitance system, measurements are made of capacitance and loss tangent as before, with the test piece in, but the electrode spacing is adjustable and after removal of the test piece the spacing is reduced until the capacitance of the empty electrodes is the same as when the test piece was in place. The equations are then

$$\varepsilon = \frac{t_s}{t_s - t_x} \tag{13.10}$$

$$\tan \delta = (\tan \delta_i - \tan \delta_o) \frac{t_o}{t_s - t_x} \tag{13.11}$$

where t_s = test piece thickness
t_x = electrode movement between test piece in and out positions
t_o = electrode separation, test piece out
tan δ_i = loss tangent, test piece in
tan δ_o = loss tangent, test piece out
ε = test piece permittivity
tan δ = test piece loss tangent.

With suitable test pieces a precision of the order of 1 part in 1000 is achievable for permittivity, such precision being required, for example, in the control of polyethylene used in submarine cables. Lynch[72] gives useful information on the use and construction of micrometer air gap electrodes with a full analysis of electrode and test piece imperfections; BS 4542 (1970)[73] is largely based on his work. This standard is complementary to BS 2067 (1953)[74] (§13.6.7) and principally covers the frequency range 1–100 kHz.

In recent years *liquid immersion electrodes* have been developed, with which very precise measurements of permittivity can be made; these are based on the fixed gap technique, but with the gap filled with a suitable non-polar liquid. Principally developed to meet the submarine cable requirement referred to above, a high accuracy is achieved by using a fluid of closely similar permittivity to that of the test piece so that only small changes in the electrode capacitance and loss occur when the test piece is inserted. If there were no change at all in these parameters on insertion, the test piece would have an identical permittivity and loss tangent to that of the liquid; otherwise the test piece properties are given by

$$\varepsilon = \varepsilon_l + \left(\frac{\Delta C}{C}\frac{t_o}{t_s}\right) \tag{13.12}$$

$$\tan\delta = \tan\delta_l + (\tan\delta_i - \tan\delta_l)\frac{t_o}{t_s} \tag{13.13}$$

where t_o = electrode separation
t_s = test piece thickness
ΔC = increase in capacitance when test piece inserted (pF) (negative sign if decrease)
C = capacitance of electrodes, empty (pF)
ε = permittivity of test piece
ε_l = permittivity of liquid
$\tan\delta$ = loss tangent of test piece
$\tan\delta_l$ = loss tangent of liquid
$\tan\delta_i$ = loss tangent with test piece inserted.

This method of measuring very precise permittivity is mainly limited to materials the permittivities of which can be matched to readily available liquids. ASTM D1531 (1962)[75] gives details of the requirements for low-density polyethylene measurement using either benzene or silicone fluid. The equations given apply only where the liquid and solid permittivities are within a few per cent of each other.

13.6.5 Measuring Techniques—General

The measurement of power factor and permittivity, over the wide band of frequencies in which these properties are important, is rendered difficult by the limitations of existing test equipment and the necessarily large variety of methods required to cover this range. Insulating materials are used as dielectrics at commercial frequencies between approximately 50 Hz and 10^{11} Hz and the variety of test methods used reflects the many differing practical techniques necessary to generate and distribute power at different parts of this frequency spectrum. The range up to 10^8 Hz, it is true, can be covered by only two types of measuring equipment, but between approximately 10^9 and 10^{11} Hz a different set of equipment is generally required for each specific frequency of interest, since the waveguides used at these frequencies are of particular dimensions directly related to the wavelength.

The two kinds of technique related to these two frequency ranges (above and below 10^8 Hz) are quite different and are sometimes loosely differentiated by the terms 'lumped circuit' and 'distributed circuit'. In the former case the circuitry involves the use of the three basic electrical components: inductors, capacitors and resistors. A passive circuit (one containing no sources of voltage), however complex, comprising these elements can be represented at a given frequency by an equivalent circuit having single values of inductance or capacitance and resistance. This notion of equivalence is a powerful aid to the solution of such complex circuits. At the higher frequencies, say 10^9 Hz and above, the three basic parameters can no longer be 'lumped' in this fashion and the waveguide behaves as though these parameters were distributed properties, so that solutions can only be found in terms of the equations relating to the associated electric and magnetic field distributions in and around the waveguide.

Dielectric measurements at u.l.f. (ultra low frequency)[31,76–78] are of some interest in view of the light they throw upon the basic structure of some materials, notably glasses, but are outside the brief of this book. Such measurements have been made down to 0·001 Hz and lower.

The power factor and permittivity of an insulator (or a capacitor) are so closely related that in practice many measuring circuits are designed to give both properties simultaneously, and as in any case the determination of one property can only be made with accuracy by adjustment of a circuit component which compensates for the presence of the second property, it is generally expedient to measure the two together.

13.6.6 Techniques for Measurements at 50 Hz–20 kHz

Bridge circuits are invariably employed in the audio frequency range (50 Hz–20 kHz) and beyond up to 100 kHz, but the upper limit is probably of the order of 10 MHz, although bridges suitable for the latter frequency have to be of rather special construction. The basic bridge circuit (Figure 13.12) consists of four arms, and the simplest case involving four resistances is that of the Wheatstone bridge for measuring resistance. The circuit is completed by the addition of a voltage source across one diagonal of the bridge and a suitable voltage detector across the other. At least one of the arms must be variable and the bridge is balanced when the detector indicates zero voltage; the relation $P/R = Q/S$ then holds. Thus one unknown component can be measured in terms of three known ones.

For a.c. measurements at least one arm other than that containing the test piece must contain a reactive component, that is one having inductance (self or mutual) or capacitance, and two examples will be considered in some detail since most present-day determinations are based on them or their variants. These are the Schering bridge and the transformer ratio-arm bridge.

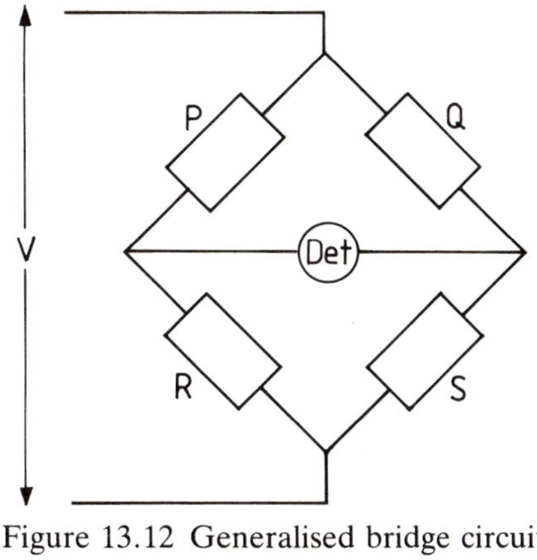

Figure 13.12 Generalised bridge circuit

Schering Bridge

The Schering bridge in its simplest form comprises two fixed resistors of equal value and two capacitance arms, one of which is a relatively loss-free variable standard capacitor and the other the unknown, i.e. the test piece. At balance the total capacitance in the unknown arm is equal to the total capacitance in the standard arm, while the loss balance is achieved by a variable capacitor across the resistance arm diagonally opposite the unknown arm, the loss tangent being a function of the capacitance and resistance of this arm and the unknown capacitance.

This arrangement is seldom used in practice, since the residual stray capacitances which are inevitably associated with each arm of the bridge are difficult to assess and allow for. Instead, a substitution method is employed using a modified Schering bridge containing two capacitive arms and two resistive arms (Figure 13.13), the bridge being initially balanced with the unknown disconnected and then re-balanced with the unknown connected across one capacitive arm containing a low-loss calibrated standard. The diagonally opposite resistive arm has a calibrated loss-balancing capacitor in parallel and the other resistive arm usually also contains an uncalibrated capacitor to facilitate initial balancing. The capacitance and loss are given by the changes in settings of the standard and the loss capacitor, provided that the remaining components are left untouched after initial balance. The virtue of this arrangement is that most of the residual capacitances and losses in the bridge remain constant during the measurement and do not enter into the bridge equations.

In Figure 13.13, C_1 is the low-loss standard capacitor of known calibration, C_2 a low-loss uncalibrated balancing capacitor, and C_3 and C_4 two capacitors of lesser quality, the latter suitably calibrated. With the unknown C_x (of loss $\tan \delta_x$) disconnected and assuming equal ratio-arms ($R_3 = R_4$), the initial balance gives $C_1 = C_2$ where these symbols denote the total capacitance in arms 1 and 2 including extraneous capacitance. Any difference in loss in arms 1 and 2 will be balanced out by the difference in settings of C_3 and C_4 (again including stray

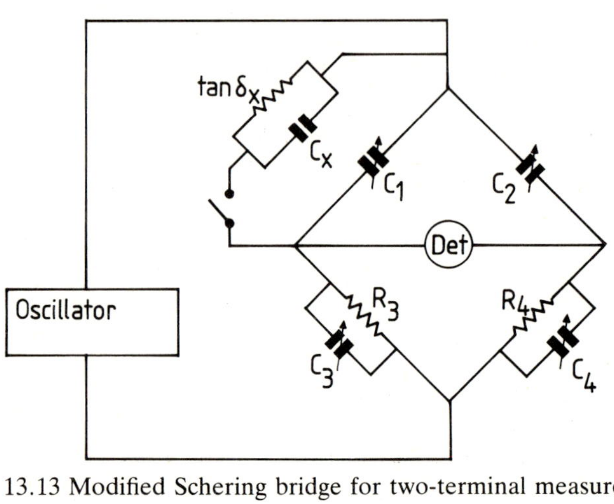

Figure 13.13 Modified Schering bridge for two-terminal measurements

capacitance). The balance equations give

$$C_x = \Delta C_1 \tag{13.14}$$

$$\tan \delta_x = \omega \frac{C_1'}{10^{12}} R_4 \frac{\Delta C_4}{\Delta C_1} \tag{13.15}$$

where $\omega = 2\pi \times$ frequency
ΔC_1 = change in capacitance of C_1 (pF)
ΔC_4 = change in capacitance of C_4 (pF)
C_1' = initial setting of C_1 (pF)
$R_4 = R_3$ = ratio-arm resistance (Ω).

It should be noted that the residual losses in C_1 and C_2, although small, are finite, but they disappear provided that the loss in C_1 is due to a constant equivalent parallel resistance, which is usually the case for good quality capacitors.

In the two-terminal example given it is important to connect the unknown via a low-capacitance switch as shown, where any alteration in the distribution of the stray capacitances from the specimen electrodes to earth between the two balances has a negligible effect. If the switch were in the upper or high voltage lead, a large stray capacitance to earth could appear across arm 3 on the initial balance, which would disappear on final balance, and the resultant change in the capacitance of this arm would be reflected as a spurious contribution to the loss balance ΔC_4. The values of C_3 and C_4 are often of the order of 500–1000 pF, while C_1 and C_2 are usually 1000 pF or less. The values of R_3 and R_4 are typically 1000–10 000 Ω; R_3 and R_4 are sometimes made unequal, the ratio giving a multiplying factor by which both capacitance and loss ranges may be extended.

Three-terminal electrode systems are accommodated on the modified Schering bridge by the provision of a further two arms, similar to one half of the existing bridge, which form a Wagner earth circuit. There are many forms of this arrangement. That shown in Figure 13.14 is given in BS 903 (1956)[33] and shows the screening in detail. It also includes an inductance L6 which may be required to balance the Wagner earth. The Wagner earth eliminates not only the effect of stray capacitances between the guard and guarded electrodes, but also of those within the bridge and may thus be useful even with two-terminal electrodes.

The main capacitance in the electrode system to be measured is connected across C_1, with the larger electrode to the high voltage side of the bridge (or to earth), and the smaller central electrode to b. The guard ring is connected to earth. The main feature of the Wagner earth is that the detector can be switched either across bc, the bridge proper, or to c and earth so that when the detector indicates a null in both positions the points b and c are both at earth potential and the guarding of the electrode system is perfect. The procedure in the substitution method is as follows:

(i) With specimen main (high voltage) electrode connected to earth, balance the main bridge using C_2 and C_3, C_1 and C_4 being meanwhile left at their zero settings (corresponding usually to maximum capacitance and minimum capacitance respectively).

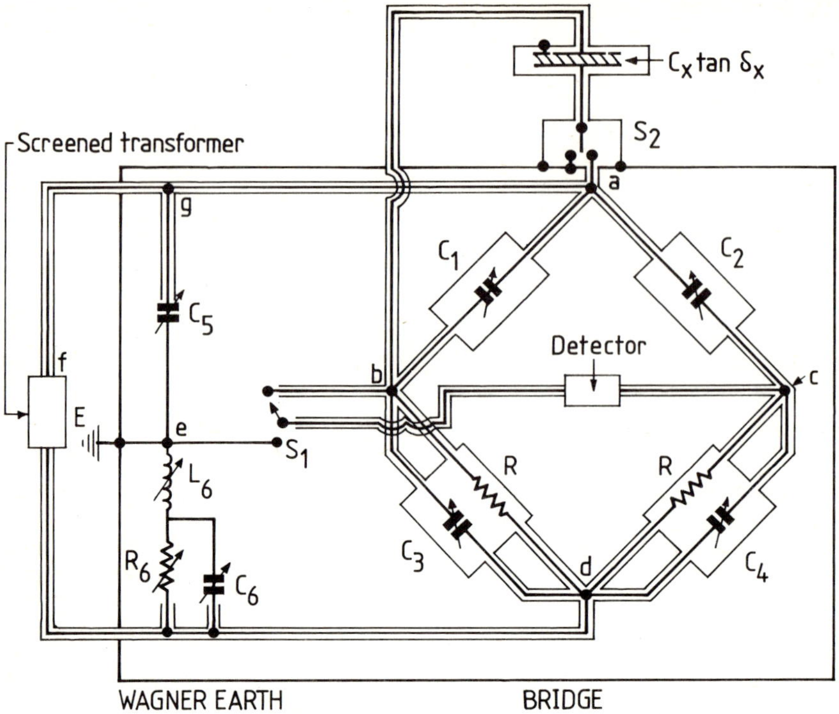

Figure 13.14 Modified Schering bridge for three-terminal measurements

(ii) Transfer detector to ce and balance the Wagner earth components C_5 and C_6 (and L_6 if necessary).

(iii) Alternate detector between bc, balancing C_2 and C_3, and ce, balancing C_5 and C_6, until both positions give a null simultaneously. Thereafter C_2 and C_3 must be left undisturbed.

(iv) Connect main electrode to a and re-balance main bridge with C_1 and C_4 only (detector at bc) followed by Wagner earth with C_5 and C_6 (detector at ce). Alternate between bridges until both are in balance simultaneously as before.

The loss tangent and capacitance are then derived as for the two-terminal arrangement.

Hague's excellent book[79] on a.c. measurements, although somewhat dated, gives an extremely detailed account of Schering's bridge and many others.

Transformer Ratio-arm Bridge

The other type of bridge circuit which will be considered in detail is the transformer ratio-arm bridge. This has been known for many years but has not been widely used until fairly recently, its development being somewhat delayed until advances in transformer design and materials enabled its virtues to be better exploited.

Figure 13.15 shows such a bridge in its simplest form, with the transformer secondary winding divided electrically exactly into two, forming two of the bridge arms, and the unknown and standard impedances (represented by C_x in parallel with R_x, and C_s in parallel with R_s, respectively) forming the other arms. The primary winding of the transformer is connected to an a.c. source and the detector completes the circuit.

For equal capacitors of equal loss tangent, at balance $C_s = C_x$ and $R_s = R_x$. For measuring low-loss capacitors it is usual to provide a series variable resistance in the standard arm instead of the parallel resistance, the values of loss tangent and capacitance of the unknown being given in terms of the appropriate equivalent circuits. For three-terminal unknowns the transformer centre tap is normally earthed and becomes the guard point. The capacitances to guard from the remaining two terminals or electrodes of the unknown are thus either a shunt on the detector, or across one half of the transformer winding. The effect is either to decrease the sensitivity of the detector (negligible for capacitances of the order of a few hundred pF) or to impose an additional loading on the transformer winding. This latter effect is in general small, with modern constructional methods and the improved magnetic core materials now available.

The simplicity of the basic circuit conceals the versatility of the transformer ratio-arm bridge and its outstanding advantages are its range and accuracy. In some bridges the single standard capacitor is replaced by a number of standards in decade steps, say 1 pF, 10 pF, 100 pF, etc. The 'standards' side of the transformer secondary is tapped at $\frac{1}{10}$, $\frac{2}{10}$, $\frac{3}{10}$ and so on of the winding and the switches are arranged so that any standard capacitor or parallel resistor can be switched to any ratio. The total capacitance of the 'standard' arm is then given by the sum of the standards multiplied by their respective ratios.

The inherent accuracy of the bridge arises from the precision with which modern transformers can be wound and tapped, combined with the use of precision fixed-value capacitors. In the best bridges the ratios can be made as accurately as 1 p.p.m. or better. The use of a number of relatively cheap fixed precision standard capacitors in place of the expensive variable or decade

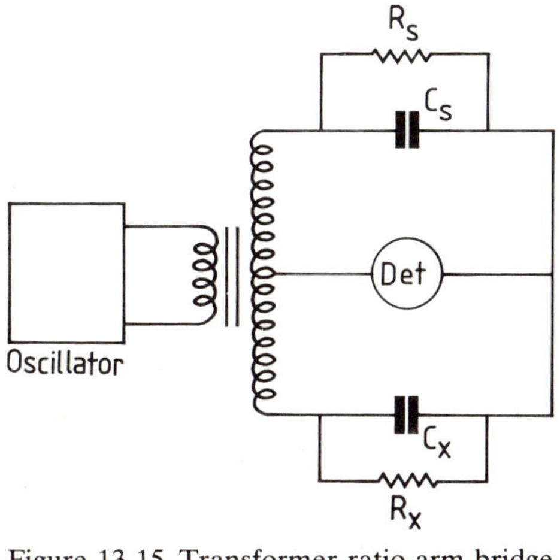

Figure 13.15 Transformer ratio-arm bridge

capacitors required in conventional bridges, results in a considerable economy. The modern transformer bridge has a discrimination of 1 p.p.m. in both loss and capacitance and an accuracy of 0·01 per cent in the latter variable. Its virtues include a self-guarding facility, simple balancing procedure and freedom from corrections. It can be used over the whole of the frequency range from 50 Hz to 10 kHz.

Automatic self-balancing bridges with digital read-outs based on transformer ratio-arms, the advantages of which hardly need extolling where rapid routine measurements are required, are now available.

13.6.7 Techniques for Measurements at 10 kHz–100 MHz

The frequency range 10 kHz–100 MHz corresponds approximately to that of radio transmissions, and telecommunication techniques based upon the tuned (or resonant) circuit are widely used for measurement purposes. The resonant circuit, comprising basically an inductance and capacitance in parallel, has the important property that it is frequency sensitive, and at one particular frequency its sensitivity to relatively slight changes in either frequency or capacitance is enormously greater than at any other frequency. These two features are exploited in the Q-meter and in the susceptance variation methods of measuring dielectric properties, whereby the addition of a capacitance in the form of a dielectric test specimen to the circuit causes a change in resonant frequency, in the first case the change being compensated by frequency adjustment and in the second by capacitance adjustment. The Q-meter will not be dealt with since it is used rather less in the UK than the second method, although it is probably capable of as great an accuracy in its refined forms.[80,81]

Hartshorn and Ward Apparatus

The susceptance variation method was originated by Hartshorn and Ward[82] in 1936 and is dealt with in BS 2067 (1953)[74] in some detail. Refinements by Barrie[83] enable it to be used to measure loss tangents of the order of 0·0001 to a few μR (1 μR = 0·000 001).

In this method a resonant circuit (see Figure 13.16) comprising an inductance (L) in parallel with a variable air capacitor (C) which also forms the electrodes for the test piece is loosely coupled electromagnetically to a stable oscillator. A valve voltmeter with a square law relationship (i.e. one whose output is proportional to the square of its input voltage) is connected in parallel with the circuit and serves both to indicate resonance and to measure the degree to which the circuit is adjusted off resonance during a measurement. The voltmeter output is connected to a reflecting galvanometer, to achieve the high sensitivity

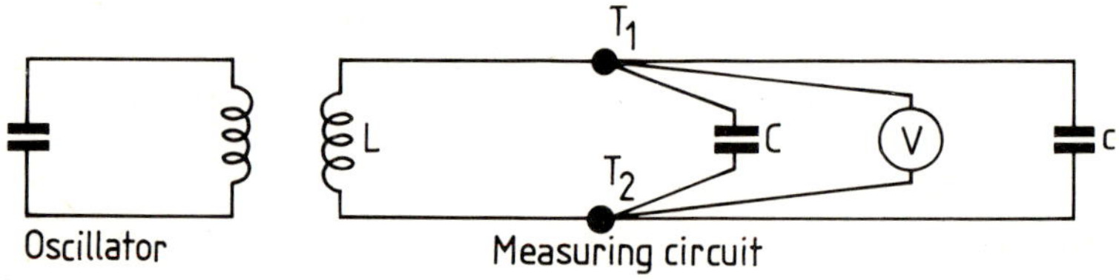

Figure 13.16 Basic circuit of Hartshorn and Ward apparatus[82]

required. A second small capacitor (c) is also connected in parallel with the main circuit and is used to produce a given degree of detuning in the circuit.

In practice a test piece is placed between the electrodes and the oscillator frequency is adjusted to resonance as indicated by maximum galvanometer deflection θ. The circuit is then de-tuned using capacitor c to produce a deflection of $\theta/2$, first on one side of resonance by increasing the value of c and then on the other side of resonance by decreasing c, the total change of capacitance being denoted by Δc_i (Figure 13.17). Finally c is adjusted back to resonance. The test piece is then removed and resonance is re-established by increasing the capacitance of C, after which the de-tuning to half deflection is repeated using c, the total change in its value being denoted by Δc_o (less than Δc_i).

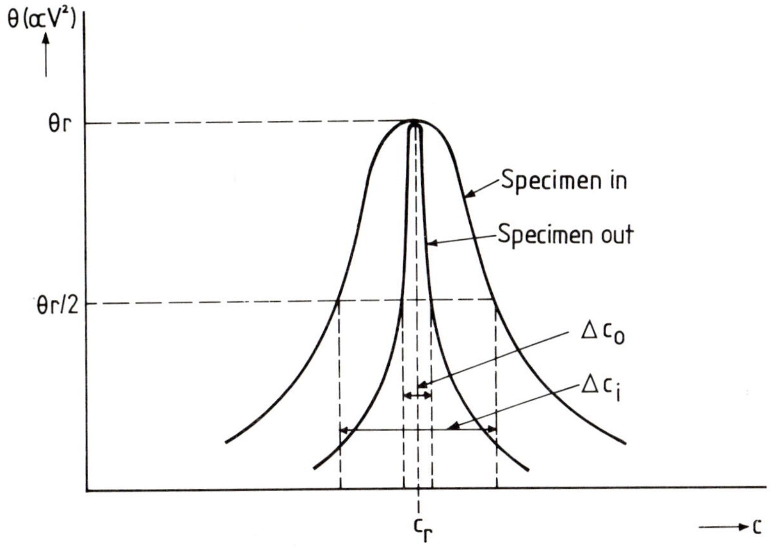

Figure 13.17 Resonance curves

The true test piece capacitance (C_s') is thus given by the changed setting of the main calibrated capacitor (C) after appropriate edge corrections, and the loss tangent is given by the expression

$$\tan \delta = \frac{\Delta c_i - \Delta c_o}{2C_s'} \tag{13.16}$$

The deflections at resonance for both conditions are shown in Figure 13.17 as equal, but this is not necessary; for accuracy they should be kept as large as possible and this is achieved by moving the oscillator nearer or away, as appropriate.

For low loss tangents, of the order of 0·0002 or less, the difference $\Delta c_i - \Delta c_o$ becomes too small to measure accurately in this way (for example, with $C_s' = 20$ pF typical values are $\Delta c_i = 0{\cdot}65$ and $\Delta c_o = 0{\cdot}64$ pF at 1 MHz for $\tan \delta = 0{\cdot}000\,25$) and an alternative method based on the ratio of deflections at resonance is used. It can be shown that

$$\frac{\Delta c_i}{\Delta c_o} = \frac{\theta_o}{\theta_i} \tag{13.17}$$

where θ_o and θ_i are the deflections at resonance corresponding to test piece 'out' and test piece 'in' respectively, *provided that the oscillator is not disturbed during the measurements* (i.e. the coupling is not altered). By re-arranging,

$$\theta c_i - \Delta c_o = \left[\left(\frac{\theta_o}{\theta_i}\right)^{1/2} - 1\right] \Delta c_o \tag{13.18}$$

so that the loss tangent becomes

$$\tan \delta = \frac{[(\theta_o/\theta_i)^{1/2} - 1]\,\Delta c_o}{2C'_s} \tag{13.19}$$

and for small losses this approximates further to

$$\tan \delta = \frac{\Delta\theta \Delta c_o}{4\theta_i C'_s} \tag{13.20}$$

where $\Delta\theta = \theta_o - \theta_i$.

Since Δc_o is a constant for a particular apparatus and frequency, by using a fixed value of θ_i a determination is reduced to a measurement of $\Delta\theta$ and C'_s only. For the example given above of a loss tangent of 0·000 25, $\Delta\theta$ would be 25 mm using a metre scale and θ_i = 800 mm so that measurements down to 0·000 01, corresponding to $\Delta\theta$ = 1 mm, are quite feasible. Some care in ensuring stability of the deflections is essential at these low levels and Barrie's paper [83] is particularly helpful.

13.6.8 Techniques for Measurements above 100 MHz

At frequencies above 10^8 Hz the two test methods of principal interest are those employing standing wave methods and cavity resonator techniques.

In the standing wave method a waveguide is employed appropriate to the particular frequency range of interest and a dielectric test piece of rectangular or cylindrical section (depending on the type of waveguide), which must be a close fit in the guide, is inserted in one end. The length of the test piece required is a function of its permittivity and the wavelength corresponding to the test frequency and it may be necessary to adjust the length by machining after an initial measurement. The frequency of a generator coupled to the opposite end of the guide is adjusted to resonance, a suitable crystal probe inserted in a slot in the guide serving as a detector. The detector may be accurately positioned along the axis of the guide by means of a vernier slide so that the positions of the standing wave nodes can be evaluated. The standing wave pattern is studied with the test piece both in position and removed, and the permittivity and loss tangent are given by rather complicated equations; these yield a number of solutions and some care is needed in obtaining the correct one.

Further details are given by Roberts and Von Hippel,[84] Westphal,[85] Gevers[86] and Dube and Natarajan.[87]

With cavity resonators the test piece is normally a disc, which forms the end of a cavity, coupled to a waveguide. As with the standing wave methods, a generator and crystal detector (in this case usually a square law type) are also coupled to the guide and the frequency is adjusted to resonance. Measurements are now made of the bandwidth of the sytem, usually at half power points, by means of a fine frequency control (accurately calibrated) in a manner somewhat analogous to that using the Hartshorn and Ward apparatus. The solutions are again complex and tedious to solve. Gevers,[86] Parry,[88] Horner *et al.*[89] and Brydon and Hepplestone[90] describe in detail various cavity resonators.

The two methods described are suitable for the frequency range from approximately 3×10^8 Hz to 3×10^{10} Hz.

13.6.9 Standard Test Methods

IEC 250 (1969),[67] relating to insulating materials in general, gives information on the form of test pieces, types of measuring equipment, electrode materials, testing procedures and calculation of results for measurements from 15 Hz to 300 MHz. It is, however, very general, and national standards for plastics generally lay down restrictions on these factors, although the results are not affected much by the measuring equipment used and sometimes 'other equipment which can be shown to give the same results' is permissible.

In BS 2782 (1970)[32] the low-frequency methods (Method 205 for 50 Hz, Method 206 for 800–1600 Hz) both specify a Schering bridge with Wagner earth as mandatory, but neither gives any details. The applied voltage is stated to be preferably 1000 V at 50 Hz, or less for thin or low electric strength specimens, and 100 V for 800–1600 Hz. Electrodes are mercury or metal foil only, although at 50 Hz plain brass discs applied with pressure are permitted, provided that they give the same results. At 800–1600 Hz, power factor (but not permittivity) measurements are allowed to be made without a guard ring. No electrode sizes are given, but a guard ring width of 25 mm and a gap of approximately 1·5 mm are stipulated at both frequencies, the only limitation on the central electrode area being that it should be sufficiently large to enable capacitance to be determined to ±5 per cent. For thin specimens the guard ring width is excessive and in practice this requirement means that the approximately 51 mm diameter Hartshorn and Ward test piece cannot also be used for low-frequency measurements. For measurements at 10 kHz to 1 MHz (Method 207) the Hartshorn and Ward method detailed in BS 2067 (1953)[74] is prescribed. BS 903 (1956)[33] Part C3 has similar methods, but the use of the Hartshorn and Ward equipment is not mandatory. BS 2782 requires duplicate test pieces, whereas BS 903 requires only one. BS 771 (1980)[91] invokes IEC Publication 250 in Part 1 and BS 2782 in Part 2.

BS 4618 (1970)[92] lays down a method of presentation of permittivity data (Section 2.1) and loss tangent data (Section 2.2). These recommendations give advice for selecting temperature intervals, etc., and graphically displaying the results in a clear, standardised way. If these documents become widely accepted, then the present confusion resulting from the use of different methods, frequencies, temperatures, stresses, etc., by manufacturers may yet disappear. As with resistivity, it may be questioned whether the measurements at less than 1 per cent r.h. have any relevance for design purposes.

The blanket ASTM specification covering dielectric measurements from 1 Hz to several hundred MHz is D150 (1978)[68] which, although extremely helpful and informative, is written in quite general terms and makes very few specific recommendations about methods or apparatus. For example, no electrode sizes are even suggested and the only recommendations made are that the guard gap should be as small as possible and the guard width should be least twice the specimen thickness. No particular type of apparatus is required to be used, but an appendix suggests details of a number of bridge and other circuits. Some details are given of the liquid immersion technique, but this method is dealt with at some length in ASTM D1531 (1962).[75] The cell is a two-terminal type with a double electrode arrangement requiring two specimens. The fluid is either very pure dry benzene or silicone oil of viscosity 1·0 cS, both of permittivity near 2·28, and only low-density polyethylene is covered by the method. The frequency range covered is between 1000 Hz and 1 MHz; although no one type of apparatus is mandatory, equations are given for direct reading bridges and Q-meters. Figure 13.18 shows the cell.

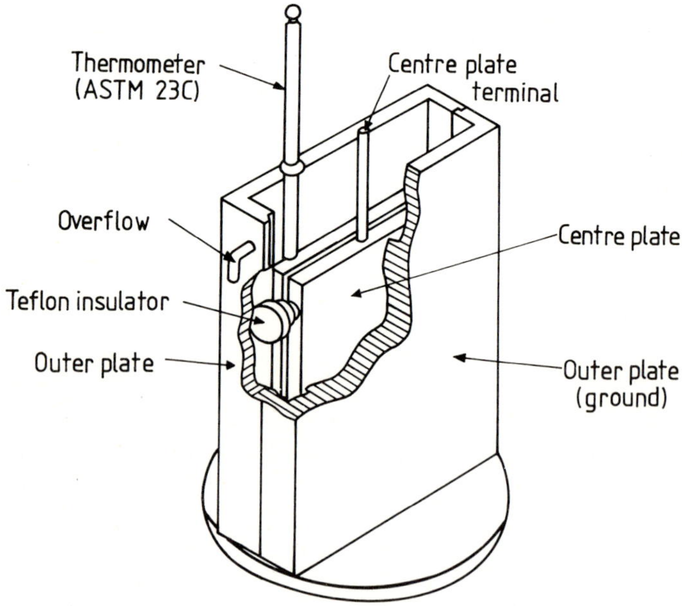

Figure 13.18 Liquid displacement cell: ASTM D1531

ASTM D1673 (1979)[93] for cellular plastics, although invoking ASTM D150 for test methods, is interesting in that it takes account of the heterogeneity of such materials and their availability in large pieces by permitting specimens of up to 400 mm square by 50 mm thickness to be tested, though at reduced maximum frequency (<1 MHz). ASTM D1674 (1967)[94] for embedding compounds is also relevant in that a simple two-terminal electrode system suitable for measurements up to 300°C is described (Figure 13.19).

The German standard for dielectric measurements is DIN 53483 (1969).[69]

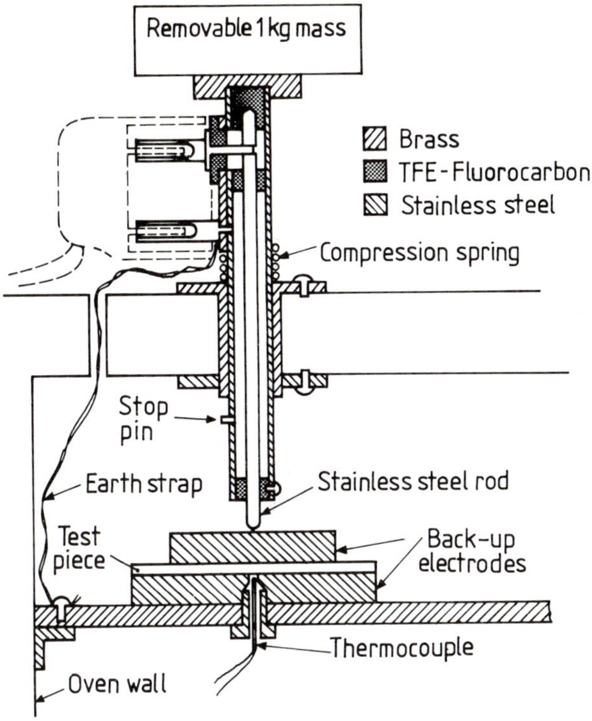

Figure 13.19 Elevated temperature apparatus: ASTM D1674

13.7 ELECTRIC STRENGTH

13.7.1 Definitions and Explanatory Notes

In alternating voltage electric strength tests the voltage applied at any instant is normally defined as the peak voltage divided by $\sqrt{2}$; for a perfectly sinusoidal waveform this is equal to the r.m.s voltage.

In some cases, especially with products, the *breakdown voltage* only is recorded. This is either (i) the voltage at which breakdown occurs, or (ii) the highest of a specified series of voltages which the test peice will withstand for a specified time without breaking down.

The *electric strength* or *breakdown strength* is the breakdown voltage divided by the test piece thickness (or the appropriate other dimension in the case of tests parallel to the plane of the sheet (§13.7.6)).

The difficulties associated with the definition and the measurement of the property arise because of the importance of a large number of variables which affect the result, and because of the number of modes in which breakdown can occur, all or some of which may contribute to failure in a given situation. Breakdown is a disruptive phenomenon and as such is very dependent upon the macroscopic, and even the microscopic, structure of the insulation. It may occur through dielectric heating causing an excessive temperature rise (and with many materials power factor increases with temperature, resulting in further heating and ultimately 'thermal runaway'); the voltage stress may be sufficiently high to accelerate electrons through the material, causing an electron avalanche; gaseous voids inside the material may ionise, causing bombardment of the material and subsequent degradation; the stress may even be high enough to break the chemical bonds in the polymer structure and cause decomposition into less resistant materials, and so on.

Breakdown is a 'weak point' phenomenon and as such it may be instructively compared with, say, a tensile strength failure on a fabric specimen where one thread invariably parts first, thus initiating rupture; similarly breakdown commences where the stress is highest, and one of the major difficulties in electric strength tests lies in the facts that normally the electric stress varies, being greatest near the electrode edges, and that discharges occurring in the ambient medium just outside the electrodes normally initiate breakdown. Irregularities in electrode surfaces or in the test piece due to inhomogeneity affect the results. It is therefore necessary to keep the electrodes clean and free from pitting (resulting from arcing) and to use homogeneous and smooth test pieces.

Under the most carefully controlled conditions, when all spurious discharges are eliminated and all secondary mechanisms of breakdown are avoided, it is possible to achieve for a material a consistent value known as the *intrinsic electric strength*, which is independent of thickness; rather surprisingly, the range of values obtained for many different homogeneous plastics materials is small and of the order of 0·5–1 MV/mm. Such measurements are of great interest to the theoretician, but in practice industrial measurements are usually made under relatively crude conditions, in order to simulate more closely practical situations, and breakdown can occur by a number of modes, sometimes operating simultaneously.

The tests to be described measure the *industrial electric strength* at power frequencies under discharge conditions (impulse and d.c. tests are specialised and will not be considered). Since breakdown under these conditions is a time-dependent phenomenon, and the rate of voltage application is analogous to rate of loading in the tensile example given earlier, it is important to standardise on the time of voltage application and even the rate of initial voltage increase; so-called 'industrial electric strength tests' are careful to specify such details. The results of such tests are very thickness-dependent in contrast to intrinsic electric strength tests, and the values obtained in general rarely approach even one-tenth of the intrinsic values.

The results are affected by the electrical (and sometimes by the thermal) characteristics of the ambient medium and by the dimensions (and sometimes by the thermal properties) of the electrodes. In addition, the results are dependent upon the frequency and waveform of the applied voltage.

The electric strength is thus dependent on so many variables that there is no 'correct' value to aim at, and consistency of results between one experimenter and another can only be achieved by mutual attention to the test variables. In other words, specifications are essential, and it should not be expected that two apparently fairly similar test methods will give similar results. Comparisons between materials are only strictly valid when experimental conditions are strictly controlled and virtually identical.

A high industrial electric strength does not necessarily mean that the material will withstand *long-term degradation processes* such as erosion or chemical deterioration by discharges (§13.8) or electrochemical deterioration in the presence of water which may cause eventual failure in service at much lower stresses. On the other hand, if discharges are eliminated in service a higher stress may be withstood.

Thus these tests are only of limited value for design, although they are useful for quality control purposes.

13.7.2 Schedule of Voltage Application

There are three types of schedule of voltage application commonly used in standards.

In the *rapidly applied voltage* test, the voltage is raised at a uniform rate until breakdown occurs. The rate may be directly specified or be such that breakdown occurs within 10–20 seconds (on average). (In ASTM D149 (1975),[95] 50 per cent of the short-time breakdown voltage may be applied initially and the voltage then raised at a slow uniform rate.) The electric strength is calculated from the voltage at which breakdown occurs.

In the *20 second step-by-step* test a trial specimen is usually tested initially, using the rapidly applied voltage schedule. The starting voltage for the test proper is then taken as 40 per cent of the failure voltage of the trial specimen and the test proceeds in defined voltage steps in the series (kV) 0·50, 0·55, 0·60, . . . , 1·0, 1·1, 1·2, . . . , 2·0, 2·2, 2·4, . . . , 4·8, and then 10 and 100 times this series up to 100 kV, each step being maintained for 20 s. Here the breakdown strength is generally calculated from the highest of these voltages withstood for the complete 20 s.

In a *proof test* the voltage is raised as rapidly as possible without overshoot to a specified value and then held at this value for a specified period (usually 1 minute) and breakdown constitutes a failure.

13.7.3 Ambient Medium

Most tests are carried out in transformer oil at room temperature, but the thermosetting plastics more usually are tested at 90°C (see BS 771 (1980)[91] for example). Some tests may also be made in air. If this is done at elevated temperatures, the oven required must have an adequate insulating bushing and be sufficiently large to prevent flashover to the oven and significant influence of the electric field in and close to the test piece.

13.7.4 Number of Tests

The number of test results required varies widely with the specification. In some cases only two are required in addition to those for any initial trial tests, while IEC 243 (1967)[96] calls for five with a further five tests being carried out if any result differs from the mean by more than 15 per cent. In the last case the mean of all 10 results is calculated. BS 2782 (1970)[32] requires the central value of five results to be calculated.

13.7.5 Voltage Supply and Measurement

Safety apart, there are a number of hazards associated with the tester itself, amongst which are errors caused by loading of the transformer due to excessive specimen leakage current or discharge currents, distortion of waveform by poor transformer design, malfunctioning of circuit breakers, and so on. Since it is the peak value of voltage which determines breakdown, whereas many indicating meters are average value instruments (although scaled in r.m.s.), it is essential to check the meter readings against a peak measuring device, such as a sphere gap. BS 358 (1960)[97] is particularly helpful here; 50 mm spheres are appropriate for the range 8–65 kV peak, which is relevant to most of the requirements in plastics specifications. A guide to high voltage testing techniques is given in BS 923 (1980)[98] and most standards detail requirements of the breakdown tester and voltage measurement methods. These requirements in most standards are very similar.

13.7.6 Standard Test Methods

The master standard relating to electric strength at power frequencies of insulating materials in general is IEC 243 (1967).[96] It is proposed to issue it as an amended version of BS 2918 (1957),[99] which was originally based on an early draft of the IEC document.

A revision of BS 2782 (1970)[32] Method 201 for plastics and BS 903 (1956)[33] Part C4 for rubbers and ebonite is being carried out to provide two common documents for both standards which are substantially in line with IEC 243; what follows is an outline of the methods which will probably be contained in the new documents. In BS 2782, they will probably be known as Methods 220 and 221, referring to the rapidly applied voltage method and the 20 s step-by-step method, respectively, but they are otherwise similar in content.

The first test in each of these documents is for sheet materials. The electrodes consist of solid brass cylinders, the lower one being 75 mm in diameter and 15 mm high and the upper one 25 mm in diameter and 25 mm high. The electrodes have radiused (3 mm) edges to reduce the local stress. The arrangement is shown in Figure 13.20. For most tests the electrodes are immersed in transformer oil (complying with BS 148 (1972)[100]) and the temperature is either 23° ± 2°C or 90° ± 2°C.

The test for narrow strips and tapes uses coaxial brass rod electrodes radiused to 1 mm, the upper being of weight 50 g. These are mounted five pairs at a time in suitable insulating blocks so that the quintuplicate tests usually specified can be rapidly carried out. Essentially the test is for thin materials tested in air and consequently it is limited to voltages of about 10 kV or less. Figure 13.21 shows the arrangement.

The third method is for rigid round or rectangular tubes. For tubes less than 100 mm in diameter it uses a closely fitting long inner electrode of rod, tube or foil, and the outer electrode is a band of foil 25 ± 1 mm wide. For tubes larger than 100 mm in diameter the inner electrode is a disc of foil 25 mm in diameter and the outer band around the tube is 75 mm wide.

The next three methods in the draft revision of BS 2782 and BS 903 cover flexible tubing. The first is essentially the same as for rigid tubing of small

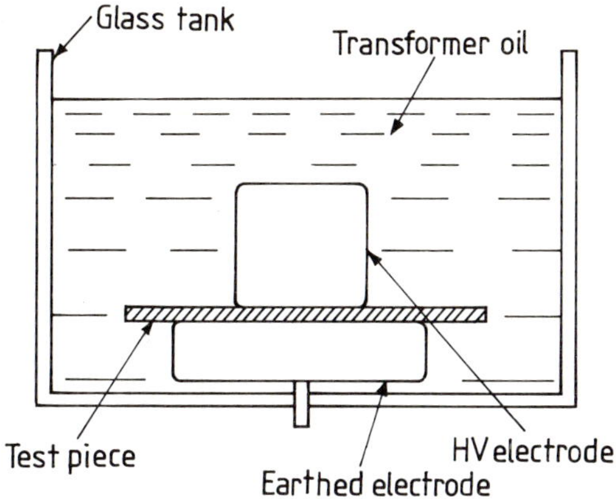

Figure 13.20 Electric strength electrodes for sheet materials

diameter and the inner electrode is straight. The second uses a similar inner electrode but bent into a U-shape, while the outer electrode is of metal shot (Figure 13.22). A subsidiary test is made to ensure adequate conductivity through the bulk of the shot. The third of these methods is for large bore flexible tubing, where the tubing is cut longitudinally, opened flat and tested as for sheet material.

The remaining method is intended principally for laminated materials whose electric strength may be much lower in the plane of the laminate than normal to it. The test is therefore conducted on test pieces 25 mm wide placed edgewise between metal plates large enough to overlap all round by at least 15 mm (Figure 13.23). It is important that the test piece has the contacting edges truly machined. To assist the mechanical stability of the arrangement, two or three test pieces may be used to support the electrodes, although of course fresh test pieces are required for replicate tests. For sheet material and for large bore tubing, a test piece about 100 mm long is cut, while for tubes up to 100 mm in diameter (or external side dimension) the complete cross-section is used.

Many recent specifications, e.g. BS 3784 (1973),[101] BS 3816 (1964),[102] BS 3924 (1978)[43] and BS 5102 (1974),[103] invoke BS 2782 directly and it is probable that others will eventually be aligned with it. BS 771 (1980)[91] however, invokes IEC 243 (1967)[96] in Part 1 and BS 2918 (1957)[99] in Part 2.

An unusually small electrode system is given in BS 3873 (1965).[104] As the title of the standard implies, various product sizes are involved and the electrodes are designed so that very small pieces can be tested. A test piece 9·5 mm diameter by 0·76 mm thickness is prepared by cutting from a moulding and it is required that the test piece has smooth square edges and plane parallel faces. The test piece is inserted in the centre of a non-rigid electrical grade sleeving 25 mm in length, which must be of such a bore as to grip it tightly; this assembly is mounted in the electrodes (Figure 13.24) which are then immersed in transformer oil. The test piece is subjected to a proof voltage of either 11·8 kV/mm or 23·6 kV/mm (depending on grade) for one minute. Five test pieces are tested.

Other special kinds of electrodes are used for finished products. For example, BS 2848 (1973),[46] which includes sleeving of PTFE, silicone rubber, PVC and

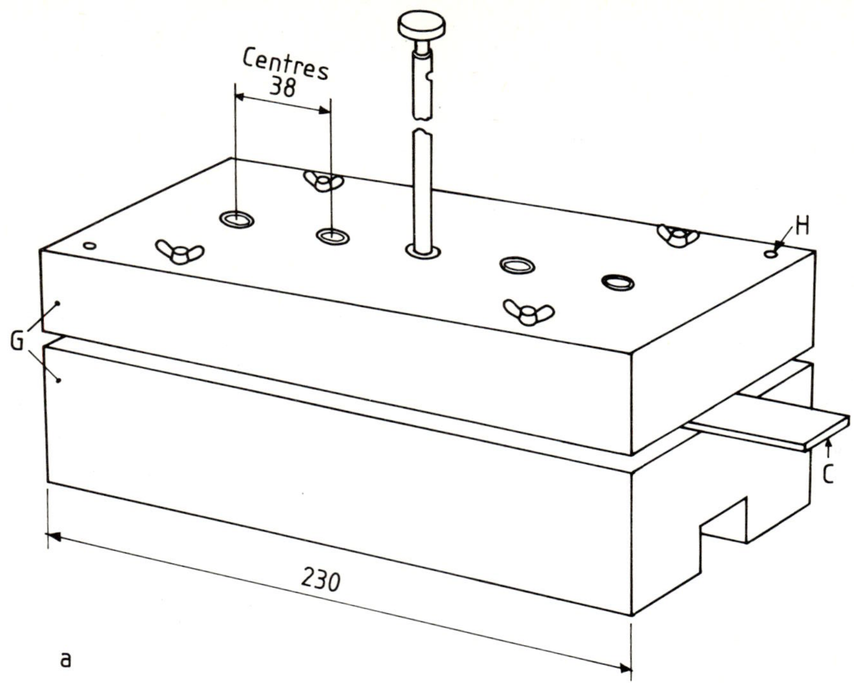

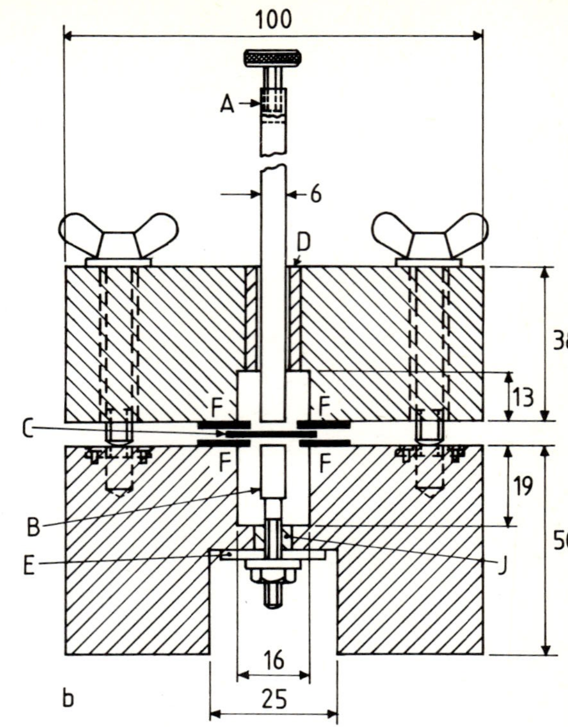

Figure 13.21 Test for tapes—dimensions in mm
(a) General arrangement of apparatus
(b) Section with top slightly raised

A Upper electrode to be an easy fit in bush (D)
B Lower electrode
C Test piece
D Brass bush inside diameter just sufficient to clear 6 mm rod
E Brass strip 25 mm wide connecting all lower electrodes
F Strips of suitable insulating material overlapping edges of sample
G Blocks of suitable insulating material, e.g. paper-filled laminate
H Dowel hole
J Brass bushing with internal thread

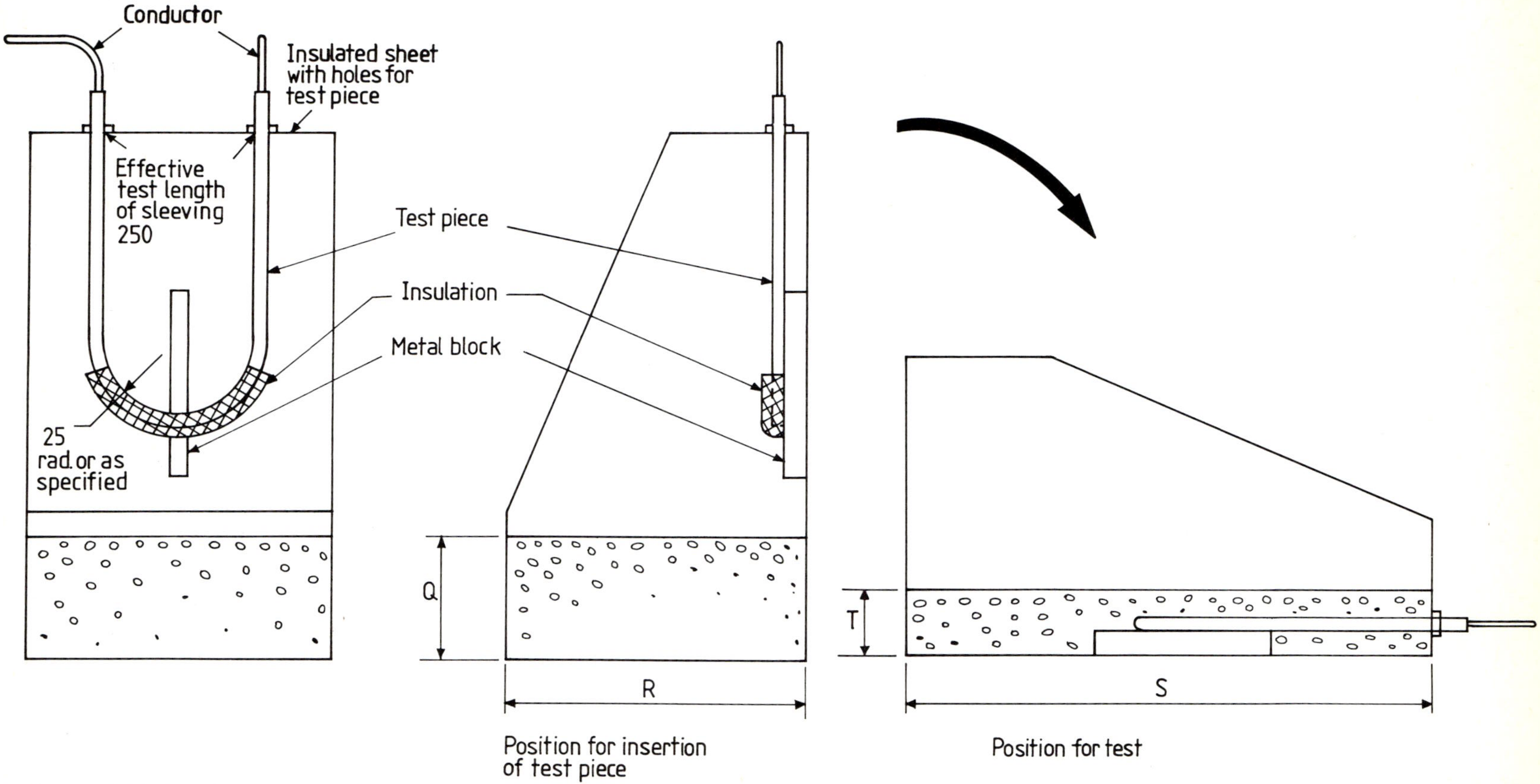

Figure 13.22 Test for sleeving—dimensions in mm

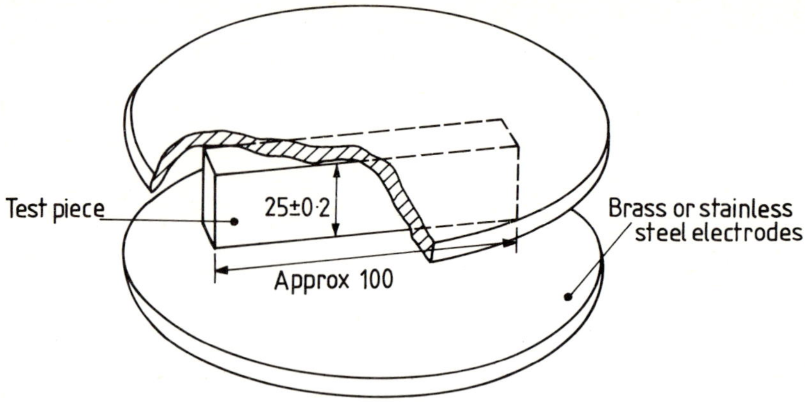

Figure 13.23 Test parallel to the surface—dimensions in mm

polyethylene, gives an *ad hoc* test in which a test piece is threaded on to a wire or rod of diameter approximately 75 per cent of the bore and wrapped round a metal mandrel. A proof voltage is applied between the mandrel and the wire or rod and the test is carried out in air or water at room temperature or in air at the maximum operating temperature.

The ASTM blanket test for electric strength is ASTM D149 (1975).[95] For sheet and film materials the upper and lower electrodes are identical and are 25 mm high. There are two different sizes, 51 mm diameter with the radius of the edges 6·4 mm, and 25 mm diameter with an edge radius of 3·2 mm. It is not very clear why the two sets of electrodes should be so close together in size, nor why and when one set is used in preference to the other. For tapes and films the electrodes are identical to the 6 mm British Standard ones. ASTM D350 (1978)[105] gives a breakdown voltage test on a test piece with foil outer and rod or wire inner electrodes.

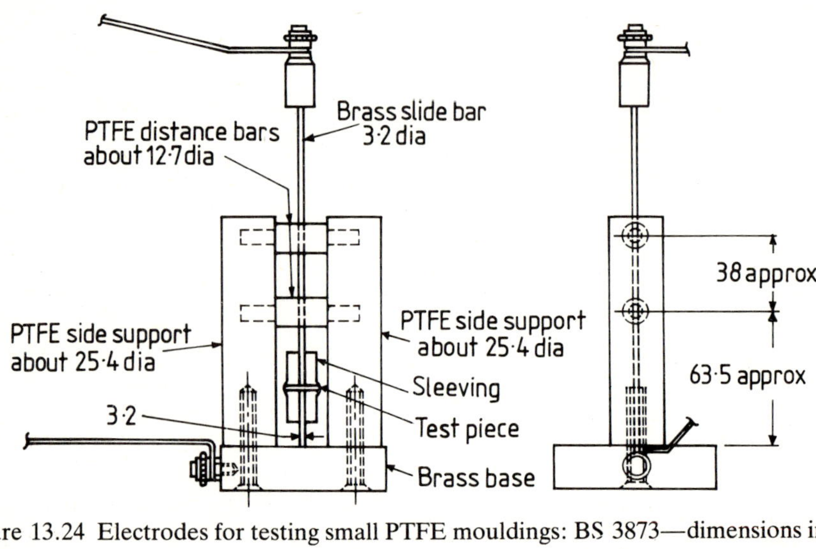

Figure 13.24 Electrodes for testing small PTFE mouldings: BS 3873—dimensions in mm

A test for determining conductive paths in tapes, such as that in BS 4893 (1973),[106] is carried out at a relatively low voltage, and the current is limited to a low value. The test piece is laid flat on a polished brass or steel plate and traversed by a roller of polished brass or steel, a voltage of 5 kV being applied between the two metal electrodes. The number of breakdowns is counted.

13.8 RESISTANCE TO SURFACE DISCHARGES, TRACKING RESISTANCE AND ARC RESISTANCE

13.8.1 General Considerations

Tests for resistance to surface discharges, tracking resistance and arc resistance are all concerned with the destruction of the surface of the material by sustained electrical discharges or arcs. The results obtained are very dependent on so many factors that (i) standard test methods need to be very tightly specified and (ii) the results of these tests, while being an aid to material selection, are not directly applicable to calculating design stresses or creepage distances.

In the surface discharge test there is no deliberate contamination of the surface; in tracking tests, the surface is wet. In both these tests, the discharges do not initially develop into a persistent arc, whereas in the arc resistance test a persistent arc is created initially.

13.8.2 Resistance to Surface Discharges

It might be thought that the effect of surface discharges from conductors to the plastics surface was dealt with in §13.7, but in the electric strength test failure is produced in periods of the order of minutes. The resistance to surface discharges is the ability of a material to withstand the effects of less intense discharges over a long period (measured from 100 hours to, say, 5000 hours). Discharges are here defined as the discontinuous movement of electric charges through an insulating medium, initiated by electron avalanches and supplemented by secondary processes. A more complete treatment of the subject of surface discharges than can be presented here is given by Mason.[107]

The initiation of the discharges occurs at a critical voltage which is determined by the properties of the ambient medium, the surface properties, thickness and permittivity of the test piece, and the configuration of the electrodes.

The method of test given in IEC 343 (1970)[108] recommends rod electrodes mounted perpendicular to the test pieces which rest on a flat plate electrode of much greater area. With soft or thin materials the rod may be separated from the test piece surface by up to 100 μm in order to avoid mechanical damage.

13.8.3 Tracking Resistance

Tracking is defined as the progressive formation of a conductive (carbonised) path across the surface of an insulator by surface discharges.

Since the presence of carbon in the insulator is a prerequisite for this type of failure, it might be expected that plastics in general would be prone to tracking, but this does not follow and some of the polymers which are richest in carbon, such as polystyrene, are non-tracking, although other adverse effects such as erosion occur in this particular case. Tracking is influenced considerably by the presence of surface films of moisture and dirt, etc., and the test methods involve the introduction of conducting media such as salt solutions on to the surface in the form of droplets or by dipping, in order to produce an acceleration of the natural processes. Although tracking may occur in a variety of practical situations—two common examples are the failure of domestic light switches and of car distributor caps—the problem is more usually associated with areas of high atmospheric pollution, and those near the coast and on board ships where salt spray is a natural hazard.

Numerous tracking resistance tests have been developed in attempts to simulate the results of practical failures in the field and two types of test have received unusually general support. The first is the 'comparative tracking index' (CTI) test, which gives a numerical result only on materials which are not highly tracking-resistant, and gives only a 'does not track' result on a number of plastics. It appears that this shortcoming has resulted in the second test, the so-called 'inclined plane test', which leaves fewer materials undifferentiated.

Comparative Tracking Index (CTI)

This most widely used test is described in IEC 112 (1979),[109] which has recently been revised. BS 3781 (1964),[110] which was based on, but not identical to, an earlier version of IEC 112, has now been withdrawn and it is proposed to issue the IEC document as a dual-numbered British Standard.

The test involves two chisel-shaped platinum electrodes 4 mm apart pressed on the surface by a force of 1 N (Figure 13.25). A 48–60 Hz voltage with a short-circuit current limit of 1 A is applied between the electrodes and drops of aqueous solution of size 20 mm^3 are allowed to fall between the electrodes at intervals of 30 seconds. Two solutions are permitted. The preferred solution (A) is less aggressive than solution (B) and consists of 0·1 per cent NH_4Cl.

The test is performed at a number of different voltage levels and the number of drops to track at each level is noted (Figure 13.26), tracking being defined arbitrarily by the current in the circuit reaching at least 0·5 A for 2 s.

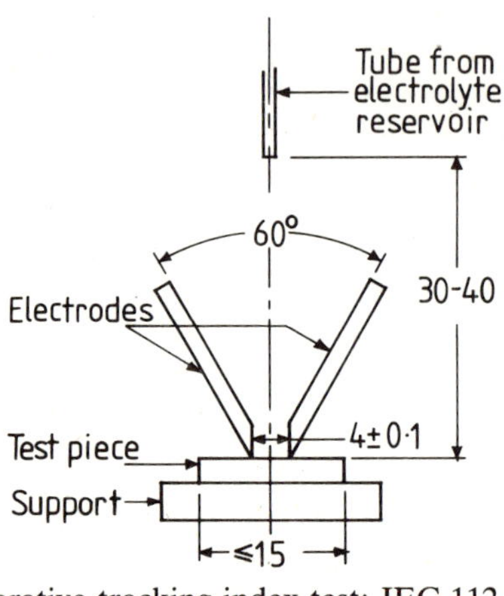

Figure 13.25 Comparative tracking index test: IEC 112—dimensions in mm

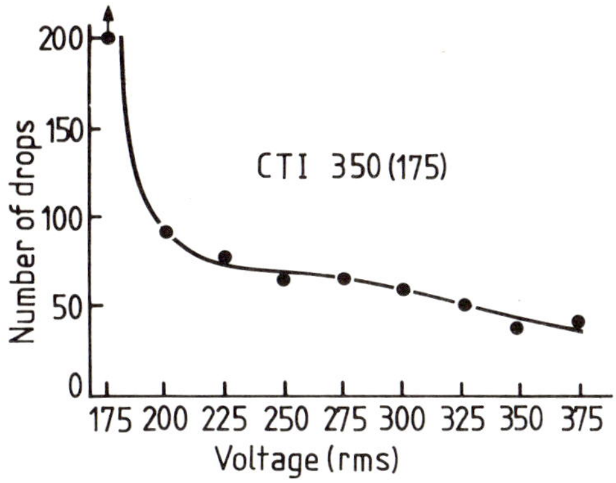

Figure 13.26 CTI curve

The CTI is the voltage which the material will just withstand for 50 drops and is quoted, for example, as CTI 350. However, if the material tracks in less than 100 drops at 25 V below the 50 drop value, the voltage corresponding to failure in 100 drops is determined and given in parenthesis after the 50 drop voltage. Additionally, the letter M is appended if solution B is used. A proof test with preferred voltages is also given.

Some care should be taken in applying the test. The error of the number of drops to track at a given voltage is high. This partly reflects the inhomogeneity of most filled plastics but it is also an inherent feature of a test which endeavours to measure such a complex phenomenon as surface failure, dependent as it must be on the formation of a random path across the insulation surface. It is often possible to obtain the odd tracking failure at a voltage below the CTI and there are cases on record of materials becoming non-tracking at voltages significantly higher than the CTI.

Some materials (e.g. poly(methyl methacrylate)) do not track, but fail by erosion, or by material between the electrodes volatilising to leave a crater. As, in service, such behaviour would be undesirable, since terminals could become loosened, erosion is recognised in the test as a mode of failure. The depth of erosion (mm) of test pieces which do not track is therefore measured.

The test described is only suitable for test voltages up to 600 V. Above this level flashovers tend to occur and the end-point becomes increasingly difficult to define. The significance of the test voltage and its relationship with practical use voltages has not been established and the comparative tracking index obtained should not be regarded as a design level below which it is safe to operate.

The test given in ASTM 3638 (1977)[111] is based on this method but differs in certain respects.

Inclined Plane Test

This test is described in IEC 587 (1977)[112] and has been published without deviation as BS 5604 (1978).[113]

A large flat specimen is inclined at 45° and electrolyte is allowed to flow down its under-surface, forming a bridge between two electrodes pressed against it. The electrolyte stream boils under the influence of the test voltage, which has preferred values of 2·5, 3·5 and 4·5 kV (tests may if required be carried out in 250 V steps from 2 to 4·5 kV) and the resulting discontinuities occurring in the conducting path produce scintillation near the lower electrode.

There are two criteria of failure: in Method A the test is terminated when a current of 60 mA flows for 2 seconds; in Method B, when a track reaches a mark 25 mm from the lower electrode. The advantage of this test is that the conditions can be carefully controlled so that steady scintillation occurs, unlike the intermittent and somewhat uncontrolled scintillation which produces failure in 'drop' tests.

Any material which fails in under 6 h at 2·5 kV is referred to as Class 0 (or Class B0 in the case of Method B).

If five test pieces withstand 2·5 kV for 6 hours, the material may be classified as 2·5 (or B2·5). Tests at 3·5 and 4 kV may be required and the material is classified by the highest voltage withstood for 6 hours (prefixed B for Method B). When no tracking occurs, the depth of erosion after 6 hours is measured.

There is a somewhat similar and more closely defined test in ASTM D2303 (1973).[114] Many of the mechanical and circuit details are identical to the IEC test but the procedures for carrying out the test are markedly different. In the *initial tracking voltage test*, the voltage is raised in 250 V steps every hour, and the end-point is when progressive tracking starts; the voltage then obtained is the initial tracking voltage. In the *time-to-track test* a specified voltage is applied and the time for tracking to proceed 25·4 mm up the test piece is recorded.

Figure 13.27 shows the type of apparatus required but even this does not indicate the complexity of the test. An erosion test may be carried out using a lower voltage (2 kV) than that causing progressive tracking. The eroded depth and eroded volume are recorded.

Dust and Fog Tracking Test

The earliest US method for tracking resistance is given in ASTM D2132 (1968).[115] The principle is that 51 × 13 × 3·2 mm brass or copper electrodes are placed flat upon the surface of a test piece, the assembly is coated with a synthetic dust and subsequently wetted by fine water spray in an appropriate chamber. A relatively high voltage (up to 1·5 kV) applied to the electrodes induces tracking and/or erosion effects along the surface or through the thickness of the test piece and the test is continued until a permanent track ensues or the test piece is penetrated, the time to such failure being recorded. Classification of tracking and erosion resistance is into several groups, a tracking resistant or erosion resistant material being defined arbitrarily as one withstanding respectively 100 h or 200 h without failure by the relevant mode. The test piece size is 127–152 mm square and normally 1·59 mm thick.

Although the test is simple in principle, the test details relating to the dust formulation, the fog deposition rate, etc., are carefully stipulated, as are the circuit details. Figure 13.28 shows the general arrangement.

Differential Wet Tracking Test

This test is given in ASTM D2302 (1975)[116] and is extremely rapid compared with the dust–fog and inclined plane tests; the test conditions are very severe and lead to failure in a matter of minutes. A test piece with a small hole drilled in it is

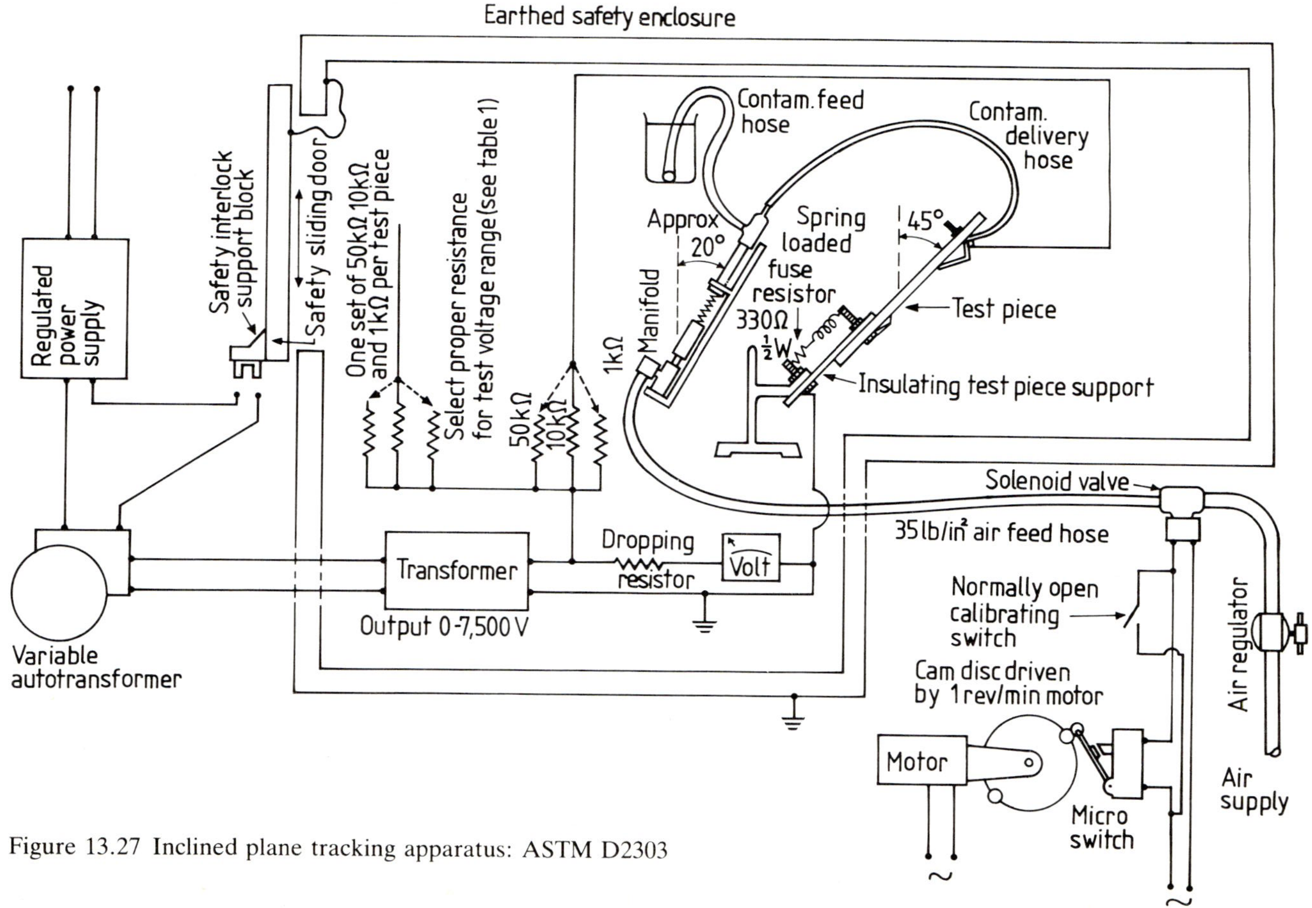

Figure 13.27 Inclined plane tracking apparatus: ASTM D2303

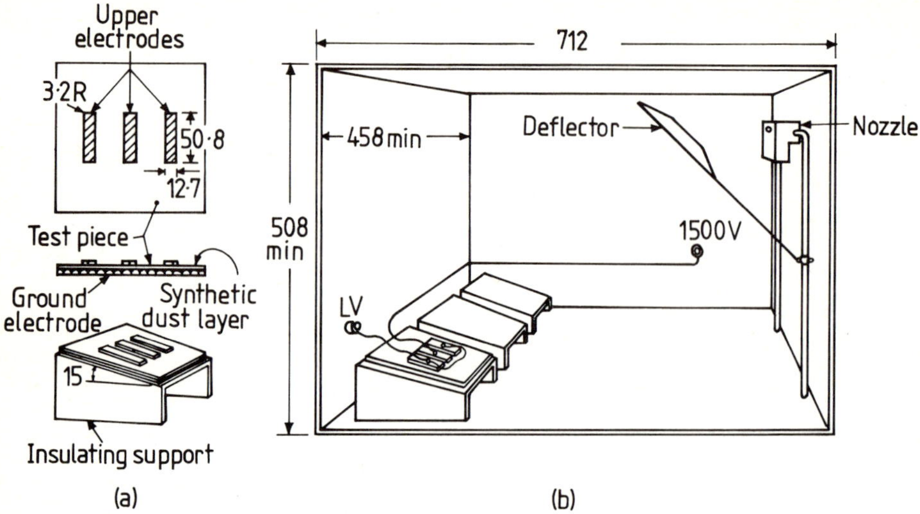

Figure 13.28 Dust–fog tracking and erosion test: ASTM D2132—dimensions in mm

immersed at an angle in a trough of 1 per cent NH_4Cl solution so that the hole just fills with liquid. A 3 kV voltage is applied at various power levels, for 1 minute each, between the liquid and a square stainless steel electrode mounted on the dry emergent portion of the test piece, oriented so that one corner is close to the hole.

13.8.4 Arc Resistance

An arc resistance test is described in ASTM D495 (1973).[117] In this test, which is of a fairly complicated nature, a high voltage arc (15 kV) is struck between two electrodes placed on the test piece, the current initially being of a few milliamps. The arc is interrupted by a timer, the on/off intervals being carefully defined. As the test progresses the 'off' period decreases until finally, in the later stages, the arc is continuous and its severity is increased by increasing the arc current. The maximum time of the test, if completed, is 420 s and materials are classified by their time to failure.

The electrodes are either tungsten rods, with their ends ground at 30° to their axes, mounted at 35° to the surface, or stainless steel strip electrodes placed corner to corner, in both cases the separation being 6·35 mm: quintuplicate tests are a minimum requirement.

The test conditions and components used are rigorously detailed. Nevertheless, some ambiguity over the interpretation of the end-point, which is defined as the extinction of the arc (i.e. the current is entirely within the test piece surface), can occur: some materials may obscure the arc by burning and the test is not applicable to materials that do not produce conductive paths under the action of the arc or which melt and float the conductive residues out of the test area.

A further arc resistance test, given in DIN 53484 (1955),[118] is very simple and involves the production of an arc between two 8 mm carbon rods inclined at 90°

to each other and resting on the test surface. The electrodes are then separated mechanically at a maximum rate of 1 mm/s and the behaviour of the material is categorised into one of six classes. The supply voltage is 220 V d.c. and the current is limited to approximately 10 A.

13.9 OTHER ELECTRICAL TESTS

The properties outlined above are considered to be the prime ones: in all applications of plastics involving electrical insulation, some knowledge of one or more of these properties is required to ensure the proper and efficient use of the appropriate material.

A more specialised test which does not fit into the above classifications is the copper corrosion test for adhesive tapes. It is an important requirement of adhesive electrical tapes that the adhesive should not corrode copper, as this is a particular hazard in high-humidity, high-temperature atmospheres. In BS 3924 (1978)[43] a test for electrolytic corrosion involves a length of tape wound round the outside of a large glass tube (Figure 13.29), with two bare copper wires wound under tension over the top of it and anchored. The wires are connected to a 120 V d.c. supply and the assembly is subjected to a humidity cycling test for three days, after which the positive wire and a 'blank' sample have their breaking loads tested in a tensile testing machine. A *corrosion liability factor* (CLF) is then calculated as the percentage (loss in strength of the tested wire divided by the strength of the untested wire).

In practice, ten test pieces are employed, half with wires in contact with the adhesive side, and half on the non-adhesive side of the tape. All positive wires are tested and the standard deviation is reported. Water-soluble impurities in either the adhesive or the backing tend to be anionic and are therefore attracted to the positive wire, which they attack and weaken.

A further test for the liability of insulating tapes to corrode metals is described in §13.4.

BS 3924 also specifies tests for the determination of conductivity of water extracts of, for example, insulating tapes.

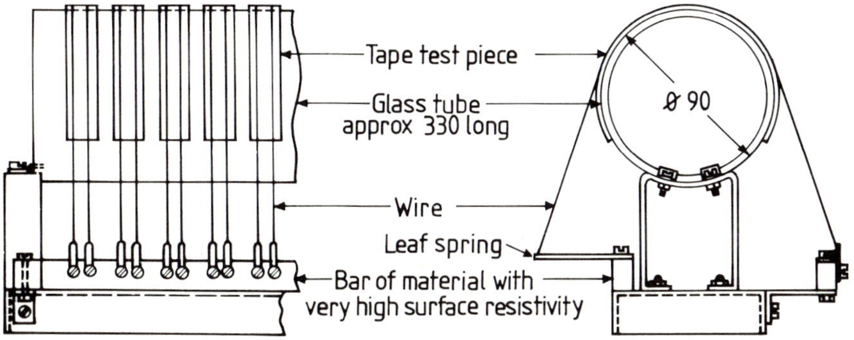

Figure 13.29 Apparatus for determining electrolytic corrosion: BS 3924—dimensions in mm

REFERENCES

1. Baird, M. E. (1973) *Electrical Properties of Polymeric Materials*, Plastics Institute.
2. Blythe, A. R. (1979) *Electrical Properties of Polymers*, Cambridge University Press.
3. Daniel, V. V. (1967) *Dielectric Relaxation*, Academic Press.
4. Moullin, E. B. (1944) *Journal of the Institution of Electrical Engineers, Part 1*, **91** (48), 448.
5. Swiss, J. and Dakin, T. W. (1954) *Westinghouse Engineer*, **14** (3), 144.
6. Hoffman, J. D. (1957) *Transactions of the British Institution of Radio Engineers*, **CP4** (June), 42.
7. Vail, C. B. (1962) *Electro-Technology*, **69** (February), 82.
8. Sharbaugh, A. H. (1962) *Electro-Technology*, **69** (February), 146.
9. Devins, J. C. and Sharbaugh, A. H. (1961) *Electro-Technology*, **68** (February), 104.
10. Mason, J. H. (1962) *Transactions of the Plastics Institute*, **30** (87), 171.
11. Stark, K. H. (1962) *Transactions of the Plastics Institute*, **30** (87), 184.
12. Baker, W. P. (1965) *Electrical Insulation Measurements*, Newnes.
13. Parkman, N. (1965) in *Physics of Plastics*, P. D. Ritchie (Ed.), Iliffe, Chapter 6.
14. Brydson, J. A. (1975) *Plastics Materials* 3rd edn, Butterworth.
15. Sharbaugh, A. H. (1965) in *Testing of Polymers* vol. 1, J. V. Schmitz (Ed.), Interscience.
16. Scott, A. H (1965) in *Testing of Polymers* (*loc. cit.*), Chapter 6.
17. Tucker, R. W. (1965) in *Testing of Polymers* (*loc. cit.*), Chapter 7.
18. Warfield R. W. (1965) in *Testing of Polymers* (*loc. cit.*), Chapter 8.
19. Dakin. T. W. (1965) in *Testing of Polymers* (*loc. cit.*), Chapter 9.
20. Brunton, J. H. (1965) in *Testing of Polymers* (*loc. cit.*), Chapter 10.
21. Kepler, R. G. (1977) in *Treatise on Materials Science and Technology* vol. 10, Part B, J. M. Schultz (Ed.), Academic Press, Chapter 4.
22. Fava, R. A. (1977) in *Treatise on Materials Science and Technology* vol. 10, Part B (*loc. cit.*), Chapter 5.
23. Geppe, A. P. (1958) *Elektrichestvo,* 3, 60.
24. Dorcas, D. S. and Scott, R. N. (1964) *Review of Scientific Instruments*, **35** (9), 1175.
25. Sazhin, B. I. and Skurikhina, V. S. (1960) *High Molecular Compounds*, **10,** 1535.
26. IEC 93 (1958) *Recommended Methods of Test for Volume and Surface Resistivities of Insulating Materials.*
27. Munick, R. J. (1956) *Journal of Applied Physics*, **27** (10), 1114.
28. France, G. (1957) *Electronic Engineering*, **29** (6), 30.
29. Hitchcox, G. (1956) *Journal of the British Institution of Radio Engineers*, **16** (June), 299.
30. ASTM D257 (1978) *Tests for DC Resistance or Conductance of Insulating Materials.*
31. Reddish, W. (1959) *The Physical Properties of Polymers*, SCI Monograph No. 5, Society of Chemical Industry, page 138.
32. BS 2782 (1970) *Methods of Testing Plastics.*
33. BS 903 (1956) *Methods of Testing Vulcanised Rubber.*
34. BS 3815 (1964) *Epoxide Resin Casting Systems for Electrical Applications.*
35. IEC 645 (1971) *Methods of Test for Electrical Resistance and Resistivity of Insulating Materials at Elevated Temperatures.*
36. DIN 53482 (1967) *Methods for Determination of Electrical Resistance Values.*
37. BS 4618 (1975) *Recommendations for the Presentation of Plastics Design Data.* Part 2 *Electrical Properties.* Section 2.3 *Volume Resistivity;* Section 2.4 *Surface Resistivity.*

38. ISO 1325 (1973) *Determination of Electrical Properties of Thin Sheet and Film.*
39. ASTM D2305 (1972) *Testing Polymeric Films used for Electrical Insulation.*
40. IEC 167 (1964) *Determination of Insulation Resistance of Solid Insulating Materials.*
41. ISO 2951 (1974) *Vulcanised Rubber. Determination of Insulation Resistance.*
42. ASTM D700 (1975) *Specification for Phenolic Molding Compounds.*
43. BS 3924 (1978) *Specification for Pressure Sensitive Adhesive Tapes for Electrical Insulating Purposes.*
44. ASTM D1000 (1978) *Testing Pressure-sensitive Adhesive Coated Tapes used for Electrical Insulation.*
45. ASTM D876 (1971) *Testing Non-rigid Vinyl Chloride Polymer Tubing used for Electrical Insulation.*
46. BS 2848 (1973) *Flexible Insulating Sleeving for Electrical Purposes.*
47. Norman, R. H. (1970) *Conductive Rubbers and Plastics*, Elsevier.
48. Morris, W. T. (1970) *Plastics and Polymers*, **38** (February), 41.
49. Davies, D. K. (1967) in *Static Electrification*, Institute of Physics/Physical Society, page 29.
50. Morris, W. T. and Norman, R. H. (1972) *RAPRA Bulletin*, **26,** 254.
51. Beesley, J. and Norman, R. H. (1977) *RAPRA Members Report No. 5.*
52. Klouda, J. C. (1978) *Journal of Cellular Plastics*, 14 (January/February), 33.
53. Cross, A. S. (1953) *British Journal of Applied Physics*, Supplement No. 2, 47.
54. Shashoua, V. E. (1958) *Journal of Polymer Science*, **33** (126), 65.
55. Langdon, S. J. (1964) *Plastics, London*, **29** (322), 43.
56. Bertein, H. (1967) in *Static Electrification*, Institute of Physics/Physical Society, page 11.
57. Moreno, R. A. and Gross, B. (1967) *Journal of Applied Physics*, **47,** 3397.
58. BS 2782, Methods 250 A-C (1976) *Antistatic Behaviour of Film.*
59. ISO 2878 (1978) *Rubber, Vulcanised—Antistatic and Conductive Products—Determination of Electrical Resistance.*
60. BS 2050 (1978) *Electrical Resistance of Conducting and Antistatic Products made from Flexible Polymeric Materials.*
61. DIN 51953 (1975) *Testing of Organic Floor Coverings; Testing of the Ability for Electrostatic Charge Dissipation in Confined Spaces Liable to Contain Explosive Mixtures.*
62. BS 3289 (1960) *Conveyor Belting for Underground use in Coal Mines.*
63. BS 2044 (1978) *Laboratory Tests for Resistivity of Conducting and Antistatic Rubbers.*
64. ISO 1853 (1975) *Conducting and Antistatic Rubbers—Measurement of Resistivity.*
65. Clark, F. M. (1962) *Insulating Materials for Design and Engineering Practice*, Wiley.
66. Norman, R. H. (1953) *Proceedings of the Institution of Electrical Engineers*, **100, IIA,** 41.
67. IEC 250 (1969) *Recommended Methods for the Determination of the Permittivity and Dielectric Dissipation Factor of Electrical Insulating Materials at Power, Audio and Radio Frequencies including Metre Wavelengths.*
68. ASTM D150 (1978) *Tests for A-C Loss Characteristics and Permittivity (Dielectric Constant) of Solid Electrical Insulating Materials.*
69. DIN 53483 (1969) *Determination of Dielectric Properties.*
70. Electrical Research Association (1958) *Technical Report L/S9.*
71. Hartshorn, L. (1941) *Radio Frequency Measurements by Bridge and Resonance Methods*, Chapman and Hall.
72. Lynch, A. C. (1965) *Proceedings of the Institution of Electrical Engineers*, **112** (2), 426.
73. BS 4542 (1970) *Method for the Determination of Loss Tangent and Permittivity of*

Electrical Insulating Materials in Sheet Form (Lynch Method).
74. BS 2067 (1953) *Determination of Power Factor and Permittivity of Insulating Materials (Hartshorn and Ward Method).*
75. ASTM D1531 (1962) *Test for Dielectric Constant and Dissipation Factor of Polyethylene by Liquid Displacement Procedure.*
76. Vince, P. M. (1964) in *IEE Conference on Dielectric Insulating Materials, London.*
77. Scheiber, D. J. (1961) *Journal of Research of the National Bureau of Standards,* **65c,** 23.
78. Barney, W. M. (1961) in *Annual Conference on Electrical Insulation*, NAS-NRC 973, page 59.
79. Hague, B. (1943) *Alternating Current Bridge Methods* 5th edn, Pitman.
80. Weston, D. (1962) *Plastics, London*, **27** (302), 105.
81. Hazen, T. (1967) in *Conference on Plastics Telecommunications Cables*, Plastics Institute, London, April.
82. Hartshorn, L. and Ward, W. H. (1936) *Journal of the Institution of Electrical Engineers,* **79,** 597.
83. Barrie, I. T. (1965) *Proceedings of the Institution of Electrical Engineers*, **112** (2), 408.
84. Roberts, S. and von Hippel, A. (1946) *Journal of Applied Physics*, **17,** 610.
85. Westphal, W. B. (1954) *Dielectric Materials and Applications* Section II A 2, MIT Technology Press/Wiley.
86. Gevers, M. (1955) in *Precision Electrical Measurements, Proceedings of the National Physical Laboratory Symposium*, HMSO.
87. Dube, D. C. and Natarajan, R. (1974) *Journal of Physics, E*, **7,** 256.
88. Parry, J. V. L. (1951) *Proceedings of the Institution of Electrical Engineers*, **98,** III, 303.
89. Horner, F., Taylor, T. A., Dunsmuir, R., Lamb, J. and Jackson. W. (1946) *Journal of the Institution of Electrical Engineers*, **93,** III (21), 53.
90. Brydon, G. M. and Hepplestone, D. J. (1964) in *IEE Conference on Dielectric Insulating Materials, London.*
91. BS 771 (1980) *Phenolic Moulding Materials.*
92. BS 4618 (1970) *Recommendations for the Presentation of Plastics Design Data.* Part 2 *Electrical Properties*, Section 2.1 *Permittivity;* Section 2.2 *Loss Tangent.*
93. ASTM D1673 (1979) *Tests for Dielectric Constant and Dissipation Factor of Expanded Cellular Plastics used for Electrical Insulation.*
94. ASTM D1674 (1967) *Testing Polymerisable Embedding Compounds used for Electrical Insulation.*
95. ASTM D149 (1975) *Tests for Dielectric Breakdown Voltage and Dielectric Strength of Electrical Insulating Materials at Commercial Power Frequencies.*
96. IEC 243 (1967) *Electric Strength of Solid Insulating Materials at Power Frequencies.*
97. BS 358 (1960) *Method for the Measurement of Voltage with Sphere-gaps (One Sphere Earthed).*
98. BS 923 (1980) *Guide on High-voltage Testing Techniques.*
99. BS 2918 (1957) *Electric Strength of Solid Insulating Materials at Power Frequencies.*
100. BS 148 (1972) *Insulating Oil for Transformers and Switchgear.*
101. BS 3784 (1973) *Polytetrafluoroethylene Sheet.*
102. BS 3816 (1964) *Cast Epoxide Resin Insulating Material for Electrical Applications at Power Frequencies.*
103. BS 5102 (1974) *Phenolic Resin Bonded Paper Laminated Sheets for Electrical Applications.*
104. BS 3873 (1965) *Polytetrafluoroethylene Moulded Basic Shapes.*
105. ASTM D350 (1978) *Testing Flexible Treated Sleeving used for Electrical Insulation.*
106. BS 4893 (1973) *Veneered Polytetrafluoroethylene (PTFE) Tape.*

107. Mason, J. H. (1971) *Insulation Engineer*, **1**, 5.
108. IEC 343 (1970) *Recommended Test Methods for Determining the Relative Resistance of Insulating Materials to Breakdown by Surface Discharges*.
109. IEC 112 (1979) *Method of Test for Determining the Comparative and Proof Tracking Indices of Solid Insulating Materials under Moist Conditions*.
110. BS 3781 (1964) *Method for Determining the Comparative Tracking Index of Solid Insulating Material*.
111. ASTM D3638 (1977) *Test Method for Comparative Tracking Index of Electrical Insulating Materials*.
112. IEC 587 (1977) *Test Method for Evaluating Resistance to Tracking and Erosion of Electrical Insulating Materials used under Severe Ambient Conditions*.
113. BS 5604 (1978) *Method of Test for Evaluating Resistance to Tracking and Erosion of Electrical Insulating Materials used under Severe Ambient Conditions*.
114. ASTM D2303 (1973) *Test for Liquid-contaminant, Inclined-plane Tracking and Erosion of Insulating Materials*.
115. ASTM D2132 (1968) *Standard Test Methods for Dust and Fog Tracking and Erosion Resistance of Electrical Insulating Materials*.
116. ASTM D2302 (1975) *Test Method for Differential Wet Tracking Resistance of Electrical Insulating Materials with Controlled Water-to-metal Discharges*.
117. ASTM D495 (1973) *Test Method for High-voltage, Low-current, Dry Arc Resistance of Solid Electrical Insulation*.
118. DIN 53484 (1955) *Testing of Insulating Materials. Determination of the Arc Resistance*.

Chapter Fourteen

Optical Properties

14.1 INTRODUCTION

In only a few plastics products, such as lenses, windows and roof lights, are light transmission and refractive index the most important optical properties. In general it is the aesthetic properties such as colour, haze, gloss and reflectance which are dominant. This is unfortunate, for light transmission and refractive index are a good deal easier to define and measure accurately than the other properties mentioned. Aesthetic properties such as surface texture, colour, haze, gloss and reflectance are difficult to measure or to correlate with visual experience. Indeed, even their definition presents some difficulty.

All these aspects are very specialised and detailed information should be sought from specialised textbooks. Thus no attempt will be made to describe, for example, all the glossmeters which have been used, but rather attention will be confined to those instruments and specifications which are in common use in the plastics industry (i.e. excluding instruments and methods employed in other industries such as paints, paper and ceramics, unless such methods are widely used for plastics also). The measurement of colour is, however, common to many different industries and the methods employed are very similar.

Fibres represent a special case. Here the optical properties of refractive index and birefringence are important not so much for their influence on the appearance of the product but as an aid to fibre identification. Although similar identification techniques are applicable to transparent plastics in general, such tests are not widely used outside the forensic science laboratory. Infrared and ultraviolet absorption are more often used for polymer identification, but these techniques and others such as X-ray scattering are outside the scope of this chapter. Similarly, methods of studying changes in the refractive index of solutions as an aid to following chemical reactions in solution, or methods of molecular weight determination, will not be considered, nor will changes of colour on exposure to light or heat such as are described in ISO R877, R878, R879 and IS 1600 be considered in detail as these are dealt with in §17.

BS 4618 Section 5.3 (1972),[1] *Optical Properties*, is a useful guide outlining the properties of refraction, transparency, haze, gloss and light transfer, together with a brief account of the available experimental methods.

ISO 31/VI (1973),[2] *Quantities and Units of Light and Related Electromagnetic Radiation*, is a table giving the international symbols and units relating to light and other radiations.

Ross and Birley[3] review optical properties and Venable[4] discusses the role of spectrophotometry in the specification and measurement of optical properties of translucent materials. A much more useful review is that given by Christie and Hunter,[5] who make the observation that when sufficient economic demand arises for more sophisticated instrumentation, new instrument development will keep pace with the need.

Bugash[6] points out that by using a modulated light source and appropriate electronics much of the uncertainty associated with conventional instruments can be eliminated.

Test pieces for optical measurements need to be prepared to a very high standard and firm recommendations are set out in ISO R869 (1968).[7] Platens of borosilicate glass without a release agent are recommended in order to obtain surfaces of adequate optical quality. It is recommended that reference be made to ISO R293 (1963)[8] for general details of compression moulding.

14.2 REFRACTIVE INDEX

14.2.1 Terms and Definitions

When a ray of light passes from one isotropic medium to another, the sine of the angle of incidence bears a constant ratio to the sine of the angle of refraction (both measured with respect to the normal) for all angles of incidence. This ratio depends not only on the two media concerned, but also on the wavelength of the light used and on temperature.

'Refractive index' is the term used for this ratio when light passes from a vacuum (or from air for less accurate work) into a more dense medium; it is always greater than unity.

The ISO definition[2] of refractive index is given in terms of the velocity of light: 'Refractive index is the ratio of velocity of electromagnetic radiation *in vacuo* to the phase velocity of electromagnetic radiation of a specified frequency in the medium'.

Refractive index is dimensionless. For anisotropic materials the state of polarisation of the light (and its direction, where appropriate) must be defined relative to a reference axis in the sample. It is then customary to quote two refractive indices; additionally, the maximum difference between two indices measured in two mutually perpendicular directions is termed the 'birefringence' of the material.

14.2.2 Methods of Measurement

General methods of measuring refractive index, not necessarily confined to plastics, can be found in standard textbooks on optics.[9] Perhaps the best known methods are those depending on matching the unknown solid with known liquids, using the Becke line method[10] or dispersion staining.[11] Other well known methods are those based on the Abbe refractometer[9] and the measurement of the ratio of real depth to apparent depth.[9]

The two methods recommended in ISO R489 (1966)[12] are the Abbe refractometer and the Becke line methods.

The test piece for the *refractometer method* should be about 12 × 6 × 3 mm with one flat face and one truly perpendicular surface, these two surfaces intersecting along a sharp line without a bevelled or rounded edge. (For anisotropic materials specimens should be made with their polished surfaces parallel and perpendicular to the direction of orientation.) The test piece is attached to the prism of the refractometer with a drop of liquid of refractive index higher than the test piece by at least 0·01 and this liquid should not soften, attack, or dissolve the plastics material. A table of suggested liquids is given. The specification states that measurement should be carried out at 20° ± 0·5°C using white light, but if it were updated it is probable that a mean temperature of 23°C would be specified. The instrument gives the refractive index for the sodium [D] line to a precision of about 0·001.

ASTM D542-50 (1977)[13] is similar and here the specified conditions are 23° ± 2°C and 50 ± 5 per cent r.h. In cases where the material has a high thermal coefficient of refractive index the temperature has to be accurately controlled at 23°C. The claimed accuracy of the method is four significant figures and where maximum accuracy is required the use of sodium light is recommended.

DIN 53491 (1955),[14] which gives useful advice on practical details relevant to refractometer measurements, is similar to ISO R489 in specifying a test temperature of 20° ± 0·5°C, but like ASTM D542 recommends the use of sodium light for maximum accuracy which is claimed to be four decimal places. All the specifications give lists of liquids suitable for different plastics.

It is convenient to consider the alternative, less preferred method in ASTM D542 at this point as this also requires a test piece with parallel faces. This method is based on the well known phenomenon that the apparent depth of a transparent object viewed through the thickness is less than the true depth, refractive index being the ratio of real depth to apparent depth. ASTM D542 outlines the technique for carrying out this measurement with a standard microscope having a magnifying power of at least 200 diameters and a means of measuring the relative movement of the objective lens to the sample with an accuracy of 0·025 mm. The claimed accuracy for this method of measurement is three significant figures.

The alternative measurement in ISO R489, the *Becke line method*, is much more useful in that it can be used with powdered or granulated transparent material or indeed with any small chip of material taken from a larger specimen. A microscope having a magnifying power of at least 200 diameters is required together with a range of liquids of known refractive index. If an Abbe refractometer is available, but only small chips of material rather than parallel-sided test pieces, then the refractometer can be used to calibrate test liquids for use with the Becke line method. The test pieces should have a thickness significantly less than the working distance of the 8 mm microscope objective and linear dimensions sufficiently small and so distributed that simultaneous observation of approximately equal areas of sample and surrounding field is possible.

The material under test is mounted in a liquid of known refractive index and examined in monochromatic light with the condenser adjusted to give a narrow axial beam. When the test pieces and the liquid have different refractive indices each particle is surrounded by a narrow luminous halo (the Becke line) which moves as focus is adjusted. If the focus is lowered then the Becke line moves towards the medium having the lower refractive index. The test is repeated with particles mounted in other immersion liquids until a match is found or until the index of the test sample lies between two known indices in the series of liquid standards. If the Becke line phenomenon does not appear then the refractive index of the material being examined is equal to that of the immersion liquid. ISO R489 recommends a test temperature of 20° ± 0·5°C, but doubtless if it were up-dated the mean temperature would be changed to 23°C.

ISO R489 does not recommend any conditioning procedure prior to the test. ASTM D542 specifies that test specimens must be conditioned for at least 40 hours at 23° ± 2°C and 50 ± 5 per cent r.h. DIN 53491 recommends that hygroscopic materials should be dried for 24 hours over anhydrous calcium chloride, phosphorous pentoxide or concentrated sulphuric acid prior to measurement.

Dispersion staining is another useful microscopical method for the determination of refractive index described in detail by McCrone in ASTM STP 348.[11]

In general, methods have changed very little since the nineteenth century so that the older references still have value. Wiley and Hobson's review[15] covers refractometer methods and immersion methods whereas Brown, McCrone *et al.*[16] deal with dispersion staining. Billmeyer,[17] Ellis[18] and Schael[19] deal specifically with polymer films where it must be borne in mind that, as in the case of fibres, birefringence may result from the processing conditions. There is no doubt that when anisotropic materials are considered these are best dealt with by means of the polarising microscope. Suitable techniques are discussed under Birefringence, below.

14.3 BIREFRINGENCE

14.3.1 Definition

Birefringence is a measure of optical anisotropy. It is defined as the maximum algebraic difference between two refractive indices measured in two perpendicular directions. Some materials are isotropic until stressed elastically, others have permanent birefringence induced by processing, e.g. drawing or rolling.

14.3.2 Methods of Measurement

The most convenient way of measuring birefringence is by means of a polarising microscope. It would be inappropriate to discuss methods of analysis in detail here, but some idea of the principle will be given.

When a beam of light falls on an isotropic material it is resolved into two components. A polariser is a device which is designed so that one of these components is totally absorbed (or internally reflected in the case of a Nicol prism). Thus light emerging from the polariser is plane polarised. A similar sheet of polarising material mounted with its plane of polarisation at right angles to the first will absorb all of the light. Only if the two directions of polarisation are parallel will light pass through both filters. This second piece of polarising material is known as the analyser. The two arrangements described above are referred to as 'crossed polars' and 'open polars', respectively.

Consider now a birefringent object mounted between crossed polars. The simplest case is that where a rectangular object, having its planes of polarisation parallel to its sides, is mounted with one of those planes parallel to the plane of the incident light (Figure 14.1a). Light from the polariser passes through the birefringent test piece and is absorbed by the analyser, so that the field is dark.

Similarly, if the test piece is rotated through 90° the field remains dark (Figure 14.1b). If the test piece is mounted at 45° to the plane of polarisation, however, the object stands out brightly on a dark field. To understand this, reference must be made to Figure 14.1c. The original plane-polarised beam may be considered to be resolved into two components A and B corresponding to the principal planes of the test piece. These two components traverse the test piece at different velocities so that the two emergent beams A′ and B′ are out of phase. On reaching the analyser each of the components A′ and B′ may be resolved into components parallel to and at right angles to the analyser direction. Components parallel to the analyser direction pass through and, since these are out of phase, interference results. Because the phase difference varies with wavelength, brilliant interference colours may be obtained. The effect on these colours of retardation wedges or compensators enables an estimate of birefringence to be made. Detailed descriptions of these methods can be found in standard textbooks.[20,21]

If a polarising microscope is available the experimental technique involved is relatively straightforward. Many of the difficulties arise from a lack of understanding of the physical principles behind the method and initial study of theoretical principles of polarising microscopy is well worth while.

Fibres sometimes present difficulty in that there is a different orientation at the skin and in the core of the fibre. In these cases it must be borne in mind that the Becke line method gives the refractive index of the skin, whereas birefringence measured by a compensator method is that of the fibre as a whole. Care is

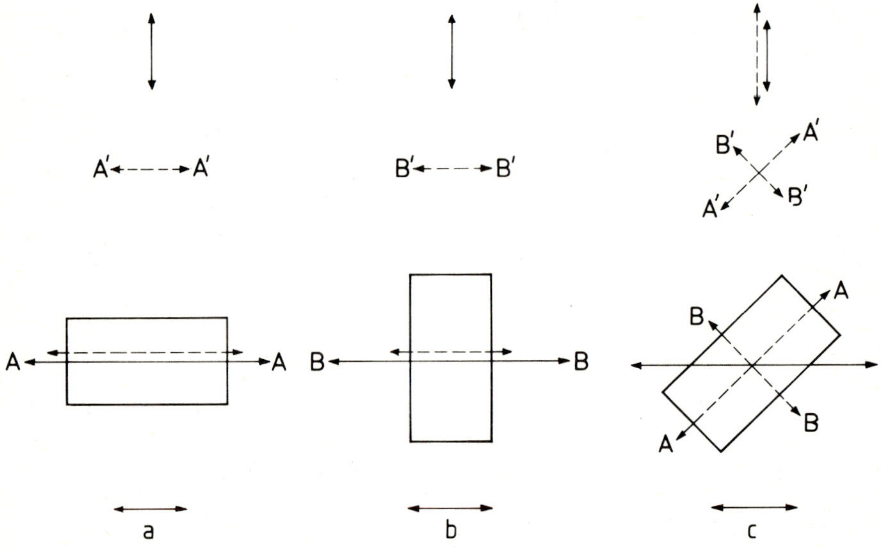

Figure 14.1 Interaction of plane polarised light with a birefringent object

needed in interpretation and advice should be sought from specialised textbooks or, better still, from specialists in the field.

14.4 TRANSPARENCY

14.4.1 General Considerations

Light transmission, haze and see-through clarity are closely linked properties which together constitute transparency. Aberration of the image formed by any optical system or the merging of points distinct in the original object is known as 'loss of clarity', whereas the scattering of light with consequent loss of contrast is termed 'haze'. Light transmission is the percentage of light transmitted without deviation.

As an illustration of these properties, glazing material for windows and diffusing material for lighting fixtures must both have a high light transmittance, but the former must be free from haze and very transparent while the latter must have maximum diffusion and minimum transparency. Film for packaging must have good see-through clarity, a property which should not be confused with haze as hazy materials often have quite good see-through clarity.

Webber[22] and Miles and Thornton[23] have expressed the view that see-through clarity is a function of light scatter over a very small forward angle, as distinct from the larger angles of scatter usually considered in haze measurements.

Haze and loss of clarity indicate discontinuities or irregularities within or at the surface of the material and these could be filler particles, impurities, minute bubbles or voids, surface roughness, or regions of different refractive indices brought about, say, by crystallisation. Abrasion, weathering, absorption of water and even temperature change can all bring about a change in the haziness of a material, so that test conditions need to be closely specified.

BS 4618 Section 5.3 (1972)[1] gives useful background material and the principles of experimental methods. Other papers covering general principles are those by Venable,[4] Christie and Hunter[5] and Willmouth.[24]

14.4.2 Light Transmission and Haze—Small-scale Test Methods

Light transmission may be defined as the percentage of light transmitted without deviation, and haze as the percentage of incident light which is transmitted with more than a certain specified angular deviation by forward scattering (say, 2·5°). Both properties vary with the geometry of the optical measuring system used which must be carefully specified.

The three principal light sources used in illumination work and photometry are CIE (Commission International de l'Eclairage) Sources A, B and C which correspond, respectively, to incandescent light, noon sunlight and overcast sky daylight. Although methods of measuring light transmission and haze are under consideration by ISO TC61 (Plastics) no International Standards have yet been agreed and the principal standards in use for small specimens are ASTM D1003-61 (1977)[25] and BS 2782 Method 515A (1970).[26] (Curiously, the British Standard does not refer to light transmission.) The ASTM method gives CIE Source C as the preferred light source, but allows Source A as an alternative, whereas the BS method specifies CIE Source C only. In other respects the methods are similar for haze measurement. The method was originally devised for low-density polyethylene film, which in practice involves a fairly narrow range of thicknesses. When applied to other thin films, or to sheets or plaques, the method can give misleading answers, so that some caution in its application is called for.

ASTM D1003 is based on a spherical hazemeter which is pivotable about a vertical axis through the entrance port, where the specimen is placed (Figure 14.2). In the normal position the collimated incident light passes straight through the sphere, leaving through the exit port which is closed by an absorbent light trap. Any light which is scattered by the instrument alone (specimen removed) or instrument plus specimen (specimen in) is reflected from the region around the edge of the exit port and finally collected by the photocell after multiple reflections from the highly reflective walls of the sphere.

When the sphere is rotated slightly so that the incident light hits the opposite highly reflecting wall of the sphere (which forms a reflectance standard) adjacent to the exit port, a measurement with and without specimen gives a measure of the total transmittance.

The three properties of interest are:

$$\text{total transmittance, } T_t = \frac{T_2}{T_1} \tag{14.1}$$

$$\text{diffuse transmittance, } T_d = \frac{(T_4 - T_3)(T_2/T_1)}{T_1} \tag{14.2}$$

$$\text{haze (per cent)} = \frac{T_d}{T_t} \times 100 \tag{14.3}$$

where T_1 = photocell output, specimen out, reflectance standard in beam
T_2 = photocell output, specimen in, reflectance standard in beam
T_3 = photocell output, specimen out, reflectance standard out of beam
T_4 = photocell output, specimen in, reflectance standard out of beam.

T_1, T_2, T_3, T_4 represent respectively the incident light, the total light transmit-

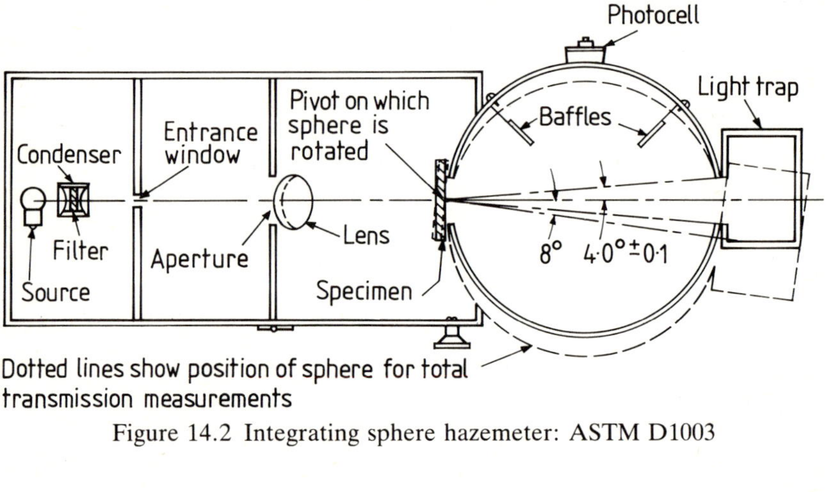

Figure 14.2 Integrating sphere hazemeter: ASTM D1003

ted by the specimen, the light scattered by the instrument and the light scattered by instrument and specimen.

The method of ASTM D1003 is also invoked by ASTM D1044 (1978)[27] to measure the diffuse transmittance of specimens after Taber abrasion tests, as a measure of the degree of abrasion (§10.3).

The wider implications of transmittance and reflectance measurements carried out on translucent plastics are discussed by Johnston *et al.*[28]

14.4.3 Light Transmission and Haze—Large-scale Test Methods

A light transmission test for panels 2 ft (0·6 m) square appears in two British Standards, BS 4154 (1967)[29] and BS 4203 (1967),[30] and in ASTM D1494–60 (1974)[31] (see also Meyer and Smith[32]). Since it is a test on a finished product rather than a material it will not be described in detail. A photocell measures the transmitted light and the transmittance is given as the ratio of photocell outputs with the specimen in and the specimen out. In BS 4154 an additional test for light diffusion is given, based on a simple slit diffusion photometer, and an appropriate correction is then calculated and applied to the total light transmission. This refinement is not required in BS 4203, presumably because only transparent material is involved.

14.4.4 See-through Clarity

There is no method universally accepted for measuring see-through clarity. The following method, outlined in BS 4618 Section 5.3 (1972),[1] is based on the work of Webber.[22]

Sets of charts (Snellen charts) are viewed through the film under test. Each chart consists of sets of parallel lines which are vertically perpendicular, each group of lines having a different separation. The angle subtended at the eye by the minimum resolvable separation between the lines is noted and compared with the angular resolution without a film. The difference between the two values gives a measure of clarity. The angular resolution with a film present depends on the distance between chart and sample and this must be specified or recorded.

14.5 GLOSS

14.5.1 General Considerations

According to Hunter,[33] 'Gloss is defined as the degree to which a surface simulates a perfect mirror in its capacity to reflect incident light'; he goes on to say that 'gloss is determined by the surface's geometric selectivity in reflecting light'. In most practical glossmeters specular gloss is the property determined and this is usually taken to be the fraction of light flux reflected in the direction of mirror reflection (the specular direction) when a specimen is illuminated by a parallel light beam.

The standard methods of measuring gloss are numerous. Superficially many of the methods are similar, but small differences in optical geometry can cause large variations in results from one type of instrument to another.

The review of Christie and Hunter[5] usefully outlines different types of instrument.

The greatest difficulty with gloss measurements lies in correlating measurements with visual impressions; this is discussed at some length by Hammond,[34] Dinsdale and Malkin,[35] Hunter[36] and Elm.[37]

Knittel[38] points out that an observer, in assessing visual gloss, turns the sample over and over using many angles of incidence and reflection, and his judgment is therefore based on a whole series of observations.

Because there is no absolute standard for gloss, existing standards (usually optically flat black glass plates) are assigned values which depend on the refractive index of the glass employed and which may vary from one test to another.

14.5.2 Standard Test Methods

ASTM D523 (1978)[39] is the recommended standard for gloss measurements on plastics (excepting films) and uses a 20° (to the vertical) incident and reflected light beam geometry for high gloss materials and 85° geometry for low gloss materials. An intermediate 60° angle is used for inter-comparative purposes and for deciding which of the other angles should be used with a given specimen.

Although the geometry of source and receiver is defined with close tolerances on the angles, the actual instrument to be used is not described in great detail: Figures 14.3 and 14.4 (taken from the standard) merely indicate the general layout. The primary standard of gloss is a highly polished, plane, black glass surface with a refractive index of 1·567 to which is assigned the arbitrary value of 100 gloss units for each of the three geometries (the reader is warned that elsewhere and in other circumstances different values may be assigned to similar gloss standards).

The specification requires that any measurements should be in accordance with CIE Source C values without actually stating that this source should be used, the inference being that the type of source is not very critical.

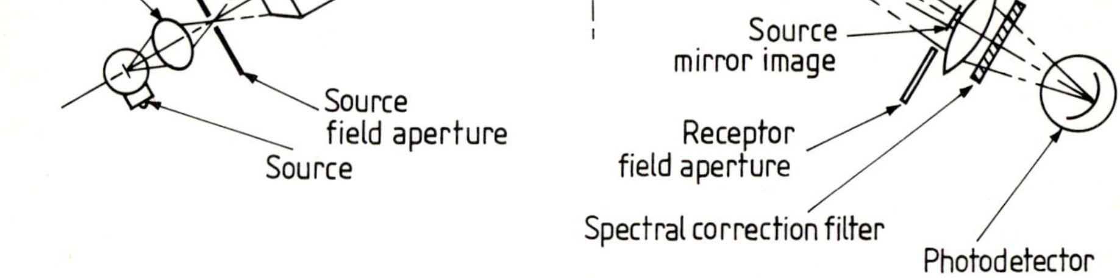

Figure 14.3 Parallel-beam glossmeter showing apertures and source mirror-image position: ASTM D523

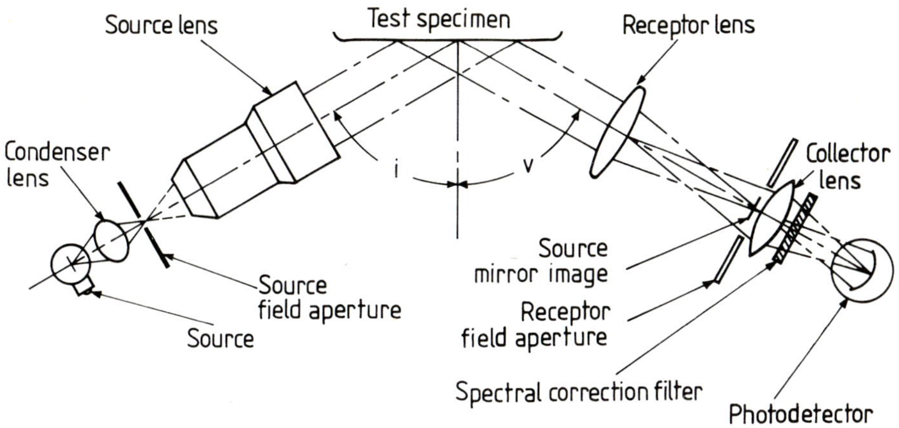

Figure 14.4 Converging-beam glossmeter showing apertures and source mirror-image position: ASTM D523

For films, ASTM D2457-70 (1979)[40] is based on ASTM D523 for the 20° and 60° angle tests, but the third angle is 45°, not 85°, and the 45° test is based on a method for ceramics, ASTM C346 (1976).[41] The same black glass standard is used at 20° degrees and 60° degrees as in the ASTM D523 test, but at 45° the refractive index is 1·540 and the assigned gloss value is 54·5. A note warns of the possibility of obtaining gloss values on clear films of more than 100 units with any one of the three geometries, because of reflection at both surfaces of the specimen. Three specimen mounting devices are described for ensuring the flatness so essential in gloss measurements.

BS 2782 Method 515B (1970)[42] uses a 45° optical system and the reference standard is highly polished black glass of refractive index 1·52, to which a gloss value of 53 units is assigned. The gloss head is not described in detail, the Evans Electroselenium Plaspec B head and the Gardner 45° specular gloss head (UXT) being quoted as examples. The source specified is CIE Source C. Film samples can be tested provided that they are held taut and free from wrinkles.

Similar gloss tests are BS 3900 Part D2 (1967)[43] and BS 3962 Part 1 (1965);[44] both are relevant to modern polymer-based paints and varnishes.

The BS 3900 test is based on a former British defence specification, DEF 1053 (1952),[45] Method No. 11, and employs the 45° geometry head, similar to BS 2782 Method 515B, described earlier. The optical arrangement is shown in Figure 14.5.

This standard is unique in quoting figures for the variation in gloss of blackened glass with refractive index:

Refractive index	1·4	1·5	1·523	1·55	1·6	1·7
Maximum specular reflection (μ of 1·523 assumed ≡ 100 units)	68	94	100	107	120	148

The importance of flatness in the reflectance standard is underlined by the tolerance permitted, namely plane to 2 fringes/cm. The standard itself is highly

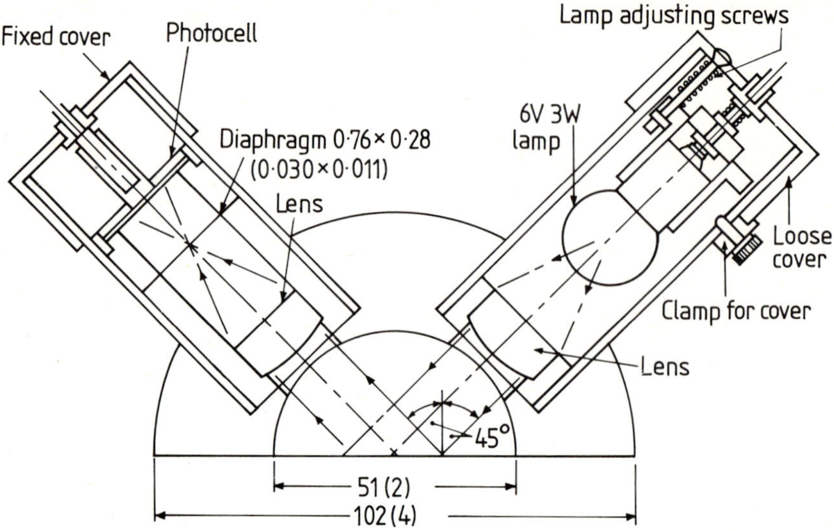

Figure 14.5 High-gloss head for paints and plastics: BS 3900 and BS 2782, Method 515B—dimensions in millimetres (inches)

polished clear glass with its underside edges roughened and coated with black paint. The refractive index is 1·523 ± 0·002 and the gloss value assigned to it is 100 units. The head is placed on it and the photocell output or galvanometer input is adjusted to 100 scale divisions. When placed on the test specimen, in this case coated plate glass, the gloss is given direct. Gloss values above 100 are evidently not impossible according to the operating procedure given.

The BS 3962 gloss head employs a 75°/75° (to the vertical) geometry. The Evans Electroselenium Ltd 'Variable-angle Glossmeter' conforms to the requirements, shown in Figure 14.6. This test is used to assess low-angle glare emanating from clear finishes used on school furniture, which may present a nuisance.

The gloss standard used is identical to that of BS 3900 and the procedure is similar.

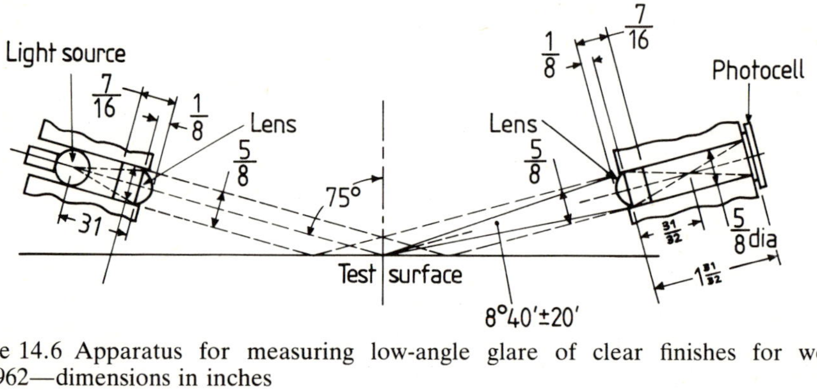

Figure 14.6 Apparatus for measuring low-angle glare of clear finishes for wood: BS 3962—dimensions in inches

14.6 COLOUR

Anyone who has watched the colour of wood change as a clear varnish is applied, or has taken cloth to the daylight to match it, knows that colour is not an absolute property but depends very much on the surface condition of the test piece and the viewing conditions. Even to specify 'daylight' or 'artificial light' is not enough, for the colour of the sun changes as it gets lower in the sky and the colours of all objects in our world change with it. Two colours which match under a particular set of lighting conditions may not match in another, and similarly two mouldings of the same plastics material may appear to be of different colours if their surface finish is different. These problems are not peculiar to the plastics industry—discussion of colour testing must of necessity cover a much wider field than plastics testing alone.

BS 4727 Part 4 (1971)[46] is relevant and Billmeyer and Saltzman[47] give valuable background information. ISO TC61 is considering the adoption of the 1931 CIE system[48] and in this regard ASTM E308-66 (1973)[49] may be considered a useful guide until the draft ISO recommendations[50] are adopted. The method is based upon spectrophotometry for the visible spectrum in the range 380 to 780 nm. Terms and symbols are described in the CIE/IEC *International Lighting Vocabulary*[51] and the instrumental evaluation of colour difference is discussed in ISO/DIS 2579.[52]

The assessment of small colour changes during weathering or of the clarity and colour of water white solutions or materials, while coming under the general heading of colour, may present less of a problem than the full definition or specification of the colour being considered. It may be easier, therefore, to deal with these first of all.

ISO recommendations R877 (1968),[53] R878 (1968)[54] and R879 (1968)[55] deal with the resistance of plastics to colour change upon exposure to daylight, carbon arc and xenon lamp, respectively, and are discussed more fully in §17.5. Similarly, BS 4618 (1974),[56] Section 4.3, uses the blue wool scale[57–59] to assess the degree of colour change which takes place when a given plastics test piece, protected by glass, is exposed to daylight. Curiously, the BS recommendation states in passing that laboratory sources such as carbon arc or xenon lamp are also allowed. BS 2782 (1976)[60] Method 530A for the determination of yellowness index and Method 530B for the determination of the colour of near-white or near-colourless materials are both based on the CIE tristimulus values; simpler methods include that of ISO 1600 (1975)[61] for the determination of light absorption of cellulose acetate before and after heating, where changes in optical density at wavelengths of 440 nm and 640 nm are reported, and ISO DIS 6271 (BS 5339) (1976),[62] which estimates the colour of clear liquids by the platinum–cobalt scale where the colour comparison is made with standard solutions by means of a simple colour comparator. The problems associated with the assessment of the colour of near-white or colourless materials are discussed in ISO/DIS 3558.[63]

In the wider context of colour in general, Willmeyer,[64] in an interesting paper, deals with the physiology of perception of colour and in the same publication Vickerstaff and Walls[65] discuss matching and measurement of colour. Adams[66] discusses basic fundamentals and Goodwin[67] is concerned with small colour differences, giving some examples in colour.

Although it is now rather old, Wright's report[68] on a symposium entitled 'Colorimetry: Its Errors and Accuracy' is still of interest because of its account of intercomparison measurements, carried out on different spectrophotometers, of standard tiles circulated to various UK laboratories. Foote[69] has reviewed the methods of measuring colour and commercial instruments available and Morse[70] and Menard[71] also describe particular instruments. ICI in conjunction with Digico Ltd[72] have described two more modern instruments. Cipolla,[73] in an Italian paper, describes the Applied Color Systems equipment. Wood,[74] in a wider-ranging paper, discusses colour control systems, colour matching equipment, colorimeters and spectrophotometers for use in plastics processing, monitoring equipment produced by specific companies.

Because colour measurement is common to many different industries, references abound. Some recent references specific to plastics are given below but these are given as a general introduction to this extensive subject and the list should not be regarded as comprehensive.

Gall,[75] in a comprehensive German review, summarises the most important elements in colorimetry and gives information on techniques and equipment for colour matching, with emphasis on quality control. Mention is also made of methods for determining colour stability, covering power and dispersion of pigments.

Rosenthal[76] lists the requirements for an ideal colour-measuring instrument and reviews the current state of the art of colour measurement and Coppock[77] in the same publication discusses the problems facing instrument manufacturers. Conner and Hammond[78] discuss both visual and instrumental approaches to measuring colour and appearance, using specific examples by way of illustration. The Gardner Colorgard system, said to have remarkable versatility, is also described.

Schreckenbach,[79] in another German paper, points out that gloss levels strongly affect colour measurement and shows that for flat unstructured surfaces mathematical relationships between colour and gloss values can be stated. Burns and Christie[80] discuss various techniques of colour measurement and consider the legal principles behind ASTM and some other US standards. Jauch[81] describes a portable spectrophotomer scanning twenty discreet wavelengths from 365 to 690 nm using an interference wedge filter and seven infrared wavelengths from 700 to 1000 nm with discrete interference filters. The 27 fully digitised values of the ratio measurements obtained in 3 seconds are recorded on casettes for data processing.

Colour measurement is not easy and those thinking of entering this field would be well advised to seek advice from laboratories with experience.

REFERENCES

1. BS 4618: Part 5, Section 5.3 (1972) *Optical Properties—Other Properties.*
2. ISO 31/V1 (1973) *Quantities and Units of Light and Related Electromagnetic Radiations.*

3. Ross, G. and Birley, A. W. (1973) *Journal of Physics D, Applied Physics*, **6**, 795.
4. Venable, W. Jr (1973) *Coloring of Plastics 7, RETEC, Cherry Hill, NJ, September*, SPE Delaware Valley Section, page 83.
5. Christie, J. S. and Hunter, R. S. (1975) *Coloring and Decorating of Plastics 9, RETEC, Cincinnati, Ohio, September*, SPE Miami Valley Section, page 33.
6. Bugash, R. S. (1973) *Coloring of Plastics 7* (*loc. cit.*), page 53.
7. ISO R869 (1968) *Preparation of Specimens for Optical Tests on Plastics Materials, Moulding Method.*
8. ISO R293 (1963) *Compression Moulding of Test Specimens of Thermoplastic Materials.*
9. Nelkon, M. (1955) *Light and Sound* 2nd edn, Heinemann.
10. James, D. I. (1974) *RAPRA Members Journal*, May, 134.
11. McCrone, W. C. (1963) *ASTM STP No.348*, page 125.
12. ISO R489 (1966) *Plastics, Determination of the Refractive Index of Transparent Plastics.*
13. ASTM D542-50 (1977) *Standard Test Methods for Index of Refraction of Transparent Organic Plastics.*
14. DIN 53491 (1955) *Testing of Plastics—Determination of Refractive Index and Dispersion.*
15. Wiley, R. H. and Hobson, P. H. (1948) *Analytical Chemistry*, **20** (6), 520.
16. Brown, K. M. and McCrone, W. C. (1963) *The Microscope*, **13** (11), 311; Brown, K. M., McCrone, W. C., Kuhn, R. and Forlini, L. (1963) *The Microscope*, **14** (2), 39.
17. Billmeyer, F. W. (1974) *Journal of Applied Physics*, **18** (5), 431.
18. Ellis, R. H. (1957) *Review of Scientific Instruments*, **28** (7), 557.
19. Schael, G. W. (1964) *Journal of Applied Polymer Science*, **8** (6), 2717.
20. Hartshorne, N. H. and Stuart, A. (1970) *Crystals and the Polarizing Microscope* 4th edn, Elsevier.
21. Stoves, J. L. (1957) *Fibre Microscopy: Its Technique and Application*, National Trade Press.
22. Webber, A. C. (1957) *Journal of the Optical Society of America*, **47** (9), 785.
23. Miles, J. A. C. and Thornton, A. E. (1962) *British Plastics*, **35** (1), 26.
24. Willmouth, F. M. (1976) *Plastics and Rubbers: Materials and Applications*, **1** (3/4), 101.
25. ASTM D1003-61 (1977) *Haze and Luminous Transmittance of Transparent Plastics.*
26. BS 2782 (1970) Method 515A *Haze of Film.*
27. ASTM D1044 (1978) *Resistance of Transparent Plastic Materials to Abrasion.*
28. Johnston, R. M., Keller, E. F., Edge, L. S. and Chang, M. (1973) *Coloring of Plastics 7, RETEC, Cherry Hill, NJ, September*, SPE Delaware Valley Section, page 39.
29. BS 4154 (1967) *Corrugated Plastics Translucent Sheets made from Thermosetting Polyester Resins* (*Glass Fibre Reinforced*).
30. BS 4203 (1967) *Extruded Rigid PVC Corrugated Sheeting.*
31. ASTM D1494-60 (1974) *Diffuse Light Transmission Factor of Reinforced Plastics Panels.*
32. Meyer, R. W. and Smith, D. (1958) *Reinforced Plastics*, **2** (5), 12.
33. Hunter, R. S. (1956) *Product Engineering*, **27** (2), 176.
34. Hammond, H. K. (1957) *American Paint Journal*, **41** (37), 95.
35. Dinsdale, A. and Malkin, F. (1955) *Transactions of the British Ceramics Society*, **54** (2), 94.
36. Hunter, R. S. (1952) *ASTM Bulletin*, 186, 48.
37. Elm, A. C. (1961) *Official Digest of the Federation of Societies for Paint Technology*, **33** (433), 163.
38. Knittel, R. R. (1962) *Materials, Research and Standards*, **2** (3), 180.

39. ASTM D523 (1978) *Specular Gloss*.
40. ASTM D2457-70 (1977) *Specular Gloss of Plastics Films*.
41. ASTM C346 (1976) *45° Specular Gloss of Ceramic Materials*.
42. BS 2782 (1970) Method 515B *Gloss (45°) of Sheet*.
43. BS 3900: Part D2 (1967) *Gloss (Specular Reflection Value)*.
44. BS 3962: Part I (1965) *Test for Low Angle Glare*.
45. DEF 1053 (1952) *Method No. 11, Gloss (Specular Reflection Value)*.
46. BS 4727 (1971) Part 4 *Terms Particular to Light and Colour*. Group 01, *Radiation and Photometry*; Group 02, *Vision and Colour Terminology*; Group 03, *Lighting Technology Terminology*.
47. Billmeyer, F. W. and Saltzman, M. (1967) *Principles of Colour Technology*, Interscience.
48. CIE (1931) *Colorimetry, Official Recommendations*, Publication No. 15.
49. ASTM E308-66 (1973) *Recommended Practice for Spectrophotometry and Description of Color in CIE 1931 System*.
50. ISO/DIS 3557 *Recommended Practice for Spectrophotometry and Calculation of Colour in CIE Systems*.
51. CIE/IEC (1970) *International Lighting Vocabulary* 3rd edn.
52. ISO/DIS 2579 *Plastics, Instrumental Evaluation of Colour Difference*.
53. ISO R877 (1968) *Determination of Resistance of Plastics to Colour Change upon Exposure to Daylight*.
54. ISO R878 (1968) *Determination of Resistance of Plastics to Colour Change upon Exposure to Light of the Enclosed Carbon Arc*.
55. ISO R879 (1968) *Determination of Resistance of Plastics to Colour Change upon Exposure to Light of a Xenon Lamp*.
56. BS 4618, Section 4.3 (1974) *Resistance to Colour Change Produced by Exposure to Light*.
57. BS 1006 (1978) *Method of Test for Colour Fastness of Textiles and Leather*.
58. ISO 105 (1978) *Textiles—Tests for Colour Fastness*.
59. BS 1006, Section A02 (1978) *Grey Scale for Assessing Change in Colour*.
60. BS 2782: Part 5, Method 530A (1976) *Determination of the Yellowness Index*; Method 530B (1976) *Determination of the Colour of Near-white or Near-colourless Materials*.
61. ISO 1600 (1975) *Plastics—Cellulose Acetate—Determination of Light Absorption Before and After Heating*.
62. ISO/DIS 6271 (BS 5339) (1976) *Method of Measurement of Colour in Hazen Units (Platinum–Cobalt Scale) of Liquid Chemical Products*.
63. ISO/DIS 3558 *Plastics, Assessment of the Colour of Near White or Near Colourless Materials*.
64. Willmeyer, E. N. (1953) *Journal of the Oil and Colour Chemists Association*, **36** (399), 491.
65. Vickerstoff, T. and Walls, I. S. M. (1953) *Journal of the Oil and Colour Chemists Association*, **36** (399), 507.
66. Adams, J. M. (1956) *Paint Manufacture*, **26** (4), 105.
67. Goodwin, W. J. (1955) *Modern Plastics*, **32** (10), 32.
68. Wright, W. D. (1957) *Nature*, **179** (4552), 179.
69. Foote, P. V. (1964) *Review of Current Literature on the Paint and Allied Industries*, **37** (259), 1.
70. Morse, M. P. (1956) *Official Digest of the Federation of Paint and Varnish Producers*, **28** (383), 1278.
71. Menard, D. F. (1959) *Modern Packaging*, **32** (7), 115.
72. ICI Fibres Limited/Digico Limited/ICI Organics Division Limited (1973) *Pigment Research and Technology*, **2** (8), 16.

73. Cipolla, B. G. C. (1978) Poliplasti, **26** (253), 51.
74. Wood, A. S. (1979) *Modern Plastics International*, **9** (4), 40.
75. Gall, L. (1973) *Kunststoffe—RD Sch.*, **20** (3), 129.
76. Rosenthal, R. D. (1973) *Coloring of Plastics 7, RETEC, Cherry Hill, NJ, September,* SPE Delaware Valley Section, page 60.
77. Coppock, W. A. (1973) *Coloring of Plastics 7* (*loc. cit.*), page 79.
78. Conner, E. T. and Hammond, H. K. (1974) *Coloring of Plastics 8, RETEC, Cleveland, Ohio, October*, SPE Cleveland Section, page 110.
79. Shreckenbach, U. (1977) *Plaste und Kautschuk*, **24** (7), 468.
80. Burns, M. and Christie, J.S. (1977) *Law and You, RETEC, Boston, Mass, October*, SPE Eastern New England Section, page 35.
81. Jauch, K. M., (1979) *Journal of Physics E, Scientific Instruments*, **12,** 1171.

Chapter Fifteen

Thermal Properties

15.1 INTRODUCTION

15.1.1 Equation of Conduction of Heat

For the flow of heat in one direction, the heat flux J_u is related to the temperature gradient $\partial\theta/\partial x$ by Fourier's law:

$$J_u = -K \frac{\partial \theta}{\partial x} \tag{15.1}$$

where K is the thermal conductivity. The minus sign indicates that the heat flows in the opposite direction to the temperature gradient. The form of equation (15.1) implies that heat conduction is a random process. If the energy were propagated without scattering then the heat flow would depend on the temperature difference between the end-faces of the specimen instead of on the temperature gradient.[1]

The general equation from which the time-dependent temperature distribution may be calculated is obtained from equation (15.1) and the equation of continuity:

$$\frac{\partial J_u}{\partial x} = -\rho c \frac{\partial \theta}{\partial t} \tag{15.2}$$

where ρ is the density, c is the specific heat (i.e. the heat capacity per unit mass) and t is the time. The equation of continuity is an expression of the conservation of energy. The heat flux can be eliminated between equations (15.1) and (15.2) to give

$$\frac{\partial^2 \theta}{\partial x^2} + \frac{1}{K}\frac{\partial K}{\partial \theta}\left(\frac{\partial \theta}{\partial x}\right)^2 = \frac{1}{\alpha}\frac{\partial \theta}{\partial t} \tag{15.3}$$

where $\alpha = K/\rho c$ is the thermal diffusivity. Equation (15.3) is the equation of conduction of heat, in the absence of heat generation and convection, for heat flow in one direction. If the conductivity is independent of temperature, it reduces to

$$\frac{\partial^2 \theta}{\partial x^2} = \frac{1}{\alpha}\frac{\partial \theta}{\partial t} \tag{15.4}$$

which is the equation usually referred to. Our results show that for rubbers the conductivity term in (15.3) is very small and equation (15.4) is adequate for most heat flow calculations.[2]

The parameter α was called the 'thermal diffusivity' by Kelvin and the 'thermometric conductivity' by Maxwell, but Kelvin's expression has been generally adopted. It measures the change in temperature which would be produced in unit volume of the substance by the quantity of heat which flows in unit time across unit area of a layer of the substance of unit thickness with unit temperature difference between its faces.[3] This is the parameter which determines the non-steady-state temperature distribution in the absence of heat generation and convection and it is therefore essential for transient heat transfer calculations.

In the above discussion the specific heat referred to is that at constant volume per unit mass, c_v, whereas in practice the sample is at constant pressure. However, a body containing temperature gradients usually also contains internal stresses and c_p is not quite correct either,[4] although more appropriate than c_v. As the difference between the two principal specific heats is small for polymers,[5] this is not a significant point. In the following sections c_v is used in theoretical arguments and c_p is used elsewhere, unless otherwise stated.

15.1.2 Units

A variety of units has been used for thermal properties which is a nuisance when different sets of results have to be compared. The two most common units for conductivity are the cal/cm s °C and the Btu in/ft^2 h °F. There are two units of length in the imperial unit, because area is measured in square feet and thickness in inches, and this inconsistency is a potential pitfall for the unwary. A self-consistent conductivity unit, Btu/ft h °F, is obtained if the temperature gradient is measured in °F/ft instead of °F/in, but this is not as common. For diffusivity the c.g.s. unit is the cm^2/s and the imperial unit is the ft^2/h. The SI unit for conductivity is the W/m K, and for diffusivity the unit is the m^2/s. In this chapter the SI unit for conductivity and a submultiple of the SI diffusivity unit, the mm^2/s, are used. Conversion factors are given in Table 15.1.

Table 15.1 Conversion Factors for Some Thermal Conductivity Units

	W/m K	**cal/cm s °C**	**W/cm °C**	**kcal/m h °C**	**Btu in/ft^2 h °F**
W/m K	1	0·002 39	0·01	0·86	6·93
cal/cm s °C	419	1	4·19	360	2900
W/cm °C	100	0·239	1	86	693
kcal/m h °C	1·16	0·002 78	0·0116	1	8·06
Btu in/ft^2 h °F	0·144	0·000 345	0·001 44	0·124	1

15.1.3 Fundamental Significance of Diffusivity

Since $\alpha = K/\rho c$, diffusivity is often regarded as just a mathematical parameter rather than as a material property. However, diffusivity is seen to be a fundamental material property if we think wholly in terms of energy.

The heat capacity at constant volume per unit volume is given by $c = (\partial u/\partial \theta)_v$, where u is the internal energy per unit volume. Obtaining from this an expression for the temperature gradient and substituting it into equation (15.1) gives

$$J_u = -\alpha \frac{\partial u}{\partial x} \qquad (15.5)$$

Thus thermal diffusivity is the parameter relating energy flux to energy gradient, whereas conductivity relates the energy flux to the temperature gradient. The origin of the units mm^2/s, which are perhaps meaningless in themselves, now becomes apparent.

15.1.4 Heat Transfer at Interface

Many heat transfer problems involve a resistance to heat transfer across the interface between the sample surfaces and the heating or cooling medium. Such a resistance may be caused by a stationary surface film (of gas or liquid) of small but indeterminate thickness through which the temperature gradually changes from that of the solid to that of the bulk fluid. Heat transfer across a solid–fluid interface depends on wetting, film thickness and temperature gradients in the film. If the fluid is moving it will also depend on whether the flow is laminar or turbulent, which in turn is determined by velocity and viscosity. Similarly, heat transfer across the interface between two solids depends on adsorbed gases, pressure, surface finishes, hardness and trapped fluids.[6]

A surface heat transfer conefficient, h, can be defined as the quantity of heat flowing per unit time normal to the surface across unit area of the interface with unit temperature difference across the interface. When there is no resistance to heat flow across the interface, h is infinite. The heat transfer coefficient can be compared with the conductivity: the conductivity relates the heat flux to the temperature gradient; the surface heat transfer coefficient relates the heat flux to a temperature difference across an unknown distance.

Table 15.2 Typical Values of Surface Heat Transfer Coefficient, *h*

System	h (W/m^2 K)
Condensing steam	∞
Rubber/mould	∞
Heating or cooling with water	280–∞
Heating or cooling with air	4·5–45
Fluid bed	110

Some theoretical work has been done on this subject[7] but, since it is rarely possible to achieve in practice the boundary conditions assumed in the mathematical formulation, it is best to regard it as an empirical factor to be determined experimentally. Some typical values are given in Table 15.2. Cuthbert[8] has suggested that values greater than about 6000 W/m^2 K can be regarded as infinite. The spread in values is caused by different fluid velocities. Heat loss by natural convection also depends on whether the sample is vertical or horizontal.

15.2 THERMAL CONDUCTIVITY MEASUREMENT

In this section the methods of measuring conductivity are described. The methods discussed have been chosen to illustrate the basic principles of conductivity measurement and the various ways of minimising the major sources of experimental error. The experimental methods can be divided into two groups: steady state and transient.

15.2.1 Steady State Methods

Under steady state conditions, i.e. when the temperature at any point does not change with time, the temperature distribution in the sample is governed by Laplace's equation. The sample geometries are chosen so that the temperature is a function of only one coordinate, and simple analytical solutions to Laplace's equation are used. There are two geometries which satisfy this condition: the parallel-faced slab with heat flow normal to the surfaces, and the hollow cylinder with radial heat flow. In the latter case, although the heat flow is now in two directions, the temperature is a function of only one coordinate, namely the radius, because of symmetry. The condition is also satisfied by the sphere but this is not relevant to the present discussion. The first case is illustrated in Figure 15.1 which shows a parallel-faced section of thickness x and cross-sectional area A.

One face is maintained at a uniform temperature θ and the other at a uniform temperature $\theta - \Delta\theta$. The rate of heat flow q is in the x direction only and is normal to the two faces. The conductivity is obtained from equation (15.1):

$$K = \frac{qx}{A\Delta\theta} \qquad (15.6)$$

The coaxial cylinder geometry is illustrated in Figure 15.2. The cylinder is of length L, outside radius r_2 and inside radius r_1. The temperature difference between the inner and outer surfaces is $\Delta\theta$. Equation (15.1) can be integrated to give

$$K = \frac{q}{2\pi L\,\Delta\theta}\log_e\left(\frac{r_2}{r_1}\right) \qquad (15.7)$$

It is necessary to construct the apparatus so that the experimental conditions agree with the boundary conditions for simple analytical solutions. Failure to achieve this perfectly gives rise to three major sources of error.

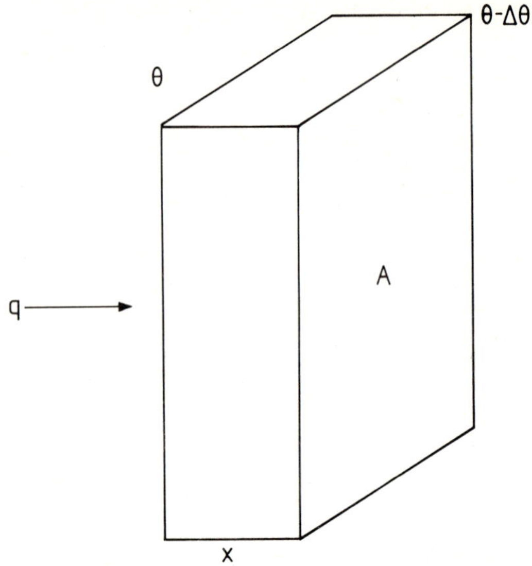

Figure 15.1 Flat slab

(i) The first major source of error is a result of a failure to obtain a normal or radial heat flow, depending on the geometry. Satisfying this boundary condition is very difficult in practice; the problem has been tackled in three ways: the flux is constrained, or the unwanted heat flow is minimised and ignored, or it is estimated and allowed for in the calculations.
(ii) The second source of error is that the surfaces of the sample are not necessarily at the same temperature as the walls of the apparatus, i.e. there is a resistance to heat flow across the interface between the sample and the apparatus. In the majority of cases the temperatures of the cell walls are measured, and this gives a conductivity value which is too low.
(iii) The third source of error results from failing to satisfy the condition that the major surfaces must be isothermals. The size of this error depends on the degree of non-uniformity.

The more important methods for measuring thermal conductivity are discussed below.

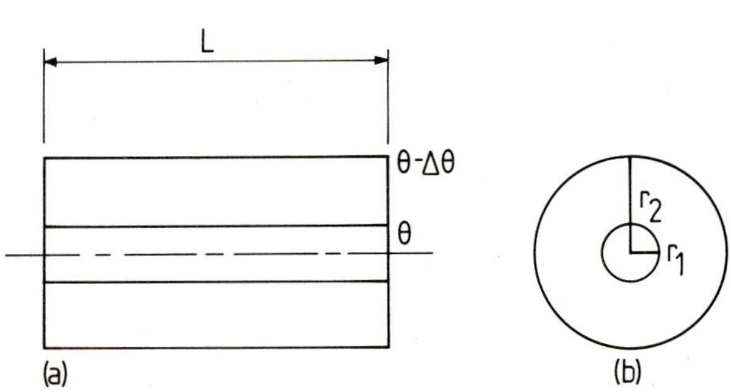

Figure 15.2 Coaxial cylinder

Unguarded Hot Plate

The modern unguarded hot plate is based on the well known Lees disc method[9] and the general arrangement is shown schematically in Figure 15.3. Recommendations are given in BS 874 (1973)[10] and ASTM C518 (1976).[11]

In this method two similar samples are placed on either side of an electrically powered heat source and sandwiched between two heat sinks. The heat sinks are controlled at a given temperature, usually by circulating a liquid from a constant-temperature bath, and a known power is supplied to the heat source. The apparatus is allowed to reach steady state and the temperature drop across the sample is measured, usually with thermocouples. The conductivity is calculated using equation (15.6).

The samples and the heat source are made as thin as possible and the apparatus is surrounded with a low-conductivity material such as vermiculite, but the side losses from the heater may still form a significant proportion of the total heat input. The heat loss may be evaluated by using a pair of specimens of low, known conductivity material, for example a cellular plastic calibrated by an absolute method. If the losses are known to, say, 10 per cent and represent only 10 per cent of the heat input, the error due to this uncertainty is only 1 per cent. The side loss calibration must be carried out over the range of specimen thickness employed and over the appropriate temperature range, and it is to be noted that it is a function of both hot and cold face temperatures and of the ambient temperature. A further refinement is to use heat flow meters on one or both sides of the sample.

By using large-area thin specimens and very thin heater plates the side losses may be reduced to negligible proportions, but of course the preparation of such specimens becomes increasingly difficult the larger the area and the smaller the thickness.

The most convenient specimen size lies usually in the range 50–200 mm across by 3–6 mm thick. The area of such specimens may easily be measured with the required precision, and the thickness similarly. The most important requirement of the specimens, however, is that the faces should be as flat as possible, since errors due to imperfect contact between specimens and heater and cold plates can be very large. As an example, an air film of thickness 0·03 mm on one side of a specimen 0·3 mm thick of a typical solid plastics material would cause an error approaching 10 per cent if the resulting film temperature drops were neglected. It is customary to reduce such errors by coating the specimens with a film of

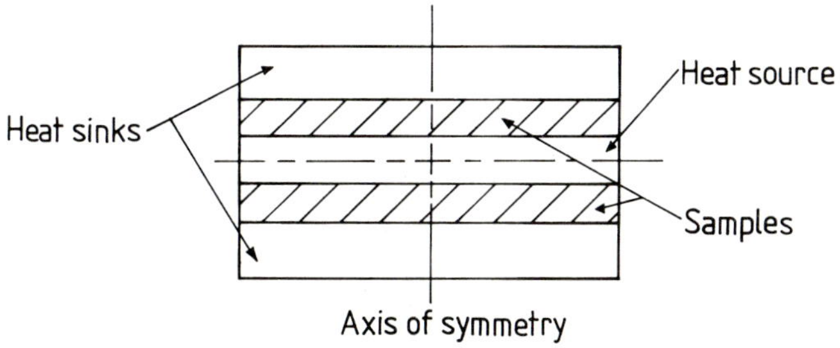

Figure 15.3 Unguarded hot plate

relatively high-conductivity liquid (compared with air) such as glycerol, or a suitable grease. Additionally, a correction for the resulting small temperature drop in such films may be obtained from a measurement with hot and cold plates in contact and specimens removed, but the contacting medium present. Plates are invariably of high-conductivity metal such as copper, brass or aluminium to ensure uniform surface temperatures. A further refinement is to make measurements on samples of different thicknesses. Assuming that the surface heat transfer coefficient is constant, it can then be eliminated from the calculations.

Temperatures (invariably of the plates rather than the specimens) are normally measured with fine wire thermocouples in conjunction with a DVM reading to 1 μV, and heat flows by measurement of heater input wattage, both measurements being capable of high precision. A stable power source is a necessity and batteries are still sometimes used, although the many admirable transistorised power supplies now available are more convenient, and do not suffer from the long-term drift associated with batteries. This is particularly important where measurements on low-conductivity materials are involved, since steady state conditions may not be reached for many hours after energising the heater.

Guarded Hot Plate

An improvement on the unguarded apparatus is the guarded hot plate. This is the most accurate method available for solid materials (including foams), and is recommended by the standards organisations (ASTM C177 (1976);[12] BS 874 (1973)[10]) for measurements on low-conductivity materials. The general arrangement is shown schematically in Figure 15.4. The standards recommend various dimension ratios for the guard width g, heater side length $2s$, and sample thickness l. If an accuracy of 1 per cent is to be achieved the limiting values of the ASTM specification should be used[13] (Table 15.3).

In this method the heat source is surrounded by a guard heater which has an independent power supply. The power to the guard is adjusted so that there is no temperature difference between the heat source and the guard. The heat from the heat source thus flows normally through the samples and the heat lost from the exposed edges comes from the guard.

The heat sinks are controlled at a given temperature, usually by circulating a liquid from a constant-temperature bath, and a known power is supplied to the heat source. The power supplied to the guard heater is adjusted to equalise the temperatures. The apparatus is allowed to reach steady state and the tempera-

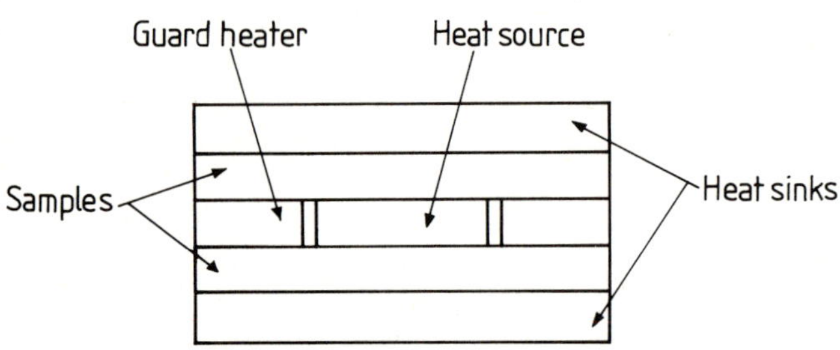

Figure 15.4 Guarded hot plate

TABLE 15.3 Dimension Ratios for Guarded Hot Plate

	BS 874	ASTM C177
$\frac{2(g+s)}{l} \geqslant$	6	9
$\frac{g}{l} \geqslant$	1	1·5
$\frac{s}{l} \geqslant$	2	3

ture drop across the samples is measured, usually with thermocouples. The conductivity is calculated using equation (15.6).

The attainment of complete equilibrium is even more important with low-conductivity materials than with solid plastics, since the long time-constant of cellular materials means that equilibrium times of the order of hours, or even a few tens of hours, are involved and the rate of change of temperature is slow and therefore difficult to detect. For this reason power supplies which are stable over long time-periods are a necessity. Since temperature differences of the order of 0·05°C between centre and guard heater plates have a significant effect on the measurement, the selection and construction of thermocouples is extremely important and it is necessary to determine the thermal e.m.f.s to within 1 μV.

As temperature differences across the specimens of as little as 10°C or less may be used, cold plate temperatures must be maintained very constant and a thermostat bath temperature control of at least ±0·1°C is required.

The apparatus dimensions are usually 300 mm square and the sample thickness range is 6 mm to 50 mm. There is a gap between the heat source and the guard heater of about 2 mm. The area used in equation (15.6) for calculating the conductivity is determined from the centre of this gap and is fractionally larger than the actual area of the heater.

In operation it is tedious to balance exactly guard and centre temperatures and it is usually more convenient to know the effect of a given amount of off-balance on the answer so that an appropriate correction can be applied. To do this, small adjustments to the guard energy are made around the balance point and the differences between guard and centre thermocouple readings are plotted against apparent conductivity. The centre energy must be maintained constant throughout and after each guard adjustment the apparatus must be left for equilibrium to be re-established. A typical plot is shown in Figure 15.5 and the correct conductivity is given, of course, by the intercept at zero difference. Over small temperature differences the plot is linear.

The virtue of this refinement in technique is that, with experience, the slope of the plot indicates whether the apparatus is responding properly or not, and also it is a measure of the efficiency of the plate, a high slope being undesirable since it indicates a high lateral conductivity between centre and guard portions of the hot plate.

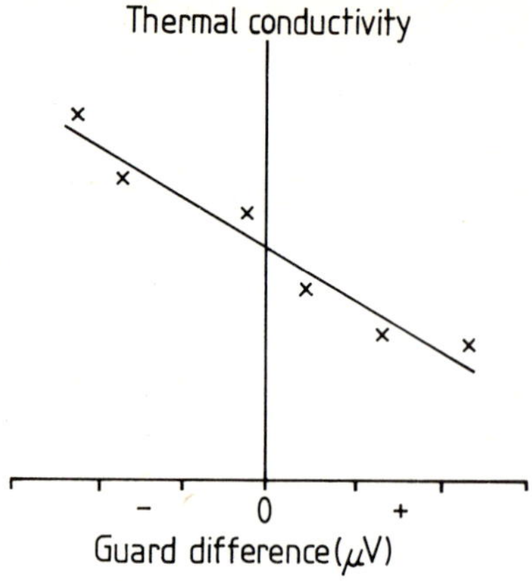

Figure 15.5 Thermal conductivity plotted against guard difference

Errors caused by a resistance to heat flow across the interface between the conductivity cell surfaces and the sample surfaces are more serious when measurements are being made on solid materials rather than on foams, because the resistance to heat flow across the interface is a larger fraction of the resistance of the sample. The errors can be minimised by wetting the surfaces with a heat transfer grease or oil. Measurements can be made on samples with different thicknesses, as described in the previous section.

Coaxial Cylinder Methods
Methods based on the coaxial cylinder geometry have the advantage that the exposed areas, from which heat can be lost, are much smaller compared with the area of the heat source than is the case with the hot plate methods. Thus the effects of lateral heat flow can be minimised without introducing guard rings. The essential features of a cylindrical conductivity cell are shown schematically in Figure 15.6.

The apparatus comprises a rod-shaped heat source surrounded by a tubular sample inside a cylindrical heat sink. The heat source and the heat sink have the same axis. The temperature of the heat sink is controlled by circulating either water or oil from a constant-temperature bath. A known power is supplied to the heat source, and the temperature difference between the heat source and heat

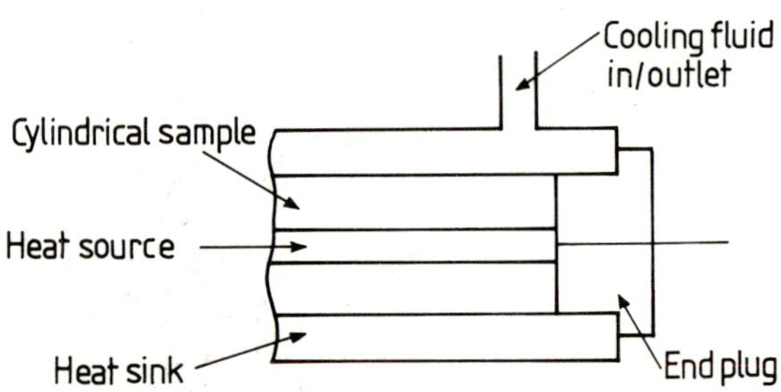

Figure 15.6 Coaxial cylinder apparatus

sink is measured after steady state conditions have been reached. The conductivity is obtained from equation (15.7).

The apparatus is constructed with a high ratio of length to diameter so that the heat lost from the ends is small compared with the heat transferred radially through the sample. This was checked for solid samples by Kline,[14] who made cells with length-to-diameter ratios of 8:1, 12:1 and 24:1. The two of higher ratios gave identical results while results from the shortest cell were 10 per cent lower. However, the greater the relative heat loss the greater should be the calculated conductivity. This shows that the major source of error is not heat loss from the ends but resistance to heat flow across the surfaces, because it is impossible to ensure that the sample is a perfect fit in the apparatus.

Quasi-stationary Method

Eiermann *et al.*[15] developed a quasi-stationary method of measuring the thermal conductivity of solid polymers. The apparatus consisted of two heat sources and two thin samples on either side of a copper heat sink, the temperature of which was monitored and which was surrounded by an adiabatic shield to minimise the heat loss. The instantaneous heat flow rate was determined from the rate of temperature rise of the heat sink:

$$q_1 + q_2 = Mc\frac{d\theta}{dt} \tag{15.8}$$

where q_1, q_2 = the heat flow rates across the two samples
M = the mass of the heat sink
c = the specific heat of the heat sink
$d\theta/dt$ = the rate of temperature rise of the heat sink.

Since $q = -KA(\Delta\theta/x)$, this gives for the conductivity

$$K = -\frac{Mcx}{A(\Delta\theta_1 + \Delta\theta_2)}\frac{d\theta}{dt} \tag{15.9}$$

The temperature range covered by the method was from −190°C to +90°C.

Enclosed Method

Most of the methods described above are for solid materials. Very little work has been done with molten polymers because the majority of present methods for use with liquids have been designed for low-viscosity liquids; loading and unloading is therefore difficult with polymers because they have high viscosities and are usually solid at room temperature. Hands and Horsfall[16] describe a method for measuring the conductivity of solid and liquid samples from ambient to 250°C. The previous steady state methods are all based on the use of simple analytical solutions to Laplace's equation. In this method the opposite design philosophy is used, i.e. the aim is towards experimental ease at the expense of mathematical simplicity.

The apparatus has an axis of symmetry and a plane of symmetry (Figure 15.7). It comprises a disc-shaped heat source supported in the centre of a short cylindrical heat sink by an insulating annulus, leaving space for a disc-shaped

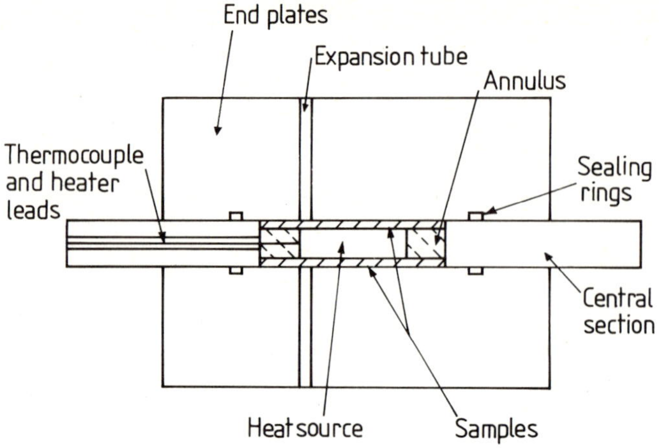

Figure 15.7 RAPRA thermal conductivity apparatus

sample on either side. The heat sink is completed by two end-plates which are bolted on, each end-plate containing an expansion tube. The samples are thus enclosed and can be melted in the apparatus.

By dimensional analysis it can be shown that the ratio of the conductivity of the polymer samples, K_p, to that of the annulus, K_A, is a function of the dimensionless power

$$\frac{K_p}{K_A} = f\left(\frac{W}{\Delta\theta h K_A}\right) \tag{15.10}$$

where W is the power supplied to the heat source, $\Delta\theta$ is the temperature difference and h is the sample thickness. Since the heat flow is three-dimensional, a numerical method is used to obtain the relationship between K_p/K_A and $W(\Delta\theta h K_A)$. For the purpose of calculating conductivity values from the experimental results, this relationship is most conveniently expressed as a polynomial. This function or instrument curve only needs to be obtained once.

The temperature of the heat sink is controlled at any desired value, a known power is supplied to the heat source and the temperature difference is measured. The conductivity is then calculated using the instrument curve. The conductivity of the annulus is needed for the calculations; this is obtained by making preliminary measurements on samples of the annulus material.

Limitations of this method are that the samples must be the correct size, since the geometry cannot be changed to accommodate samples of varying thickness; the samples must also be isotropic in their heat transfer characteristics, and the scale of any inhomogeneities must be small compared with the thickness. The major source of error is the deviation from the assumed isothermal boundary condition on the source and sink surfaces.

15.2.2 Transient Methods

The time-dependent temperature distribution in a transient experiment is governed by equation (15.4) and usually the related parameter, thermal diffusivity, is obtained. However, under certain circumstances the solution to the heat equation contains the thermal conductivity as well as the diffusivity, and by choosing a suitable method the diffusivity can be eliminated from the answer. The more important methods are the line and plane source heater methods.

Line Source Method
The most common transient technique which gives the thermal conductivity directly is the continuous line source method. For a line source heater of infinite length in an infinite mass of sample the temperature at a distance r from the line source is given by[17]

$$\theta = \frac{q}{4\pi K} \int_a^\infty \frac{e^{-u}}{u} \, du \tag{15.11}$$

where $a = r^2/4\alpha t$, and q is the rate of heat generation per unit length. For small values of a the integral reduces to $\log_e (4\alpha t/r^2)$ − const. Thus, if the diffusivity and conductivity are constant over the temperature range $\Delta\theta$, then

$$\theta_2 - \theta_1 = \Delta\theta = \frac{q}{4\pi K} \log_e \frac{t_2}{t_1} \tag{15.12}$$

where θ_1 is the temperature at r at time t_1 and θ_2 is the temperature at time t_2. Rearrangement gives

$$K = \frac{q}{4\pi \, \Delta\theta} \log_e \frac{t_2}{t_1} \tag{15.13}$$

Hence the conductivity can be obtained directly by measuring q and the change of temperature with time.

It is necessary to make small corrections for finite heater wire diameter, finite wire length, and finite sample mass. These have been discussed in detail for liquids by Horrocks and McLaughlin,[18] who found that the corrections were only significant for short heating times. This technique has been used with polymers,[19,20] and has also been developed in the form of a conductivity probe[21–23] for making single point measurements on, for example, foams.

Plane Source Method
An alternative to a line source is to use a plane course of constant flux, J. If the sample is assumed to be a semi-infinite solid ($0 \leqslant x < \infty$) then the temperature at a point x and time t is given[24] by

$$\theta = \frac{2J}{K} \sqrt{(\alpha t)} \operatorname{ierfc} \frac{x}{2\sqrt{(\alpha t)}} \tag{15.14}$$

At the point $x = 0$

$$\theta = \frac{2J}{K} \left(\frac{\alpha t}{\pi} \right)^{1/2} \tag{15.15}$$

The temperature at the surface, $x = 0$, and at a point in the sample, $x = 1$, say, are measured as functions of time, and from this both K and α can be determined. The method was developed by Harmathy[25] for measurements on building materials, and in his paper he also discussed the limitations on the technique imposed by the finite size of the samples. The method has been used for polymers by Steere.[26,27] However, to obtain a sufficiently thick sample he stacked thin films together but ignored the resistance to heat flow at the interfaces. A further difficulty with Steere's technique is that thin films are invariably oriented.

15.3 THERMAL DIFFUSIVITY MEASUREMENT

15.3.1 Introduction

To measure diffusivity it is only necessary to know the change in temperature with time at three co-linear points in the direction of the heat flow. For conductivity, although the temperature gradient can easily be obtained, the heat flux cannot be measured at a point; the total power to the heat source is known but it is difficult to control or calculate the movement of all the energy. Although diffusivity is easier to measure, and although it is the essential parameter for transient heat flow calculations, it has received less attention than conductivity. The scatter in the diffusivity data in the literature is not as great as for conductivity (possibly because not as many people have tried to measure it!).

15.3.2 Basic Principles

Boundary Conditions

For certain boundary and initial conditions, analytical solutions to equation (15.4) can be obtained. Apart from the latest method,[2] which is based on a numerical calculation, diffusivity measurement methods are all based on such solutions. The experimental conditions are matched to these mathematical conditions as closely as possible and the appropriate solution is used to give a value for the diffusivity. The experiment can be repeated at different temperatures in order to obtain the temperature dependence of the diffusivity. This experimental procedure has been criticised[28] because, if the diffusivity changes with temperature, then almost invariably the conductivity is also temperature-dependent and equation (15.3), which would not have given an analytical solution, should have been used instead of (15.4). However Hands and Horsfall[29] have shown that, except near melting transitions, thermocouples sensitive to 0·002°C would be needed to detect the effect of the conductivity term in equation (15.3). Hence generally the simpler equation is adequate for diffusivity measurement and for heat flow calculations.

Diffusivity measurement methods based on analytical solutions to equation (15.4) have all had the same initial condition that the whole sample is at a constant uniform temperature. But three different types of boundary condition have been employed: first the sample surface is subjected to a step change in temperature; second the surface is subjected to a linear rate of temperature rise; and third the surface is subjected to a periodic temperature fluctuation.

The solutions to equation (15.4) for these three types of boundary condition are usually presented in terms of dimensionless groups and are applicable to infinite solids. Before proceeding to discuss the experimental methods it is necessary to consider these points in more detail and also the application of the solutions to finite solids. The word 'solid' here is a mathematical term and does not refer to the phase of the sample.

Dimensionless Groups

The solutions to equation (15.4) may be presented graphically.[30,31] In order to give the greatest amount of information with the minimum number of curves, the graphs are presented in terms of four dimensionless groups.

In problems involving a step change in surface temperature the unaccomplished temperature change of a point in the sample is related to the maximum possible temperature change. Hence the dimensionless temperature is given by

$$Y = \frac{\theta_1 - \theta}{\theta_1 - \theta_0} \tag{15.16}$$

where θ_0 is the initial uniform temperature, θ_1 is the temperature of the heating or cooling medium, and θ is the temperature at any point in the sample for $t > 0$. All values of Y lie between 0 and 1. The extreme values indicate the heating or cooling temperature $\theta = \theta_1$, and the initial temperature $\theta = \theta_0$, respectively.

Some authors use an alternative dimensionless temperature

$$\frac{v}{V} = 1 - Y = \frac{\theta - \theta_0}{\theta_1 - \theta_0} \tag{15.17}$$

and this is the ratio of the actual temperature change to the maximum possible temperature change.

The times required for similar temperature changes in similar bodies are proportional to the square of the dimensions and inversely proportional to the thermal diffusivity. It is therefore convenient to use a dimensionless group given by

$$\phi = \frac{\alpha t}{a^2} \tag{15.18}$$

where a is the radius of a sphere or cylinder, or the semi-thickness of a slab.

At any time, the temperature at a point in the sample is related to the ratio of the distance of the point from the centre of the radius or semi-thickness. Hence the dimensionless length is given by

$$\xi = \frac{r}{a} \tag{15.19}$$

where r is the distance measured from the centre along a radius of a sphere or cylinder, or the distance measured from the centre plane of a flat slab normal to the surfaces.

The fourth dimensionless group must be introduced for problems involving a resistance to heat transfer across the interface. The heat transfer coefficient can be combined with the conductivity and the radius or semi-thickness to give a dimensionless resistance ratio

$$m = \frac{K}{ah} \tag{15.20}$$

When h is infinite, m is zero and there is no resistance to heat flow across the interface.

It should be pointed out that although the reciprocal of m looks like a Nusselt number it is not one. For a Nusselt number the conductivity would be that of the heating or cooling fluid and not, as in this case, the conductivity of the sample.

Infinite and Non-infinite Solids

The analytical solutions to the heat equation are for one-dimensional problems; this implies that the heat flow is in one direction only or that the equation reduces to one dimension because of symmetry, i.e. the temperature is a function of one length variable only. This condition is satisfied for three sample geometries: the infinite flat slab, the infinite circular cylinder and the sphere. An infinite slab is a parallel-faced slab of finite thickness and infinite length and breadth; an infinite cylinder is a cylinder of finite diameter and infinite length.

Real samples are obviously not infinite and the practical definition of an infinite solid depends on how much error can be tolerated. For example, if one lateral dimension of the slab is reduced from infinity to five times the thickness, or the length-to-diameter ratio of a cylinder is 5:1, then a 1 per cent error is introduced if lateral heat flow is ignored.

The solutions to equation (15.4) for infinite solids are of the form

$$Y = f(\phi, \xi, m) \qquad (15.21)$$

and are in the form of infinite series. For samples in the form of cubes, rectangular parallelepipeds (bricks) and short right circular cylinders, where lateral heat flow cannot be ignored, the infinite solid solutions can still be used.[32] For a brick, by considering heat transfer through each pair of faces in turn (imagining the lateral dimensions to be infinite) three values of the dimensionless temperature are obtained. The final solution is the product of these three, $Y = Y_x Y_y Y_z$. Similarly, the solution for a short cylinder is given by the product of the solutions for an infinite slab and an infinite cylinder, $Y = Y_z Y_r$.

This approach is based on separation of variables and it can be shown by substitution that a product of solutions is itself a solution. Separation of variables is only valid for certain boundary conditions and imposes restrictions on geometry which are, however, satisfied by the examples under consideration.

Sources of Error

In the experimental techniques described below there are a number of recognised sources of error. These have not always been avoided and although the measurement is basically straightforward the results in the literature are not always reliable. The significance of the errors will be discussed in connection with the methods of measurement and only a summary is given here. Error can arise:

(i) in assuming an infinite surface heat transfer coefficient between the sample and the heating or cooling medium;
(ii) in ignoring lateral heat flow when it is significant;
(iii) in conduction along thermocouple leads;
(iv) in assuming that diffusivity is temperature-independent;
(v) in ignoring the thermal expansion of the samples.

15.3.3 Quenching Methods

Thermal diffusivity has usually been measured using a quenching method, i.e. the solid sample at a uniform temperature is immersed in a temperature-controlled bath at a different temperature. The rate of change of temperature at the centre is then monitored with an embedded thermocouple. The sample dimensions are usually chosen so that lateral heat flow can be ignored and regular sample geometries, i.e. 'infinite' flat slabs, 'infinite' cylinders, or spheres, are used.

Quenching methods are based on the assumption that the thermal diffusivity is constant over the experimental temperature range. To obtain the temperature-dependence of the diffusivity, the temperature range of interest is covered in a series of steps which are small enough for this assumption to be valid. The size of the step temperature change is a compromise between this requirement and the need to make accurate temperature difference measurements. The value which is usually chosen is about 5°C. In some experiments this point has been ignored and the measurement made with a much larger step. For example, the sample may be conditioned at room temperature and then immersed in boiling water. An average value for the diffusivity is obtained and the error depends on the temperature-dependence of the diffusivity and hence changes from material to material.

Two methods have been used to obtain a value for the diffusivity from the experimental results. As was shown above, for the geometries in question the analytical solutions to the heat equation, which are in the form of infinite series, can be conveniently represented graphically. The first such collection of graphs was published by Williamson and Adams.[30] By reference to such graphs a value for ϕ and hence α can be obtained.

The second method is based on the rapid convergence of the series solutions. For example, after about 10 per cent of the time for the centre temperature to reach 99 per cent of the surface temperature has elapsed, the second term in the series is about $\frac{1}{2}$ per cent of the first term. Thus after a certain time only the first term is relevant. Now Y is always an exponential function of ϕ; for example, consider an infinite slab of thickness $2a$: the first term of the series for the dimensionless temperature at the centre is given by

$$Y = \frac{4}{\pi} \exp\left(-\frac{\pi^2}{4}\phi\right) \tag{15.22}$$

For long time-scales this is the only significant term and the diffusivity can be obtained from the slope of a graph of $\log_e Y$ against t since $\phi = \alpha t/a^2$.

15.3.4 Linear Heating Method

The second type of boundary condition to be considered is that in which the sample surface is subjected to a linear rate of temperature rise. A method based on this has been developed by Shoulberg[33] for diffusivity measurements on polymer melts. He used two discs of his material with a thermocouple sandwiched between them; the diameter-to-thickness ratio was such that his sample could be regarded as an infinite flat slab. The sample completely filled a cavity in an aluminium block and was melted in the apparatus. The aluminium block was heated electrically and the power was adjusted to give an approximately linear rate of temperature rise. Under his experimental conditions this lasted for about 30°C.

For this boundary condition the solution to the heat equation for the temperature at the centre is[34]

$$\theta = kt - \frac{ka^2}{2\alpha} + \frac{16ka^2}{\alpha\pi^3}\sum_{n=0}^{\infty}\frac{(-1)^n}{(2n+1)^3}\exp\left(\frac{-(2n+1)^2\pi^2\phi}{4}\right) \qquad (15.23)$$

where k is the linear rate of temperature rise. After a long time (in Shoulberg's case about 12 min) the summation term is negligible and the temperature difference between the surface and the centre becomes constant. The diffusivity is then given by

$$\alpha = \frac{ka^2}{2\Delta\theta} \qquad (15.24)$$

where $\Delta\theta$ is the temperature difference between the surface and the centre of the sample, and a is the semi-thickness of the sample, i.e. the thickness of one of the discs. The temperature range of interest was covered in a series of 30°C steps, and using equation (15.24) Shoulberg obtained an average value for the diffusivity for each step.

15.3.5 Periodic Heating Method

A method for measuring the diffusivity of solid polymers based on this type of boundary condition has been developed by Berlot.[35,36] A disc sample of thickness $2a$ is held at a uniform temperature and then a sinusoidal temperature fluctuation of angular frequency ω is imposed on the outer surfaces. The amplitude ratio and phase of the temperature at the centre are monitored with a thermocouple. Under these conditions the amplitude ratio, A, and phase, ϕ, are given by[37]

$$A = \left(\frac{2}{\cosh 2ka + \cos 2ka}\right)^{1/2} \qquad (15.25)$$

$$\phi = \arg\left(\frac{1}{\cosh ka\,(1+i)}\right) \qquad (15.26)$$

$$k = \left(\frac{\omega}{2\alpha}\right)^{1/2} \qquad (15.27)$$

k can be obtained from equation (15.25) or (15.26) and hence α from equation (15.27).

15.3.6 Continuous Heating Method

If the temperature-dependence of diffusivity is taken into account then the temperature range from ambient up to, say, 250°C can be covered in one experiment. With a quenching method a true step change in temperature is difficult to achieve because liquids capable of withstanding high temperatures tend to have high viscosities and this results in large temperature gradients close to the sample surface. Another objection to a quenching method is that the predominance of the quenching temperature throughout the experiment, coupled with the extremely large initial temperature difference, could give rise to computational difficulties. These problems, inherent in the quenching method, do not occur in a continuous heating method.

A method based on this principle has recently been developed by Hands and Horsfall.[2] The apparatus is shown schematically in Figure 15.8. Two disc-shaped samples are placed together and a thermocouple is sandwiched between them to monitor the centre temperature. The outside temperatures are measured with two further thermocouples which are in intimate contact with the surfaces. The overall thickness is maintained by a brass ring and two brass end-plates which contain expansion holes.

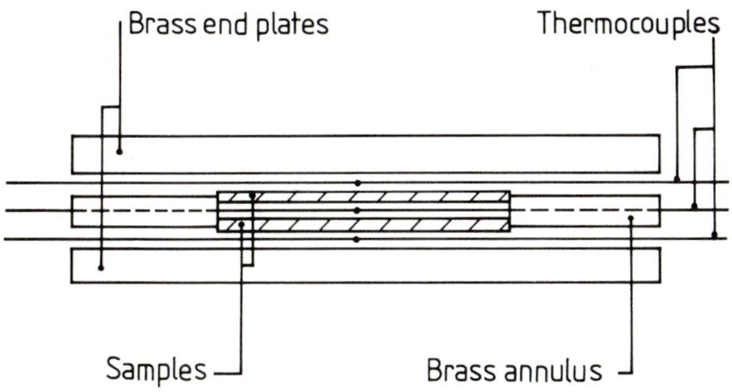

Figure 15.8 Schematic diagram of apparatus for continuous heating method

Enclosing the samples eliminates errors caused by ignoring thermal expansion. In quenching methods, for example, the appropriate sample dimension is measured at room temperature and this is the value which is used in the calculations. Since the equations involve terms containing the square of the dimensions, any error caused by ignoring expansion is automatically doubled. The error can be a few per cent depending on the temperature range covered.

The apparatus is immersed in a small oil bath which is heated continuously by an electric immersion heater, and the oil is stirred vigorously. No temperature control is needed; the only requirement is that the temperature increases smoothly. The whole temperature range is covered in about 40 minutes.

The heat equation is solved numerically[38] to give thermal diffusivity as a function of temperature. To allow for the change in conductivity at a melting transaction, equation (15.3) is solved in preference to (15.4).

15.3.7 Pulse Heating Method

If a disc-shaped sample is irradiated on one surface for a short time using a flash tube or a pulsed laser, then the temperature–time curve for the back face depends on the thermal diffusivity of the sample and the heat losses. If the pulse time τ is very much shorter than the time for the pulse to pass through the sample, then the heat losses can be ignored. Cape and Lehman[39] give a criterion for this condition:

$$a^2/\pi^2\alpha \geqslant 10\tau$$

where a is the thickness of the disc. Under these conditions the ratio of the temperature to the maximum temperature for the back face is given by[40]

$$\frac{\theta}{\theta_{max}} = 1 + 2 \sum_{n=1}^{\infty} (-1)^n \exp\left(-\frac{n^2\pi\alpha t}{a^2}\right) \tag{15.28}$$

The temperature as a function of time on the back face is usually measured with a radiation pyrometer.[41] Analysis of the results has been discussed by Parrott and Stukes.[42]

15.4 SPECIFIC HEAT

15.4.1 Introduction

Specific heat, c, is the heat capacity, C, per unit mass or per unit volume; usually the term refers to the mass specific heat. Heat capacity is defined by

$$C = \frac{dQ}{dT} \tag{15.29}$$

where Q is the heat energy absorbed and T is the temperature. The heat capacities of greatest interest are those at constant pressure, C_p, and constant volume, C_v. Substituting into equation (15.29) for the heat energy absorbed from the first law of thermodynamics, assuming that the work done is of a mechanical nature only, the heat capacities are given by

$$C_p = \left(\frac{\partial U}{\partial T}\right)_p + P\left(\frac{\partial V}{\partial T}\right)_p \tag{15.30}$$

and

$$C_v = \left(\frac{\partial U}{\partial T}\right)_v \tag{15.31}$$

where U is the internal energy of the system, P is the pressure, and V is the volume.

The heat capacity at constant volume is of more interest from a theoretical point of view because it is directly related to the internal energy of the system. However, it is almost impossible to measure C_v directly. To obtain a value for C_v it is necessary to measure C_p and to calculate C_v from the formula[43]

$$C_p - C_v = TVB\beta^2 \tag{15.32}$$

where B is the bulk modulus, and β is the volume expansion coefficient. In terms of the principal specific heats this is

$$C_p - C_v = \frac{TB\beta^2}{\rho} \tag{15.33}$$

Putting in some typical values, the difference between the principal specific heats for polymers at room temperature is of the order of 0·004 J/g K, which is usually small enough to be ignored.[5] Typical specific heat values for solids, including plastics, range from 0·4 J/g K to 4 J/g K. Data are given in the review by Wunderlich.[44]

15.4.2 Adiabatic Calorimeters

One of the most precise way of measuring specific heat is by means of an adiabatic calorimeter; it is also the most relevant and direct method since a small measured amount of heat is applied to the specimen and the resultant temperature rise recorded. It is adiabatic since the temperature of the system is allowed to rise, but no heat exchange with the surroundings is permitted; this is normally contrived by surrounding the calorimeter proper by a jacket maintained at the same temperature to within very close limits (of the order of 0·01°C or less). All the heat applied to the specimen, less that used in raising the temperature of the calorimeter itself, is therefore employed in raising the specimen temperature; the specific heat is simply calculated. The thermal capacity of the calorimeter is determined in a separate experiment.

A number of workers have described adiabatic calorimeters and the constructional and experimental difficulties.[45–58]

15.4.3 Drop Calorimeters

Drop calorimeters are widely used on account of their simplicity. A specimen, often contained in a metal capsule, is heated to some appropriate constant temperature in an oven or furnace and allowed to drop into liquid in a stirred calorimeter. The temperature of the calorimeter is plotted against time and from the curve the temperature rise, allowing for heat losses or gains to or from the environment, is obtained. The specific heat is calculated from this rise after applying corrections for the water equivalent of the calorimeter, etc., determined separately.

ASTM C531 (1961)[59] is typical of this method, although not intended specifically for plastics.

Sometimes the drop calorimeter and adiabatic methods are combined, the heated capsule being dropped into an adiabatic container. Griskey and Hubbell[60] described such a method as applied to measurements on methacrylic polymers in the range 120°–300°C.

15.4.4 Scanning Calorimeters

The usual techniques are differential thermal analysis (DTA)[61] and differential scanning calorimetry (DSC).[62–64] In these methods it is assumed that the heat loss from the calorimeter is a function of temperature only. By comparing the rate of heat input and temperature rise for a polymer sample with that of a standard, usually synthetic sapphire, the specific heat of the polymer can be calculated.

DTA measures the difference in temperature between the sample and a standard for the same rate of heat input. DSC compares the rate of heat inputs for the same rate of temperature rise. The latter is easier to analyse as it gives a direct measure of the rate of heat input. The method is based on the assumption that the samples are so small that thermal equilibrium is obtained almost immediately. For polymers this is not correct; errors from this source are discussed by Strella and Erhardt.[65]

Richardson[66] discusses methods of calibrating the DSC to improve the accuracy of the results.

15.4.5 Other Methods

One other method is worthy of brief mention. The thin heater calorimeter was developed by Hager[67] from his thin foil thermal conductivity apparatus.[68] Results obtained with this apparatus on PVC are reported by Dunlap.[69]

If a mass of material is uniformly heated, the central section experiences adiabatic conditions for a short time and Hager's method takes advantage of this. He achieves uniform heating by laying up alternate layers of polymer film and metal foil heater and then tightly winding them to form a cylinder. Electrical connections are made to the central and outer edges of the foil and thermocouples are inserted in suitable positions to measure the small temperature rise.

REFERENCES

1. Kittel, C. (1966) *Introduction to Solid State Physics* 3rd edn, Wiley, page 168.
2. Hands, D. and Horsfall, F. (1977) *Rubber Chemistry and Technology*, **50,** 253.
3. Carslaw, H. S. and Jaeger, J. C. (1959) *Conduction of Heat in Solids* 2nd edn, Oxford University Press, page 9.
4. Parrott, J E. and Stukes, A. D. (1975) *Thermal Conductivity of Solids*, Pion, page 5.
5. Choy, C. L. (1975) *Journal of Polymer Physics*, **13,** 1263.
6. Howard, J. R. (1975) *Engineering*, **215,** 220.
7. McAdams, W. H. (1954) *Heat Transmission* 3rd edn, McGraw-Hill.
8. Cuthbert, C. (1954) *Transactions of the Institution of the Rubber Industry*, **30,** 16.
9. Lees, C. H. (1898) *Philosophical Transactions of the Royal Society*, **A191,** 399.
10. BS 874 (1973) *Methods for Determining Thermal Insulating Properties*.
11. ASTM C518 (1976) *Steady-state Thermal Transmission Properties by Means of Heat Flow Meter*.
12. ASTM C177 (1976) *Steady-state Thermal Transmission Properties by means of the Guarded Hot Plate*.
13. Parrott, J. E. and Stukes, A. D. (*op. cit.*), page 19.
14. Kline, D. E. (1961) *Journal of Polymer Science*, **50,** 441.
15. Eiermann, K., Hellwege, K. H. and Knappe, W. (1961) *Kolloidzeitschrift*, **174,** 134.
16. Hands, D. and Horsfall, F. (1975) *Journal of Physics, E, Scientific Instruments*, **8,** 687.
17. Carslaw, H. S. and Jaeger, J. C. (*op. cit.*), page 261.
18. Horrocks, J. K. and McLaughlin, E. (1963) *Proceedings of the Royal Society*, **A273,** 259.
19. D'Eustachio, D. and Schreiner, R. E. (1952) *Journal of the American Society of Heating and Ventilating Engineers*, **58,** 331.
20. Vos, B. H. (1956) *Applied Scientific Research*, **A5,** 425.
21. Mann, G. and Forsyth, F. G. E. (1956) *Modern Refrigeration*, June, 188.

22. Hooper, F. C. and Lepper, F. R. (1950) *Heating, Piping and Air Conditioning*, **20** (August), 129.
23. Hooper, F. C. and Chang, S. C. (1952) *Heating, Piping and Air Conditioning*, **22** (October), 125.
24. Carslaw, H. S. and Jaeger, J. C. (*op. cit.*), page 75.
25. Harmathy, T. Z. (1964) *Journal of Applied Physics*, **35,** 1190.
26. Steere, R. C. (1966) *Journal of Applied Physics*, **37,** 3338.
27. Steere, R. C. (1966) *Journal of Applied Polymer Science*, **10,** 1673.
28. Martin, B. (1970) *Polymer*, **11,** 287.
29. Hands, D. and Horsfall, F. (1971) *Polymer*, **12,** 145.
30. Williamson, E. D. and Adams, L. H. (1919) *Physical Review*, **14,** 99.
31. Hands, D. (1971) *RAPRA Technical Review No. 60.*
32. Carslaw, H. S. and Jaeger, J. C. (*op. cit.*), page 33.
33. Shoulberg, R. H. (1963) *Journal of Applied Polymer Science*, **7,** 1597.
34. Carslaw, H. S. and Jaeger, J. C. (*op. cit.*), page 104.
35. Berlot, R. (1966) *Plastiques Modernes et Elastomères*, **18,** 231.
36. Berlot, R. (1967) *Plastiques Modernes et Elastomères*, **19,** 117.
37. Carslaw, H. S. and Jaeger, J. C. (*op. cit.*), page 105.
38. Horsfall, F. (1976) MSc thesis, University of Bradford.
39. Cape, J. A. and Lehman, G. W. (1963) *Journal of Applied Physics*, **34,** 1909.
40. Parker, W. J., Jenkins, R. J., Butler, C. P. and Abbott, G. L. (1961) *Journal of Applied Physics*, **45,** 2321.
41. Taylor, R. (1965) *British Journal of Applied Physics*, **16,** 509.
42. Parrott, J. E. and Stukes, A. D. (*op. cit.*), page 37.
43. Zemansky, M. W. (1957) *Heat and Thermodynamics* 4th edn, McGraw-Hill, page 251.
44. Wunderlich, B. and Baur, H. (1970) *Advances in Polymer Science*, **7,** 151.
45. Southard, J. C. and Brickwedde, F. G. (1933) *Journal of the American Chemical Society*, **55,** 4378.
46. Bekkedahl, N. and Matheson, H. (1935) *Journal of Research of the National Bureau of Standards*, **15,** 503.
47. Scott, R. B., Myers, C. H., Rands, R. D. Jr, Brickwedde, F. G. and Bekkedahl, N. (1945) *Journal of Research of the National Bureau of Standards*, **35,** 39.
48. Williams, J. W. and Daniels, F. (1924) *Journal of the American Chemical Society*, **46,** 903.
49. Dole, M., Hettinger, W. P. Jr, Larson, N. R., Wethington, J. A. and Worthington, A. E. (1951) *Review of Scientific Instruments*, **22,** 812.
50. Worthington, A. E., Marx, P. C. and Dole, M. (1955) *Review of Scientific Instruments*, **26,** 698.
51. West, E. D. and Ginnings, D. C. (1958) *Journal of Research of the National Bureau of Standards*, **60,** 309.
52. Richardson, M. J. (1965) *Transactions of the Faraday Society*, **61,** 1876.
53. Aukward, J. A., Warfield, R. W., Petree, M. C. and Donovan, P. (1959) *Review of Scientific Instruments*, **30,** 597.
54. Tautz, H., Gluck, M., Hartmann, G. and Leuteritz, R. (1963) *Plaste und Kautschuk*, **10,** 648.
55. Hellwege, K. H., Knappe, W. and Wetzel, W. (1962) *Kolloidzeitschrift*, **180,** 126.
56. Wunderlich, B. and Dole, M. (1957) *Journal of Polymer Science*, **24,** 201.
57. Bowring, R. W., Garton, D. A. and Norris, H. F. (1960) *UKAEA Report AEEW-R38*, HMSO.
58. Dainton, F. S., Evans, D. M., Hoare, F. E. and Melia, T. P. (1962) *Polymer*, **3,** 263.
59. ASTM C351 (1961) *Mean Specific Heat of Thermal Insulation.*

60. Griskey, R. G. and Hubbell, D. O. (1968) *Journal of Applied Polymer Science*, **12,** 853.
61. Slade, P. E. Jr and Jenkins, L. T. (Eds.) (1966) *Techniques and Methods of Polymer Evaluation* vol. 1, *Thermal Analysis*, Dekker.
62. Watson, E. S., O'Neill, M. J., Justin, J. and Brenner, N. (1964) *Analytical Chemistry*, **36,** 1233.
63. O'Neill, M. J. (1964) *Analytical Chemistry*, **36,** 1238.
64. O'Neill, M. J. (1966) *Analytical Chemistry*, **38,** 1331.
65. Strella, S. and Erhardt, P. F. (1969) *Journal of Applied Polymer Science*, **13,** 1373.
66. Richardson, M. J., and Burrington, P. (1974) *Journal of Thermal Analysis*, **6,** 345.
67. Hager, N. E. Jr (1964) *Review of Scientific Instruments*, **35,** 618.
68. Hager, N. E. Jr (1960) *Review of Scientific Instruments*, **31,** 177.
69. Dunlap, L. H. (1966) *Journal of Polymer Science, A2*, **4,** 673.

Chapter Sixteen

Effect of Temperature

16.1 INTRODUCTION

Plastics materials as a whole, and thermoplastics in particular, are sensitive to temperature; while those that are ductile or flexible may demonstrate shortcomings in becoming brittle or rigid at some low temperature, far more serious as a rule is the softening and even degradation that may occur when the temperature is raised to only moderate levels—that is, 'moderate' in the eyes of the designer, engineer or artisan more familiar with traditional materials such as metal, wood and stone.

Much effort has been expended in designing tests to measure the upper and lower temperature limits of performance of plastics, for most of them do not demonstrate sharp melting points or *first order* transitions. On the contrary, they change their physical condition over a range of temperature, the extent and position of which is dependent on a host of variables such as physical form, processing cycle, moisture content, time-scale of measurement and physical parameter of interest.

For those few plastics which show well defined melting points suitable methods of tests are described in this chapter.

Standardised procedures for assessing the softening characteristics of plastics are invariably *ad hoc* in nature, selecting some mechanical property—usually bending stiffness or hardness—which is characterised indirectly by a measurement of deflection, the end-point being taken when some arbitrarily chosen value of deflection is reached. Rate of heating, and level of stress are equally arbitrarily chosen. These tests are described in this chapter, but it will be realised at the outset that such tests yield figures *which have no absolute significance*; they are only intended as quality control methods and the data they yield should not be regarded as anything else. In the ultimate, the only satisfactory way of determining the upper or lower limits of usage temperature is to identify the property or properties of significance for the intended use (they may be electrical rather than mechanical) and then to make appropriate fundamental measurements to follow the deterioration of value with change in temperature. Some hints on this subject are given in the following sections.

Prior to all this, however, in considering *reversible* effects of temperature, the methods for measuring thermal expansion and the related shrinkage will be described together with a brief description of methods of finding the glass transition temperature.

Most plastics materials, being based on a carbon–carbon chain backbone or containing a high content of this bond and of carbon hydrogen bonding, have very limited resistance to actual degradation by heat. Depending on the temperature, the nature of the polymer and the presence or absence of oxygen, so initially breakdown or crosslinking (of thermoplastics) may result; ultimately, and invariably, total breakdown occurs into a variety of products ranging from carbon through to virtually pure monomer—again depending on the conditions, but particularly on the nature of the polymer. The last sections of this chapter are therefore devoted to methods of test and investigational methods for the *irreversible* effects of temperature, but do not include fire tests which are dealt with in Chapter 17.

16.2 THERMAL EXPANSION

The coefficients of linear and volume expansion are defined respectively as

$$\alpha = \frac{1}{l}\left(\frac{\partial l}{\partial T}\right)_P, \qquad \text{and} \qquad \beta = \frac{1}{V}\left(\frac{\partial V}{\partial T}\right)_P \qquad (16.1,2)$$

where l is length, V is volume and T is temperature. For isotropic and homogeneous materials the two are related by $\beta = 3\alpha$. However plastics are not often isotropic and homogeneous and if accurate values are needed it is necessary to measure the linear expansion in orthogonal directions. A useful review of methods of measuring the thermal expansion of polymers has been given by Griffiths.[1]

ASTM D696 (1979)[2] gives a method for determining the coefficient of linear thermal expansion using a quartz dilatometer and a dial gauge. Linear expansion can also be measured with the commercially available thermomechanical analysers (TMA). These instruments usually cover a wider temperature range and use a linear displacement transducer for measuring the displacement.

Volume expansion is usually measured with a liquid-in-glass dilatometer (ASTM D864-52 (1978)[3]). A detailed account of this technique has been given by Bekkedahl.[4] The sample is placed in the dilatometer and covered with a known mass of liquid which must not react with the sample. The liquid which is most commonly used is mercury. The dilatometer is suspended in a temperature-controlled bath and allowed to reach equilibrium. The position of the liquid meniscus in the capillary tube is determined with a cathetometer. The temperature is changed and after thermal equilibrium has been achieved the change in the liquid height is measured. The temperature range of interest is covered in a series of small steps. After correction for the expansion of the containing liquid and the dilatometer, the volume expansion of the sample can be calculated for each step. Great care has to be taken in calibrating and operating the apparatus, although the dilatometer method is essentially simple and can yield very accurate results.

16.3 SHRINKAGE

Closely related to the coefficient of linear thermal expansion is the property known as 'shrinkage'; this general-sounding term is normally taken to apply to the difference between the linear dimensions of a moulding and the corresponding dimensions of the mould cavity from which the moulding was produced. This property is of importance to the processor, as has already been mentioned in Chapter 5. In addition there is the shrinkage which takes place when a resin polymerises, as, for example, when using an epoxy resin in a casting application, and shrinkage due to heat is also an important property for some plastics film products.

ISO 2577 (1975)[5] is the International Standard for the determination of shrinkage of compression moulded test specimens in the form of bars. In this method, test pieces in the form of bars, length greater than 80 mm, width 10 to 15 mm and thickness 4 to 10 mm, are moulded in a positive or semi-positive compression mould. After careful cooling to 23° ± 2°C in an atmosphere of 50 ± 5 per cent r.h. for not less than 16 h and not more than 72 h, their length is measured to the nearest 0·02 mm (L_1) and compared with the length of the mould cavity measured to the nearest 0·01 mm (L_0). The moulding shrinkage (MS) is then calculated as a percentage by the relationship:

$$MS = \frac{L_0 - L_1}{L_0} \times 100 \tag{16.3}$$

It is permissible to use engraved lines on the mould for the measurements and the standard requires that the test pieces are examined for warpage after moulding by testing them on a flat surface (e.g. glass). Test pieces showing significant warpage are rejected.

The same standard also allows the determination of post-moulding shrinkage. Here the moulded and measured test pieces are conditioned at 80° ± 2°C for urea formaldehyde moulding materials and 110° ± 3°C for all other thermosetting moulding materials. For the normal determination the heating period is 168 ± 2 h, but for a rapid determination 48 ± 1 h is used. After the heating period the test pieces are allowed to cool and remeasured. The post-shrinkage (PS) is calculated as a percentage by subtracting the length after heating from the length before heating, dividing by the length before heating and multiplying by 100.

BS 2782 Method 640A (1979)[6] follows ISO 2577 and replaces BS 2782 (1970) Methods 106A to 106E which use a circular test piece, the change in diameter of which is used to calculate mould shrinkage and after-shrinkage as a change in length per unit length.

ASTM D955 (1979)[7] covers the subject somewhat more comprehensively in making specific reference to compression and transfer moulding and injection moulding; for the last two a longitudinal bar is used. The mode of expression of results is in terms of change of length per unit length.

ASTM D1299-55 (1979)[8] describes a method for finding the shrinkage of moulded and laminated thermosetting plastics at elevated temperature. Discs 100 mm in diameter or 100 mm squares, both 3·2 mm thick, are used as test pieces. A temperature of 70°C, 90°C, 105°C, 130°C, 180°C or 230°C is used, as appropriate, and the shrinkage is expressed as change in dimension per unit dimension.

DIN 53464 (1962),[9] which makes specific reference to phenolic, urea formaldehyde and melamine formaldehyde moulding materials, lays down precise moulding conditions for those materials and uses test bars 120 ± 2 mm long by 15 ± 0·5 mm wide by 10 ± 0·5 mm thick.

ASTM D702-68 (1974),[10] Section 6.1.8, gives a method for measuring shrinkage due to exposure for 30 min at 125°C or 140°C according to the grade of material. Sheets 305 mm (12 in) square are used, and the separation of gauge marks placed thereon is measured to the nearest 0·25 mm (0·01 in).

There are several other standardised methods of tests for shrinkage of thermoplastics which are specific for certain types of product. BS 2782 (1970), Method 106F,[11] details the determination of the dimensional stability at 100°C of flexible PVC sheet (this method may be withdrawn in the later edition of BS 2782). A strip of the sheet 250 mm long by 6·4 mm wide is cut and its length measured. The length is again measured after immersion for 15 ± 1 min in a water bath at 100° ± 1°C. The result is expressed as a percentage change of length on the original length of test piece. ASTM D2732-70 (1976)[12] details a broadly similar test but uses test pieces 100 mm square. BS 2782 (1970), Method 106H,[13] is a test method for finding the dimensional change of decorative laminated sheet. It is claimed to give an indication of the dimensional change that could occur with aminoplastic faced phenolic decorative sheet as the result of an increase in humidity of the surrounding atmosphere from a desiccated condition to a very high humidity. Test pieces 140 mm long and 12·7 mm wide are measured for length to the nearest 0·02 mm and tested at 70° ± 2°C for 24 ± 1 h, then cooled in a desiccator and the length is measured. Other measured test pieces are conditioned at 40° ± 2°C and 90–95 per cent r.h. for 7 days and the length is measured within one minute of removal from the humid condition. The percentage decrease in length (shrinkage $\bar{s}$) is calculated after the oven treatment and the percentage increase in length ($\bar{d}$) is calculated after the humidity treatment. The dimensional change of the sample under test is then expressed as the sum of $\bar{s}$ and $\bar{d}$.

BS 2782 (1976) Method 643A[14] and ASTM D2838-67 (1975)[15] describe test methods for determining the shrink properties of heat-shrinkable plastics film. Method 643A is a simple 20 ± 1 s immersion in a heating bath maintained at a specified temperature with a change in length determined and expressed as a percentage of the original length of the test piece. D2838 details the determination of shrink tension and orientation release stress of plastics film and thin sheeting. This test recommends the use of a commercially available apparatus and it is claimed to be useful to predict the appearance and performance of the film in shrink packaging applications. It does not determine the shrinkage on heating of film intended for shrink wrapping applications, as does Method 643A of BS 2782. Method 643A is quite adequate for normal quality control purposes, while ASTM D2838 is more appropriate for research and development.

Finally, two published standards for determining the shrinkage of casting resins during cure, ISO 3521 (1976)[16] and ASTM D2566 (1979),[17] give two quite different methods: ISO 3521 determines the volume change by comparing the density of the cured resin with that of the uncured resin, while ASTM D2566 uses the more conventional method of comparing the length of a cured casting with the length of the mould used to make the casting.

BS 2782 Method 644A (1979)[18] is identical to ISO 3521.

16.4 TRANSITION TEMPERATURE

Unlike crystalline materials, amorphous materials do not melt when they are heated, but go through a glass transition. In the glass transition temperature range many physical properties of amorphous materials undergo a more or less drastic change. The most obvious change is that from a viscous or flexible material above T_g to a hard, sometimes brittle, solid below. On a molecular level, as the temperature is raised to the glass temperature and above, segments of the molecules are able to rotate. The actual value of T_g for a polymer is clearly method- and rate-dependent; T_g is that temperature at which the molecular relaxation times are of the same order as the experimental times. Brydson,[19] for example, quotes values for a polyoxacylobutane polymer which ranged from 7°C to 32°C for dilatometric and electrical methods where the frequencies were 10^{-2} Hz and 100 Hz respectively. For a detailed discussion of the glass transition the reader is referred to the review articles by Shen and Eisenberg.[20,21]

A crystalline solid melts when it is heated and this change of phase from solid to liquid is known as a first order transition, because there is a discontinuity in the first order derivatives of Gibbs function, such as the volume.[22] In a second order phase transition there is a discontinuity at the transition temperature in the second order derivatives of Gibbs function, such as the specific heat.[23] For a true second order transition in the Ehrenfest sense the phase change must take place at constant temperature and pressure with no change of entropy or volume. A glass transition, therefore, is not a second order transition, although it approximates to one.[24] Gibbs and DiMarzio[25,26] developed a theory of the glass transition by assuming that there is an underlying second order transition,[27] and that the value of the glass transition temperature obtained from an experiment done at an infinitely slow rate would be the same as the thermodynamic second order transition temperature. This theory is not without its critics.[28]

From the structural point of view the molecular arrangements in an amorphous material (or glass) below T_g possess the permanence of crystals and the randomness of liquids. As a crystalline arrangement of the molecules has a lower energy, for a glass to form on cooling from the melt there must be some barrier to crystallisation. This may be caused by steric hindrances in the molecule such as bulky side groups, or, if the material is cooled quickly enough, by insufficient thermal energy, or by a combination of both. T_g is always less than T_m (melting temperature) since a crystal has a lower energy than an amorphous structure.

The glass transition does not occur at a single temperature but takes place over a short temperature range, because real polymeric materials have a molecular weight distribution. The single point temperatures which are quoted are obtained by interpolation. T_g can be measured in a number of ways because most material properties change on going through it. It is not sufficient to just quote a value for T_g; it is important to give the method and the rate of measurement.

There are no standard methods for measuring T_g, but a number of methods have been reported in the literature. T_g can be determined by measuring the expansion, either linear or volume, since the expansion coefficient above T_g is greater than that below.

The change in the dynamic mechanical properties, as discussed in Chapter 9, can also be used, since the dissipation factor plotted against temperature shows a pronounced peak. A similar effect is observed with the electric loss factor as described in Chapter 13.

Optical properties also change when the sample is heated or cooled through the glass transition range. A value for T_g can be obtained by measuring refractive index[29] or birefringence.[30]

Thermodynamic properties such as the specific heat can also be used because there is a step change at the transition temperature. A differential scanning calorimeter or a differential thermal analyser allows the temperature range of interest to be scanned quite quickly, making this a very convenient method to use. The value of T_g depends on the scan rate and it may be necessary to make a number of measurements at different scan rates and extrapolate to zero rate.

16.5 MELTING POINT

16.5.1 General Considerations

Under this heading only those tests will be considered which are suitable for materials with a fairly sharp change of state (solid to liquid), i.e. those which behave somewhat like an ordinary low-molecular-weight organic compound. In the characterisation of these, most of the standard methods depend on a visual observation of the transition. These melting point tests are generally unsuitable for the majority of plastics, or more correctly thermoplastics, which soften slowly over a temperature range which may be quite extensive and frequently never reach a free-flowing condition before thermal decomposition sets in. With such an indefinite change of state there is no melting point in the usual sense of the term and it is necessary, if it is required to put a temperature figure to such an indefinite property, to follow the change in some parameter—usually mechanical—with rise of temperature and ascribe a softening point to that temperature at which the property being observed reaches some prescribed value. This type of test is discussed more fully in §16.6 below.

There are several tests available for melting point determination and attention is given below to those which have been standardised. However, mention must be made of the simple and useful Durrans method, where 3 g of resin is gently melted in the bottom of a test-tube, and the bulb of a thermometer is inserted in the resin which is then cooled to solidify. 50 g of mercury is then poured on top of the resin and the whole assembly is heated so that the resin temperature rises at a rate of 2°C/min until the resin melts and floats on the mercury. The temperature at which this first occurs is recorded as the melting point of the resin.[31]

16.5.2 'Ring and Ball' Method

This method appears as Method 103A of BS 2782 (1970)[32] and is generally used for synthetic resins of the novolak type (Figure 16.1).

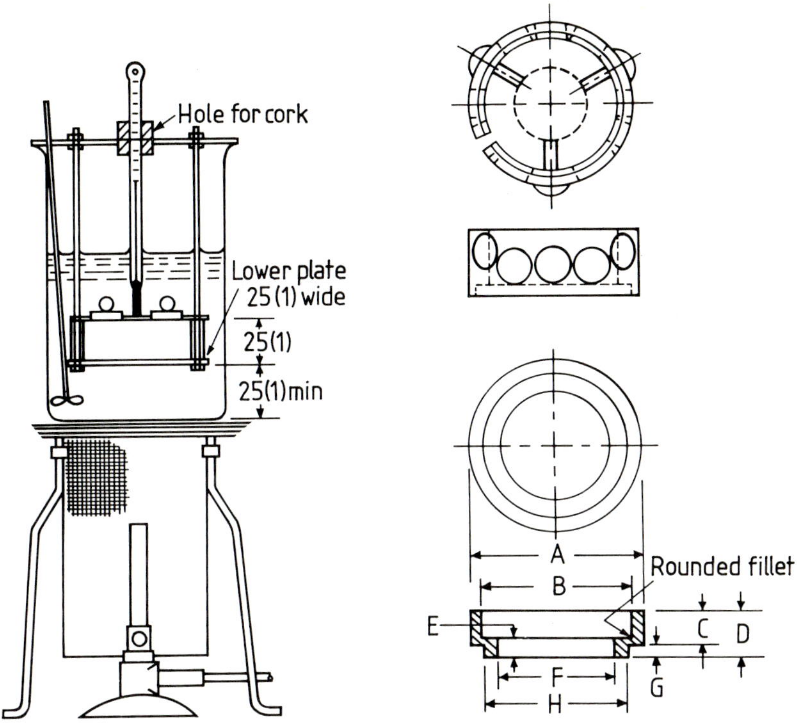

Figure 16.1 'Ring and ball' softening point apparatus: BS 2782—dimensions in millimetres (inches)

Ring dimensions:	A	23 ± 0·1	(0·906 ± 0·004)
	B	19·9 ± 0·1	(0·781 ± 0·004)
	C	4·4 ± 0·1	(0·172 ± 0.004)
	D	6·4 ± 0·1	(0·250 ± 0·004)
	E	2·8 ± 0·1	(0·109 ± 0·004)
	F	15·9 ± 0·1	(0·625 ± 0·004)
	G	2·0	(0·08)
	H	19·0 ± 0·1	(0·750 ± 0·004)

The sample under examination is reduced to small pieces (not powder, as bubbles may result on melting) and 1·6 g is weighed into a ring (as shown) placed on a metal plate, of suitable finish to avoid sticking. This assembly is then placed in an oven at a temperature 10°–20°C above the expected melting point until the resin has melted (time should not exceed 15 min). The assembly is then removed, placed on a flat metal surface, and any excess resin over the top of the ring is immediately removed by scraping with a hot knife. Having been allowed to cool for at least 15 min, the ring filled with resin is removed from the plate, if necessary using gentle tapping.

A steel ball, 9·53 mm ($\frac{3}{8}$ in) diameter and mass 3·45–3·55 g, is placed centrally on the cast resin, using a centering guide (as illustrated) which rests on the ring.

Two such specimens are prepared and placed in the apparatus illustrated, using as heat transfer medium a liquid without effect on the resin (glycerol will often be suitable). The heating bath is raised in temperature by 5° ± 0·5°C per min, with constant mechanical stirring, and the melting point is recorded as that temperature at which the ball or surrounding resin first touches the lower plate. The agreement between the duplicate results must be within 1°C.

16.5.3 Capillary Tube Method

ISO 1218 (1975)[33] Method A, for the determination of the melting point of polyamide, and BS 2782 (1976) Method 123A[34] are technically similar except that Method 123A allows the use of the well known Thiele tube as well as the metal heating block for polymers which show only a sinter point, e.g. phenol formaldehyde solid resins in the uncured state.

The sample under examination is ground to a fine powder with a pestle and mortar and filled to a depth of about 10 mm into a thin-walled capillary tube, length about 60 mm, and internal diameter about 1 mm, which is sealed at one end. The sample is compacted as much as possible by tapping the sealed end of the tube sharply on a hard surface, and then the open end of the tube is sealed in a Bunsen flame. For materials not in powder form, or those that cannot be ground in this way, it is permissible to cut a thin sliver of the material about 5 mm long, by means of a razor blade, and to seal this in the capillary tube.

Two such specimens are used; one or both are positioned in the apparatus (Figure 16.2). The temperature is then raised rapidly to between 10°C and 20°C below the expected melting point. The heat input is then adjusted to give an increase in temperature of 2° ± 0·5°C/min. As the melting point is approached, the powdered sample first contracts away from the tube wall and coalesces (the 'sinter point'). Immediately afterwards it becomes translucent and subsequently becomes transparent. Finally, if the sample truly melts, it runs down the tube.

The melting point is defined as the temperature at which the powder specimen first begins to transmit light through about half its bulk. In some cases, the resin does not reach a translucent stage, and it is not possible to observe a melting point as defined above. In such cases, the sinter point is reported instead. For samples which are not in powder form the melting point is defined as the temperature at which the sharp edge of the specimen disappears.

The determination is carried out in duplicate and if the difference between the two results exceeds 3°C two additional test portions have to be tested.

DIN 53736 (1973)[35] is similar to ISO 1218 Method A.

16.5.4 Hot Plate Method

ISO 1218 (1975)[33] Method B and BS 2782 (1976) Method 123B[36] are technically similar but ISO 1218 gives a more precise description of the hot plate apparatus. The method is described as specifically for polyamides, but is presumably suitable for any material with similar melting characteristics.

The test sample is a single granule of carefully defined size (related to a certain sieve size) which is placed on a microscope slide resting across the centre of a hot plate. A few drops of silicone fluid are also placed on the slide, adjacent to the

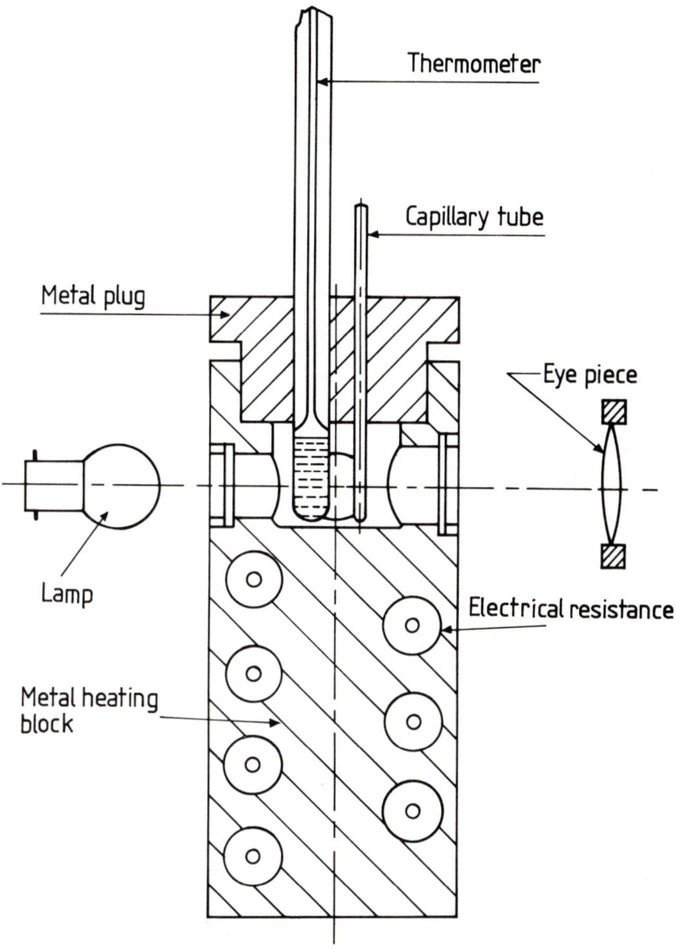

Figure 16.2 Melting apparatus with capillary tube

granule, and the spacing is such that a cover glass placed on the latter is slightly tilted towards the fluid which makes contact with only a part of the underside of the glass (Figure 16.3).

The temperature of the hot plate, which has a thermometer inserted in a cavity immediately below its centre, is raised at a rate of 2° ± 1°C/min and the boundary line of the silicone fluid is carefully observed. The melting point is taken as that temperature at which the boundary line moves rapidly across the cover glass, i.e. when the granule no longer supports the glass. Two such tests are carried out.

This is the basis of the methods described in ASTM D789 (1978),[37] a specification for nylon materials, and in ASTM D2133 (1978)[38] for acetal resin materials, where the use of a commercial apparatus, the Fisher-Johns melting point apparatus, is specified. Using the same apparatus, appearance changes are used to measure the melting points of polytetrafluoroethylene resins

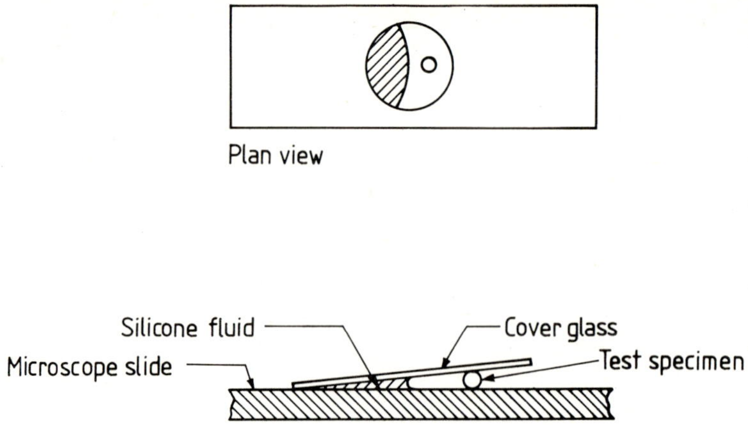

Figure 16.3 Specimen mounting for hot plate method: BS 2782

(ASTM D1457 (1978)[39]) and fluorinated ethylene propylene copolymer (ASTM D2116 (1979)[40]).

16.5.5 Optical Method

This method is applicable only to crystalline polymers in that it depends on their birefringent, or double refracting, property. Thus, when crystalline, such a material transmits light when viewed between crossed Nicol prisms, i.e. it is visible against the dark background, but when molten it has become non-crystalline and becomes invisible in the 'blackness' of the crossed prisms.

The specimen is prepared as a thin layer between a microscope slide and cover glass. If it is a moulded or tableted material that is being examined, it may simply be microtomed to 0·01–0·05 mm thickness; if it is in film form of this thickness it may be used direct. To avoid light scattering and ensure good thermal contact it is advisable to immerse the specimen in a liquid, preferably of similar refractive index, which of course must not affect the material. Powdered material, or thicker film or sheet, is placed between the microscope slide and cover glass, which together are put on a hot plate and heated to a temperature about 25°C above the melting point. This cover glass is pressed down to reduce the specimen to a thickness of 0·01–0·05 mm and, having been kept at elevated temperature for about 10 min, the assembly is then allowed to cool slowly, to induce optimum crystallinity. If the material is liable to thermal decomposition under these conditions the temperature may have to be lowered and the cooling started immediately.

A microscope is used, fitted with analyser and polariser and capable of giving ×50 to ×100 magnification. It is also fitted with a hot stage, which is mounted just above the microscope stage and consists of an insulated metal block with a central hole for the passage of light, and with a recess close to the hole for the insertion of a thermometer or thermocouple. The hot stage is electrically heated at controlled rates and should be designed so that it can be enclosed in an atmosphere of nitrogen if necessary.

At the start of the test, the polariser and analyser are adjusted for complete extinction of light and the specimen assembly is placed on the hot stage. The polariser and analyser are then rotated again, if necessary, to obtain maximum brightness of the specimen and thus overcome the effects of strain in the latter. The hot stage is then heated at a (linear) controlled rate, usually 1°± 0·25°C/min, and the melting point is taken as that temperature at which the field becomes completely dark. For high melting point materials, the test may be conveniently started 50°C below the expected softening point.

A test of this type is given in ISO 3146 (1974)[41] and is essentially similar to that described in ASTM D2117-64 (1978).[42] BS 2782 (1976) Method 123C[43] is technically similar to ISO 3146 but differs in respect of rate of heating and gives more precise instructions on the preparation of the test sample.

16.6 SOFTENING POINT

16.6.1 General Considerations

The results of the *ad hoc* tests, which are referred to in §16.1 and which have found wide favour as standardised methods, flourish under a number of informal names, particularly: 'softening point', 'plastic yield', 'heat distortion point', 'deformation under load' and 'temperature of deflection under load'. They measure the temperature response of the material under test to a variety of stress conditions, especially in bending or under a penetrative load. There are two methods of obtaining an end-point; the most popular is to select some arbitrary stress condition and raise the temperature until the strain has reached some arbitrary level. This is essentially a measure of the temperature at which the modulus deteriorates to some prescribed level (but without reference to the value of the modulus at normal ambient conditions). The other principal variant is to select the test temperature and measure the results in terms of the degree of yield resulting, at this temperature, from standardised conditions of stress.

Crofts and Brown[44] have comprehensively reviewed methods for measuring softening point and provide a useful bibliography.

The size of the test piece required for the various types of test are quite different: for instance, a penetration-type test requires only a small area and thickness whereas a cantilever bending method may need a test piece of substantial cross-section. Since the temperature of the surrounding heat exchange medium (water bath, air or whatever) must reflect accurately the temperature of at least the stressed area of the specimen, the rates of temperature rise of the heat exchange media may have to be quite different for the various methods.

While acknowledging the value of softening point determinations and the like as *quality control* tests, it is necessary to stress just how *ad hoc* and arbitrary they are and how incapable of comparison their various results.[45–48] Table 16.1 serves to emphasise this point: not only will the wide differences between results in each column be noted, but also the inconsistent differences and, in some cases, the changed order of ranking.

More recently, however, Abolins[49] has demonstrated that a linear relationship exists between the Vicat softening point and the temperature of deflection under load as measured by ASTM D648 (1972),[50] particularly for annealed test pieces.

Table 16.1 Softening Points (°C) Determined by Various Standard Methods

Material	Cantilever BS 2782 102C	DTI (ASTM D648) 455 kPa	DTI (ASTM D648) 1826 kPa	Vicat BS 2782 120A
Polystyrene	95	90	97	98
Toughened polystyrene	84	72	85	86
ABS	94	84	96	95
Poly(methyl methacrylate)	95	80	97	90
Cellulose acetate	76	64	77	72
Rigid PVC	78	70	82	82
Polyethylene (LD)	Too flexible	Too flexible	45	85
(MD)	90	35	69	105
(HD)	115	45	75	125
Polypropylene	145	60	140	150
Nylon 66	180	75	183	185
Acetal	170	120	165	175

16.6.2 Cantilever Techniques

BS 2782 (1978) Method 115A[51] determines the 'plastic yield' and requires two test pieces each 200 ± 1 mm long by a square cross-section of side 15 mm + 0, −0·3 mm. A notch is moulded or machined into one long face of each test piece, the root of the notch being 5 mm from one end. The notched surface is one of those perpendicular to the direction of moulding pressure.

The test is carried out in an oven capable of being maintained throughout the test within the specified tolerances of the specified test temperature; this latter is normally quoted in the relevant product specification but if not it is to be one of 40°C, 55°C, 70°C, 85°C, 105°C, 125°C, 150°C, 175°C or 200°C, the temperature being selected so that the result ('plastic yield') does not exceed 6 mm. The test set-up is shown in Figure 16.4.

The clamped test piece, without weight *W* but with stirrup plus attachment (combined weight not exceeding 20 g), is placed in the oven at the test tempera-

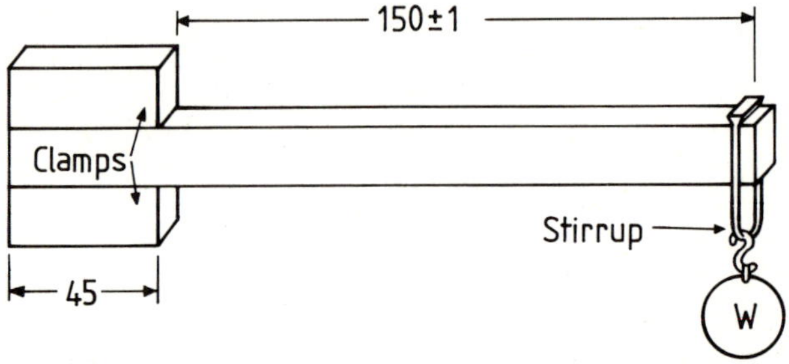

Figure 16.4 Plastic yield test: BS 2782—dimensions in millimetres

ture; 15 min afterwards the height of the stirrup is taken, with respect to some suitable datum point, to the nearest 0·1 mm and then the weight W is attached to the stirrup (W + stirrup + attachment = 450 g).

After a further 6 h ± 10 min the height of the stirrup with respect to the same datum is again taken. The difference between the two height values (mm) is the plastic yield; the results for the two test pieces are averaged.

Method 102B of BS 2782 (1970)[52] describes a somewhat similar test to that of Method 115A, but it is termed 'Deformation in bend under load at elevated temperature of laminated sheet'. Again two specimens are used, each approximately 200 mm long and 25·4 ± 0·5 mm wide; the thickness is that of the sheet unless this exceeds 12·7 mm, in which case *one* face is machined to yield a thickness of 12·7 ± 0·5 mm. A shallow notch is machined in one original face across the width, approximately 3 mm from one end.

The test is carried out at 90° ± 2°C, but otherwise generally as described above. The distance between the edge of the clamp and the notch root is 152 mm. The weight added is to be such as to give a maximum surface stress in bend in the test piece of 6·9 MPa within 2 per cent, calculated from the formula

$$W = \frac{700BT^2}{6L} = 0{\cdot}77BT^2 \qquad (16.4)$$

where W = requisite mass (g)
B = width of specimen (mm)
T = thickness of specimen (mm)
L = effective length of cantilever (152 mm).

The second stirrup height reading is taken after 60 min in this test, and the result is expressed in millimetres.

This method is likely to be withdrawn in the latest edition of BS 2782.

Finally, in cantilever methods, there is Method 102C of BS 2782 (1970).[53] Although this method will be withdrawn from the latest edition of BS 2782, it is worth describing as it enjoyed many years of popularity and was often used to define the softening point of thermoplastics in the trade literature. It has now been rendered obsolete in the UK by the Vicat test (§16.6.6). The test piece is shown in Figure 16.5.

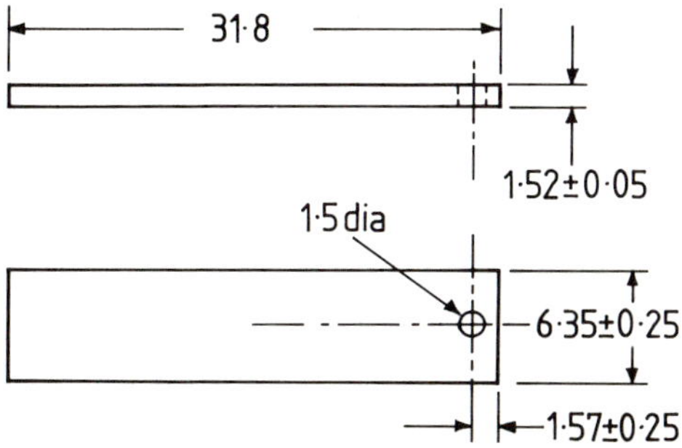

Figure 16.5 Softening point test piece: BS 2782, Method 102C—dimensions in millimetres

The test piece is mounted horizontally in a clamp (Figure 16.6). Initially the 20 g weight is supported so that there is no load on the test piece. The assembly is immersed in a bath of appropriate liquid, at a temperature 25°–30°C below the expected softening point. (The 'appropriate liquid' must be without effect on the material under test: liquid paraffin is suitable for cellulose acetate and glycerol for polystyrene, toughened polystyrene and rigid PVC.) The support is then removed from under the load and the temperature of the bath is raised at 1·0° ± 0·2°C/min with adequate stirring.

The temperature of the bath is noted at which the upper edge of the free end of the test piece coincides with a 30° graduation marked on the quadrant plate.

The mean of the temperatures obtained from two specimens is taken as the softening point.

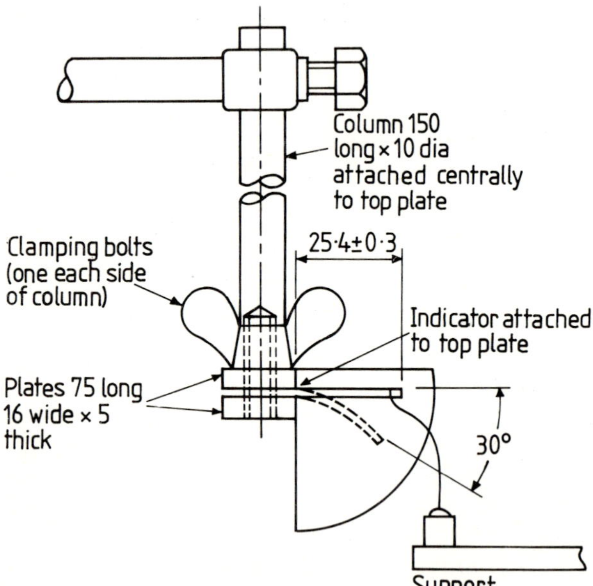

Figure 16.6 Test assembly for softening point test: BS 2782, Method 102C—dimensions in millimetres

16.6.3 Three-point Bending Techniques

ISO 75 (1974)[54] describes the determination of temperature of deflection under load of plastics and ebonite. This test method was directly derived from ASTM D648-72 (1978).[50] BS 2782 (1976) Methods 121A and 121B[55] are in general agreement with ISO 75. DIN 53461 (1969)[56] is similarly related to the ISO method. As the various national and international standards are so closely related, only the BS method will be described.

To accommodate materials of different initial (room temperature) stiffness, two variants of the method are available, one applying a maximum surface stress of 1·8 MPa (Method 121A), and the other one of 0·45 MPa (Method 121B).

Test pieces are as follows:

Moulding material and extrusion compounds: rectangular bar of minimum length 110 mm, width 9·8–12·8 mm and thickness 3·0–4·2 mm, moulded as appropriate to the material in question with the direction of pressure perpendicular to the largest faces of the test piece.

Sheet: length and width as above, but with thickness that of the sheet (which is to be 3–13 mm).

Casting and laminating resin systems: length and width as above but thickness 3–7 mm.

The apparatus used is shown in Figure 16.7. The parallel cylindrical metal blocks which form the outer supports for the test piece are 100 ± 2 mm apart; these and the loading block, which rests centrally across the width of the test piece, all have contacting radii of 3 ± 0·2 mm. All vertical members of the supports should have the same coefficient of expansion so that expansion of the apparatus cancels out and no error is thus introduced into the reading of the deflection of the test piece through uncompensated expansion in the apparatus; even so, it is as well to check this by carrying out a 'control' run beforehand using a test piece of very low expansion coefficient, for example Invar steel or borosilicate glass.

The heating bath into which the above assembly is immersed is filled with a liquid which has been shown to have no effect on the material under test (liquid

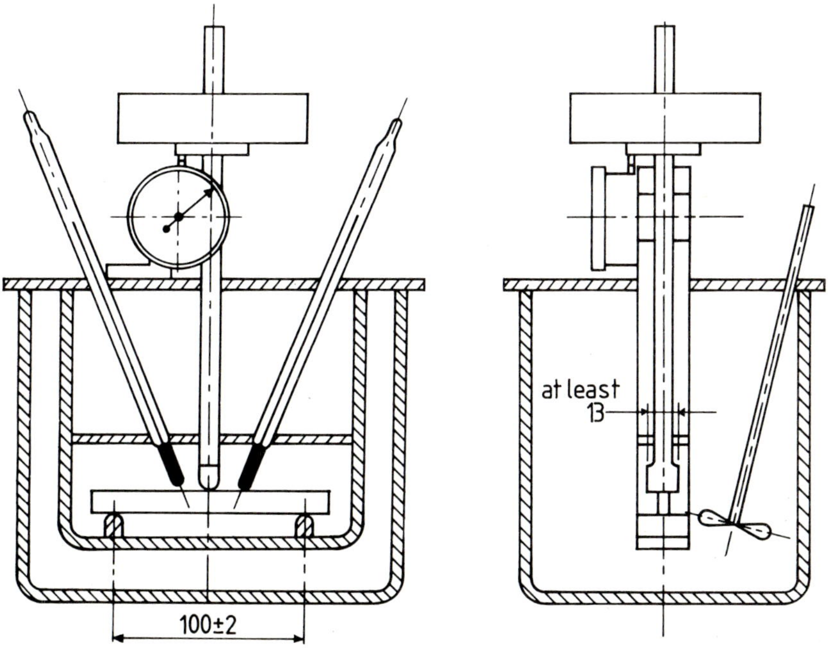

Figure 16.7 Apparatus for determination of temperature of deflection under load: BS 2782—dimensions in millimetres

paraffin, transformer oil, glycerol and silicone fluid are suggested for trial). The bath must be capable of being heated so that the temperature can be raised at a rate of 2°C/min, with the temperature at any time being not more than 1°C from the value corresponding to this rate.

To undertake the test, the cross-sectional dimensions of the test piece are measured to the nearest 0·02 mm and then it is placed in the apparatus so that the direction of application of load is parallel to the plane of the largest faces. In this position, the vertical height is designated depth *d* and horizontal width from one large vertical face to the other is designated breadth *b*. The test piece is loaded under a bending stress of 1·8 MPa or 0·45 MPa (as appropriate) by applying a force as follows (making due allowance for any force exerted by the spring of the dial gauge):

$$F = \frac{2abd^2}{3l} \tag{16.5}$$

where *F* is the force (N)
a is the bending stress (1·8 MPa for Method A, 0·45 MPa for Method B)
b is the breadth of the test piece (mm)
d is the depth of the test piece (mm)
l is the span between supports (mm).

The initial temperature of the bath is generally 23°C, but if it can be proved that the use of a higher initial temperature is without effect on the test result, then this may be used, provided that it is not less than 30°C below the expected temperature of deflection under load. 5 min after the load has been applied the deflection-measuring device is set to zero (or its reading is recorded) and the temperature of the bath is raised as described above. The temperature (°C) is noted at which the deflection increase is reached which corresponds to that specified in a table relating deflection to depth *d* (0·33 mm deflection for 9·8 to 9·9 mm, falling linearly to 0·25 mm for 12·8 mm).

The mean of the results on the two test pieces is taken.

The higher alternative applied stress is too low for high-modulus materials, particularly glass fibre reinforced resins, and as a result this test does not yield deflection temperatures low enough to be measured conveniently, say below 200°C. This is not only a question of experimental convenience, however, as it is most probable that at some temperature within the normal test range such materials will have lost a significant proportion of that initial (high) value which was the very reason for their selection. In an endeavour to overcome this drawback, a British Standard product specification for GFRP (BS 3534 Part 2 (1964)[57]) has extended the test by applying a bending stress of 35·3 MPa on a test piece approximately 150 mm long and 12·5 mm wide by the prepared thickness of 3·2 ± 0·1 mm. The span is at least 50 mm.

Deflection is recorded against temperature to a value of the latter at least 30°C above that at which the 'critical deformation' is achieved; best straight lines are drawn through the two sections of the curve (below and above the 'critical' point) and the point of intersection is designated the temperature of deflection. One test piece is used for a trial run and the test result is taken as the average of the figures from the second and third.

The extension of the technique is taken one stage further by BS 2782 Method 121C (1976)[58] for rigid thermosetting resin bonded laminated sheet, by applying a stress equivalent to 10 per cent of the (measured) ultimate crossbreaking strength. The test piece is as in BS 3534 Part 2 and the technique of obtaining the deformation point is precisely as before. This test is somewhat inelegant and the choice of 10 per cent of the normal temperature crossbreaking strength rather than some fraction of the bending modulus is surprising.

16.6.4 Four-point Bending Techniques

The Martens test is described in DIN 53458 (1968)[59] and is an ingenious way of converting four-point bending into magnified vertical movement of the end of an arm (Figure 16.8).

The apparatus is carefully specified and a value of weight *G* is given for each of the three different sizes of test piece permitted. The rate of heating is 50° ± 1°C/h and the Martens temperature is taken when the deflection at the end of the lever reaches 6 mm. The values of *G* are selected to give the same maximum surface stress to each of the three alternative forms of test piece, but the maximum strain is necessarily different. The ratio of the two (stress–strain) is stiffness, which is a function of temperature and therefore test results differ according to which of the standard test pieces has been used.

Gohl[60] discusses the limitations of the Martens test and concludes that it is unsuitable for materials of modulus less that 4902 MPa.

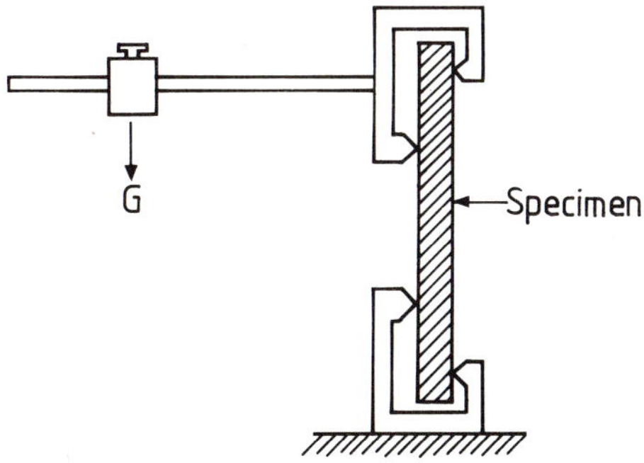

Figure 16.8 Martens test: DIN 53458

16.6.5 Tensile Techniques

For sheets and films of thickness from 0·025 mm to 1·5 mm and of modulus greater that 69 MPa at 23°C, a test for tensile heat distortion temperature is described in ASTM D1637-61 (1976).[61] It measures the temperature at which thermoplastic sheeting begins to deform appreciably (either to stretch or shrink) under a small tensile force.

Two forms of apparatus are shown in Figure 16.9. The oven or heating bath must be capable of being raised in temperature at a constant rate of 2° ± 0·2° C/min. The weight(s) required must be appropriate to apply a stress of 0·345 MPa to the test piece.

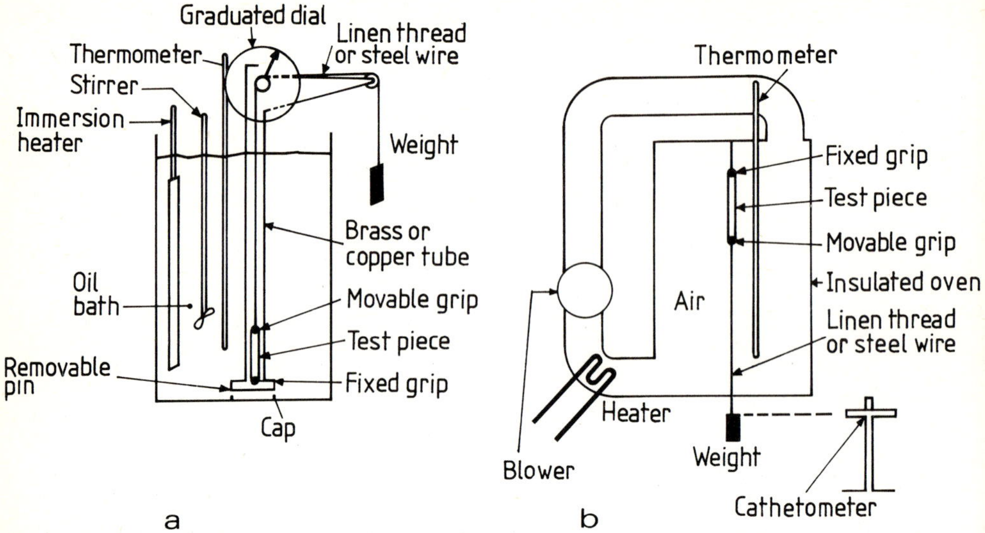

Figure 16.9 Tensile heat distortion apparatus: ASTM D1637
(a) Heat bath method (b) Air oven method

The test pieces are strips 6 to 25 mm wide and 50 to 180 mm long. The initial distance between the grips is between 25 and 125 mm. Materials of less than 0·075 mm in thickness and capable of being readily cemented together (or otherwise joined) are formed into a loop, with an overlap of not more than 6 mm, and tested in this manner.

After the width and thickness have been carefully measured, the test piece is preconditioned and then mounted in the apparatus with the appropriate weights attached to the moving grip to give the specified stress of 0·345 MPa. The assembly is placed in the oven or bath held at a temperature approximately 20°C below that at which the test piece is expected to begin to deform. The temperature is then raised to 2° ± 0·2°C/min and extension is recorded against temperature until either the test piece has elongated by 50 per cent or the temperature has reached 250°C. It is suggested that a temperature correction (which should not exceed 3°C) be applied to allow for any temperature lag between heating medium and test piece, which correction can be determined once and for all for each apparatus by attaching a thermocouple to a test piece and comparing the heating curve for the thermocouple with that from the thermometer in the oven or bath.

The tensile heat distortion temperature is determined from a plot of percentage change in length against temperature and is the temperature at which a 2 per cent shrinkage or a 2 per cent extension occurs. Bourgault[62] has shown that with properly designed apparatus for this test it is a useful tool in the investigation of properties of heat shrinkable and crosslinked polymers.

In the product specification ASTM D1430 (1978),[63] *Polychlorotrifluoroethylene (PCTFE) Plastics*, a test is included for 'zero strength time' (ZST) which it is stated is useful for control of molecular weight of this type of polymer. The test piece is a notched strip (Figure 16.10) cut and punched out of

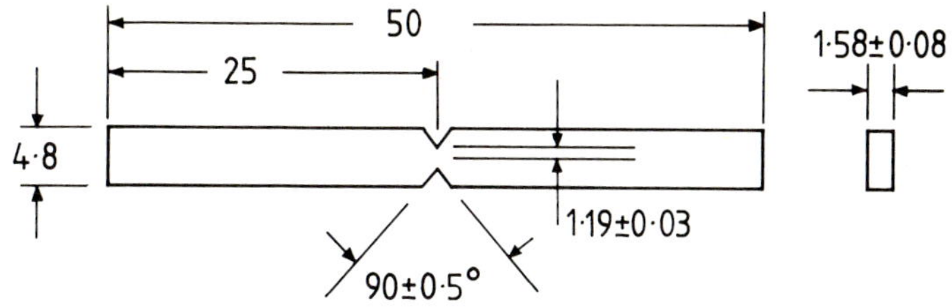

Figure 16.10 ZST test piece: ASTM D1430—dimensions in millimetres

sheet 1·58 ± 0·08 mm thick preformed and moulded from the powder in specified form.

One end of the test piece is hung from a specimen holder and a weight of 7·5 ± 0·1 g is attached to the other. Two such weighted test pieces are inserted one each into two cylindrical holes in a brass 'thermostat' at 250° ± 1°C and timers are started. The time is noted when each specimen breaks, denoted by the weight dropping through the bottom of the 'furnace' (Figure 16.11). The average of two readings in seconds is taken as the ZST; if the two values differ by more than 10 per cent the test is repeated.

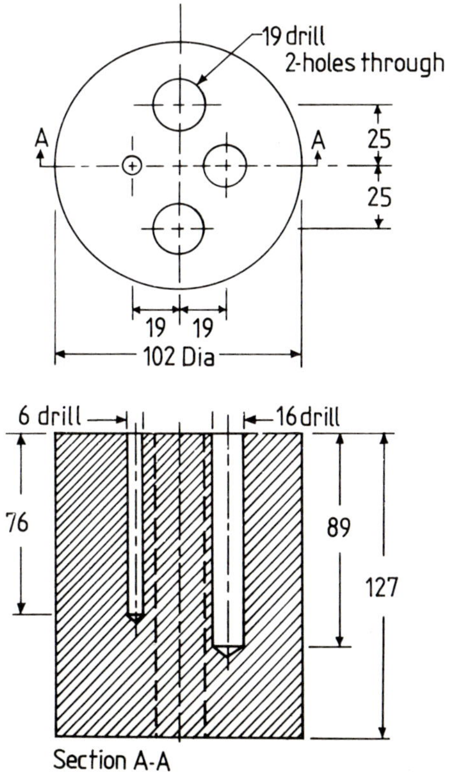

Figure 16.11 ZST thermostat: ASTM D1430—dimensions in millimetres

Sandiford and Buckingham[64] describe the tensile heat distortion test in some detail, which they term the 'TDT' test (tensile deformation versus temperature), give details of single-test-piece and six-test-piece apparatus and present some typical values for thermoplastics films.

16.6.6 Penetrometer Techniques

The most important penetrometer method and one which has gained almost universal acceptance is the *Vicat softening point* test. A test piece at least 10 mm square and not less than 2·5 mm thick is subjected to a round indentating tip of cross-section area 1·00 ± 0·015 mm^2. The lower surface of the indentating tip is free of burrs and is plane and perpendicular to the axis of the rod to which it is attached (Figure 16.12). A specified load is applied to the indentor tip and the degree of movement of the tip is indicated by a dial gauge. The test piece and tip are immersed in an oil bath and the temperature is raised at a specified rate. The softening point is defined as the temperature at which the indentor tip penetrates the test piece to a depth of 1 mm.

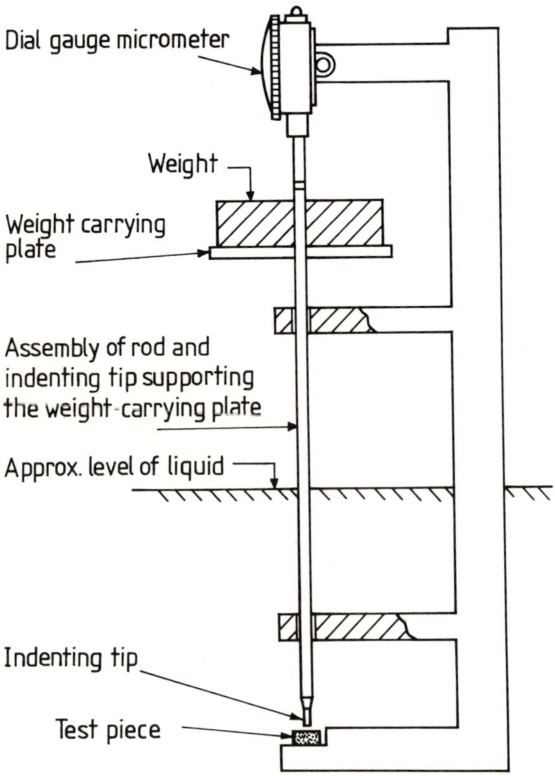

Figure 16.12 Vicat softening point apparatus: BS 2782
It is recommended that the frame and indenting rods are constructed from low-expansion alloy

ISO 306 (1974),[65] BS 2782 (1976) Methods 120A to 120E,[66] ASTM D1525 (1976)[67] and DIN 53460 (1976)[68] are all similar in describing the Vicat softening point test for thermoplastics. All allow the use of a force of 9·81 N on the indentor (this is rounded to 10 N in BS 2782, while ASTM D1525 specifies the force in terms of a total mass of 1000 g applied to the indentor) but BS 2782, ISO 306 and DIN 53460 offer alternative forces of 49 N. The standard rate of temperature rise during the test is 50 ± 5°C/h, but for some materials a rate of 120° ± 5°C/h is permissible.

Test pieces are between 2·5 and 6·5 mm thick (BS 2782) or between 3 and 6 mm thick (ISO 306 and ASTM D1525) but for samples, say, of sheet, of thickness less that 3 mm the plying up of samples to give a composite thickness of 2·5 mm to 3 mm is allowed, provided that the stack is composed of not more than three pieces. For samples of thickness greater than 6 mm (6·5 mm for BS 2782), the test piece is reduced to a thickness of 3 mm by machining one surface only prior to testing. Two test pieces of each sample are tested and if the results show a difference greater than 2°C the sample must be retested.

BS 2782 Method 120C describes the so-called 1/10 Vicat softening point. Here a force of 10 N is applied to the indentor and the softening point is defined as the temperature at which there is a penetration of 0·1 mm when the rate of heating is 50° ± 5°C/h. The respective uses of the normal and 1/10 Vicat softening point tests are discussed by Willbourn and Stephenson.[46]

With wide adoption of the Vicat method it is a pity that the various national and international standards differ in minor details, although the difference in results obtained is probably insignificant. Crofts[69] has reported the effect on Vicat softening point of the rate of heating and of the force applied to the indentor. By extrapolating curves to the hypothetical softening point at zero force and zero rate of heating he obtained good agreement with the glass transition temperature of certain thermoplastics.

Several workers have been prompted to refine the Vicat softening point apparatus. Automatic indication of the end-points of Vicat softening point (and deflection temperature under load) tests is achieved by Graves and Loveless[70] using a lever actuating a mercury switch to cause a panel light to glow; they also describe VSP tests at rates of temperature increase of both 50°C/h and 120°C/h (i.e. 2°C/min), quoting results for seven thermoplastics to justify the use of the faster rate.

Ehlers and Powers[71] use a transducer attached to the top of the penetrometer shaft of the Vicat apparatus to record depth of penetration against temperature fed to the recorder from a thermocouple.

BS 2782 (1976) Method 122A[72] describes a penetrometer method entitled 'Determination of deformation under heat of flexible polyvinyl chloride extrusion compound'. A test piece 13 mm in diameter and 1·27 ± 0·07 mm thick is accurately measured for thickness under specified conditions and placed on a flat horizontal base. A force of 9·8 N is applied to the test piece with a vertical plunger, 3 mm in diameter and flat, and the whole assembly is placed in an oven for 24 h at 70° ± 1°C. After this the assembly is removed from the oven and allowed to cool for one hour at room temperature. Finally the thickness of the test piece in the deformed area is remeasured between three and five minutes after removal of the load.

The deformation under heat of the test piece is taken from the difference between the initial and final thicknesses calculated as a percentage of the initial thickness, and the result is expressed as the mean of two readings.

Another test which determines deformation under load at both room and elevated temperatures is described in ASTM D621-64 (1976).[73] This contains two test methods: Method A is for rigid plastics and is designed to assess the ability of such materials to withstand compressive load, e.g. when held by bolts, without yielding and loosening the assembly; Method B is intended to determine the ability of non-rigid plastics to return to their original shape after having been deformed.

Method A

The essentials of the apparatus are that it shall be capable of exerting constant forces of 113 kg, 227 kg and 454 kg, all ± 10 per cent, and that it is equipped with a dial gauge measuring the relative movement of the faces of the anvils to 0·02 mm or less.

It will be realised that many compression testing machines (Chapter 8), if fitted with temperature controlled ovens, are adaptable to meet the specification requirements.

Test pieces are 12·7 mm cubes, either solid or composite, machined down or built up as necessary. Surfaces must be plane and parallel.

After preconditioning, if necessary, the test piece is placed between the anvils and the load is applied without shock. After 10 s the dial reading is taken and again after 24 h at the test temperature of 23°C, 50°C or 70°C, all ±1°C. The test piece is removed, its thickness is measured and the original height (B) is calculated from this plus the change in dial reading (A).

Deformation, per cent = $A/B \times 100$.

Method B

This is basically the same as Method A but the stress is 0·69 MPa ±1 per cent and the period of test only 3 h.

The test pieces are 28·67 mm diameter by 12·7 mm thick. After preconditioning, the procedure of Method A is followed with the exception given above and a difference in the dial reading sequence. The zero point is obtained at the beginning of the test by determining the dial reading with the anvils together under full load and the thickness of the test piece (H_0) is measured by micrometer, with low-pressure ratchet attachment, to the nearest 0·0002 in (0·005 mm).

After the thickness of the test piece has been determined at the end of three hours (H_1), it is removed from the apparatus and left in the test chamber for one hour. Finally the test piece is removed from the chamber, kept at room temperature for $\frac{1}{2}$ hour and then its thickness is redetermined by micrometer (H_2).

Deformation per cent = $[(H_0 - H_1)/H_0] \times 100$.

Recovery, per cent = $[(H_2 - H_1)/(H_0 - H_1)] \times 100$.

The reader will have noticed the curious mixture of units given in this standard. Although ASTM D621 was reapproved in 1976 it needs editing to bring it into line with the current practice of using SI units.

16.7 LOW-TEMPERATURE BRITTLENESS AND FLEXIBILITY TESTS

16.7.1 General Considerations

Most applications of rigid plastics make use of their stiffness, even if this is low by comparison with metals and stone. It is therefore logical that in the majority of quality control tests to check the thermal properties of such rigid plastics it is loss of stiffness that is used as the parameter of measurement. Likewise, as many of the applications of flexible plastics utilise their characteristic 'give', the corresponding quality control for these follows their loss of flexibility with decrease in temperature. Amongst the standardised tests, however, will be found those which are slow in action, for example plotting torsional stiffness against temperature, and those which are fast acting, such as those which assess the temperature at which impact shock causes shattering. Early background to both types of test is described by Clash and Berg.[74]

16.7.2 Torsional Stiffening Methods

Welding[75] critically reviewed torsion tests before some of them had been standardised and concluded that they could be simple and accurate. The Clash and Berg type of test has been standardised by ISO, BS, ASTM and DIN and is commonly used for testing plastics. The apparatus is characterised by the torque to the test piece being applied by a pulley, cords and two weights (Figure 16.13). The Gehman test uses an apparatus where the torque is applied by calibrated torsion wire. Williamson,[76] in an early paper, found the Clash and Berg type of apparatus too insensitive at low values of modulus, and described a simple torsion apparatus based on calibrated torsion wires; an automated version of this type of equipment has been described by Wheeler.[77] The Gehman test is used extensively for testing rubbers and has been standardised by ISO, BS and ASTM, (see Brown,[78] page 251).

ISO R458 (1975,[79] BS 2782 (1976) Method 150A,[80] ASTM D1043-72 (1978)[81] and DIN 53447 (1966)[82] are similar; all are for the determination of stiffness in torsion as a function of temperature. The apparatus is shown diagramatically in Figure 16.13. The weights and pulley of the apparatus are such that a torque in the range 0·01 Nm to 0·12 Nm can be applied, through clamps, to the test piece which is immersed, in a Dewar flask, in a suitable heat transfer medium. Good stirring is essential during heating and cooling and commercially available apparatus usually incorporates an efficient electrically driven stirrer in the Dewar flask. For cooling, mechanical refrigeration or dry ice is recommended and an immersion heater in the flask is commonly used for heating.

Test pieces are rectangular strips and various sizes are allowed as given in Table 16.2. The ASTM method allows the use of different size test pieces where the span (i.e. distance between the grips of the machine) is variable. Spans of 38 to 100 mm may be used provided that the span to width ratio of the test piece is between 6 and 8.

In operation the width and thickness of the test piece are measured to an accuracy of ±0·02 mm and ±0·01 mm respectively and then mounted in the grips of the apparatus with the torque pulley clamped at the zero position. The heat transfer medium is brought to the required temperature (which is somewhat

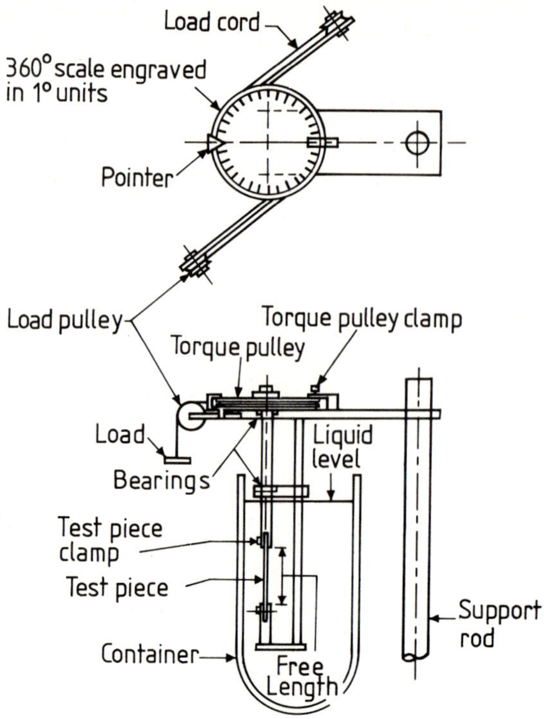

Figure 16.13 Torsion tester: BS 2782

below the lowest temperature of the range to be studied) and the test piece is immersed. After 3 min +15, −0 s the pulley clamp is released and the reading of angle of movement (θ) of the pulley is taken after an interval of 5 s. The pulley is then returned to zero and the procedure is repeated at successively higher temperatures, the value of θ for each temperature being recorded. Two test pieces are used for each test.

The apparent modulus of rigidity for each test temperature is calculated by the relationship

$$G = 9 \times \frac{Tl}{b(d^3) \times \mu\theta} \tag{16.6}$$

Table 16.2 Test Piece Dimensions for Stiffness in Torsion as a Function of Temperature

Standard method	Length (mm)	Width (mm)	Thickness (mm)
ISO R458	70 ± 10	6·0 to 6·3	1 to 5
BS 2782 Method 150A	60 to 80	6·0 to 6·3	1 to 5
ASTM D1043	63·5 ± 0·025	6·35 ± 0·025	1 to 3
DIN 53447	60 to 65	6·0 to 6·3	1 to 5

where G is the apparent modulus of rigidity (Pa)
T is the applied torque (Nm)
l is the span (mm)
b is the test piece width (mm)
d is the test piece thickness (mm)
μ is a constant depending on the ratio b/d
θ is the angle of rotation of the pulley (degrees).

Tables giving μ against b/d are included in all the standard test methods.

A plot of the mean value of G on a logarithmic scale is made against temperature on a linear scale and the value of G is read off at any desired temperature from the curve. ASTM D1043 specifies a value, designated T_F, as the temperature at which G is equal to 310·3 MPa. This standard also gives appendices which include comprehensive instructions for calibrating the equipment and for modifying to eliminate friction in the load pulleys and to automatically take up slack in the test piece when it expands during heating.

A modification of the test is used to specify the cold flex temperature of flexible polyvinyl chloride compound in BS 2782 (1976) Method 150B.[83] Here a test piece of closely specified dimensions (65 mm long, 6·2 to 6·4 mm wide and 1·30 ± 0·08 mm thick) is tested in the Clash and Berg apparatus at three temperatures chosen so that an angular deflection of 200° occurs between the highest and lowest temperatures. A graph of angular deflection versus temperature is plotted and the temperature at which a deflection of 200° occurs is read off from the graph. The mean of two test results is reported as the cold flex temperature of the sample under test.

16.7.3 Extensibility Method

Resistance to low temperature of flexible PVC sheet is measured by the extensibility in tension at a specified temperature according to Method 104C of BS 2782 (1970),[84] which will appear in its present form in the new edition of BS 2782.

A constant-rate-of-loading instrument is required, capable of increasing stress on a rectangular test piece (180 mm long × 6·3 ± 0·1 mm × sheet thickness) uniformly from zero at a rate of 20·7 ± 0·4 MPa per minute. The free (unclamped) length of test piece is 125 mm and an extensometer device is needed to measure extension to the nearest one per cent. A cooling bath at −5 ± 0·5°C, comprising one volume of industrial methylated spirit and three volumes of water, to which is added solid carbon dioxide as necessary, is provided for immersing that part of the apparatus which contains the clamped test piece.

The load required to give a stress of 10·3 MPa on the test piece is calculated from the width and thickness obtained gravimetrically (Chapter 7). With the grips set 127 ± 1·3 mm apart, the test piece is mounted symmetrically in them, first in the upper one and then, with a 20 g weight attached to its lower end to keep it taut, in the lower one. The mounted test piece is then completely immersed in the bath, for 30 to 60 s according to thickness, and after this the load is applied uniformly so that 10·3 MPa is reached in 30 s. When this stress is reached the extension of the test piece is read (all the extended specimen to be in the bath) to the nearest 1·3 mm.

Three such determinations are carried out and their results are averaged.

16.7.4 Flex Cracking Methods

The 'cold bend temperature' of flexible PVC extrusion compounds, described in Method 104A of BS 2782 (1970)[84] measures the lowest temperature, measured in multiples of 5°C, at which none of a set of three test pieces fractures or cracks when wound onto a standard mandrel. (This test will be republished in the later edition of BS 2782.)

At least six test pieces, each 100 mm long and 4·8 mm wide, are cut from specially prepared sheet of thickness 1·27 ± 0·08 mm. They are tested in an apparatus of the type shown in Figure 16.14 in which all the essential features are given.

Three test pieces are placed in the guides and one end of each is secured by a clamp to the mandrel. Test pieces and winder are then immersed in a cooling bath of industrial methylated spirit maintained at the required temperature by solid carbon dioxide. After 10 min at the temperature, and while still immersed, the test pieces are wound tightly around the mandrel for three complete helical turns at the rate of one revolution per second. They are then withdrawn from the bath and examined for signs of failure—fracture or surface cracking. The test is repeated, on fresh sets of test pieces, at various temperatures until two temperatures, differing by 5°C, are found such that at the higher no test piece fails and at the lower one or more fail.

The cold bend temperature is the higher of these two temperatures.

In the course of an investigation into low-temperature failures of handbag materials, Wormald[85] developed a flexing tester operated at −5°C in which the test result was the number of flexings required to cause failure of a standard test piece when repeatedly flexed under controlled conditions, particularly incorporating in each cycle a rapid fold followed by an appreciable waiting period to simulate service conditions.

16.7.5 Impact Methods

ISO R974 (1969)[86] and ASTM D746 (1979)[87] both describe the determination of brittleness temperature by impact of plastics. They determine the temperature at which 50 per cent of the test pieces would fail under the conditions of the test (but see below).

Test pieces are die punched either as 6·35 ± 0·51 mm wide pieces or as shown in Figure 16.15. In either case they are 1·91 ± 0·25 mm thick, of suitable length to be clamped in the apparatus as shown in Figure 16.16. The alternative test piece is clamped so that the entire tab is inside the jaws for a minimum distance of 3·18 mm.

Sharp dies must be used. Hoff and Turner[88] have conducted an investigation into the effect of mode of specimen preparation on test results and found up to 50°C difference between razor-cut and die-cut test pieces on low-density polyethylene of MFI 20.

The striking edge moves relative to the test piece at a linear speed of 1·8 to 2·1 m/s at impact and for at least 6·4 mm thereafter. A cooling bath is required, of liquid heat transfer medium without appreciable effect on the test pieces, and held within ±0·5°C of the desired value.

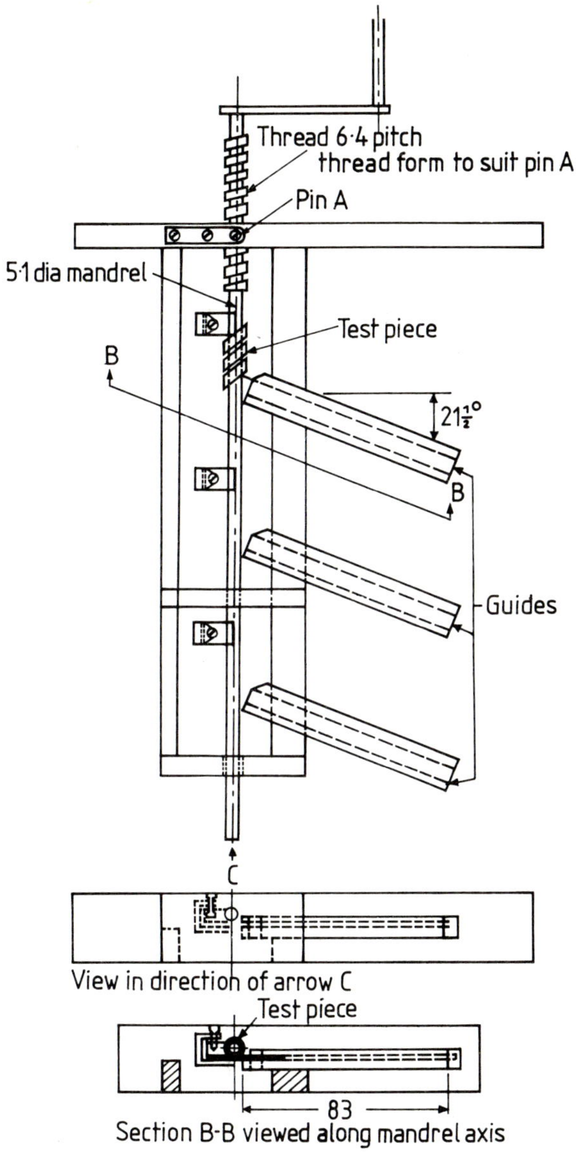

Figure 16.14 Cold bend test: BS 2782—dimensions in millimetres

The test piece is mounted in the apparatus and immersed in the bath at the required temperature for 3 ± 0·5 min. After this the test piece is impacted as described above and examined for failure—fracture or visible crack (as shown by bending through 90°). A new test piece is used for each test.

The complete procedure is started at a temperature at which a 50 per cent chance of failure is expected, and at least ten test pieces are tested at this

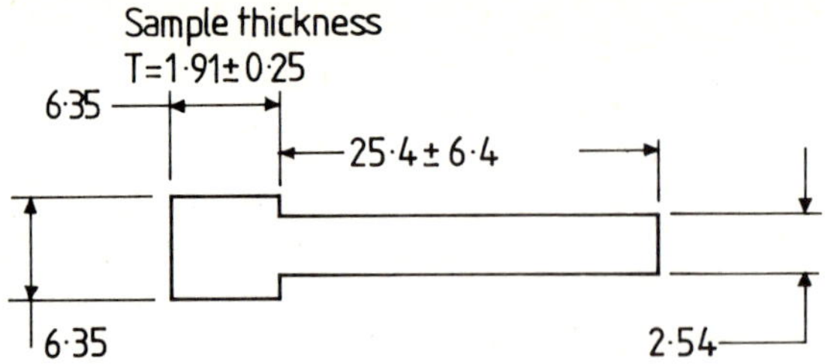

Figure 16.15 Alternative test piece for brittleness temperature test: ASTM D746—dimensions in millimetres

temperature. If all the test pieces fail or do not fail, the bath temperature is increased or decreased (respectively) by 10°C and the test repeated. When temperatures have been found where all failures and non-failures are recorded, the temperature is changed by steps of 2° or 5°C, testing ten test pieces each time and including the 'all fail' and 'no fail' temperatures.

$$\text{Brittleness temperature, } T_B = T_h + \Delta T\left(\frac{S}{100} - \frac{1}{2}\right) \tag{16.7}$$

where T_h = highest temperature at which failure of all the specimens occurs
Δ = temperature increment (°C)
S = sum of the percentage breaks at each temperature (from a temperature corresponding to 'no fail' down to and including T_h).

Alternatively a graphical method may be used for determining T_B or, for quality control purposes, the test may be carried out at one standard temperature and the material deemed to pass if not more than five test pieces fail out of the ten.

Although the title of ASTM D746 refers to 'plastics and elastomers', the brittleness at low temperature for rubber is defined differently in the test

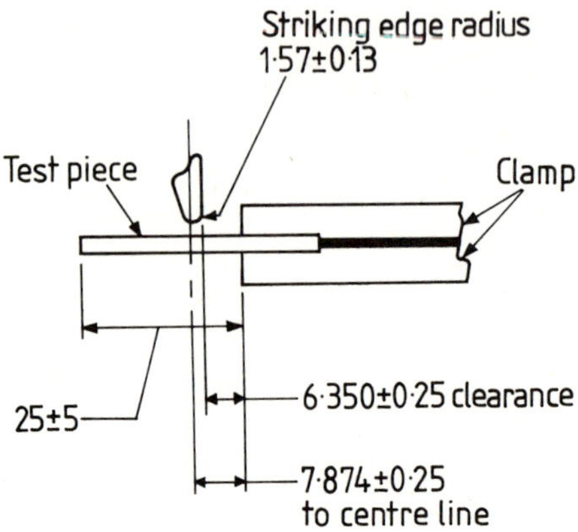

Figure 16.16 Clamp and test piece for brittleness temperature test: ASTM D746—dimensions in millimetres

methods for rubber, Part A25 of BS 903 (1968),[89] yet the apparatus is essentially the same (see Brown,[78] page 257).

The development and study of the ASTM D746 method is described by Webber.[90] A number of authors[91–93] recommend notching the test pieces to obtain more reproducible results; while this departs from the ASTM procedure, it has been recognised as a useful supplementary technique by ISO R974 and is permitted by this International Recommended Test Method. R974 also includes an Appendix which specifies three procedures for preparing test pieces from polyolefin samples.

Two brittleness impact tests at low temperature for plastics film and coated fabrics have been standardised. BS 2782 (1976) Method 150D,[94] *Cold Crack Temperature of Film and Thin Sheeting,* and ASTM D1790-62 (1976)[95] are basically similar to a test described by Williams.[96] In the BS test a test piece 32 mm long and 6·5 mm wide is folded into a loop, the free end being clamped, and the looped test piece is immersed in a bath of coolant. The test piece is struck with a flat-ended striker (25 mm in diameter) at a velocity of 2 m/s and a nominal energy of 2·6 kJ. Tests are carried out at increments of temperature of 2°C on at least 12 test pieces and the cold crack temperature is calculated as the temperature at which 50 per cent of the test pieces would have cracked. The BS test in a less sophisticated form also appears in BS 3424 (1973).[97]

16.8 HEAT AGEING

16.8.1 General Considerations

So far in this chapter we have been dealing mainly with *reversible* processes, measurements designed to study or check the influence of change of temperature on plastics materials wherein any change of property so induced may be cancelled by return to the original datum temperature: i.e., no 'permanent' damage has been done. We now consider the test for measuring the *irreversible* effects of change of temperature—thermal stability and degradation; as a consequence, we are only concerned with effect of heat, since lowering the temperature reduces the energy in a system and of itself causes no permanent change, though, depending on circumstances, the process of returning to the original state may be slow. Elevation of temperature causes loss of volatile constituents, such as an evaporation of plasticiser if present, as a first irreversible step; however, considering the polymer alone, since most polymers are based on a wholly or partially carbon backbone they are susceptible to bond scission at relatively modest temperatures—perhaps as low as 100°C depending on polymer type, environment and duration.

The mode of polymer degradation varies widely with polymer type; for instance, polyethylene in oxygen initially crosslinks, PVC degrades by losing hydrogen chloride and poly(methyl methacrylate) depolymerises almost totally into monomer. It is not therefore surprising that standardised tests for thermal stability tend to differ from polymer to polymer: clearly, with an unknown material it is not sufficient, say, only to measure weight loss due to heating—there may be in fact a gain.

A general standard on the subject is ISO 2578 (1974),[98] *Plastics—Determination of Time–Temperature Limits After Exposure to Prolonged Action of Heat.* This defines a time–temperature limit as the highest temperature to which the particular plastics material can be subjected for a determined time before the numerical value of a chosen characteristic reaches a predetermined critical (threshold) value. For general purposes a period of 20 000 h is recommended but the choice of the characteristic and the threshold value of that characteristic is left to the user, or is obtained from the specification of the material under test. This standard is, at best, a guide only; selection of the temperature for exposure involves predicting, or knowing beforehand, the approximate temperature range in which the time–temperature limit of the material to be tested is located, otherwise exploratory tests must be made. It would have been helpful if a list of the common plastics materials had been included, with a guide to the temperature limit at 20 000 h. IEC publication 216,[99] *Guide for the preparation of Test Procedures for Evaluating the Thermal Endurance of Electrical Insulating Materials*, was taken into account during the preparation of ISO 2578. DIN 53446 (1962)[100] and ASTM D3045-74 (1979)[101] are similar in content to the International Standard. IEC publication 216 was also used as a basis for BS 4618 (1974) Section 4.6[102] on the presentation of plastics design data in thermal endurance tests. Underwriters Laboratories Inc. Standard No. UL 746 (1975)[103] on long-term property evaluation is similar to ISO 2578 but is expanded to give recommended test methods and it also gives a list of thermosetting materials and their approximate thermal indices, i.e. the maximum temperature for which they have indefinite service life.

Caution needs to be used in interpreting the results of these tests in terms of actual service life at elevated temperatures. Mair and Wiebusch[104] have shown that it is unwise to use time and temperature limits to predict permanent service temperatures as this requires the taking into account of all properties relevant to the application, such as mechanical or electrical loading throughout the service life of the material. Parthum[105] has investigated the ageing behaviour of various moulding grades of phenol-formaldehyde-based materials under prolonged heating to determine the limiting temperature at 25 000 h (three years). The results obtained were not in agreement with practical experience in the use of the various materials. More recently Meacham[106] has made a plea for a 'standard' heat resistance test and has proposed the use of ASTM D794-68 (1972)[107] for this purpose. He also proposes the criterion of heat resistance to be retention of flexural strength. Unfortunately, ASTM D794 is rather vague, does not closely specify the oven to be used and, for continuous heat tests, specifies the temperature time limit be determined without giving precise instructions as to how to do so. It would be better, therefore, to use ISO 2578 as the basis of Meacham's proposal.

16.8.2 Standard Test Methods

The procedures mentioned in §16.6 come into this category in certain instances, in that some of them may be carried out at temperatures which cause irreversible changes. ASTM D794-68 (1977)[107] specifies general requirements for conducting any suitable standard tests at high temperature; for example

ASTM D700 (1975)[108] for phenolic moulding compounds refers to ASTM D794 under the requirements for heat resistance and requires the determination of flexural strength after seven days' ageing at the specified temperature followed by reconditioning to normal temperature. If the flexural strength so determined is less than 75 per cent of the initial flexural strength the material is deemed to have failed.

Similarly, Methods 131C and 131D of BS 2782 (1978)[109] determine the crushing strength after heating ('heat resistance') of, respectively, moulding material and laminated sheet. The test is basically a compression strength measurement (Chapter 8) after the test pieces have been cooled following heating in an air oven for 17 ± 1 h at 135° ± 5°C, followed by 6 h ± 15 min at 170° ± 5°C and total immersion in fusible metal at 400° ± 10°C for 30 ± 2 min. The test pieces are either cylinders (Method 131C) of length 10 ± 0·2 mm and similar diameter or cubes (Method 131D) of side of the same dimension; the latter may be a composite specimen made by building up thin sheet. The compression load on the (heat treated) test pieces is increased steadily so that failure occurs within 30 ± 15 s.

Method 108B of BS 2782 (1970)[110] for the ageing of polyethylene by hot air oven uses a specimen approximately 76 × 76 × 1·3 mm which is heated for 48 ± 2 h at 140° ± 1°C in an oven with fan circulation. The test piece rests on a sheet of polytetrafluorethylene or a sheet of metal coated with that polymer. This method of ageing is particularly used for examination of polyethylene used for cable insulation or sheath and is usually followed by determination of power factor at 1–20 MHz (Chapter 13) or, for black compounds, by determination of melt flow index (Chapter 5). This method is being revised and will be issued as Method 130 of BS 2782.

Methods Based on Weight Loss

There is a considerable number of 'loss in weight on heating' tests which have been standardised. Most, however, are not ageing tests, but rather control tests of solids content of resin solutions and monomer or resin content of laminating resins; as such they are not considered to fall within the scope of this book (being more analytical in character). Nevertheless this chapter would not be complete without mention of certain of the tests.

ISO 1137 (1975),[111] for the determination of behaviour in a ventilated tubular oven, allows for the loss in mass (and change in mechanical properties) after heating for a specified period of time and temperature in a tubular oven with an air velocity of 100 m/min. ASTM D1870-68 (1978)[112] is similar but allows a greater freedom of choice of temperatures and air velocities. Both of these methods are useful where it is desired to determine the long-term effect of heat on plastics under carefully controlled and contamination-free conditions.

ISO 176 (1976),[113] *Determination of Loss of Plasticisers—Activated Carbon Method*, is used as the basis for Method 465A of BS 2782 (1979).[114] ISO 176 requires the use of three circular test pieces 50 ± 1 mm in diameter and 1 ± 0·1 mm thick, cut from compression moulded sheet. Activated carbon with a grain size of about 4 to 6 mm, free from powder, is required and, for Method B, cylindrical cages constructed from bronze gauze having apertures of about 600 μm, 60 mm in diameter and 6 mm high are used for keeping the test pieces out of contact with the activated carbon. The test is carried out by loading the test pieces in metal cases (about 100 mm diameter and 120 mm high) fitted with a non-airtight cover, at a temperature between 50° and 150°C, in a thermostatically controlled bath or oven capable of maintaining the temperature to within ±1°C of the test temperature.

For Method A a layer of about 120 cm^3 of activated carbon is spread evenly over the bottom of a container and a test piece is placed on top. A further 120 cm^3 of carbon is placed on top of this, followed by two further test pieces each covered by 120 cm^3 of carbon. After the lid of the container is secured, the whole is heated for 24 h at 70° ± 1°C. For Method B the procedure is the same except that each test piece is put into a gauze cage to avoid direct contact with the carbon and the heating is carried out for 24 h at 100° ± 1°C. By weighing each test piece to within 0·001 g before and after heating, the change in mass is calculated as a percentage of the original mass and the result is expressed as the arithmetic mean of the values obtained from the three test pieces.

ASTM D1203-67 (1974)[115] is similar to ISO 176.

ISO 3671 (1976),[116] for aminoplastics moulding materials, determines the loss in mass (assumed to be predominantly water) after heating for 3 h at 55° ± 1°C a 5 g sample of the moulding powder. Method 107G of BS 2782 (1970),[117] for cellulose acetate moulding material, uses moulded test pieces 50·8 mm long, 25·4 mm wide and 1·52 mm thick which are heated in an oven of specified dimensions for 3 h at 70° ± 1°C; the loss in mass is expressed as a percentage of the original mass. In addition, there are many mass loss tests given in the various materials specifications of Part 36 of the ASTM Series.[118]

In ASTM D1457 (1975),[39] *TFE-Fluorocarbon Resin Moulding and Extrusion Materials*, there is a thermal instability test, the index of which is obtained by multiplying by 100 the difference in specific gravity between (a) discs moulded by sintering compressed powder at 380°C for 30 min followed by cooling to 300°C at a rate of 1·2° ± 5°C per minute in the oven and then to room temperature in air, and (b) discs similarly prepared but sintered for 2 h.

In ISO 177 (1976),[119] mass loss of plasticised material at 70°C is determined when a sample is placed between specified 'absorbing discs', to simulate certain end-uses.

Methods Specifically for Vinyl Chloride Polymers

ISO R182 (1970)[120] Method A, BS 2782 (1976) Method 130A[121] and DIN 53381 (1971),[122] Test 1, all assess the time taken for the vinyl chloride polymer or copolymer to evolve sufficient hydrogen chloride to change the colour of a piece of Congo Red paper placed with its lower edge 25 mm above the top of the polymeric sample, the upper end of the paper being held in position by a plug of glass wool inserted in the mouth of the tube. The assembly is placed in a constantly-stirred oil bath at 180° ± 1°C to the level of the upper surface of the material in the tube. The heat stability (mean of two determinations) is taken as the time in minutes between insertion of the tube in the hot oil bath and the first signs of a change of the indicator paper from red to blue.

PVC compositions discolour markedly when degradation sets in, as a result, it is believed, of the creation of conjugated doublebond systems by loss of hydrogen chloride. Oven ageing tests which depend on visual assessment of such degradation are therefore popular. ISO 305 (1976),[123] ASTM D2115-67 (1974)[124] and DIN 53381 (1975),[125] Test 2, are three such standard tests. There is nothing particularly profound in these methods and the reader is referred to the original standards for details.

ASTM D793-49 (1976),[126] *Short Term Stability at Elevated Temperatures of Plastics Containing Chlorine*, actually involves titration of the hydrogen chloride liberated when 10 g of material, cut so that no dimension is greater than $\frac{1}{16}$ in (1·6 mm), is heated in the bottom of an Erlenmeyer flask. The flask is placed in a constant-temperature bath at 180° ± 2°C to within 30 mm of the top of the flask and is fitted with a rubber stopper holding two glass tubes. One of these is a glass inlet tube and extends nearly to the bottom of the flask (Figure 16.17). The tube is connected to the gas preheater, in the bath, which in turn is connected to a nitrogen cylinder. The outlet tube is connected to an absorption tube containing 40 ml of 0·1 N sodium hydroxide solution.

The nitrogen gas flow is adjusted to a rate of 2–4 bubbles/second in the absorbing tube and allowed to flow for 30 min. The absorbent solution is washed into a beaker, acidified with nitric acid and the chloride present is precipitated with 50 ml of 0·02 N silver nitrate solution. The excess silver nitrate is back-titrated with 0·02 N ammonium or potassium thiocyanate solution, using ferric nitrate as indicator. A 'blank' is performed using 40 ml of the 0·1 N sodium hydroxide solution diluted with distilled water to approximately the same volume as used for the test.

$$M = [(A - B) - C] \times \text{normality of silver nitrate solution} \times 3{\cdot}65 \quad (16.8)$$

where M = Short-term stability expressed as mg hydrogen chloride evolved per g test specimen (under the test conditions)
A = ml silver nitrate solution
B = ml thiocyanate solution
C = ml silver nitrate solution consumed in the 'blank'.

More recently, the use of pH change has been used to determine the time of heating of a PVC material to cause the first evolution of hydrochloric acid gas as

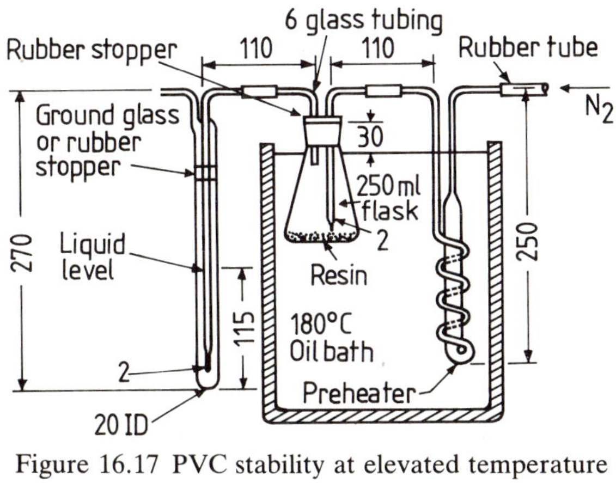

Figure 16.17 PVC stability at elevated temperature

evidence of decomposition. ISO R182 (1970)[120] Method B, BS 2782 (1976) Method 130B[127] and DIN 53381 (1971),[128] Test 3, are all similar and based on this method of measurement. A weighed portion (1 g) of the material is placed in a test tube (150 mm long and 17 mm diameter) fitted with glass inlet and outlet tubes. The outlet tube is taken to a pH-measuring cell fitted with calomel and glass electrodes and filled with 60 ml of 0·1 N potassium chloride solution of pH 6·0. The inlet tube is connected to a gas supply (either air or nitrogen), which must be dry and free of carbon dioxide, at a rate of 6·0 l/h. To carry out the test, the gas flow is started and after one minute the test tube is immersed in an oil bath at 180° ± 1°C and the time is noted. The test is continued until the pH of the potassium chloride solution in the pH cell reaches 3·9±0·1, when the time is again noted. The thermal stability is expressed as the time, in minutes, from insertion of the tube containing the test portion in the oil bath until the pH has reached 3·9. Two test portions are used and the arithmetic mean of the two results is recorded as the thermal stability. When the two values are more than ±10 per cent apart from their average the test is repeated.

The pH method has the advantage over the Congo Red method that different gases can be used to simulate oxidative (air) and neutral (nitrogen) conditions of use. The end-point is also more precise; the Congo Red paper sometimes bleaches before turning blue, or only changes colour slowly, making the determination of the end-point difficult.

Other Standard Tests for Plastics

A spectacular method included in ASTM D701-49 (1976),[129] *Cellulose Nitrate (Pyroxylin) Plastic Sheets, Rods and Tubes,* is called the 'fuming-off temperature'. Shavings not thicker than 0·025 mm are prepared and heated for 2 h in an oven at 50°C to drive off moisture and solvents. Approximately 0·1 g of the dried shavings is placed in a test tube, 150 × 13 mm, lightly stoppered with a perforated or notched cork, and the tube is immersed 50 mm into a liquid bath at 100°C. The bath is heated at a rate between 3° and 5°C per minute and the temperature is noted at which the sample decomposes explosively. At least five such tests are run.

In the dimensional stability test for extruded acrylic plastic sheet in ASTM D 1547-75 (1979),[130] the presence of bubbles is studied after specimens 6 in (152 mm) square have been heated at either 143° ± 6°C or 154° ± 6°C for 15 or 30 mins according to, respectively, grade and thickness of material. The specimens, after preconditioning at 50° ± 6°C, are rested on a preheated glass plate sprinkled with talcum powder.

For testing the thermal oxidative stability of propylene plastics (ASTM D2445-69 (1974)[131]), a quantity of pellets or granules is placed in a U-tube which is partially immersed in an oil bath at 150° ± 0·5°C. Oxygen is metered in at one end of the tube at a rate of 10 cm^3/min, and the pellets or granules are inspected at regular intervals for signs of failure as manifested by the appearance of surface crazing. Embrittlement is confirmed by removing a few of the granules and crushing them; if they are readily reduced to powder, embrittlement is taken as complete.

Finally, ISO 1599 (1975),[132] for the determination of viscosity loss on moulding of cellulose acetate, requires the test sample to be compression moulded under a specified moulding cycle. The moulding is then ground up, the granulate is dried and its viscosity in dilute solution is determined according to ISO 1157 (1975).[133] The percentage viscosity loss is calculated from the values of viscosity before and after moulding.

16.9 PHYSICAL TESTS AT NON-AMBIENT TEMPERATURES

16.9.1 General Considerations

If more meaningful data on temperature effects are required, more useful for design purposes than the arbitrary values obtained from the *ad hoc* thermal yield, brittleness and stiffening tests described in §§16.6 and 16.7, it is essential to perform relevant 'absolute' tests at the temperature or temperatures of interest. By and large, their execution follows the description of the tests at standard temperatures, the variation being in creating an atmosphere of the appropriate temperature around the specimens—and perhaps the whole apparatus—with the minimum of variation from point to point and at any one point. There may, of course, be practical difficulties in enclosing an apparatus, or even the essential part of it, in a temperature-controlled atmosphere, or, if this is successfully achieved, of actually operating it in such an enclosure. Carrying out impact tests with a pendulum machine is a particular case in point but, depending on the case in question, simple solutions are often available. Thus, impact testing of this type is a fast operation, and it has been found[134] that appropriate precooling of the test pieces, followed by testing immediately after removal from the cooling cabinet, is a satisfactory alternative. There may, however, be unexpected pitfalls with otherwise simply accommodated tests—e.g. ice formation on the guides of compression chucks causing apparently high results, or on the surface of test pieces, leading to low results in electrical resistivity measurements.

ISO 3205 (1976)[135] and ASTM D618-61 (1977)[136] give lists of preferred test temperatures from −269°C to 1000°C, but there are differences between these two standards (see Chapter 4).

16.9.2 Standard Test Methods

ISO 2578 (1974),[98] BS 4618 (1974) Section 4.6,[102] ASTM D3045 (1974),[101] DIN 53446 (1962)[100] and UL 746B (1975)[104] have already been described in §16.8. The product specification BS 3534 (1964),[137] *Epoxide Resin Systems for Glass Fibre Reinforced Plastics* Part 2 *Pre-Impregnating Systems*, requires crossbreaking strength of prepared laminates to be measured, by Method 335B of BS 2782 (1978),[138] at 10° ± 2°C below the 'lower classification temperature' after conditioning at that temperature for two hours; these temperatures range from 70°C to 200°C. No details or recommendations are given on how to carry out the test at elevated temperatures. The related BS 4045 (1966),[139] replaces the high-temperature test by one at room temperature after exposure for long periods at elevated temperature—and terms it a 'thermal endurance' test. The abilities to resist deterioration due to increased temperatures, as measured by the two tests, are of course quite different.

Specifications ASTM D759-66 (1976)[140] and D2733-70 (1976)[141] give instructions for carrying out mechanical tests at above or below ambient temperatures.

ASTM D2733 does not recommend the test temperatures to be used but ASTM D759 cross refers to ASTM D618-61 (1971)[136] which gives a list of preferred test temperatures from 170°C to 600°C.

16.9.3 Test Equipment

Clearly the design of test equipment and test chambers for use at very high and very low temperatures is very important if satisfactory results are to be obtained. Account must be taken of problems which are likely to arise with the use of testing machines which operate satisfactorily at ordinary temperatures when they are exposed to subnormal and supernormal temperatures (see §16.9.1 above). Thermal insulation of test chambers is all-important if adequate temperature control is to be achieved. Many suppliers of test machines market ovens and chambers suitable for total enclosure of the machine or, more often, of the area immediately surrounding the test piece and some of these are capable of working from well below ambient temperatures up to a few hundred degrees centigrade.

For very low-temperature testing recourse usually has to be made to the use of liquefied gases; the relative merits and disadvantages and dangers of the various gases have been summarised by Mathes.[142] Liquid nitrogen vapour is often used and, properly handled, is reasonably safe. Following the ideas of Wessel and Olleman,[143,144] Ives and Mead[145] describe a cryostat for low-temperature mechanical testing using the vapour from liquid nitrogen which gives good temperature control up to near room temperature.

The liquid nitrogen moves from its storage flask to the cold chamber of the apparatus through a dip tube which passes from the bottom of the vessel up to a well insulated pipe (as short as possible) and thence to the chamber; no valves or other restrictions are included in this line. The liquid nitrogen container is otherwise sealed, except for one vent pipe connected to the vapour above the liquefied gas; the increase in pressure of this vapour, caused by continuous slow evaporation, forces liquid or cold vapour into the working chamber (Figure 16.18).

Temperature is controlled by releasing or reducing the excess pressure in the storage flask and to this end the vent pipe is connected directly to an adjustable pressure relief valve (No. 1) and by way of a solenoid valve to relief valve No. 2. In operation, the settings of the valves are adjusted so that valve No. 1 permits the driving force in the flask to be high enough to ensure an adequate supply of coolant to the chamber, but the pressure is such that the supply is not excessive and overshooting is avoided. The second pressure relief valve is brought into action by the opening of the solenoid valve, which in turn is actuated by a temperature controller with a sensing element in the cold chamber. This second valve is adjusted to control the pressure at a value slightly below that necessary to ensure a supply of coolant to the working chamber, and thus an enhanced demand for coolant, initiated by a rise in chamber temperature, can be met almost instantaneously.

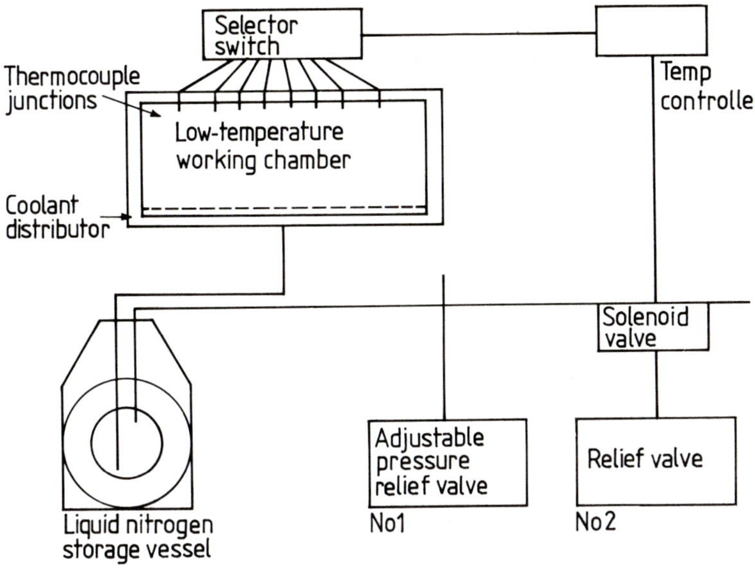

Figure 16.18 Schematic diagram of low-temperature apparatus[145]

The subject of testing at low temperatures has been reviewed by Lieb and Mowers.[146]

Comments on the types of chamber commercially available are given by Brown.[147] General directions for achieving elevated or sub-normal temperatures are given in ISO 3383 (1976)[148] and ASTM D832 (1979)[149] may also be useful to the reader contemplating low-temperature testing. ISO 3383 is a useful document but is of an elementary nature and its shortcomings are highlighted by Brown,[78] page 66.

When testing at non-ambient temperatures it is necessary to condition the test piece for a time long enough for it to reach the test temperature, although excessive conditioning will cause heat ageing. Attention is drawn to the conditioning times given in the appendix to Chapter 4.

REFERENCES

1. Griffiths, M. D. (1974) *RAPRA Members Journal*, July, 187.
2. ASTM D696 (1979) *Coefficient of Linear Thermal Expansion of Plastics.*
3. ASTM D846-52 (1978) *Coefficient of Cubical Thermal Expansion of Plastics.*
4. Bekkedahl, N. (1949) *Journal of Research of the National Bureau of Standards*, **42** (Rp 2016), 145.
5. ISO 2577 (1975) *Plastics—Determination of Shrinkage of Compression Moulded Test Specimens in the Form of Bars.*
6. BS 2782, Method 640A (1979) *Mould Shrinkage of Thermosetting Moulding Materials.*
7. ASTM D955 (1979) *Measuring Shrinkage from Mold Dimensions of Molded Plastics.*

8. ASTM D1299-55 (1979) *Shrinkage of Molded and Laminated Thermosetting Plastics at Elevated Temperatures.*
9. DIN 53464 (1962) *Testing Plastics, Determination of Shrinkage Properties of Moulded Thermosetting Materials.*
10. ASTM D702-68 (1974) *Cast Methyacrylate Plastics Sheets, Rods, Tubes and Shapes.*
11. BS 2782 (1970) Method 106F *Dimensional Stability at 100°C of Flexible PVC sheet.*
12. ASTM D2732-70 (1976) *Unrestrained Linear Thermal Shrinkage of Plastic Film and Sheeting of 0·76 mm (0·030 in) Thickness or Less.*
13. BS 2782 (1970) Method 106H *Dimension Change of Decorative Laminated Sheet* (to be republished as Method 642A in the new edition of BS 2782).
14. BS 2782, Method 643A (1976) *Shrinkage on Heating of Film Intended for Shrink Wrapping Applications.*
15. ASTM D2838-67 (1975) *Shrink Tension and Orientation Release Stress of Plastic Film and Thin Sheeting.*
16. ISO 3521 (1976) *Plastics—Polyester and Epoxy Coating Resins—Determination of Total Volume Shrinkage.*
17. ASTM D2566 (1979) *Standard Test Method for Linear Shrinkage of Cured Thermosetting Casting Resins During Cure.*
18. BS 2782 Method 644A (1979) *Total Volume Shrinkage of Polyester and Expoxy Coating Resins.*
19. Brydson, J. A. (1975) *Plastics Materials* 3rd edn, Butterworth.
20. Shen, M. C. and Eisenberg, A. (1970) *Rubber Chemistry and Technology*, **43,** 95.
21. Eisenberg, A. and Shen, M. C. (1970) *Rubber Chemistry and Technology*, **43,** 156.
22. Zemansky, M. W. (1957) *Heat and Thermodynamics* 4th edn, McGraw-Hill, page 318.
23. Zemansky, M. W. (*op. cit.*), page 332.
24. Gordon, M. (1965) in *Physics of Plastics*, P. D. Ritchie (Ed.), Iliffe.
25. Gibbs, J. H. and DiMarzio, E. A. (1958) *Journal of Chemical Physics*, **28,** 373.
26. DiMarzio, E. A. and Gibbs, J. H. (1958) *Journal of Chemical Physics*, **28,** 807.
27. Staverman, A. J. (1968) *Rubber Chemistry and Technology*, **41,** 544.
28. Rehage, G. and Borchard, W. (1973) in *The Physics of Glassy Polymers*, R. N. Haward (Ed.), Applied Science.
29. Wiley, R. H. (1947) *Journal of Polymer Science*, **2,** 10.
30. Sheldon, R. P. (1962) *Journal of Applied Polymer Science*, **6,** 343.
31. Durrans, T. H. (1929) *Journal of the Oil and Colour Chemists Association*, **12** (108), 173.
32. BS 2782 (1970) Method 103A *Softening Point of Synthetic Resin* (*Ring and Ball Method*) (to be published in the new edition of BS 2782).
33. ISO 1218 (1975) *Plastics—Polyamide—Determination of Melting Point.*
34. BS 2782 Method 123A (1976) *Determination of the Melting Point of Synthetic Resins.*
35. DIN 53736 (1973) *Testing of Plastics; Determination of the Melt Temperature of Semi-Crystalline Plastics.*
36. BS 2782 Method 123B (1976) *Determination of the Melting Point of Polyamides.*
37. ASTM D789 (1978) *Nylon Injection Molding and Extrusion Materials.*
38. ASTM D2133 (1978) *Acetal Resin, Injection Molding and Extrusion Materials.*
39. ASTM D1457 (1978) *TFE-Fluorocarbon Resin Molding and Extrusion Materials.*
40. ASTM D2116 (1979) *FEP-Fluorocarbon Molding and Extrusion Materials.*
41. ISO 3146 (1974) *Determination of Melting Point of Semi-crystalline Polymers—Optical Method.*
42. ASTM D2117-64 (1978) *Melting Point of Semi-crystalline Polymers.*
43. BS 2782 Method 123C (1976) *Melting Point of Crystalline Polymers with Polarised Light.*

44. Crofts, D. and Brown, R. P. (1971) *RAPRA Technical Review No. 61*, August.
45 Tordella, J. P., Webber, A. C. and Cooper, E. R. (1953) *ASTM STP No. 14.*
46. Stephenson, C. E. and Willbourn, A. H. (1959) *ASTM STP No. 247*, page 169.
47. Sherr, A. E. (1965) *SPE Journal*, **21** (1), 67.
48. Horsley, R. A. (1965) *Transactions of the Plastics Institute, London*, **33** (106), 119.
49. Abolins, V. (1975) *Materials Engineering*, **81** (4), 52.
50. ASTM D648-72 (1978) *Test for Deflection Temperature of Plastics under Flexural Load.*
51. BS 2782 Method 115A (1978) *Plastics Yield.*
52. BS 2782 (1970) Method 102B *Deformation in Bend under Load at Elevated Temperature of Laminated Sheet.*
53. BS 2782 (1970) Method 102C *Softening Point of Thermoplastic Moulding Material (Bending Test).*
54. ISO 75 (1974) *Plastics and Ebonite—Determination of the Temperature of Deflection under Load.*
55. BS 2782 Methods 121A and 121B (1976) *Determination of Temperature of Deflection under a Specified Bending Stress of Plastics and Ebonite.*
56. DIN 53461 (1969) *Testing of Plastics—Determination of Temperature of Deflection under Load according to ISO 75.*
57. BS 3534: Part 2 (1964) *Epoxide Resin Systems for Glass Fibre Reinforced Plastics—Pre-impregnating Systems.*
58. BS 2782 Method 121C (1976) *Determination of Temperature of Deflection under a Bending Stress of Rigid Thermosetting Resin Bonded Laminated Sheet.*
59. DIN 53458 (1968) *Testing of Plastics: Determination of Temperature of Deflection under Load (Martens Method).*
60. Gohl, W. (1959) *Kunststoffe*, **49** (5), 228.
61. ASTM D1637-61 (1976) *Tensile Heat Distribution Temperature of Plastics Sheeting.*
62. Bourgault, C. V. (1974) *Plastics Design and Processing*, **14** (7), 17.
63. ASTM D1430 (1978) *Polychlorotrifluoroethylene (PCTFE) Plastics.*
64. Sandiford, D. J. H. and Buckingham, K. A. (1961) *British Plastics*, **34** (11), 594.
65. ISO 306 (1974) *Determination of the Vicat Softening Temperature of Thermoplastics.*
66. BS 2782 Methods 120A to 120E (1976) *Determination of the Vicat Softening Temperature of Thermoplastics.*
67. ASTM D1525-76 *Vicat Softening Temperature of Plastics.*
68. DIN 53460 (1976) *Determination of the Vicat Softening Temperature of Thermoplastics.*
69. Crofts, D. (1972) *RAPRA Research Report No. 200.*
70. Graves, F. L. and Loveless, H. S. (1963) *Materials, Research and Standards*, **3** (1), 33.
71. Ehlers, G. F. L. and Powers, W. H. (1964) *Materials, Research and Standards*, **4** (6), 298.
72. BS 2782 Method 122A (1976) *Deformation under Heat of Flexible Polyvinylchloride Extrusion Compound.*
73. ASTM D621-64 (1976) *Deformation of Plastics Under Load.*
74. Clash, R. F. Jr and Berg, R. M. (1944) *ASTM Symposium on Plastics*, page 54.
75. Welding, G. N. (1955) *Plastics, London*, **20** (214), 158.
76. Williamson, I. (1950) *British Plastics*, **23** (256), 87.
77. Wheeler, A. (1969) *Plastics and Polymers*, **37** (131), 469.
78. Brown, R. P. (1979) *Physical Testing of Rubbers*, Applied Science.
79. ISO R458 (1965) *Determination of Stiffness in Torsion as a Function of Temperature.*
80. BS 2782 Method 150A (1976) *Stiffness in Torsion as a Function of Temperature.*
81. ASTM D1043-72 (1978) *Stiffness Properties of Plastics as a Function of Temperature by means of a Torsion Test.*

82. DIN 53447 (1966) *Determination of Stiffness in Torsion (Clash-Berg Method).*
83. BS 2782 Method 150B (1976) *Cold Flex Temperature of Flexible Polyvinyl Chloride Extrusion Compound.*
84. BS 2782 (1970) Method 104A *Cold Bend Temperature of Flexible Polyvinyl Chloride Extrusion Compound*; Method 104C *Low Temperature Extensibility of Flexible Polyvinyl Chloride Sheet.*
85. Wormald, D. (1958) *British Plastics*, **31** (9), 392.
86. ISO R974 (1969) *Method of Determining the Brittleness Temperature by Impact.*
87. ASTM D746 (1979) *Brittleness Temperature of Plastics and Elastomers by Impact.*
88. Hoff, A. E. W. and Turner, S. (1957) *ASTM Bulletin*, 225, 58.
89. BS 903: Part A25 (1968) *Brittleness Temperature by Impact.*
90. Webber, A. C. (1958) *ASTM Bulletin*, 227, 40.
91. Bestelinki, P. N. and Turner, S. (1958) *ASTM Bulletin*, 231, 68.
92. Birks, A. M. and Rudin, A. (1960) *ASTM Bulletin*, 246, 49.
93. Turner, S. (1958) *British Plastics*, **31** (12), 526.
94. BS 2782 Method 150D (1976) *Cold Crack Temperature of Film and Thin Sheeting.*
95. ASTM D1790-62 (1976) *Brittleness Temperature of Plastic Film by Impact.*
96. Williams, H. O. (1958) *British Plastics*, **31** (3), 107.
97. BS 3424 (1973) *Methods of Test for Coated Fabrics.*
98. ISO 2578 (1974) *Plastics: Determination of Time–Temperature Limits after Exposure to Prolonged Action of Heat.*
99. IEC Publication No. 216 (1966) *Guide for the Preparation of Test Procedures for Evaluating the Thermal Endurance of Electrical Insulating Materials.*
100. DIN 53446 (1962) *Testing Plastics, Determination of Temperature–Time Limits.*
101. ASTM D3045-74 (1979) *Heat Ageing of Plastics without Load.*
102. BS 4618 Section 4.6 (1974) *Recommendations for the Presentation of Plastics Design Data; Thermal Endurance Test.*
103. UL 746B (1975) *Polymeric Materials—Long Term Property Evaluation* (Underwriters Laboratories Inc.).
104. Mair, H. J. and Wiebusch, K. (1977) *Kunststoffe*, **67** (9), 530.
105. Parthum, W. (1975) *Plaste und Kautschuck*, **22** (4), 334.
106. Meacham, S. E. (1978) *Plastics Technology*, **24** (1), 83.
107. ASTM D794-68 (1977) *Determining Permanent Effect of Heat on Plastics.*
108. ASTM D700 (1975) *Phenolic Molding Compounds.*
109. BS 2782 Methods 131C and 131D (1978) *Crushing Strength after Heating (Heat Resistance) of Thermosetting Moulding Materials and Laminates.*
110. BS 2782, Method 108B (1970) *Ageing of Polythene by Air Oven.*
111. ISO 1137 (1975) *Plastics—Determination of Behaviour in a Ventilated Tubular Oven.*
112. ASTM D1870–68 (1978) *Elevated Temperature Ageing Using a Tubular Oven.*
113. ISO 176 (1976) *Determination of Loss of Plasticisers—Activated Carbon Method.*
114. BS 2782 Method 465A (1979) *Determination of Loss of Plasticisers— Activated Carbon Method.*
115. ASTM D1203–67 (1974) *Loss of Plasticisers from Plastics (Activated Carbon Methods).*
116. ISO 3671 (1976) *Determination of Volatile Matter of Aminoplastics Moulding Materials.*
117. BS 2782 (1970) Method 107G *Loss in Mass on Heating of Cellular Acetate Moulding Material* (this test may not appear in the latest edition of BS 2782).
118. ASTM (annual) *ASTM Standards*, Part 36 *Plastics, Materials, Film, Reinforced and Cellular Plastics, Fibre Compounds.*
119. ISO 177 (1976) *Plastics—Determination of Migration of Plasticisers.*

120. ISO R182 (1970) *Determination of the Thermal Stability of Polyvinyl Chloride and Related Co-polymers and their Compounds by Splitting off of Hydrogen Chloride.*
121. BS 2782 Method 130A (1976) *Thermal Stability of Polyvinyl Chloride Compound (Congo Red Method).*
122. DIN 53381 (1971) Test 1 *Determination of the Thermal Stability of Polyvinyl Chloride and Related Copolymers and their Compounds: Congo Red Method.*
123. ISO 305 (1976) *Determination of Thermal Stability of Polyvinyl Chloride, Related Chlorine Containing Polymers, and their Compounds—Discolouration Method.*
124. ASTM D2115–67 (1974) *Oven Heat Stability of Poly(Vinyl Chloride) Compositions.*
125. DIN 53381–Test 2 (1975) *Determination of the Thermal Stability of Polymers and Copolymers of Vinyl Chloride and their Compounds. Discolouration Method.*
126. ASTM D793–49 (1976) *Short Term Stability at Elevated Temperatures of Plastics containing Chlorine.*
127. BS 2782 Method 130B (1976) *Thermal Stability of Polyvinyl Chloride Compounds (pH Method).*
128. DIN 53381–Test 3 (1971) *Determination of the Thermal Stability of Polyvinyl Chloride and Related Copolymers and their Compounds: pH Method.*
129. ASTM D701–49 (1976) *Cellular Nitrate (Pyroxylin) Plastic Sheets, Rods and Tubes.*
130. ASTM D1547-75 (1979) *Extruded Acrylic Plastics Sheet.*
131. ASTM D2445–69 (1974) *Thermal Oxidative Stability of Propylene Plastics.*
132. ISO 1599 (1975) *Cellulose Acetate—Determination of Viscosity Loss on Moulding.*
133. ISO 1157 (1975) *Cellulose Acetate in Dilute Solutions—Determination of Viscosity Number and Viscosity Ratio.*
134. Ives, G. C. and Mead, J. A. (1959) *SCI Monograph no. 5*, Society of Chemical Industry, page 80.
135. ISO 3205 (1976) *Preferred Test Temperature.*
136. ASTM D618–61 (1977) *Conditioning Plastics and Electrical Insulating Materials for Testing.*
137. BS 3534: Part 2 (1964) *Pre-impregnating Systems for Epoxy Resin Glass Reinforced Plastics.*
138. BS 2782 Method 335B (1978) *Crossbreaking Strength of Rigid Materials.*
139. BS 4045 (1966) *Epoxy Resin Pre-impregnated Glass Fibre Fabrics.*
140. ASTM D759–66 (1976) *Determining the Physical Properties of Plastics at Subnormal and Supernormal Temperatures.*
141. ASTM D2733–70 (1976) *Methods of Test for Interlaminar Shear Strength of Structural Reinforced Plastics at Elevated Temperatures.*
142. Mathes, K. N. (1964) *SPE Journal*, **20** (7), 634.
143. Wessels, E. T. and Olleman, R. D. (1953) *ASTM Bulletin*, 187, 56.
144. Wessel, E. T. (1956) *ASTM Bulletin*, 211, 40.
145. Ives, G. C. and Mead, J. A. (1961) *Materials, Research and Standards*, **1** (3), 194.
146. Lieb, J. H. and Mowers, R. E. (1966) *Testing of Polymers* vol. 2, J. V. Schmitz (Ed.), Interscience, Chapter 3.
147. Brown, R. P. (Ed.) (1979) *RAPRA Guide to Rubber and Plastics Test Equipment.*
148. ISO 3383 (1976) *General Directions for Achieving Elevated or Sub-normal Temperatures.*
149. ASTM D832 (1979) *Conditioning of Elastomeric Materials for Low Temperature Testing.*

Chapter Seventeen

Environmental Resistance

17.1 INTRODUCTION

The subjects considered in this chapter have one factor in common—they are test methods designed to evaluate the resistance of plastics materials to change as a result of exposure to some environment other than high or low temperature. Many are accelerated tests and invariably these have at least one shortcoming in common which is the very result of their aim to provide a forecast of durability within a finite period of time: unless they are 'accelerated' they offer nothing over a field trial, but to achieve this acceleration the stringency of test conditions may be so enhanced as to scale an energy barrier and create an effect which would never occur in normal usage. Even if this does not happen, the equation of the results of accelerated ageing tests into likely service life is difficult if not impossible because the accelerated conditions are usually so idealised that they may not truly simulate the sequence of events in the field; this, and the 'energy barrier' consideration, may lead to orders of merit from the accelerated ageing tests which are not fulfilled in practice.

Temperature plays a part in all environmental tests. The effects of elevated temperature alone have already been dealt with in Chapter 16, except for that of fire which is considered to be more of an environmental effect than simply one of exposure of plastics to very high temperatures.

It cannot be claimed that all the degradative influences which plastics materials are likely to encounter are covered in this chapter, but it is the intention to consider the more important ones.

17.2 MOIST HEAT AND STEAM TESTS

There would appear to be no great concern over the long-term effects of moist heat on plastics, although conditioning of certain materials at specified relative humidity is very important (see Chapter 4) if consistent and meaningful results are to be obtained from certain test procedures.

There is, however, a Draft International Standard, ISO/DIS 4611,[1] which provides for damp heat tests at 40°C and 93 per cent r.h., both static and cycling, with or without water or salt mist spray.

Perhaps the most comprehensive standardised test method which is relevant to this subject is ASTM D756 (1975),[2] *Standard Practice for Weight and Shape*

Changes of Plastics under Accelerated Service Conditions. In the words of the defined scope of this standard: 'This practice covers the determination of the weight and shape changes occurring in plastics under various conditions of use, not where exposure to direct sunlight, weathering, corrosive atmospheres or heat alone is involved, but where changes in atmospheric temperature and humidity are encountered. This embraces the interior of buildings, and the interior of transport facilities such as motor vehicles, airplane cargo spaces or wing interiors, holds of ships and railroad cars'.

Notwithstanding this explanation, Procedure B of ASTM D756 involves dry heat ageing only. Others specified are:

Procedure A	24 h at 60° ± 1°C at 88–89 per cent r.h. followed, within 2 h, by 24 h at 60° ± 1°C in an oven.
Procedure C	24 h at 70° ± 1°C at 70–75 per cent r.h. followed, within 2 h, by 24 h at 70° ± 1°C in an oven.
Procedure D	24 h at 80° ± 1°C over distilled water in an oven followed, within 2 h, by 24 h at 80° ± 1°C in an oven.
Procedure E	24 h at 80° ± 1°C at 70–75 per cent r.h. followed, within 30 min, by 24 h at −40° ± 2°C or −57° ± 2°C, then, within 2 h, by 24 h at 80° ± 1°C in an oven and, finally, within 30 min by 24 h again at −40° ± 2°C or −57° ± 2°C.
Procedure F	24 h at 38° ± 1°C at 100 per cent r.h. followed, within 2 h, by 24 h at 60°C in an oven.
Procedure G	As Procedure F except that the temperature is 49° ± 1°C in both humid and dry heat periods.

Desiccator conditioning is used between each part of the various cycles.

Specimens are weighed, their dimensions measured and visual changes noted after each of the distinct parts of the various cycles.

Also of interest, because of the use of plastics in electrical insulation applications, is the series BS 2011 dealing with components for telecommunication and allied electronic equipment, particularly Part 2D (1977),[3] *Damp Heat, Cyclic*, which specifies a 24 h cycle of the following:

1. 25° ± 10°C ('laboratory temperature') to 55° ± 2°C in $1\frac{1}{2}$–$2\frac{1}{2}$ h during which period the relative humidity must be between 80 and 100 per cent and condensation must occur on the components.
2. 55° ± 2°C for 16 h, during which a periodical excursion of 2°–3°C variation in temperature must occur at least four times per hour. The relative humidity during this part of the cycle is 95–100 per cent and condensation will occur.
3. Cool to 'room temperature', the relative humidity remaining at 80–100 per cent to complete the 24 h cycle. During this period droplets of water must not appear on the components.

Similarly interesting, from a related industry, is Part F2 (1973) *Resistance to Humidity under Condensation Conditions* of BS 3900,[4] *Methods of Test for Paints* (basically derived from Ministry of Defence Specification DEF 1053). Here the cycle involves the temperature cycling continuously from 42°C to 48°C and back to 42°C in 60 ± 5 min, the heating and cooling periods being approximately equal. Heating is effected through an open water bath so that 100 per cent r.h. is achieved and copious condensation occurs on the test panels.

17.3 EFFECTS OF LIQUIDS AND CHEMICALS

There is a number of published test methods for determining the matter extractable from a plastics material by a specific chemical reagent (for example, ISO 118 (1976) *Polystyrene—Determination of Methanol Soluble Matter*). These are considered to be chemical tests and will not be discussed as being outside the scope of this book. By far the greatest number of published test methods involve the use of distilled water as the test medium and even for chemicals these are mostly offered as a solution in water as the test medium.

17.3.1 Water Absorption and Water-soluble Matter

The resistance of plastics materials to water varies significantly according to their chemical nature and, if they are filled or reinforced, the nature of additives or reinforcements present. The sensitivity to moisture is quite different for the relatively hydrophilic secondary cellulose acetate and the decidedly hydrophobic polystyrene. In practical terms of influence on mechanical properties, reference has already been made (Chapter 4) to the 20-fold increase in impact strength on nylon 6 between dryness and moisture saturation.

A particularly interesting effect is described by Uijlenburg,[5] who studied the water absorption of unplasticised PVC at various temperatures up to 140°F and then measured certain mechanical properties, particularly related to the material in pipe form. It was found that at 140°F effects were produced by water absorption which were not encountered at more normal temperatures and therefore a proposal for a speedy measurement of the long-term bursting pressure of unplasticised PVC pipe, by testing at 140°F, was likely to yield very misleading results as to performance at more normal ambient temperatures.

Braden[6] has analysed the processes involved in the absorption of water by plastics materials, which he concluded depend on only two parameters: the diffusion coefficient and the equilibrium uptake. The former governs the kinetics of water absorption and is highly temperature-dependent. Equilibrium uptake is virtually unaffected by temperature. Blank[7] examined five equations proposed for relating water absorption to time, concluded that none was universally applicable and deduced an equation for obtaining equilibrium absorption (Q) from values of absorption (q_1, q_2 and q_3) measured at equally spaced short intervals:

$$Q = \left[\frac{q_2^4 - q_1^2 q_3^2}{2q_2^2 - (q_1^2 + q_3^2)}\right]^{1/2} \qquad (17.1)$$

Further experimental study of this equation is described by Bland.[8]

The effect of fillers is well illustrated by the differing requirements for maximum water absorption specified in BS 771 (1980),[9] *Phenolic Moulding Materials*.

The standard water absorption test is designed as no more than a quality control test, for use purely as a guide to water sensitivity, by measuring water take-up under a specified but arbitrary set of experimental conditions with no pretence to absolute significance or even to an assessment of the effect of mositure attack on, say, electrical or mechanical properties (for this see, for example, Hauck).[10] Even so, great care is needed to obtain reproducible and comparable test data from such water absorption tests. Control of temperature is obviously of paramount importance in a diffusion or rate process. The presence of grease, even finger marks, on the surface may influence results. The physical state of the actual surface must be specified, for instance whether 'as moulded' (with a resin skin) or as left after machining. For hydrophobic materials in particular, the precise mode of removal from the water immersion bath and of drying, and the time taken to weigh, may all significantly influence the test results. The head of water could have an effect and, since all plastics materials are relatively resistant to water, water absorption tests generally must take heed of surface-to-volume ratio of test pieces.

In practically all standard tests, since they are essentially short term, equilibrium is not obtained throughout the test piece thickness (this could take months, or even years) and therefore the effect measured is basically a surface one. Thus, the total surface area of the test piece is of paramount importance in a test measuring water absorption by simple weight increase, as is the ratio of surface to volume in tests measuring water absorption on a wt/wt basis. Finally, a constituent may well be present which is significantly soluble in water, i.e. it will be leached out in the test, or the material may even hydrolyse. A quantity of the test piece lost by solution in the water environment will obviously affect the results of a test which is designed to assess attack of water by weighing water uptake; the loss of soluble matter may cause a much lower nett gain in weight, may cancel out the gain or even exceed it. If water solubility is anticipated it should be measured in the test.

The work of Braden[6] demonstrates that two materials of the same equilibrium uptake of water could, if their diffusion coefficients were different, give quite dissimilar water absorption test results after some arbitrarily selected short test period of, say, 24 h.

Standard Test Methods

ISO R62 (1958) with Amendment 1 (1965)[11] and ISO R117 (1959)[12] use similar test pieces, 50 mm diameter discs, 3 mm thick; each has two methods, A and B. ISO R62 specifies immersion in water at 50°C for 24 hours while ISO R117 specifies boiling water for 30 min. In each case Method A is for finding the water absorption and Method B for water absorption where the material under test has significant water-soluble matter. Method B requires the heating of the test pieces after immersion for 24 h at 50°C. The results are expressed by:

$$\text{water absorption (Method A)} = W_2 - W_1 \text{ (mg)} \qquad (17.2a)$$

$$\text{water absorption (Method B)} = W_2 - W_3 \text{ (mg)} \qquad (17.2b)$$

where W_1 is the original mass of the test piece
W_2 is the mass after immersion in water
W_3 is the mass after post-immersion drying.

In addition to the standard size test pieces these standards allow the use of sheet and tube test pieces.

BS 2782 (1970),[13] *Methods of Testing Plastics*, contains no less than ten methods for determining water absorption:

Method 502A	Water absorption of phenolic moulding material
Method 502B	Cold water absorption of aminoplastic moulding material
Method 502C	Water absorption and water soluble matter of polyvinyl chloride extrusion compound
Method 502D	Water absorption of laminated sheet
Method 502E	Water absorption of laminated tube and rod
Method 502F	Water absorption. Procedure A of ISO R62
Method 502G	Water absorption. Procedure B of ISO R62
Method 503A	Boiling water absorption of aminoplastic moulding material
Method 503B	Boiling water absorption. Procedure A of ISO R117
Method 503C	Boiling water absorption. Procedure B of ISO R117

Table 17.1 Standard Water Absorption Tests

Method	Immersion Time (h)	Immersion Temperature (°C)	Post-immersion heating Time (h)	Post-immersion heating Temperature (°C)
ISO R62	A 24 ± 1	23 ± 0·5	–	–
(≡ BS 2782 Methods 502F and G)	B 24 ± 1	23 ± 0·5	24 ± 1	50 ± 2
ISO R117	A 24 ± 1	100	–	–
(≡ BS 2782 Methods 503B and C)	B 24 ± 1	100	24 ± 1	50 ± 2
BS 2782				
502 A	168 ± 2	25 ± 0·5	–	–
502 B	24 ± 1	23 ± 0·5	24 ± 1	50 ± 3
502 C	48 ± 1	50 ± 1	Dry to constant mass over Ca Cl_2	
502D, 502E	24 ± 1	23 ± 0·5	–	–
503 A	24 ± 1	100	–	–

ASTM D570

Procedure

1. 24 −0, +½ h in distilled water at 23° ± 1°C on edge.
2. As 1, but for 120 ± 4 min.
3. As 2 followed by 1, total immersion period being 24 h.
4. After 1, reimmerse, weigh at end of first week and thereafter at two-weekly intervals until equilibrium is substantially obtained (as defined).
5. 120 ± 4 min in boiling distilled water, on edge; cool in distilled water at room temperature for 15 ± 1 min.
6. As 5, but for 30 ± 1 min.
7. As 5, but in distilled water at 50° ± 1°C for 48 ± 2 h.

Test temperatures and period of immersion are summarised in Table 17.1, but it is necessary to refer to the individual tests for details of the methods.

It should be noted that all of the above test methods are currently under review, but publication of the revised methods for BS 2782 will await the outcome of the ISO revisions. ISO R62 and ISO R117 are at an advanced stage of review, in a combined version, as ISO DIS 62,[14] in which results may be expressed as mass change (mg), as mass change per unit area (mg/cm^2) or as a percentage change of mass based on the original mass of the test piece.

Seven procedures are included in ASTM D570 (1977),[15] *Water Absorption of Plastics*, which differ in their severity of duration and temperature; different test pieces are used according to the physical form of the sample under test. Again, the basic immersion periods and temperatures are given in Table 17.1. This test method prescribes the calculation of the results as a percentage change in mass.

BS 2782 Method 502C and ASTM D570 differ from the other test methods in that they permit the determination of water-soluble matter as well as water absorption as follows: if W_1 = original mass of test piece, W_2 = mass of test piece after immersion and W_3 = mass of test piece after post-immersion conditioning,

for BS 2782 Method 502C,

$$\text{water absorption} = W_2 - W_3 \text{ (mg)} \tag{17.2}$$

$$\text{water-soluble matter} = W_1 - W_3 \text{ (mg)} \tag{17.3}$$

for ASTM D570,

$$\text{water absorption} = \frac{W_2 - W_3}{W_1} \times 100 \text{ (per cent)} \tag{17.4}$$

$$\text{water-soluble matter} = \frac{W_1 - W_3}{W_1} \times 100 \text{ (per cent)} \tag{17.5}$$

DIN 53471 (1976)[16] and DIN 53475 (1962)[17] follow ISO R117 and R62 respectively. DIN 53473 (1979)[18] employs similar test pieces but the absorption is from an atmosphere of 92–93 per cent r.h. at 20° ± 2°C for 1, 2, 4, 7 or a multiple of seven days. Mention must also be made of BS 2782 (1970) Method 504A,[19] 'Water vapour absorption of cellulose acetate moulding material'. This uses rectangular test pieces of specified thickness (1·52 ± 0·5 mm), exposed for 48 hours at 23° ± 1°C and 50 ± 5 per cent r.h. The text of this test has been revised and will be issued as Method 431A of BS 2782.

17.3.2 Effect of Chemicals

Water absorption and water solubility are in reality but one aspect of the general subject of *chemical resistance* and probably the only excuse for treating the effect of water as a separate entity—and before all others—is the profusion of the medium. By their very nature the basic polymers as such would not be expected to be subject to chemical attack except by specific compounds or classes of compounds. The most aggressive media will be the organic solvents since the polymers themseves are mainly organic and, subject to the influence of molecular weight and crystallinity, non-polar polymers are generally susceptible to the attack of non-polar solvents and polar polymers to polar solvents.

Organic plastics are generally resistant to the attack of aqueous media, for the reasons briefly outlined in §17.3.1 above, but if the polymer chain is prone to oxidative attack, concentrated nitric acid for instance may have a pronounced degradative effect. Much may be gained from the *ad hoc* approach of immersion of small strip specimens in the medium, preferably at various temperatures, and observation of the effects—discoloration, crazing, cracking, complete degradation, etc. Partial immersion has much to recommend it, as the attack at an air–liquid interface, with oxygen accessibility, may be much more severe. 'Monitoring' the attack, by measurement of the change of some important physical property such as hardness and transparency, is a valuable refinement: Jessup[20] used flexural strength, hardness and electrical properties when examining the effects of dilute sulphuric acid and certain organic solvents on a range of thermosetting plastics.

Standard Test Methods

ISO R175 (1961)[21] is an empirical test method only suitable for the comparison of different plastics after immersion in chemical substances, in liquid form or in solution. ISO R462 (1965)[22] for the determination of change of mechanical properties specifies the same reagents as ISO R175 at temperatures of 20° ± 2°C or 23° ± 2°C or 27° ± 2°C or 70° ± 2°C, for a period of 7 days; results are judged by the change in a mechanical property after immersion.

Both of these standards are under review and will be issued in one standard which already exists as ISO DIS 175.[23] This proposes the use of 38 laboratory reagent quality chemicals and 8 industrial chemicals. This standard is more precise than the two which it will replace (R175 and R462) and states preferred test temperatures (23°C and 70°C) and test durations (24 h, 1 week, 16 weeks) but allows a wide range of other temperatures up to 150°C and test durations up to 5 years. Details are given for the determination of change in dimensions, mass and appearance but the choice of other properties for determination of changes in a material after immersion in a chemical is left to the discretion of the user. This standard would undoubtedly be more useful if more precise guidance were given about the physical properties to use, as in BS 4618, Section 4.1 (1972),[24] *Recommendation for the Presentation of Data on the Chemical Resistance of Plastics to Liquids*. This lays down the test temperature, recommends the exposure period and offers a list of reagents based on those of ISO R175. The criteria for assessing the effect of the chemical liquid are change of dimensions, change in appearance and change of mechanical properties; a list of recommended tests is given.

BS 2782 (1970) Method 505A[25] (currently under review) gives a method for the determination of resistance to concentrated sulphuric acid of rigid PVC compounds. This method has the merit of being quantitative because it is based on the comparison of shear strength (see Chapter 8) before and after immersion in concentrated sulphuric acid at 95°–100°C for 24 hours. This test was originally devised to ensure that unplasticised PVC compounds were truly unplasticised, as inclusion of quite small quantities of ester plasticisers in rigid PVC results in a marked loss of chemical resistance.

ASTM D543-67 (1978)[26] includes some useful early references to chemical resistance of plastics and lays down general procedures for the examination of 50 specified reagents and solvents by following weight and dimensional changes and alteration in mechanical properties.

ASTM D2299-68 (1977)[27] describes how to carry out tests for determining relative stain-resistance of plastic, not so much by chemicals as by the hazards of everyday life such as coffee, tea, lipstick, shoe polish and so on. The specific hazard of sulphide staining is covered by ASTM D1712 (1965).

DIN 53476 (1979)[28] follows ISO/DIS 175 in laying down procedures for measuring the resistance of plastics to some 50 liquids.

In marine atmospheres, salt spray may be present in the atmosphere to a significant degree. It is not reckoned a serious hazard to plastics generally and there is no British Standard or ASTM test designed to assess degradation of plastics by salt spray. BS 4618, Section 4.4 (1973), *Presentation of Plastics Design Data—The Effect of Marine Exposure*, categorically states: 'Plastics do not suffer chemical corrosion in diluted (3 per cent) salt solution, the effects of which are certainly no greater than those of distilled water. There is no evidence of any effect on plastics of evaporated salt.' Despite this, provision for a salt spray test on plastics has been included in the Draft International Standard ISO/DIS 4611[1] (see §17.2).

17.3.3 Environmental Stress Cracking

So far, consideration of the effects of liquids on plastics has been confined to absorption, extraction and chemical attack (in the sense that they cause a change in physical properties), when 'ideal', unstressed test pieces are used. Unfortunately, in real life applications of plastics it is not uncommon for the material to be used under conditions of stress induced as a result either of the end-use or of the moulding or fabrication technique. Plastics may be resistant to a chemical in the unstressed state but fail by cracking when exposed under stress to the same chemical. Before this phenomenon was recognised considerable problems were caused in the industry, particularly with injection moulded polyethylene housewares which cracked in contact with household detergents, and with the early vacuum-formed toughened polystyrene refrigerator liners which cracked when they came into contact with vegetable fat and fruit juices. Even today the exact mechanism of such environmental stress cracking, as it is called, is not fully understood but we now know that most, if not all, thermoplastics can show environmental stress cracking given the right conditions of stress and aggressive medium. The aggressive medium need not be a liquid; semi-liquids (e.g. greases) and gases are now also known to cause environmental stress cracking.

Clearly, the determination of resistance to environmental stress cracking is of great importance both for the designer and for quality control purposes. The polymer chemist has learned how to reduce the tendency to stress crack by various means and therefore the polymer manufacturer needs to know how to check his production to ensure that the product he is making is maintaining the designed level of environmental stress crack resistance.

A review of environmental stress cracking test methods is given by Brown.[29] Originally, the available test methods consisted of exposing a strained test piece to a liquid environment and measuring the time to failure as the environmental stress crack resistance. The failure might be defined as appearance of crazing or of physical cracks or, in some cases, complete rupture. Test methods have now been extended to include exposure at either constant stress or constant strain.

Simple Bent Strip Tests

ASTM D1693 (1975),[30] *Environmental Stress Cracking of Ethylene Plastics*, was developed from the well known Bell Telephone test and is generally accredited to the work of De Coste *et al.*[31] The test piece is a strip, nominally 38 mm long and 13 mm wide, in which a controlled cut (imperfection) is made with a blade held in a jig. The test piece is bent into a 'U' shape, placed in a simple retaining jig, exposed to the aggressive environment and examined at intervals for failure. For the purpose of the test, stress crack failure is definded as 'any crack visible to the observer with normal eyesight'; extension of the controlled imperfection is not to be construed as failure. The preparation of the test pieces is carefully defined and the standard aggressive liquid is Igepal CO–630 which is nonylphenoxy poly(ethylene-oxy) ethanol. Ten replicate test pieces are used and in the absence of other instructions the duration of the test is 48 hours.

The standard clearly states that the method is intended for quality control purposes and the results should not be used for direct application to engineering problems. Roe and Gienewsky[32] have made a detailed and careful study of the method to identify sources of error and have suggested modification to improve variability. In particular they mention the need to control closely the moulding conditions, the pretest conditioning and the length of time the test piece stands between moulding and the test. These authors also confirmed that variation in test results occurs when the test piece is bent without backing or guide so that its precise deformed shape and the strains realised are not closely defined or controlled.

Other variations of the simple bent strip have been proposed, all being characterised by the absence of any restraining member to define and control the degree of bending. Steinle and Pflasterer[33] used a channel to hold strips of material in an arc to give a much lower strain than the ASTM method and hence a more uniform and better defined strain. They also introduced a procedure of measuring the change of a physical property (elongation at break) as a function of exposure time to quantify the measure of degradation as well as observing cracking and fracture. Leghissa and Salvatore[34] used a simple bent strip and devised an optical system for measuring the degree of cracking or crazing, but this was only suitable for transparent material.

Defined Curvature Methods

The most obvious improvement to the ASTM test is to define precisely the imposed curvature by clamping the test piece to a rigid curved jig or, perhaps less satisfactory, by subjecting it in a fixed deflection jig to three-point loading. A logical extension of this is to bend the test piece over a jig having various radii over its length; both parabolic and elliptical jigs have been used in this way. Dempsey,[35] Ruhnke and Britiz;[36] and Bergen,[37] among others, have proposed tests and while they undoubtedly offer some improvement over ASTM D1693 they have not yet found favour with the standardisation authorities.

Insert Methods
Instead of by bending the test piece, the strain can be introduced by impressing an oversize ball or pin into a hole drilled in the test piece. This method has been described by Pohrt[38–39] and others on several occasions. Carefully drilled and reamed holes are prepared, balls (or pins) of various known oversizes are inserted and the assemblies are exposed to the aggressive medium. Either the oversize equivalent to the critical strain can be noted from visual observation or changes in tensile or flexural properties can be measured. When measuring physical properties there are no problems because the test pieces are not bent after removal from the exposure jigs.

The method has been standardised as DIN 53449, (1970)[40] and an International Standard version has reached the draft standard stage (ISO/DIS 4600[41]) which will almost certainly be published as a method of BS 2782. This method will clearly differentiate between visual detection of cracking and measurement of mechanical properties, because the latter procedure is much more time-consuming and hence less convenient for routine use. In fact, a major advantage of the ball insert test with visual observation is its usefulness for testing without first having to cut the specimen into test pieces.

Gnauck[42] evaluated the impressed ball method for use with PMMA and concluded that it was not satisfactory for materials at that level of modulus because the strains could not be maintained with enough precision. Both steel and rubber balls were embedded in polystyrene samples by Tsuey *et al.*[43] to examine the crazing around relatively soft and hard spherical inclusions. The test pieces were then strained in tension but the procedure was not intended to be used as a test method in the general sense.

Constant Stress Methods
The use of a dumbbell stressed in direct tension by a constant load was proposed by Lander[44] as an alternative to the Bell Telephone test. He used an ASTM Type A tensile impact dumbbell, cut from 1 mm sheet, loaded through a lever arm; the time was noted to complete failure of the test piece. Herman and Biesenberger[45] used effectively the same method. The essential difference between these tests and the bent strip test is that the latter operates at a constant strain as opposed to a constant stress. O'Connor and Turner[46] have tried methods using test pieces subjected to constant tensile strain but these were abandoned on the grounds that at constant strain the development of a crack changes the strain distribution near it and the stress relaxation continuously reduces the severity of the test.

There are many references in the literature to environmental stress cracking tests based on test pieces subjected to both constant tensile stress and constant tensile strain. Some workers have used holes drilled in the test pieces to provide a preferential site for failure.[47,48] Haslett and Cohen[49] used a constant load applied to a three-point loading configuration (rather than in tension) and measured the time to complete failure of the test piece. They also considered the degradation which had occurred prior to complete failure by subjecting test pieces to the static flexural loading, followed by measurement of cyclic fatigue life using a universal tensile machine. Apart from showing that fatigue life has been significantly decreased before visible signs of failure, they found evidence of a 'characteristic environmental stress' i.e. a stress, not necessarily the maximum stress, which induced maximum damage.

Judging by the interest being shown in the constant load direct tension methods by the standardisation bodies, this method is finding increased favour in the testing laboratory. An international method is at the draft proposal stage and uses the usual tensile properties dumbbell test piece (see Chapter 8) but with all dimensions halved, cut from 2 mm thick sheet. A constant load is applied, the test piece immersed in the aggressive medium and the method requires either the time to rupture for a given stress or the stress loading to rupture after 100 hours to be reported. In the latter case the result is obtained by interpolation of a graph of time to rupture against stress.

BS 4618, Section 1.3.3 (1976),[50] *Recommendations for the Presentation of Plastic Design Data—Environmental Stress Cracking*, gives a constant tensile load method using two well tried test piece geometries which are not standardised elsewhere. It also recommends the use of both notched and unnotched test pieces.

ASTM D2552-69 (1974) *Environmental Stress Rupture of Type III Polyethylenes under Constant Tensile Load*, is restricted to this particular type of material (as specified in ASTM D1248 (1978)[52]). It uses a dumbbell similar to that of Lander but with wider tab ends, presumably to improve gripping, and cut from 1 mm sheet.

It is clear that there are advantages and disadvantages in the types of test which have been described. Tests at constant strain (e.g. bent strip tests) are useful for quality control purposes and require simple apparatus, whereas tensile constant stress tests may give a better insight into the mechanism of environmental stress cracking by separating the effect of the environment on crack initiation and crack propagation. Nevertheless, it cannot be emphasised too strongly that if consistent results are to be obtained from any environmental stress cracking test great attention must be paid to all the experimental conditions. Figure 17.1 shows a typical apparatus for determining environmental stress cracking under constant tensile stress.

17.4 EFFECT OF GASES

In comparison with the effects of liquids on plastics, very little testing is carried out on the effect of gases (permeability is dealt with in Chapter 18), with the notable exception of air or oxygen which has been covered in §16.8 on heat ageing.

The effect of ozone on elastomers is well known and is important in rubber technology since many elastomers are adversely affected by exposure to ozone to a greater or lesser degree. Saturated polymers, which form the bulk of plastics materials, are quite resistant to ozone and there are no standard tests for plastics which specifically examine this property separately from its contribution to the general degradative influence of weathering (§17.5). Weiss[53] has reviewed ozone tests on plastics and Brown [54] has given a comprehensive account of ozone exposure apparatus for elastomers.

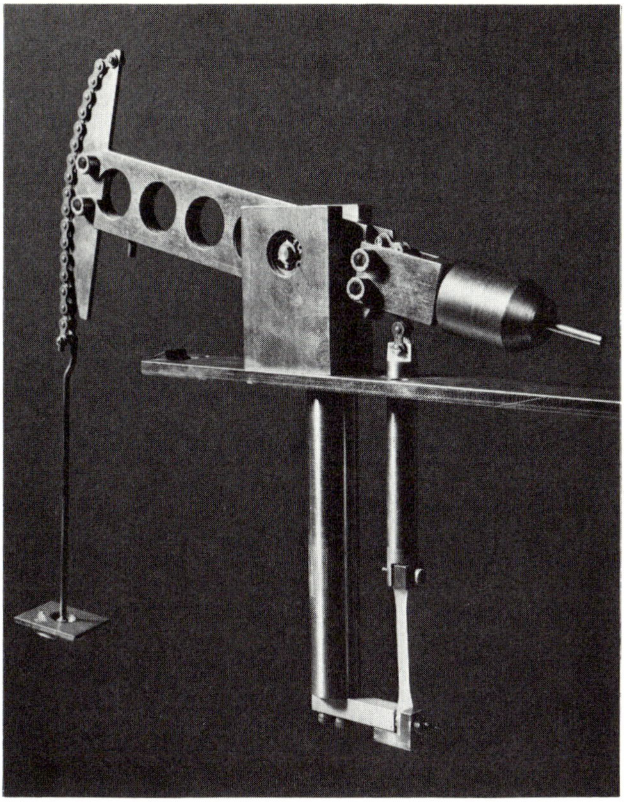

Figure 17.1 Essential parts of constant stress environmental stress cracking apparatus

17.5 WEATHERING

17.5.1 General Considerations

Weathering effects are a combination, *inter alia*, of those due to heat (cf. Chapter 16) including infrared radiation, to damp (see §§17.2, 17.3), to ultraviolet radiation, to oxidation, to ozone formed by the action of u.v. on atmospheric oxygen and perhaps to radiation from active fall-out or emitting sources. The last mentioned is conveniently separated for consideration (§17.8) and those already discussed (i.e. heat and damp) may well be encountered as singular or combined degradative influences. However, u.v., oxidation and ozonolysis are unlikely to be hazards except as constituents of general weathering attack and are therefore dealt with together. The reason for this non-classical approach is that in general weathering the separate degrading influences affect the changes or rate of change brought about by each other. To this synergistic phenomenon must be added the varying intensity of the degradative influences, which variation may be more severe in action than a continuous high level of the influence. To take a simple example, repeated drying out caused by occasional saturation with water in a high-temperature environment may be far more damaging than continuous immersion at the same temperature.

Estevez[55] analyses the weathering of plastics into its separate components and lists the major objections to accelerated tests:

(i) the degree of acceleration in relation to 'natural' ageing is unknown;
(ii) the order of merit of materials produced by an accelerated test may be quite different from that obtained in use;
(iii) phenomena may occur in accelerated tests which never occur naturally—and *vice versa.* (Point (ii) could easily result at least in part from point (iii).)

With regard to point (i), the very variability of the weather at any one geographical location, quite apart from the vast differences in average climate between various parts of the world, obviously militates against the chances of any direct correlation between accelerated weathering test results and the results of natural weathering.

The problem of predicting weathering of plastics, the methods and many test results are to be found in a set of symposium papers edited by Kamal.[56]

17.5.2 Natural Weathering and Ageing

If one can afford the time to wait, evaluation of resistance to weathering, and of associated effects such as fading, is most reliably undertaken using the natural elements. Even then the results obtained must be carefully related to the geographical location, time of the year and climatic conditions prevailing at the time. The degree of resistance to natural weathering can then be judged by the change in physical properties. This technique is standardised by ISO 877 (1976)[57] and by BS 2782 (1977) Method 540A,[58] which are both entitled *Plastics—Determination of Resistance to Change upon Exposure under Glass to Daylight.*

Test pieces, 20 mm wide and of convenient length, are used for determining colour change but other sizes of test piece are allowed for determining changes in other physical properties. The exposure case is closely specified and is set at an angle of 45° facing the equator (i.e. in the northern hemisphere it faces south) and is covered with 3 mm thick glass having specified transmittance. The exposure case is so placed that there is no obstruction in an easterly, southerly or westerly direction subtending a vertical angle greater than 20° nor any northerly obstruction subtending an angle greater than 70°. The exposure case consists essentially of an open-bottomed shallow square box, mounted at 45° to the horizontal and supported on legs so that its lower front edge is 760 mm from the ground. A framed wire screen fits snugly into the 'case' box and on the upper surface of the wire mesh rests a rack which also fits snugly in the box. The rack comprises a frame with spaced horizontal battens across it, there being a 50 mm gap between each batten; on the battens 50 mm wide wooden flaps are hinged so that the lower 50 mm of the depth of each batten may be covered. Over this rack rests a lid, kept at a distance of 75 mm from the upper surface of the rack by a distance piece mounted in the centre of the rack. This lid contains the glass cover mentioned above.

Each test piece is mounted so that it is partly covered by the batten flaps and there is a space of not less than 6 mm between each test piece or between the test piece and the edge of the frame.

The standard allows the use of instrumental means for measuring the radiation dosage under the glass lid but recommends the use of the blue dyed wool standards to ISO 105 (1978)[59] Part 11, examined for colour change by the grey scale of ISO 105 Part 2. The blue dyed wool standards range from No. 1 to No. 7, No. 1 being of very low lightfastness and No. 7 of very high lightfastness, each standard being approximately twice as lightfast as the one below it. The grey scale ranges from No. 1 (greatest degree of contrast) to No. 5 (zero contrast), having two patterns of identical colour. In use, a set of the blue standards is exposed under the glass along with the test pieces. The blue standards are examined at intervals for change of colour between the exposed and unexposed portions, the change of colour to be defined as a difference in contrast equal to that shown by No. 4 on the grey scale. Each stage of the exposure is recorded as each wool standard fades to this degree of contrast. In the event of very long exposure which exceeds the time to fade No. 7 standard, then further No. 7 standards are successively exposed as required. Test pieces are examined at each stage of exposure as required and the results expressed as change in properties or colour at each stage. In British Standard practice it is common to examine the test pieces for colour change to grey scale No. 4 and then to report the blue standard which has faded to the same degree as being the colourfastness of the material under test. A similar procedure is followed by DIN 53388.[60]

In recent years there has been an increasing interest in determining the effect of total outdoor exposure ('natural weathering') on the properties of plastics and ISO, BS and ASTM have published standards covering this subject.

ISO 4607 (1978)[61] describes procedures for exposing plastics to natural weathering in order to assess changes produced after specific stages of such exposure. It specifies a series of preferred exposure times (ranging from 1 week to 6 years) and recommends methods for measurement of total radiation received by the exposed test pieces. The use of the dyed wool standards and grey scale is permitted for this purpose but instrumental means are recommended for finding the total energy and the energy in specified wavelength intervals.

BS 4618 Section 4.2 (1972)[62] lays down the procedures for presenting design data on the resistance of plastics to natural weathering. The standard provides for total exposure to the natural elements under five standard climatic conditions and gives procedures for reporting biological effects (§17.6) and changes in dimensions, appearance and mechanical properties.

Full exposure to the natural elements is allowed in ASTM D1435 (1979).[63] The general design of the weathering racks and their siting are covered and the method of storage of control samples is laid down (23° ± 1°C and 50 ± 2 per cent r.h.) so that final control values for the material under test may be determined as well as initial values.

In many instances, change of physical appearance may be a sufficient criterion upon which to assess the effect of weathering (or other ageing); obviously this is particularly true if the intended application is basically decorative. When it comes to standardising techniques, however, degrees of colour change, darkening or fading, and changes in opacity need less subjective techniques. Alternatively, if it is a mechanical property (or several), or electrical performance which is of importance, changes in these as a result of weathering should be determined, following, for example, the guidelines laid down by the above standards. It should be noted that ISO 4607 refers to ISO 4582[64] (currently in draft form) for the specified procedures for the determination of changes in colour and appearance and variations in mechanical and other properties of plastics upon exposure to natural or artificial weathering. This standard details the methods of assessing changes in colour and appearance and gives the procedure for determining the changes in other properties.

17.5.3 Accelerated Natural Ageing

The time factor involved in performing outdoor exposure tests on plastics (exposure periods of at least two years are not uncommon) and the difficulties of accurately predicting natural ageing from (accelerated) laboratory ageing tests (§17.5.4) has prompted several attempts to accelerate natural ageing. One way of doing this is to employ exposure sites which give severe weather conditions because of their geographical location. Ellinger[65] describes the use of the Florida climate for accelerating the weathering of paints in particular; South Florida has a consistent climate of a combination of warm temperatures, high daily relative humidity, heavy annual rainfall, heavy dews and high values of solar radiation. There are undoubtedly other locations enjoying similar weather conditions; however, apart from the problem of inaccessibility, the degree of acceleration is probably not very great.

Garner and Papillo[66] have described a device called 'EMMA' which is a combination of ten aluminium mirrors so arranged as to intensify natural sunlight tenfold, while the test pieces are prevented from becoming too hot by a stream of cooled air. Results obtained indicate acceleration of degradation ranging from threefold for unsaturated polyesters to ninefold for rigid PVC. Further description, including the addition of water spray (EMMAQUA), and results from such accelerated tests have been provided by Caryl.[67])

More recently, Binder *et al.*[68] have re-opened this subject and have claimed a new procedure for accelerated weather testing by the use of stainless steel mirrors to magnify natural sunlight. Unfortunately this paper is short, does not adequately describe the apparatus and, apparently, overlooks the earlier work of Garner *et al.*

17.5.4 Accelerated Weathering and Ageing

In general, natural weathering and ageing take too long to achieve results in an industry as fast developing as plastics. Furthermore, the results will predict service life only for a time period similar to that of the weathering trial as, in the present state of our knowledge, extrapolation must largely be based on guesswork and may bear no relation to actual longer-term effects.

Many authors report work on the problem of correlating actual outdoor ageing with accelerated exposure tests in the laboratory using artificial light sources. A comprehensive survey is outside the scope of this book but a brief summary is given below.

In the past, four types of lamp source have been used for artificial weathering in the laboratory. Low-pressure mercury vapour lamps give an emission which comprises a series of sharply defined bands including a very strong emission at 2537 Å. When this is compared with natural sunlight, of which all wavelengths below 2900 Å are filtered out by the earth's atmosphere, it is obvious that the source of 2537 Å might easily cause a degradation or chemical change that would not be caused by sunlight. High-pressure mercury arc lamps give a better spectral distribution and are reasonably suitable for test purposes if the lower wavelengths are filtered out. Carbon arc lamps are still widely used and specified (particularly in the paint industry) but their spectrum bears no relation to that of sunlight. The fourth type of lamp source, the Xenon arc, has become widely used in the plastics industry during the past 15 years. With the appropriate filter this source simulates sunlight fairly closely, though the degree of acceleration compared with natural sunlight is not great.

Kamal[69] provides a comprehensive survey of the complexity of weatherability studies using outdoor exposure, carbon arc 'Weather-Ometers', Xenon arc 'Weather-Ometers' and fluorescent sunlamps. He points out that his own results, and those of other workers, show that the degree of correlation between natural outdoor weathering and artificial light exposure depends to a great extent on the polymer type as well as the type of light source. For example, better correlation between the enclosed carbon arc and outside exposure was achieved when the carbon arc was modified by incorporating eight fluorescent sunlamps in the case of cellulose acetate butyrate, yet good correlation was obtained with daylight with an unmodified carbon arc when testing polypropylene filaments.

Despite the closeness of the Xenon arc spectrum to that of natural sunlight there is no clear correlation between the two light sources which enables a positive statement to be made that x hours exposure to the Xenon lamp is equivalent to kx hours outdoor exposure where k is a constant. Kamal[69] quotes results by various workers indicating that 500 h in the Xenotest with high humidity and water spray is equivalent to 1 year of outdoor exposure as measured by colour change in a number of plastics. The manufacturers of the Xenotest apparatus suggest that exposure for 1 h to the lamp source is roughly equivalent to 10 h exposure to sunshine in temperate climates; however, Ruhnke and Biritz[70] claim that, for ABS plastics, exposure for 1 h to the Xenon arc corresponds to 5 h sunshine.

An exhaustive bibliography compiled by the Building Research Establishment[71] on weathering tests followed the RAPRA review[72] of the subject up to 1968. The references which may be found in these reviews cover detailed information on weathering tests both natural and artificial. Some work has also been done on development of instrumental methods of monitoring ultraviolet radiation[73,74] and on the use of plastics film as detectors.[75] More recently, a number of polymer standards have been developed, in place of the dyed wool blue standards, for monitoring ultraviolet radiation; these are being considered by ISO committee TC61 with a view to standardisation.

Although the problems of accelerated weathering, as outlined above, mean that such tests can only give arbitrary results and that great care is needed in their interpretation, there are a number of standardised test methods, some of which have been used for many years.

Standard Test Methods

ISO R878 (1968)[76] and ISO R879 (1968)[77] are for the determination of resistance of plastics to colour change upon exposure to light of the enclosed carbon arc and the Xenon lamp, respectively. Both are currently under revision and it is likely that they will be ultimately replaced by ISO 4892,[78] which may also be published as a British Standard under the dual numbering system. This draft standard offers the choice of carbon arc lamps, Xenon lamps and fluorescent lamps. It lays down the characteristics of the lamps to be used and the exposure conditions in terms of humidity, water sprays and temperature. It also includes an appendix on the correlation of the effects of exposure to artificial light sources and exposure to natural daylight. This appendix is only advisory and draws attention to some of the problems already mentioned.

The only published British Standard that refers to the exposure of plastics to artificial light sources is BS 4618, Section 4.3 (1974).[79] This, like ISO 4892, offers the choice of carbon arc, Xenon and fluorescent lamps.

ASTM have a number of accelerated ageing tests. ASTM D1501 (1971)[80] details the use of five fluorescent sunlamp bulbs with operating characteristics closely specified. The bulbs are mounted against a flat polished aluminium reflector. Test pieces are mounted on a turntable rotating at 30 to 40 r.p.m. in such a way that there is an air space of at least 5 mm between the bottom of the test piece and the top of the turntable. Provision is made for the test to be carried out at a temperature either of 21°–25°C (Method A) or of 55°–60°C (Method B). Results are reported as a change in appearance (colour, chalking, crazing, etc.) but changes in other physical properties (not specified) can also be used for assessment of effect of exposure.

ASTM D1499–64 (1977)[81] and ASTM D2565 (1979)[82] are similar in content to D1501 except that they specify carbon arc lamps and Xenon arc lamps, respectively, and make reference to the general standards for recommended practices for exposure of non-metallic materials G23 and G26, respectively. While D1501 makes specific reference to change in colour, D1499 requires the change in the exposed test piece to be evaluated by the applicable ASTM specification and D2565 reports only the appearance of the exposed test specimen. None of these standards uses the dyed wool blue standards for assessing the level of light dosage; ASTM D1920–69 (1976)[83] uses pieces of poly(methyl methacrylate) containing a specified type and quantity of yellow dye so that its transmittance at 420 nm wavelength on a test piece 3·15 mm thick is 17 ± 1·5 per cent. The light dosage is determined for any given carbon arc apparatus by the rate of change of light transmittance with time of exposure of these yellow test pieces.

DIN 53389 (1974)[84] is similar to ISO R879.

17.6 BIOLOGICAL ATTACK

Synthetic polymers are not usually attacked by micro-organisms (bacteria and fungi) but the picture is confused by the presence of additives, reinforcements, etc., which may themselves be attacked. The resistance of plastics to biological attack has been extensively reviewed by Heap[85] and Heap and Morrell.[86] More recently, Wienert and Hillard[87] have discussed the errors in interpreting accelerated mould growth tests in terms of service life, using plastics films to demonstrate the limitations; this work has been substantiated by Dalton,[88] using PVC films.

Undoubtedly the subject of 'biological ageing' is best entrusted to the experts, although a number of published standards deal with the subject of microbiological attack on plastics.

ISO 846 (1978)[89] specifies the use of a mixture of five strains of fungi over a period of at least 28 days at 30° ± 2°C and 95–100 per cent r.h. and measures the results of the attack by visual examination for mould growth. Two test procedures are included: Method A for finding the ability of plastics to act as nutritive medium and Method B to determine the fungitoxic properties of plastics.

BS 4618, Section 4.5 (1974)[90] broadly outlines the variables in exposure conditions for soil burial and suggests ways of determining the results of exposure. This standard makes reference to ISO R846 and also gives a short bibliography, but it can only be regarded as a guide as it does not lay down precise procedures for exposure to biological attack.

At present there is no other British Standard that offers test methods for determining biological attack on plastics but BS 1204 Parts 1 and 2 (1964/65)[91] for, respectively, gap-filling adhesives and synthetic resin adhesives for wood both contain a test for measuring resistance to micro-organisms. BS 1982 (1968)[92] measures the effect of fungi, cellulose-attacking microfungi and moulds or mildew on manufactured building materials by visual examination and weight change.

ASTM presents two standard test methods. ASTM G21-70 (1975)[93] specifies the use of five fungi at 28° to 30°C and 85 per cent r.h. for a minimum period of 21 days and provides for measuring the effects on appearance and on optical, physical or electrical properties. ASTM G22 (1976)[94] specifies the use of the bacterium *Pseudononos aeruginosa*, incubated at 35° to 57°C for a minimum of 21 days, the methods of assessing the result being similar to those of G21.

17.7 FIRE

17.7.1 General Considerations

In recent years this subject has received enormous attention because of the alarm (sometimes unjustified) which has arisen from the flammability of plastics materials, particularly foams. The importance of this subject to the industry, worldwide, can be judged by ISO TC61 committee deeming it necessary to set up a Working Group (now a Sub-Committee) exclusively to examine the subject of flammability testing of plastics.

It has long been recognised that there must be some form of agreed philosophy of fire testing; it is relatively easy to devise laboratory tests relating to the flammability of plastics but these are often confusing in terms of what they actually measure and, most important, they do not and cannot relate to what actually happens in a real fire. Brown[54] summarises the problems in applying this philosophy to rubber and plastics and emphasises the dangers of attempting to predict the performance of a polymeric material in a large fire from a small-scale laboratory test. A very comprehensive account and comparative study of existing fire tests as applied to plastics, by Eagles and Scott,[95] runs to 247 pages and more than 200 references and covers both large-scale and small-scale tests. The philosophy of fire testing receives helpful comment from Malhotra[96,97] and the complete proceedings of the 1975 Symposium of Fire Safety of Combustible Materials[97] form a comprehensive review of fire problems. There is a considerable number of published flammability test methods and a detailed review of all of them is outside the scope of this book. This study is therefore confined to those tests which are currently used for plastics materials.

17.7.2 Small-scale Tests

ISO Recommendations

ISO R181 (1961),[98] *Incandescent Resistance of Rigid Self-extinguishing Thermosets*, is similar to BS 2782 (1970) Method 508E,[99] ASTM D757 (1977)[100] and DIN 53459 (1975).[101] It is currently under review by ISO TC61 Committee and its fate is, as yet, undecided. A test piece, not less than 80 mm long, 10 mm wide and 4 mm thick, is weighed and clamped horizontally (thickness in the vertical plane) so that its 10 mm wide edge can be brought into contact with an incandescent silicon carbide rod, 8 mm in diameter, maintained at a temperature of 950°C. The incandescent bar is held in contact with the test piece with a force of 295 mN for a period of three minutes. The bar is removed, any flame is dry-extinguished and the test piece is reweighed. It is then cleaned and polished so that the limit of the flame spread is apparent and the length of the test piece is measured to this point. The difference between the original and final lengths is termed the 'flame spread'. The incandescent resistance *IR* is then calculated by

$$IR = \log_{10} \frac{100\ 000}{ps} \tag{17.6}$$

where p is the average mass loss on three test pieces (mg)
s is the average flame spread of six determinations (2 measurements on 3 test pieces) (cm).

ISO R871 (1968)[102] is a much simpler version of ASTM D1929 (1977)[103] (q.v.). It sets out to find the temperature at which plastics begin to decompose to flammable gaseous products. 1 g of plastics material is introduced into a chamber fitted with a nozzle set in a heated metal block. An igniter flame is played over the nozzle and, if after 5 min no ignitable product issues from the nozzle, the test is repeated with a fresh sample at a temperature 10°C higher until ignitable gas is detected at the nozzle. The decomposition temperature is reported to the nearest 10°C.

ISO R1210 (1970)[104] is a loosely defined test utilising a test piece in the form of a bar 80 mm long, 10 to 15 mm wide and 3 to 5 mm thick. This is clamped horizontally and a 'standard' luminous bunsen burner flame, 100 mm long, is applied in a specified way to the test piece for 60 s. A stop clock is used to determine the time of burning after the flame is removed. 'Flammability' is then assessed to be in one of three categories: Category 1, if the test piece does not burn after removal of the flame; Category 2, if it burns for less than 15 s, some of it remaining unburnt; Category 3, if the test piece has burnt away or is still burning after 15 s. Five test pieces are subjected to test and the overall category of the sample is determined as the most flammable category found amongst the five test pieces.

ISO R1326 (1970)[105] for plastics film uses a strip of film 25 mm wide and 450 mm long, hung vertically and ignited at the bottom by a bunsen burner. Again, the 'flammability' is expressed as one of three categories according to how much burns away during the test. If the test piece burns away entirely, the rate of burning is reported in mm^2/min.

British Standards

BS 2782 (1970) Method 508A,[106] 'Rate of burning (laboratory method)'—currently under review— is similar to ASTM D635 (1977)[107] (q.v.). A test piece in the form of a bar 150 mm long, 13 mm wide and 1·5 mm thick is scribed with reference lines across the test piece at 25 mm and 125 mm from one end. It is then clamped horizontally and with the plane of the width at an angle of 45°, and so mounted that the lower edge of the specimen is 6 mm above a sheet of metal gauze 130 mm square. A flame, 13 to 19 mm in height, is applied to the free end of the test piece for 10 s and the time taken for the edge of the flame of the burning test piece to travel the distance of 100 mm between the two reference lines is measured with a stopwatch. At least three test pieces are used and the result (reported for each test piece) is expressed as:

(a) the rate of burning (mm per minute); or
(b) that the flame does not reach the first mark; or
(c) that the flame does not reach the second mark; or
(d) for test pieces where the flame does not reach the first mark, the duration of flame or afterglow after removal of the burner.

Method 508E of BS 2782 (1970)[99] is, as mentioned above, similar to ISO R181. The method is under review but will not be revised until the fate of the ISO standard is decided.

BS 2782 Method 140D[108] for thin polyvinyl chloride sheeting will be issued as a revision of BS 2782 (1970) Method 508C.[106] A strip of material of the prescribed dimensions is clipped to a semi-circular rack and one end of the strip is subjected to the flame from a specified volume of ethanol. The test result is reported as the distance over which the strip is burned or charred under these conditions.

BS 2782 Method 140E[108] will be issued as a revision of BS 2782 (1970) Method 508D,[106] the alcohol cup test. The test piece of prescribed dimensions is supported on a simple brass and wire frame so that its major plane is at 45° to a horizontal. The flame from burning a specified amount of alcohol is allowed to impinge on the centre of the sheet and the result is expressed in terms of the amount of charring of the surface of the test piece and the duration of flaming or glowing.

BS 2782 (1978) Methods 141A to 141D[109] cover the oxygen index of combustion determined on four different test pieces:

141A A bar of 10 × 4 mm nominal cross-section (rigid materials)
141B A bar of 6·5 × 3 mm nominal cross-section (rigid materials)

141C A bar 10 mm wide cut from rigid sheet 1·2–6·5 mm thick
141D A bar of 6·5 × 3 mm nominal cross-section of electric cable insulation or sheathing material

The oxygen index test has received much attention during the past few years. It has been discussed by many workers[110–113] and it is claimed to give good repeatability and is therefore useful for quality control purposes. It does not purport to predict performance of a material in real fire conditions and indeed must not be used for this purpose, nor does it necessarily indicate relative merits of, say, flame retardant additives in a polymer under conditions other than those specified in the test method.

Basically, the test determines the minimum oxygen concentration necessary to support flaming combustion of the material under certain conditions. The test piece is clamped vertically at its base and supported in a glass chimney of specified dimensions. Arrangements are provided to ignite the top of the test piece with a propane flame and a mixture of nitrogen and oxygen (of known purity) is metered into the bottom of the chimney. Tests are then made to find the oxygen concentration in the gas mixture to cause the test piece, after ignition, to burn over a specified distance and time (at least 50 mm and at least 180 s). Fresh test pieces are used for each determination and the test is continued until the critical oxygen concentration for each burn is within 0·3 per cent on total volume of gas. Three such determinations are carried out and the mean value is reported as the oxygen index of the material under test.

ASTM D2863 (1976)[114] is similar and it is worth noting that an International Standard of this test is in draft form but may not be published for several years.

ASTM Tests

ASTM D757 (1975)[100] has already been mentioned as similar to ISO R181 and BS 2782 Method 508E.

ASTM D635 (1977)[107] determines rate of burning in a horizontal position and is similar to Method 508A of BS 2782. The ASTM test more precisely specifies the method than the BS test and requires the result to be reported in terms of burning rate (cm/min), average time of burning (s), and average extent of burning (mm), together with any special effects noted during the test.

ASTM D2863 (1977)[114] covering minimum oxygen concentration is similar to BS 2782 Method 141B.

ASTM 1929 (1977),[103] *Ignition Properties of Plastics*, uses a similar principle of tests to that of ISO R871 except that a much more closely specified apparatus is used and a forced air flow is introduced into the heating chamber. Three criteria can be measured:

(i) *Flash ignition temperature*—the lowest initial temperature of air passing around the specimen at which a sufficient amount of combustible gas is evolved to be ignited by a small external pilot flame.
(ii) *Self-ignition temperature*—the lowest initial temperature of air passing around the specimen at which, in the absence of an ignition source, the self-heating properties of the specimen lead to ignition or ignition occurs of itself, as indicated by an explosion, flame or sustained glow.
(iii) *Self-ignition by temporary glow*—a special case of self-ignition temperature

where, in some cases, slow decomposition and carbonisation of the specimen result only in glow of short duration at various points therein without general ignition.

ASTM D568 (1977)[115] for rate of burning in a vertical position has a test piece 25 mm wide and 450 mm long which is marked with gauge marks to define a 380 mm length over which the burning rate is to be measured. The test piece is suspended vertically and ignited at the bottom with a specified bunsen flame. The result is expressed in terms of burning rate (cm/min), average time of burning (s) and average extent of burning (mm).

ASTM D1433 (1977)[116] determines rate of burning of flexible thin plastics sheeting supported on a 45° incline. A test piece 76 mm wide and 228 mm long is supported on a metal frame inclined at 45° to the horizontal with 25·4 mm of the test piece hanging vertically over the lower end of the frame. A specified butane gas flame is applied to the centre of the hanging vertical portion and the results are expressed as the burning rate of the test piece (mm/min). In the test apparatus specified, the burn is timed automatically by microswitches activated by the burning through of nylon sewing threads set at a distance of 152 mm apart across the test piece.

ASTM D2843 (1977),[117] *Density of Smoke from the Burning or Decomposition of Plastics*, sets out a procedure for burning a test piece 25·4 mm square and 6·2 mm thick under specified conditions in a closed chamber and measuring the density of the smoke by means of a light source and photoelectric cell set 260 mm above the burning test specimen. The chamber is also fitted with a white backing sheet and a glazed front to allow visual observation of the test and of the degree to which an 'EXIT' sign on the white backing sheet becomes obscured by the smoke from the burning test piece. Three test pieces are used for each test. The test is started and the percentage light absorption is determined at 15 s intervals up to a total of four minutes. The average values for the three test pieces are plotted against time and the resultant graph is used to compute the maximum smoke density and the smoke density rating. The latter is expressed as a percentage by dividing the area under the curve by the total area of the graph, all multiplied by 100.

Other Tests

Over the years many tests have been, and still are being used for the testing of plastics, although originally designed for other materials. For example, BS 476 Part 7 (1971),[118] for surface spread of flame tests for building materials, includes a small-scale test suitable for development and quality control. A test piece 95 mm wide and 300 mm long and not more than 25 mm thick is mounted so that the width is vertical and the long axis is virtually at right angles to a gas-fired radiant panel so that the radiant intensity reaching the test piece is within the limits specified by the test method. As soon as the test piece is in position a gas flame is applied to the hotter end for one minute. Measurements of the time of the flame spread along the test piece are made over a period of ten minutes or until the flame has reached the far end of the test piece, whichever is the shorter. The flame spread so measured is used to classify the test material into one of four categories, depending on flame spread after $1\frac{1}{2}$ min and final flame spread.

One of the more important standards for flammability of plastics materials is that of the Underwriters Laboratories Subject 94[119] (usually abbreviated to UL94). It covers six test methods, five of which, listed below, refer to solid plastics and the sixth to foamed materials.

1. *Horizontal burning test* for classifying materials as 94HB. This is a similar test to ASTM D635 (or BS 2782 Method 508A).

2. *Vertical burning test* for classifying materials as 94V-0, 94V-1 or 94V-2. This is somewhat similar to ASTM D568 but requires a pad of dry absorbent surgical cotton to be placed 12 in (305 mm) below the test piece. Classifications 94V-0 and 94V-1 require that the ignited test pieces have flaming particles which will ignite the cotton. 94V-2 is a lower classification allowing some flaming particles to ignite the cotton. This classification test of UL94 has found favour in recent years in trade and insurance specifications in both the UK and USA.

3. *Vertical burning test* for classifying materials as 94-5V. This is similar to Method 2 above but the use of a plaque 6 in (152 mm) square which is tested in various positions is also allowed. The test flame is applied for 5 s and then removed for 5 s, this being repeated for five applications of the flame. To receive the 94-5V classification no test specimen must (a) burn for more than 60s, (b) drip or (c) be destroyed to any significant degree in the area of the test flame.

4. *Flame spread index* using radiant panel. The material is classified according to ASTM E162, *Test for Surface Flammability of Materials*, using a radiant heat energy source. This test has some similarity with that of the BS 476 test mentioned above.

5. *Ignition temperature test.* Pieces of material approximately $\frac{1}{4}$ in (6.5 mm) square are dropped into a heated glass conical flask, 114 mm high, 60 mm diameter at the bottom and 28 mm diameter at the top. The pieces are dropped into the flask at intervals of 5°C and it is observed whether or not ignition takes place (without application of flame) within 2 minutes. A small amount of air may be introduced into the flask by means of a rubber bulb and bent glass tube. The ignition temperature is the lowest temperature at which flaming or glowing combustion takes place within 2 minutes.

It is important to note that UL94 requirements do not cover plastics when used as building materials.

None of the tests described in this section must be used to attempt to predict the behaviour of the materials in real fire situations and all make this point in their introduction. The BS and ASTM make this very clear by carrying suitably worded caveats either in the report section (BS) or in the scope section (ASTM).

17.7.3 Large-scale Tests

As yet, there are no published national or international standards for large-scale fire tests for plastics. However, use is made of the large-scale tests which have been published for building materials. Typical examples of these are BS 476, Part 3 (1975),[118] *External Fire Exposure Roof Test*, and Part 8 (1972),[118] *Test Methods and Criteria for the Fire Resistance of Elements of Building Construction*. Part 3 requires test pieces not less than 1·5 × 1·2 m and Part 8 test pieces 2·5 m square. Thus the equipment necessary for such tests (Part 8 requires a furnace which can give a temperature rise of up to 1193°C) is beyond the normal used in the average physical testing laboratory and such testing is better left to the specialist organisation. Similarly, to test the burning characteristics of a complete product, e.g. furniture, requires a specially built room.

For these reasons detailed discussion of large-scale fire tests is outside the scope of this book and those requiring any such fire tests to be carried out are advised to get in touch with a specialist organisation such as The Fire Research Station (Boreham Wood, England), Rubber and Plastics Research Association (Shawbury, England) or their equivalent organisations in other countries. A useful appraisal of large-scale tests is given by Taylor.[120]

17.8 RADIATION

Radiation is taken here to mean bombardment by atomic and nuclear particles, i.e. gamma rays, electrons, neutrons, etc. The intensity of such radiation at the Earth's surface is not high enough significantly to affect plastics and tests are only required in connection with applications in atomic or nuclear plants or perhaps where radiation is used to induce crosslinking. Not surprisingly, such a specialised subject has not given rise to widespread standardisation of test methods. ASTM D1672 (1966), *Exposure of Polymeric Materials to High Energy Radiation*, specifies a recommended practice for exposure of polymeric materials to various types of radiation and a detailed account of testing polymers for radiation resistance has been given by Metz.[121]

REFERENCES

1. ISO/DIS 4611 *Plastics—Determination of the Effects of Exposure to Damp Heat, Water Spray and Salt Mist.*
2. ASTM D756 (1978) *Standard Practice for Determination of Weight and Shape Changes of Plastics under Accelerated Service Conditions.*
3. BS 2011 *Basic Methods for the Climatic and Durability Testing of Components for Telecommunications and Allied Electronic Equipment*, Part 2.1Db (1977) *Damp Heat, Cyclic.*
4. BS 3900 Part F2 (1973) *Determination of Resistance to Humidity under Condensation Conditions.*
5. Uijlenburg, H. (1960) *Plastics, London*, **25** (275), 359.
6. Braden, M. (1963) *Transactions of the Plastics Institute, London*, **31** (94), 83.
7. Blank, F. (1962) *Plaste und Kautschuk*, **9** (8), 391.
8. Blank, F. (1965) *Plaste und Kautschuk*, **12** (11), 657.
9. BS 771 (1980) *Phenolic Moulding Materials.*
10. Hauck, J. E. (1966) *Materials in Design Engineering*, **64** (5), 93.
11. ISO R62 (1958) with Amendment 1 (1965) *Plastics—Determination of Water Absorption.*
12. ISO R117 (1959) *Plastics—Determination of Boiling Water Absorption.*
13. BS 2782 (1970) Methods 502A–502G and Methods 503A–503C *Water Absorption of Plastics.*
14. ISO/DIS 62 *Plastics—Determination of Water Absorption.*

15. ASTM D570 (1977) *Water Absorption of Plastics.*
16. DIN 53471 (1976) *Testing of Plastics—Determination of the Water Absorption after Storage in Boiling Water.*
17. DIN 53475 (1962) *Determination of Water Absorption According to ISO R62* (now withdrawn).
18. DIN 53473 (1979) *Testing Plastics—Determination of Water Absorption after Storage in Humid Air.*
19. BS 2782 (1970) Method 504A *Water Vapour Absorption of Cellulose Acetate Moulding Material.*
20. Jessup, J. N. (1967) *Modern Plastics*, **44** (7), 174.
21. ISO R175 (1961) *Determination of the Resistance of Plastics to Chemical Substances.*
22. ISO R462 (1965) *Determination of Change of Mechanical Properties after Contact with Chemical Substances.*
23. ISO/DIS 175 *Plastics—Determination of the Effects of Liquid Chemicals, Including Water.*
24. BS 4618: Section 4.1 (1972) *Recommendation for the Presentation of Data on the Chemical Resistance of Plastics to Liquids.*
25. BS 2782 (1970) Method 505A *Determination of Resistance to Concentrated Sulphuric Acid of Rigid Polyvinyl Chloride Compound.*
26. ASTM D543 (1978) *Resistance of Plastics to Chemical Reagents.*
27. ASTM D2299-68 (1977) *Determination of the Relative Stain Resistance of Plastics.*
28. DIN 53476 (1979) *Testing of Plastics—Determination of the Behaviour against Liquids.*
29. Brown, R. P. (1979) *RAPRA Members Report No. 42.*
30. ASTM D1693-70 (1975) *Environmental Stress Cracking of Ethylene Plastics.*
31. De Coste, J. B., Malm, E. S. and Wallder, V. T. (1951) *Industrial and Engineering Chemistry*, **43,** 117.
32. Roe, R. J. and Gienewski, C. (1975) *Polymer Engineering Science*, **6** (2), 127.
33. Steinle, H. and Pflasterer, H. (1967) *Kautschuk und Gummi Kunststoffe*, **20** (9), 516.
34. Leghissa, S. and Salvatore, O. (1966) *Polymer Engineering Science*, **6** (2), 127.
35. Dempsey, L. T. (1967) *Polymer Engineering Science*, **7** (2), 86.
36. Ruhnke, G. M. and Biritz, L. F. (1970) *Plastics and Polymers*, August, 265.
37. Bergen, R. L. (1962) *SPE Journal*, June, 667.
38. Pohrt, J. (1973) *Kunststoffe*, **63** (3), 163.
39. Pohrt, J. (1976) *Kunststoffe*, **66**(8), 481.
40. DIN 53449 (1970) *Testing of Plastics—Evaluation of Resistance of Thermoplastics to Environmental Stress Cracking; Steel Ball Impression Method.*
41. ISO/DIS 4600 *Determination of Environmental Stress Cracking Resistance (ESCR) (Ball or Pin Impression Method).*
42. Gnauck, B. (1977) *Gummi und Asbest Kunststoffe*, **10,** 708.
43. Tsuey, T., Wang, M. and Kwei, T. K. (1971) *Journal of Applied Physics*, **42** (11), 4188.
44. Lander, L. L. (1960) *SPE Journal*, **16** (12), 1329.
45. Herman, J. N. and Biesenberger, J. A. (1966) *Polymer Engineering Science*, October, 341.
46. O'Connor, A. and Turner, S. (1962) *British Plastics*, September, 452; October, 526.
47. Hittmair, A., Mark, H. F. and Ullman, R. (1958) *Journal of Polymer Science*, **33** (126), 505.
48. Sternstein, S. S., Ongchin, L. and Silverman, A. (1968) *Applied Polymer Symposia*, **7,** 175.

49. Haslett, W. H. and Cohen, L. A. (1964) *SPE Journal*, **20** (3), 246.
50. BS 4618, Section 1.3.3 (1976) *Recommendations for the Presentation of Plastics Design Data—Environmental Stress Cracking.*
51. ASTM D2552-69 (1974) *Environmental Stress Rupture of Type III Polyethylenes under Constant Tensile Load.*
52. ASTM D1248 (1978) *Polyethyelene Molding and Extrusion Materials.*
53. Weiss, E. (1966) *Testing of Polymers* vol. 2, J. V. Schmitz (Ed.), Interscience, Chapter 9.
54. Brown, R. P. (1979) Physical Testing of Rubbers, Applied Science, page 207.
55. Estevez, J. H. J. (1965) *Transactions of the Plastics Institute, London*, **33** (105), 89.
56. Kamal, H. R. (1967) Weatherability of plastics materials, in *Applied Polymer Symposium No. 4*, Interscience.
57. ISO 877 (1976) *Plastics—Determination of Resistance to Change upon Exposure under Glass to Daylight.*
58. BS 2782, Method 540A (1977) *Plastics—Determination of Resistance to Change upon Exposure under Glass to Daylight.*
59. ISO 105 (1978) *Textiles—Tests for Colour Fastness.*
60. DIN 53388 (1974) *Testing of Plastics—Testing of Stability under Natural Light (Global Radiation behind Glass).*
61. ISO 4607 (1978) *Plastics—Methods of Exposure to Natural Weathering.*
62. BS 4618: Section 4.2 (1972) *Recommendations for the Presentation of Plastics Design Data—Resistance to Natural Weathering.*
63. ASTM D1435-75 (1979) *Outdoor Weathering of Plastics.*
64. ISO 4582 (Draft) *Determination of Changes in Colour and Variation in Properties after Exposure to Daylight under Glass, Natural Weathering or Artifical Light.*
65. Ellinger, M. I. (1963) Paint Technology, **27** (12), 40.
66. Garner, H. I. and Papillo, P. J. (1962) *ACS Division of Organic Coatings and Plastics Chemistry, Atlantic City Meeting*, **22** (2), 110.
67. Caryl, C. R. (1967) *SPE Journal*, **23** (1), 49.
68. Binder, K., Vukovich, S. and Tschamler, H. (1978) *Chemie Kunststoffe Aktuell*, **32** (5), 209.
69. Kamal, H. R. (1972) *Coloring of Plastics 6, RETEC, Philadelphia, Pa, October*, SPE Philadelphia Section, page 4.
70. Ruhnke, G. M. and Biritz, L. F. (1972) *Plastics and Polymers*, **40** (147), 118.
71. BRE (1975) *BRE Library Bibliography*, No. 247, February.
72. Matthan, J., Scott, K. A. and Wiechers, M. (1970) *Ageing and Weathering of Plastics*, RAPRA.
73. Harris, P. B. (1973) *BRE Current Paper no. 17/73*, June.
74. Capron, E. and Crowder, J. R. (1975) *BRE Current Paper No. 17/75*, February.
75. Davis, A., Gordon, D. and Howell, G. V. (1973/74/75) *ERDE Technical Report No. 141, No. 186, No. 190.*
76. ISO R878 (1968) *Determination of the Resistance of Plastics to Colour Change upon Exposure to Light of the Enclosed Carbon Arc.*
77. ISO R879 (1968) *Determination of the Resistance of Plastics to Colour Change upon Exposure to Light of a Xenon Lamp.*
78. ISO 4892 (Draft) *Plastics—Method of Exposure to Laboratory Light Sources.*
79. BS 4618: Section 4.3 (1974) *Resistance to Colour Change Produced by Exposure to Light.*
80. ASTM D1501 (1971) *Exposure of Plastics to Fluorescent Sunlamps.*
81. ASTM D1499-64 (1977) *Operating Light and Water Exposure Apparatus (Carbon Arc Type) for Exposure of Plastics.*
82. ASTM D2565 (1979) *Operating Xenon Arc Type (Water Cooled) Light and Water Exposure Apparatus for Exposure of Plastics.*

83. ASTM D1920-69 (1976) *Determining Light Dosage in Carbon Arc Light Ageing Apparatus*.
84. DIN 53389 (1974) *Testing of Plastics: Short Test of Light Stability (Simulation of Global Radiation behind Glass by Filtered Xenon Arc Radiation)*.
85. Heap, W. H. (1965) *RAPRA Information Circular No. 476*.
86. Heap, W. H. and Morrell, S. M. (1968) *Journal of Applied Chemistry*, **18** (7), 189.
87. Wienert, L. A. and Hillard, M. W. (1975) *Australian Plastics and Rubber Journal*, **26** (10), 15.
88. Dalton, D. L. (1977) *Polymer Paint and Colour Journal*, **167** (3964/5), 987/990.
89. ISO 846 (1978) *Determination of Behaviour under the Action of Fungi and Bacteria—Evaluation by Visual Examination or Measurement of Change in Mass or Physical Properties*.
90. BS 4618: Section 4.5 (1974) *Recommendations for the Presentation of Plastics Design Data—The Effect on Plastics of Soil Burial and Biological Attack*.
91. BS 1204 (1979): Part 1 *Gap Filling Adhesives*; Part 2 *Synthetic Resin Adhesives for Wood*.
92. BS 1982 (1968) *Methods of Test for Fungal Resistance of Manufactured Building Materials Made of, or Containing, Materials of Organic Origin*.
93. ASTM G21-70 (1975) *Determining the Resistance of Polymeric Materials to Fungi*.
94. ASTM G22 (1976) *Determining Resistance of Plastics to Bacteria*.
95. Eagles, A. E. and Scott, K.A. (1974) *RAPRA Technical Report on Fire Tests*, November.
96. Malhotra, H. L. (1973) *Plastics and Polymers*, **41** (151), 44.
97. Malhotra, H. L. (1975) in *Symposium—Fire Safety of Combustible Materials*, University of Edinburgh.
98. ISO R181 (1961) *Incandescent Resistance of Rigid, Self-extinguishing Thermosets*.
99. BS 2782 (1970) Method 508E *Incandescent Resistance of Rigid Thermosetting Plastics*.
100. ASTM D757 (1977) *Incandescent Resistance of Rigid Plastics in a Horizontal Position*.
101. DIN 53459 (1975) *Testing of Plastics; Evaluation of Behaviour During and After Contact with the Glow Rod*.
102. ISO R871 (1968) *Determining of the Temperature of Evolution of Flammable Gases from Plastics*.
103. ASTM D1929 (1977) *Ignition Properties of Plastics*.
104. ISO R1210 (1970) *Determination of Flammability of Plastics in the Form of Bars*.
105. ISO R1326 (1970) *Determination of the Flammability and Burning Rate of Plastics in the Form of Fibres*.
106. BS 2782 (1970) Method 508A *Rate of Burning—Laboratory Method*; Method 508C *Degree of Flammability of Thin Polyvinyl Chloride Sheeting*; Method 508D *Flammability (Alcohol Cup Test)*.
107. ASTM D635 (1977) *Rate of Burning and/or Extent and Time of Burning of Self-supporting Plastics in a Horizontal Position*.
108. BS 2782 Method 140D (1980) *Flammability of a Test Piece 550 mm × 35 mm of Thin Polyvinyl Chloride Sheeting (Laboratory Method)*; Method 140E *Flammability of a Test Piece 150 mm Square, not exceeding 50 mm in Thickness, of a Plastics Material (Alcohol Cup Method) (Laboratory Method)* (to be published).
109. BS 2782 (1978) Methods 141A–141D *Oxygen Index Combustion Test*.
110. Day, A. G. (1975) *Plastics and Polymers*, **43** (164), 64.
111. Walker, A. G. (1973) *Europlastics Monthly*, **46** (11), 72.
112. Abbot, C. (1973) *Europlastics Monthly*, **46** (11), 68.
113. Charsley, E. L. and Schulz, R. A. (1975) *Journal of Physics, E*, **8** (2), 147.

114. ASTM D2863 (1977) *Measuring the Minimum Oxygen Concentration to Support Candle-like Combustion of Plastics* (*Oxygen Index*).
115. ASTM D568 (1976) *Rate of Burning and /or Extent and Time of Burning of Flexible Plastics in a Vertical Position.*
116. ASTM D1433 (1977) *Rate of Burning and/or Extent and Time of Burning of Flexible Thin Plastics Sheeting Supported on a 45° Incline.*
117. ASTM D2843 (1977) *Density of Smoke from the Burning or Decomposition of Plastics.*
118. BS 476: Part 3 (1975) *External Fire Exposure Roof Test*; Part 7 (1971) *Surface Spread of Flame Test for Materials*; Part 8 (1972) *Test Methods and Criteria for the Fire Resistance of Elements of Building Construction.*
119. Underwriters Laboratories Subject 94 (1973) *Standard for Tests for Flammability of Plastics Materials.*
120. Taylor, W. (1973) *RAPRA Members Journal*, **1,** 167.
121. Metz, D. J. (1966) *Testing of Polymers* vol. 2, J. V. Schmitz, (Ed.), Interscience, Chapter 5.

Chapter Eighteen

Permeability

18.1 INTRODUCTION

Gas and vapour transport in polymeric materials is of great importance in the packaging and container industries. It is in these industries that plastics in the form of film and sheet find extensive usage.

No plastics material forms a complete barrier to the movement of gas or vapour molecules. This is due to one, or a combination, of three structural features of the material, namely pin-hole defects, porosity and intermolecular space. In this chapter we shall, in theory, be concerned with intermolecular space and the term 'permeation' will apply to the movement of gas and vapour molecules through such molecular scale voids.

Permeation of gases and vapours takes place in two stages. In the first stage the permeant dissolves into the surface of the material. The second stage is formed by the diffusion of the molecules through material under the action of a concentration gradient.

Strictly, the solubility of the permeant in a polymer is defined as that volume which dissolves in unit volume of material under applied unit pressure of permeant. It is expressed as Henry's law:

$$C = SP \tag{18.1}$$

where C = the volume (STP) of dissolved permeant
S = the solubility coefficient
P = applied pressure.

The diffusion process is described by Fick's first and second laws:

Fick's first law $$\frac{\mathrm{d}q}{\mathrm{d}t} \times \frac{1}{A} = -D\,\frac{\mathrm{d}C}{\mathrm{d}x} \tag{18.2}$$

where $\mathrm{d}q/\mathrm{d}t$ = rate of flow of permeant
A = permeated area
D = the diffusion constant
$\mathrm{d}C/\mathrm{d}x$ = concentration gradient across an infinitesimally thin layer of polymer.

Fick's second law $$\frac{\mathrm{d}C}{\mathrm{d}t} = D\,\frac{\mathrm{d}^2C}{\mathrm{d}x^2} \tag{18.3}$$

Fick's first law describes the steady state diffusion of a gas or vapour through a polymer; the second law, which is derived from the first, describes the non-steady state where the concentration gradient is a function of time.

Taking a polymer sheet of finite thickness L and assuming D to be independent of concentration, equation (18.2) can be integrated across this thickness, giving

$$\frac{dq}{dt} \times \frac{1}{A} = \frac{D(C_1 - C_2)}{L} \tag{18.4}$$

where C_1 and C_2 are the concentrations at the two surfaces ($C_1 > C_2$). Expressing concentrations as pressures using equation (18.1) and re-arranging, equation (18.4) becomes

$$\frac{dq}{dt} = DSA\,\frac{(P_1 - P_2)}{L} \tag{18.5}$$

(at constant temperature). This is the equation describing steady state permeation. The product DS is known as the permeability coefficient, Q. Hence if any two coefficients are known the third may be obtained from

$$Q = DS$$

It is essential when measuring the permeability coefficient that steady state gas transfer exists. This is achieved when equal quantities of permeant enter and leave the polymer. If measurement is made before the attainment of this condition then a smaller numerical value will be obtained. In this non-steady state period more permeant dissolves into the polymer than evaporates from it. The first stage of the permeation process (solubility) is dominant initially but, as the concentration gradient increases, the second stage (diffusion) becomes the major contributor. The attainment of steady state may take anything from a few hours to many days. The recognition of the steady and non-steady states will be dealt with in the following sections.

The units in which the permeability coefficient is expressed are varied. With the adoption of the SI system the units of the permeability coefficient are $m^4\ N^{-1}\ s^{-1}$.

This is a rather abstract way of expressing the quantity but clarity may be improved by rearrangement with N = Pa m^2: m^3(STP) m $m^{-2}\ s^{-1}\ Pa^{-1}$.

Other units which are extensively used in current literature include cm^3 (STP) cm/cm^2 s cm Hg. The barrer, which has been adopted by the ASTM, is also used. This is expressed as

$$1 \text{ barrer} = 10^{-10}\ \text{cm}^3\ (\text{STP})\ \text{cm/cm}^2\ \text{s cm Hg}$$

The magnitudes of these units are more realistic when small quantities of permeant are measured, which is usually the case. Also, comparisons of the coefficients between materials are facilitated.

For vapours, the quantity of permeant is expressed as a mass: 10^{-10} g cm/cm^2 s cm Hg.

If there is doubt as to the homogeneity of a material, reference of the rate of flow of permeant to unit thickness may be omitted and a gas transmission rate

$(G.T.R.)$ preferred:

$$\frac{\mathrm{d}q}{\mathrm{d}t} = (G.T.R.)\,A(P_1 - P_2) \tag{18.6a}$$

where $G.T.R. = Q/L$. Rearranging equation (18.6a),

$$G.T.R. = \frac{\mathrm{d}q}{\mathrm{d}t} \times \frac{1}{A(P_1 - P_2)} \tag{18.6b}$$

the thickness being stated. The gas transmission rate may also be used if a product is tested. The units are $m^3\ N^{-1}\ s^{-1}$ or cm^3/m^2 24 h atm. As in the case of the permeability coefficient, the vapour transmission rate is expressed in terms of mass (see §18.3.1).

The steady state equation for gas transfer only applies to gases and vapours which are sparingly soluble in the polymer and with which they do not interact, i.e. chemical association. This is true of the air gases but often not of water and organic vapours, which show pressure dependence. Vapours may also act as plasticisers and so increase diffusion. For all permeant–polymer systems the permeability coefficient has a dependence on temperature which is expressed by the Arrhenius relationship:

$$Q = Q_o \exp\left(-E_p/RT\right) \tag{18.7}$$

where Q_o = constant for a particular polymer
E_p = activation energy for the permeation process (E_p = heat of solution + activation energy of the diffusion process)
R = gas constant
T = absolute temperature.

For both vapour and gases the diffusion constant, D, increases with increasing temperature. The solubility constant, S, increases with temperature for gases and decreases for vapours.

Molecular size and polarity of the permeant also influence permeation. The morphology of the polymer (e.g. the degree of crystallinity and crosslinking) plays an important role in the diffusion stage.

The use of polymer laminates in packaging is common and so it is useful if the permeability coefficient of the laminate can be derived from the coefficients of its constituent layers. This may be carried out using the following expression:

$$\frac{1}{Q_L} = \left(\frac{x_1}{L}\frac{1}{Q_1}\right) + \left(\frac{x_2}{L}\frac{1}{Q_2}\right) + \left(\frac{x_3}{L}\frac{1}{Q_3}\right) + \cdots \text{ etc.}$$

or

$$\frac{1}{Q_L} = \frac{1}{L}\sum_{i=1}^{i=n}\frac{x_i}{Q_i} \tag{18.8}$$

where Q_L = permeability coefficient of the laminate
Q_i = permeability coefficient of the ith layer
x_i = thickness of the ith layer
L = thickness of the laminate.

It should be noted, however, that the use of this equation is limited to permeability coefficients which are independent of pressure. If this is true, the ordering of the layers is unimportant; if not, then differing permeability coefficients will be obtained for the laminate depending on the positioning of the layers.

18.2 GAS PERMEABILITY

In this section well established techniques used for the measurement of the permeability coefficient and gas transmission rate will be described, together with some more recent developments.

18.2.1 Manometric Method

The manometric method forms the basis of test standards ISO 2556 (1974),[1] BS 2782 Part 8 (1979) 821A,[2] ASTM D1434 (1975)[3] and DIN 53380 (1969).[4] It measures the quantity of gas which has permeated through a test piece in a given time, as a change in pressure and volume. The test piece forms a barrier between two chambers in a gas transmission cell (Figure 18.1). A constant high pressure (usually 1 atmosphere) is maintained in one chamber and a low pressure (usually vacuum) is initially established in the other chamber. The test piece is supported against the high pressure by a porous substrate. A manometer is coupled to the low-pressure chamber to measure the variation in pressure and volume over a specified test time. A plot of mercury displacement versus time gives the characteristic curve illustrated in Figure 18.2. The initial build-up to steady state gas transfer is shown, with the inflection in the curve marking the start of steady state gas transfer.

The differential equation for the instantaneous steady gas transmission rate is given in terms of pressure and volume as:

$$G.T.R. = \frac{K}{AT\,\Delta P}\left(P\frac{\mathrm{d}V}{\mathrm{d}t} + V\frac{\mathrm{d}P}{\mathrm{d}t}\right) \tag{18.9}$$

where K = constant = $273 \times 24 \times 10^4$ if the $G.T.R.$ is expressed in $\mathrm{cm^3\ (STP)/m^2\ 24\ h\ atm}$
A = permeated area
T = absolute test temperature
ΔP = pressure difference across the test piece.

The above expression may be written in terms of mercury displacement:

$$G.T.R. = \frac{K}{AT\,(P_{\mathrm{H}} - P_0 - h)}\left[(P_0 + h)a\frac{\mathrm{d}h}{\mathrm{d}t} + (V_0 + ah)\frac{\mathrm{d}h}{\mathrm{d}t}\right] \tag{18.10a}$$

Rearranging gives

$$G.T.R. = \frac{K}{AT\,(P_{\mathrm{H}} - P_0 - h)}[V_0 + aP_0 + 2ah]\frac{\mathrm{d}h}{\mathrm{d}t} \tag{18.10b}$$

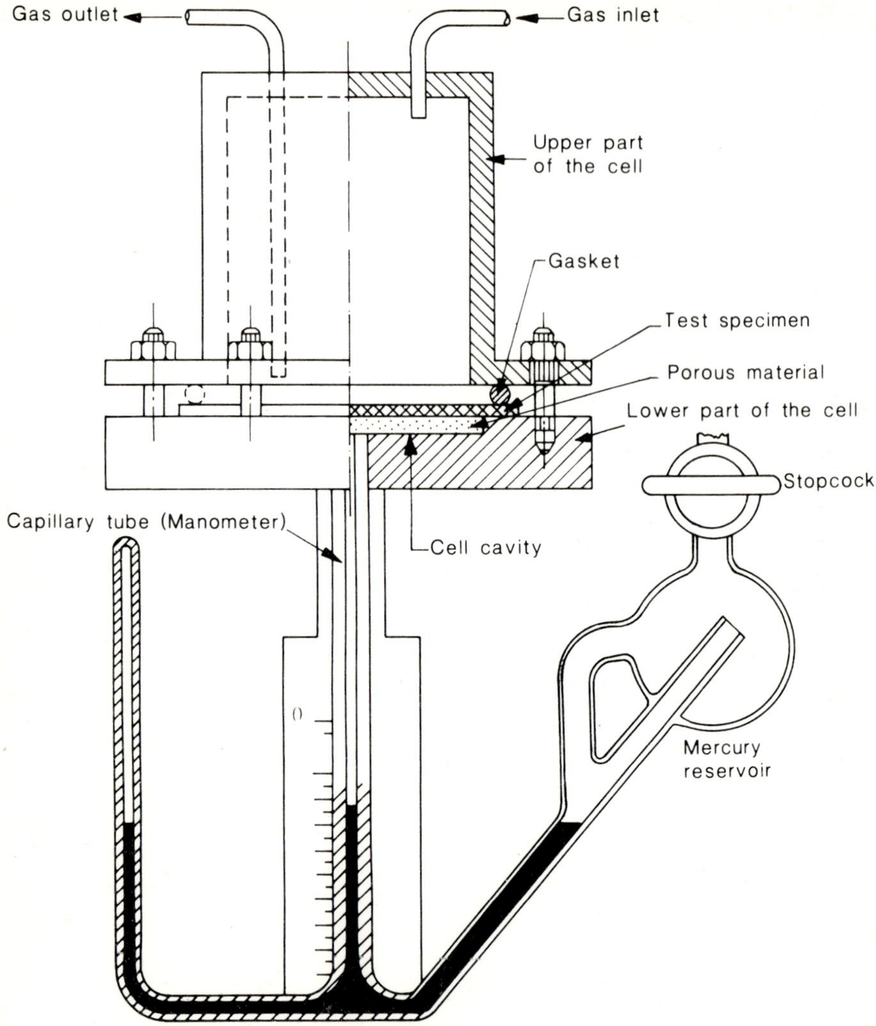

Figure 18.1 Manometric apparatus

where V_0 = initial volume between test piece and datum line in the manometer
P_0 = initial pressure in the manometer (0·2 mm Hg)
P_H = high pressure
h = mercury displacement from the datum line
dh/dt = rate of change of mercury displacement
a = cross-sectional area of the measuring capillary.

ISO 2556 and BS 2782 adopt equation (18.10) to measure the gas transmission rate at a single point chosen on the steady state portion of the plot of mercury displacement versus time. ASTM D1434 uses an approximation of the integrated form of equation (18.10) for an extended time interval.

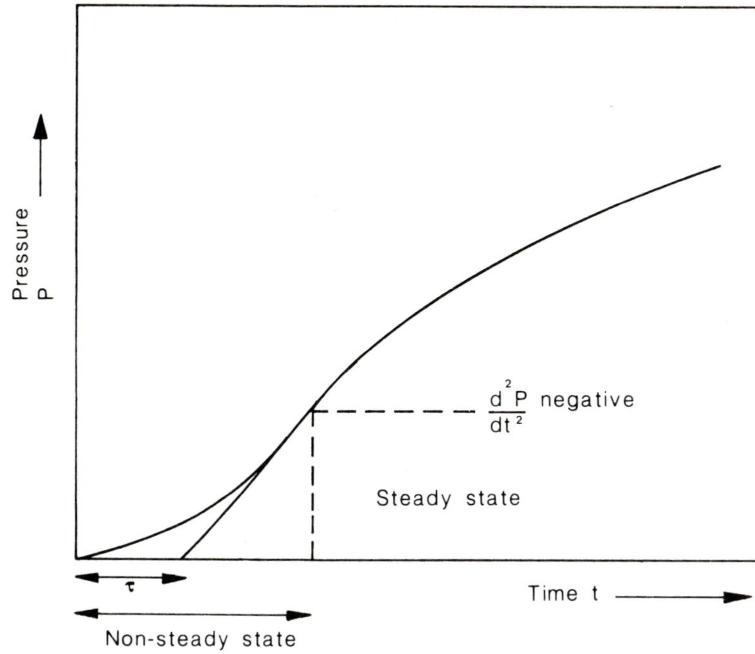

Figure 18.2 Plot of pressure versus time

In practice it is essential that the manometer and mercury are thoroughly clean to reduce stick–slip effects in the movement of the mercury meniscus. This may be further reduced by adoption of a larger bore capillary and a vibrator.

Leaks in the high-pressure chamber are unimportant if a large reservoir of gas is used. They are more important in the low-pressure chamber, either at the interface between test piece and metal case of the transmission cell, or at the several joints in the manometer. Some simple procedures for detecting leaks are given in the relevant standards.

The volume of the low-pressure chamber can usually be varied to alter the sensitivity of the apparatus. This may also be achieved by choice of a larger permeated area or by the use of an inclined measuring capillary.

A tipping manometer is frequently employed to transfer the mercury from the reservoir into the arms of the manometer after evacuation. This has been subject to modification with a direct connection between measuring capillary and reservoir.[5] This permits the mercury to remain in the manometer while it is evacuated or aerated.

Readings of mercury displacement may be made from a scale mounted behind the capillary or with a travelling microscope. ASTM D1434 also makes provision for the use of a calibrated resistance wire inserted into the capillary, thus allowing a continuous measurement of displacement as a function of resistance.

A close tolerance on test temperature is required, therefore the apparatus is usually sited in a temperature controlled chamber.

For full details of test procedures the relevant standards should be consulted.

18.2.2 Constant Volume Method

Although no current plastics test standard (ISO, BS, ASTM or DIN) is believed to use this method, its use in the field of research has been reported (it is also used in a test standard for rubbers[6]).

If the volume on the low-pressure side of the test piece is kept constant, equation (18.9) becomes:

$$G.T.R. = \frac{K}{AT\,\Delta P}\left(V\frac{\mathrm{d}p}{\mathrm{d}t}\right) \qquad (18.11)$$

where V is the volume between the test piece and the measuring device. The steady state portion of the plot of pressure versus time may be near linear, hence the calculation of the gas transmission rate can be carried out either at a single point or over an extended time interval. The permeability coefficient may then be obtain from $G.T.R. = Q/L$ where L is the thickness. An estimate of the time required to reach steady state is obtained from the relationship $t = L^2/2D$ if the value of the diffusion constant, D, is approximately known.

The use of a differential pressure transducer instead of a manometer to measure pressure change has been demonstrated in an automated apparatus[7] (Figure 18.3). The reference side of the transducer is held at some arbitrary chosen pressure in the low-pressure chamber of the gas transmission cell at the start of a test. This low pressure can be in the high vacuum range, i.e. $<0{\cdot}07$ Pa ($<5 \times 10^{-4}$ mm Hg). The sensitivity range of the transducer depends on the choice of diaphragm thickness. The instrument is robust and permits operation of the apparatus at high pressures in safety. In this case high pressures from

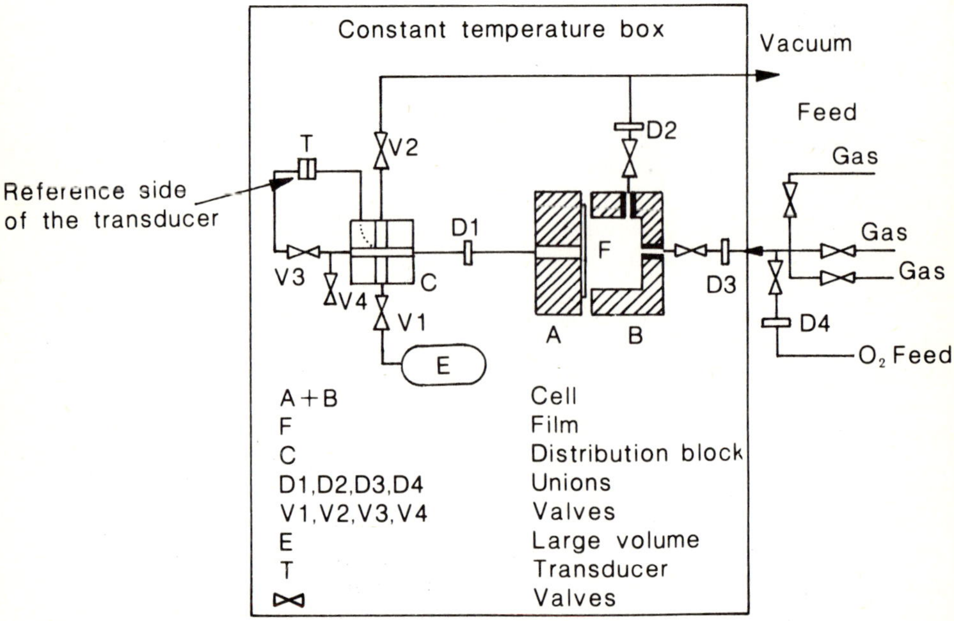

Figure 18.3 Constant volume apparatus[7]

approximately 0·2–7 MPa have been employed. The volume change due to diaphragm deflection is considered insignificant in comparison with the volume of the low-pressure chamber, which is quoted at 12 cm^3. The volume of the latter can be increased by coupling with a larger vessel, to allow highly permeable materials to be tested over sufficiently long experimental times. Conversely, the high driving pressures reduce experimental times for materials of low permeability. However, it should be remembered that a material may behave differently under differing hydrostatic pressures.

An interesting practical detail of the transmission cell is the incorporation of a magnetic stirrer when the permeation rate of a gas mixture is investigated. This is used to prevent concentration polarisation effects at the surface of the test piece exposed to the high pressure.

The high-pressure chamber is flushed out with test gas prior to pressurisation. In addition, if low pressures other than vacuum are used then the low-pressure chamber and ancillary pipework must also be adequately purged with test gas to prevent bidirectional gas permeation. Leaks in this chamber manifest themselves as a falling off in the gradient of the pressure–time curve. The test piece is supported on a laminate of coarse and fine wire mesh.

The pressure transducer is usually calibrated against a Macleod gauge.

A temperature controlled cabinet houses the transmission cell and low-pressure circuitry.

18.2.3 Constant Pressure Method

ASTM D1434 (1975)[3] provides for a second method to measure the gas transmission rate of plastics, namely that of constant pressure (this technique also forms the basis of a rubber test standard[8]). In this case the pressure is kept constant in the low-pressure chamber and the volumetric change in permeated gas is measured. Equation (18.9) reduces to

$$G.T.R. = \frac{K}{AT\,\Delta P}\left(P\,\frac{\mathrm{d}v}{\mathrm{d}t}\right) \tag{18.14}$$

where P = atmospheric pressure.

An example of a typical apparatus[9] is given in Figure 18.4. The same basic design of gas transmission cell is used as in the manometric and constant volume methods. The test piece forms a barrier between high- and low-pressure chambers. The test gas is maintained at atmospheric pressure on the low-pressure side of the test piece and at 0·2 MPa or greater on the high-pressure side. The volumetric change is registered by the displacement of a slug of manometer fluid in a vertically mounted measuring capillary. This displacement may be measured on a scale or by a travelling microscope. The bore of the capillary must be uniform to a close tolerance. Three diameters are suggested, 1 mm, 0·5 mm and 0·25 mm, for use with materials ranging from high to low permeability.

The volumetric displacement is plotted against time and the slope of the linear steady state portion is used in the calculation of gas transmission rate and permeability coefficient.

As in the previous method, it is essential that both test chambers and associated pipework be flushed out with test gas unless air is to be used. This method is

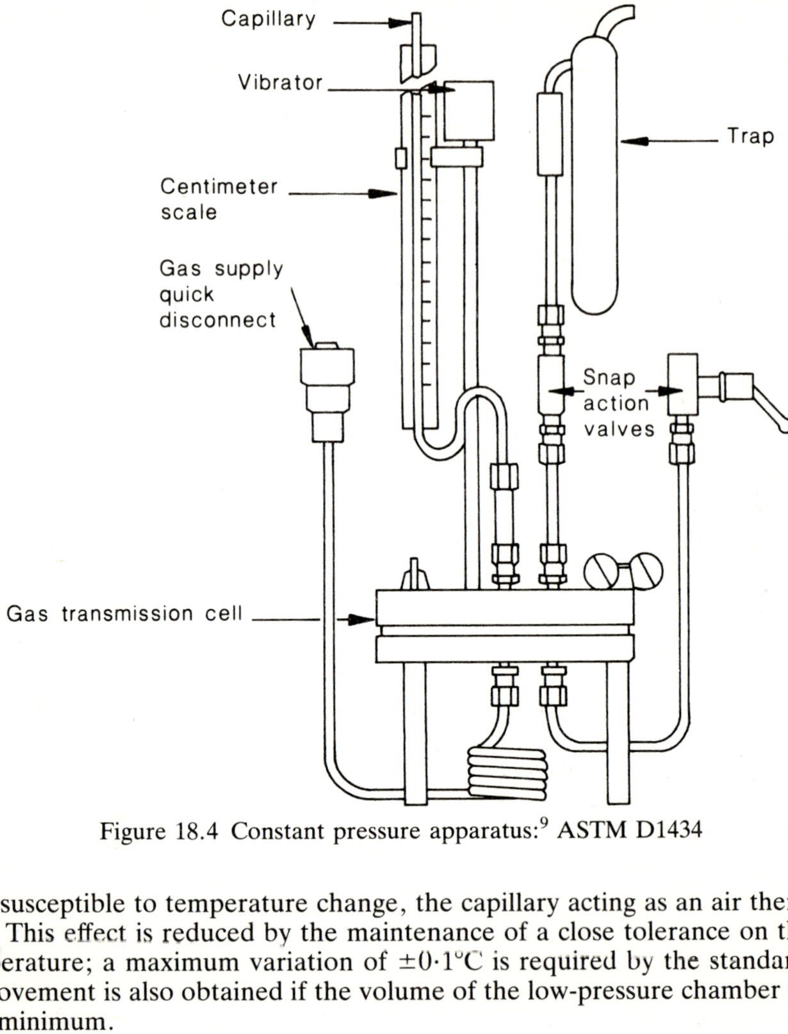

Figure 18.4 Constant pressure apparatus:[9] ASTM D1434

very susceptible to temperature change, the capillary acting as an air thermometer. This effect is reduced by the maintenance of a close tolerance on the test temperature; a maximum variation of ±0·1°C is required by the standard. An improvement is also obtained if the volume of the low-pressure chamber is kept to a minimum.

An alternative apparatus (Figure 18.5) is also suitable. The main difference is that the measuring capillary is replaced by an adjustable manometer. The volumetric change is registered as a change in the height of the balanced liquid columns.

18.2.4 Carrier Gas Methods

Carrier gas techniques are commonly referred to as 'dynamic' because gas is permitted to flow across each side of the test specimen, at equal pressure. The gas transmission cell is similar to those of the previous methods, the test piece forming a barrier between two chambers. In the most basic of carrier gas methods the test gas flows at a constant rate through one chamber and a second gas, the carrier, flows through the other chamber at a constant rate. Test gas

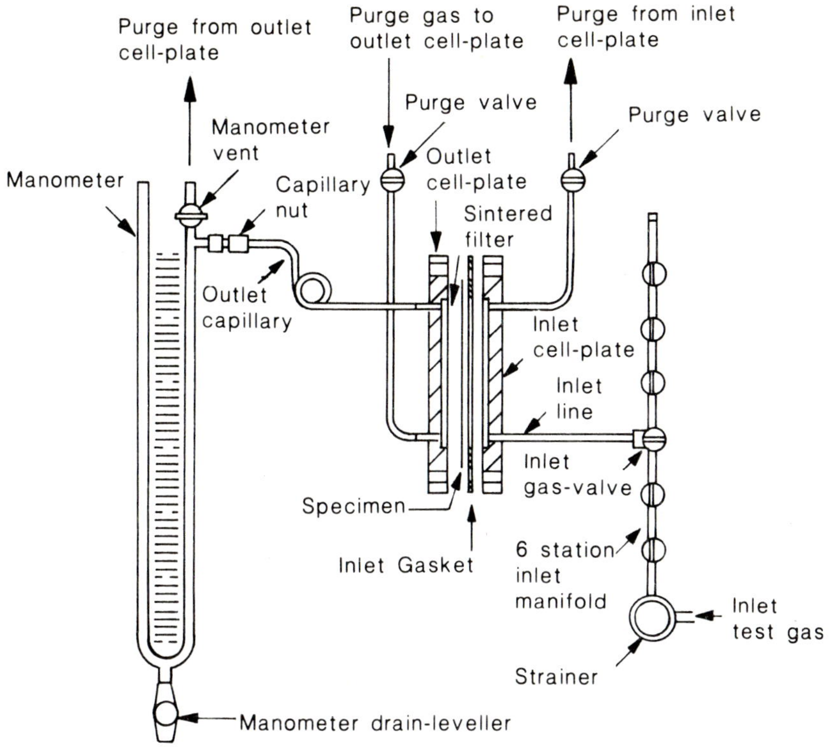

Figure 18.5 Constant pressure apparatus using adjustable manometer: ASTM D1434

which permeates through the polymer is swept away to a detector which may be of the absorptiometric[10] or thermal conductivity type.[11,12]

An important feature is the lack of need for mechanical support of the test piece, as both gases are held at the same pressure, usually one atmosphere. The polymer experiences the same hydrostatic pressure as in the vacuum manometric method. The difference in test gas partial pressure (the driving pressure) across the test piece is also the same, because the concentration of permeant in the carrier gas is kept small by suitable choice of flow rate.

Another feature of the technique is that small differences in partial pressure can be established across the polymer membrane while still maintaining a hydrostatic pressure of one atmosphere. Thus the service conditions of some polymers used in the packaging industry (e.g. for food) can be approximated in the laboratory. These small partial pressure differences are achieved by passing the carrier gas through both chambers with the permeant introduced as an impurity into one of the gas streams. As an example, the permeation of sulphur dioxide at partial pressure differences of less than 100 Pa through a polymer membrane has been demonstrated[13] using an ultraviolet detector.

As illustrated above, detectors other than those of the thermal conductivity type can be used, depending on the nature of the permeant. However, the thermal conductivity detector is satisfactory for most gases.

Gas leakage is less of a problem with the use of a carrier gas, although the problem of diffusion of air gases through seals is still present when small permeation rates of such gases are being measured. Suitable choice of sealing materials should minimise this.

A logical development of the carrier gas technique has been the introduction of a gas chromatographic column between transmission cell and detector.[14–16] This permits gas mixtures to be used for the permeant and their individual transmission rates to be measured. It should be noted that no single gas chromatographic column–detector combination is suitable for all gases. This must be chosen to suit the permeant and the carrier gas used. In addition, a pulsed input to the column is required if separation of components is achieved by elution development (this is the most common and versatile method of separation).

18.2.5 Other Methods

Several techniques have been developed which have not been included in the previous sections. Perhaps the most important of these from the research point of view is the use of a mass spectrometer[17] as the detector. In this case a pressure differential may be maintained across the test piece by coupling the low-pressure chamber of the transmission cell directly to the inlet manifold of the instrument. The mass spectrometer offers high sensitivity and the facility to measure the permeation rates of several gases simultaneously. The cost of such an instrument naturally limits its use for such measurements, but where it serves several analytical functions such a use is a natural extension of its capabilities.

A colorimetric method[18] utilises the change in colour intensity of ammonia solution due to the formation of cupric ammonium ions by the permeation of oxygen. In this case the test piece is a product (package) containing the solution and strands of copper. The intensity is measured with a colorimeter giving an electrical output which is a function of oxygen concentration. This method is suitable for quality control testing.

Two other techniques are based on the sorption of the permeant by the test material and hence operate under non-steady state conditions. It will be recalled from §18.1 that the non-steady state is described by Fick's second law. Of the two methods, one uses a highly sensitive torsion balance[19] to measure weight change in the polymer as gas is sorbed. The other is based on the phenomenon of radio-thermoluminescence of polymers,[20] as a function of oxygen content. These last-named methods determine the diffusion constant, but if the solubility constant is known or can be measured then the permeability constant may be obtained.

A radioactive isotopic method has also been used.[21] In this case the permeabilities of elastomeric membranes to carbon dioxide, labelled with carbon-14, were measured. The experimental set-up is basically the same as the constant volume method but with a Geiger counter substituted for the pressure-measuring instrument.

18.3 VAPOUR PERMEABILITY

Methods used to measure the permeability of plastics to vapours fall broadly into two categories, namely those for water vapour and for volatile liquids. In both cases the most common measuring technique is the weight change or gravimetric technique on which several test standards are based. Other methods include techniques similar to those used for gases.

18.3.1 Water Vapour

Gravimetric Method
The transmission cell used in the gravimetric method is relatively simple, consisting of a lightweight metal dish with lid (Figure 18.6). Depending on the procedure, a dessicant, such as calcium chloride, or water is introduced into the bottom of the dish and the test piece is sealed onto the lip of the dish using wax, which has a very low permeation rate to water vapour. The assembly is then placed in an atmosphere of controlled temperature and humidity. The weight gain or loss is measured at regular intervals of time and, when this becomes approximately constant, the rate of weight change is used to calculate the quantity of water vapour which has permeated through unit area of the test piece in unit time.

ISO R1195 (1970),[22] BS 3177 (1959)[23] and BS 2782 Part 5 (1970),[24] Methods 513A/B, provide for two sets of environmental conditions, (a) temperate—25°C, 75 per cent (90 per cent ISO) r.h. and (b) tropical—38°C, 90 per cent r.h. All these standards use the dessicant technique.

ASTM E96-66 (1972)[25] offers six different procedures including both dessicant and water techniques, with environmental conditions ranging from 23°C, 50 per cent r.h. to 37·8°C, 90 per cent r.h.

In all cases the dessicant in the dish effectively reduces the water vapour pressure to zero on one side of the test piece, while the water methods raise the vapour pressure to a maximum for a given test temperature. The water contact method specified in ASTM E96 also tests for capillary transfer as well as diffusion.

Thicker sheet materials and rigid cellular materials are accommodated in separate test standards: ASTM C355-64 (1973),[26] BS 4370 Part 2 (1973),[27] Method 8, and ISO R1663 (1970).[28]

In all the previously mentioned standards for water vapour transmission, the use of wax to seal the test piece into the dish requires care and dexterity if minute leaks are to be avoided. Such leaks are difficult to detect and it is useful to test the sealing technique using a metal template in place of the polymer.

An alternative to the dish method is that which uses the test material in the form of a sachet. This is suitable for thermoplastic materials in thin sheet form capable of being heat sealed. The transmission of vapour through the seals must be considerably less than through the sheet. BS 2782 Part 5 (1970),[24] Methods 513C/D provide for such a test using the dessicant technique.

Several definitions are applied to the transfer of water vapour through a polymer film. BS 3177 and BS 2782 define it as follows:

permeability is the mass transfer rate of water vapour per unit area (g/m^2 24 h) (see §18.1).

BS 3177 extends this definition, when the material is homogeneous and the transfer rate is inversely proportional to thickness, as an equivalent permeability

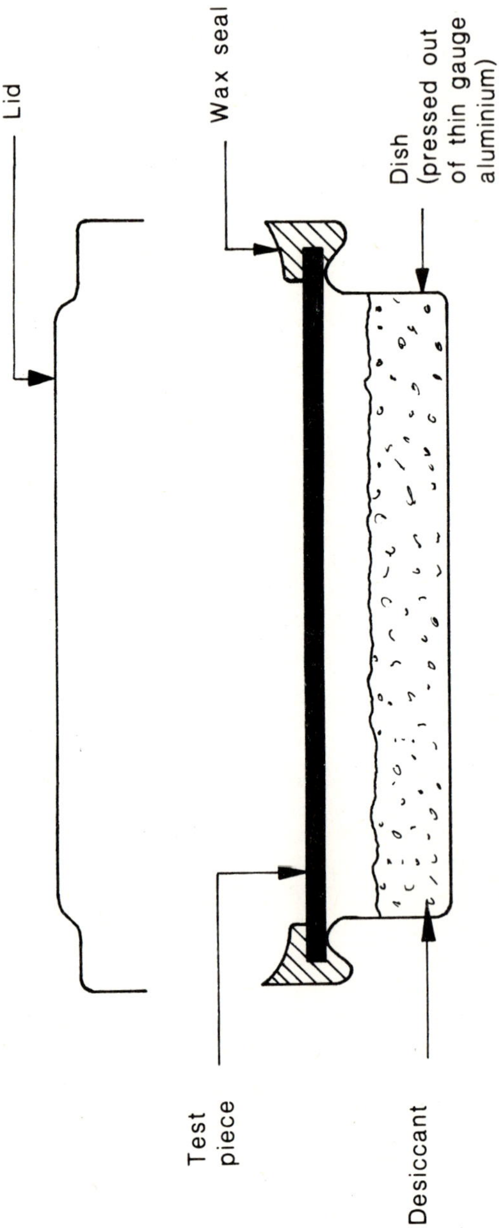

Figure 18.6 Water vapour transmission apparatus—dish method: BS 2782, Methods 513A,B

defined as follows:

equivalent permeability is permeability × the thickness of the test piece in thousandths of an inch (g mil/m^2 24 h).

The ASTM standards adopt the following definitions:

water vapour transmission rate is the mass transfer rate of water vapour per unit area (g/m^2 24 h);

permeance is the ratio of the water vapour transmission rate to the difference in vapour pressure betwween the surfaces of the test piece (g/m^2 24 h mm Hg = metric perms). This is equivalent to the gas transmission rate.

permeability is the product of the permeance and the thickness of the test piece, assuming that the permeance is inversely proportional to thickness for homogeneous materials (g cm/m^2 24 h mm Hg = perm-centimetre).

Since the adoption of SI units the water vapour permeability may also be expressed in the units of microgramme metre per newton hour ($\mu g\ m\ N^{-1}\ h^{-1}$ or $\mu g\ m\ m^{-2}\ Pa^{-1}\ h^{-1}$).

Other Methods

Refinements of the weighing technique have been made by enclosing the dish and a sensitive weighing balance in a chamber whose temperature and relative humidity are controlled.[29] This eliminates the need to remove the cell from its controlled environment for weighing. However, this is still a time-consuming test.

Electrical and chemical detection systems have also been employed.[29] An interesting variant which forms the basis of ASTM F372 (1973)[30] uses an infrared detector. In addition, radioactive labelling of water and its subsequent detection by liquid scintillation counter have been demonstrated.[31]

18.3.2 Volatile Liquids

The vapour transmission rates of other volatile liquids may also be measured from weight change data. This technique is also being considered for an International Standard; a suitable weighing vessel is illustrated in Figure 18.7. The test piece is sealed into the vessel by a screw-on lid having an aperture which defines the permeated area. To prevent distortion of the material, a metal ring separates the lid from the specimen. The ring and lid are free to rotate with respect to one another on ball bearings. The vessel is fitted with inlet and outlet ports to allow replenishment of the test liquid. This is particularly important if the liquid is made up of more than one volatile component having differing permeation rates. Thus the composition may be maintained constant throughout the test. When testing, the vessel is usually inverted so that polymer and liquid are in contact.

The result is quoted as a transmission rate with the test conditions stated. It should be remembered that some polymers swell when in contact with organic liquids, hence the thickness and the permeability are not constant throughout

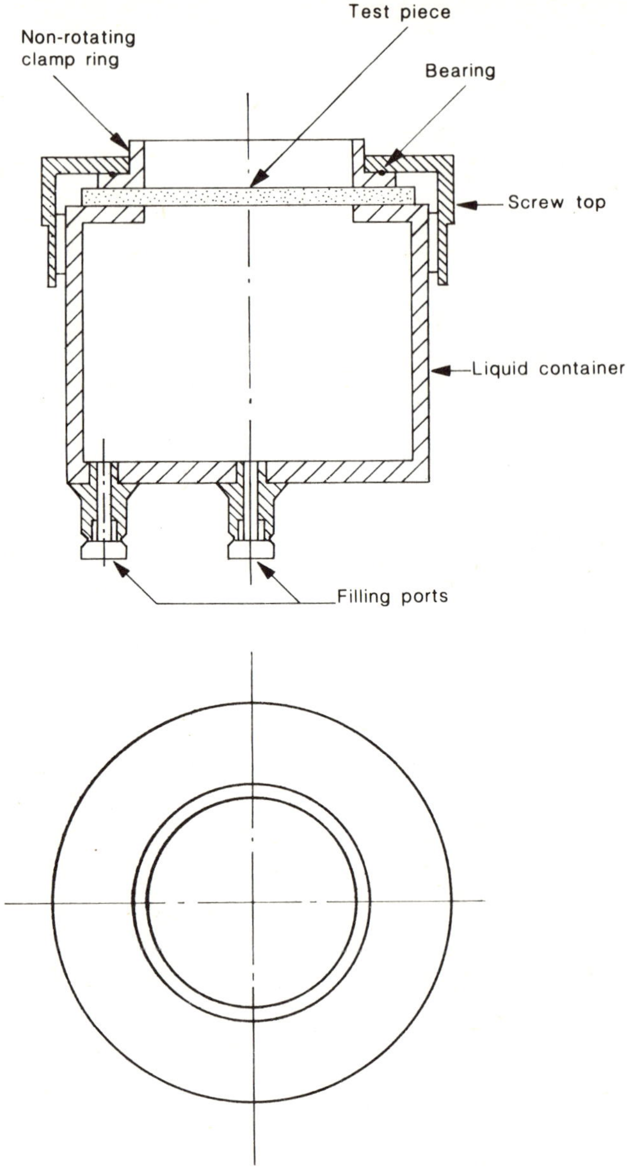

Figure 18.7 Vapour transmission cell

the test. Steady state conditions must also have been achieved before the transfer rate is calculated.

Other methods make use of the manometer[32] and gas chromatograph[33] (see also gas chromatography tests, §18.2.4[14–16]). The mass spectrometer perhaps offers the greatest potential in a single instrument for both water and other volatiles.

18.4 FURTHER READING

For a more detailed treatment of the theory of gas and vapour transport in polymers, the reader should consult some of the standard literature on the subject.[34,35] In addition a more comprehensive review of measurement techniques has been published.[36]

REFERENCES

1. ISO 2556 (1974) *Plastics—Determination of the Gas Transmission Rate of Films and Thin Sheets under Atmospheric Pressure—Manometric Method.*
2. BS 2782: Part 8 (1979) Method 821A *Determination of the Gas Transmission Rate of Films and Thin Sheets under Atmospheric Pressure (Manometric Method).*
3. ASTM D1434 (1975) *Standard Test Methods for Gas Transmission Rate of Plastic Film and Sheeting.*
4. DIN 53380 (1969) *Testing of Plastics Film; Determination of the Gas Transmission Rate.*
5. Linovitski, V. (1977) *Kunststoffe,* **67** (8), 433.
6. BS 903: Part A17 (1973) *Determination of the Permeability of Rubber to Gases* (*Constant Volume Method*).
7. Pye, D. G., Hoehn, H. H. and Panar, M. (1976) *Journal of Applied Polymer Science*, **20,** 1921.
8. BS 903, Part A30 (1975) *Determination of the Permeability of Rubber to Gases* (*Constant Pressure Method*).
9. Sweeting, O. J. (Ed.) (1971) *The Science and Technology of Polymer Films* vol. II, Wiley-Interscience.
10. Landrock, A. H. and Proctor, B. E. (1952) *Modern Packaging*, **25,** 131.
11. Pasternak, R. A., Schimscheimer, J. F. and Heller, J. (1970) *Journal of Polymer Science, Part A-2*, **8,** 467.
12. Ziegel, K. D., Frensdorff, H. K. and Blair, D. E. (1969) *Journal of Polymer Science, Part A-2 Polymer Physics*, **7** (5), 809.
13. Davis, E. G. and Rooney, M. L. (1975) *Journal of Applied Polymer Science*, **19,** 1829.
14. Caskey, T. L. (1967) *Modern Plastics*, **45** (4), 148/153/154.
15. Kapanin, V. V., Lemanik, O. B. and Reithninger, S. A. (1974) *Vyscokomol. Soyed.,* **A16** (4), 911.
16. Pye, D. G., Hoehn, H. H. and Panar, M. (1976) *Journal of Applied Polymer Science*, **20,** 287.
17. Invashchenko, D. A., Krotov, V. A., Talakin, O. G. and Fuks, Ye. V. (1972) *Vyscokomol. Soyed.*, **A14** (9), 2109.
18. Speas, C. A. (1972) *Package Engineering*, October, 78.
19. Volabuyev, P. V., Kupryazhkin, A. Ya. and Suetin, P. Ye. (1972) *Vyscokomol. Soyed.*, **A14** (2), 489.
20. Kiryushkin, S. G. and Gromov, B. A. (1972) *Vyscokomol. Soyed.*, **A14** (8), 1715.
21. Kirshenbaum, A. D., Streng, A. G. and Dunlap, W. B. Jr (1954) *Rubber Age*, March, 903.
22. ISO R1195 (1970) *Plastics—Determination of the Water Vapour Transmission Rate of Plastics Films and Thin Sheets—Dish Method.*
23. BS 3177 (1959) *Method for Determining the Permeability to Water Vapour of Flexible Sheet Materials used for Packaging.*

24. BS 2782: Part 5 (1970) Methods 513, A, B, C and D *Methods of Testing Plastics—Permeability to Water Vapour.*
25. ASTM E96-66 (1972) *Standard Test Methods for Water Vapor Transmission of Materials in Sheet Form.*
26. ASTM C355-64 (1973) *Standard Test Methods for Water Vapor Transmission of Thick Materials.*
27. BS 4370: Part 2 (1973) Method 8 *Methods of Test for Rigid Cellular Materials—Measurement of Water Vapour Transmission.*
28. ISO R1663 (1970) *Plastics—Determination of Water Vapour Transmission Rate of Rigid Cellular Plastics.*
29. Lelie, H. J. (1973) *Plastica,* **26** (12), 544.
30. ASTM F372-73 (1978) *Standard Test Method for Water Vapor Transmission Rate of Flexible Barrier Materials using an Infra-red Detection Technique.*
31. Matsui, E. S. (1970) *US NCEL Technical Report R674*, April, page 33.
32. Izydorczyk, J., Podkówka, J., Salwinski, J, and Grzywna, Z. (1977) *Journal of Applied Polymer Science*, **21**, 1835.
33. Duskova, D. (1974) Plasty a Kaucuk, **11** (3), 72.
34. Barrer, R. M. (1941) *Diffusion In and Through Solids*, Cambridge University Press.
35. Hennessy, B. J., Mead, J. A. and Stening, T. C. (1966) *Permeability of Plastics Films*, Plastics Institute.
36. Lomax, M. L. (1979) *RAPRA Members Report No. 33.*

Chapter Nineteen

Non-destructive Testing

19.1 INTRODUCTION

Non-destructive testing (NDT) is invariably performed on finished or semi-finished products. It is applied where sampling and subsequent destructive testing do not provide sufficient assurance of the quality of an individual component and where destructive testing of large, expensive components is impractical. NDT is also useful for in-service inspection of critical components to detect the onset of failure.

As the applications of plastics expand to fill roles traditionally the preserve of metals, so NDT takes on more importance. The metals industry has long used NDT on a routine basis and it is frequently called for in product specifications. It is not unreasonable to assume that, as development continues, NDT requirements for plastics products will become commonplace.

A great proportion of the work on NDT of plastics centres around the use of composites, particularly in the aerospace industry. The application of glass and carbon fibre laminates to critical structures in civil and military aircraft has provided the impetus for the development of techniques particularly suited to these materials. Plastics welding and adhesive jointing have also provided areas of interest.

Although techniques exist for the NDT of most plastics products, the major problem in their application is one of cost. For many products the cost of testing each item is often far higher than that of the product. Automatic test equipment can often reduce cost, but again the capital outlay is very high.

When applying NDT, it is essential to know the nature and likely location of the fault to be found. Faults tend to fall into one of three classes: firstly, voids or holes of macroscopic dimensions; secondly, discontinuities where the separation of the surfaces is extremely small; and finally, areas where there is no discontinuity, but where the mechanical properties of the component have been degraded.

Of the first class, typical examples are voids in the thick sections of mouldings and gaps in thick adhesive layers. Examples of the second class are lack of 'wetting out' in GRP components and incomplete fusion of welds. In the third class, orientation of thermoplastics and lack of cure of thermosets are typical. A fault not included in the above classes is the misplacement of reinforcing members, e.g. resin-rich areas in GRP.

To a greater or lesser extent, all components have faults; it is only where they are of a size or number to detract significantly from the service performance of the component that it is necessary to detect them. This critical flaw size must be calculated, considering the service conditions and the design safety factor. This figure will set the sensitivity requirement for the NDT technique to be used.

In the following sections brief descriptions of the major NDT techniques applicable to plastics are given, together with sources of further details and examples of their application.

A number of tests previously described in this book that are essentially non-destructive—notably dimensional measurement, hardness and electrical tests—are not described further here.

19.2 ULTRASONIC TESTING

Ultrasonic testing comprises a range of methods which make use of mechanical oscillation at frequencies above 20 kHz. These waves behave in a similar manner to light: they obey the laws of reflection and refraction. Krautkramer[1] gives a detailed account of the physics and applications of ultrasonics when applied to materials testing.

Ultrasonics is used both to find imperfections in a component and to measure mechanical moduli. In flaw detection, pulses of ultrasound are reflected from boundaries to build up a 'picture' of the interior of a component. These pulses are produced and received by transducers which are normally piezoelectric crystals. With this type of transducer, it is possible to use the same transducer for both transmission and reception. Figures 19.1 and 19.2 show various configurations of transducers for flaw detection.

There are two types of sound wave: (i) longitudinal (compression), where the mechanical energy is transmitted through the material, exciting the molecules in the direction of the sound wave; and (ii) transverse (shear), where the excitation is normal to the wave.

For a given material the two types of wave have different values of velocity and attenuation. The shear wave has a lower velocity and receives much higher attenuation.

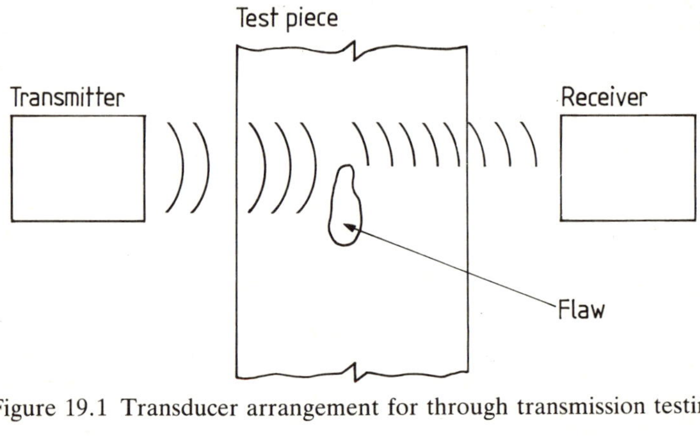

Figure 19.1 Transducer arrangement for through transmission testing

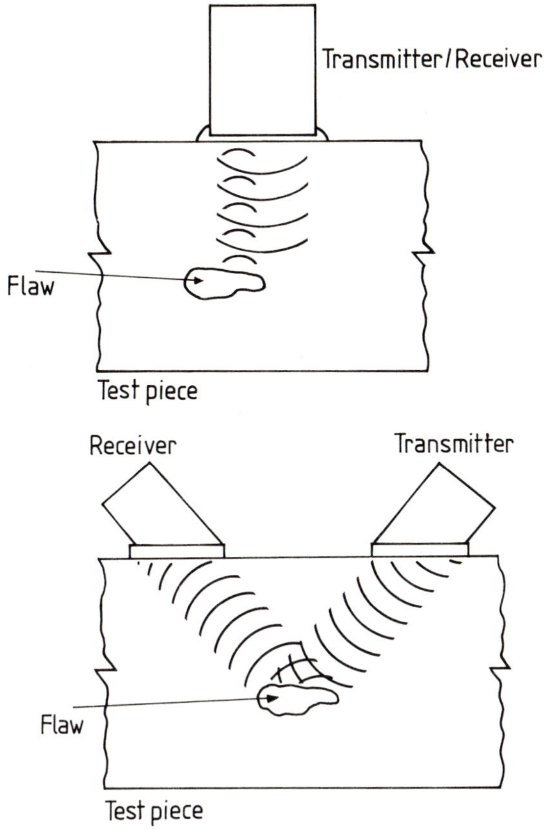

Figure 19.2 Transducer arrangements for pulse echo testing—one and two transducers

Compared with metals, plastics have low sound velocities. Table 19.1 gives values for some common materials. Attenuation of the shear wave in plastics is very high, hence it is rarely used for flaw detection. Some of the more flexible materials, such as LDPE and plasticised PVC, will not support a shear wave.

An important property is acoustic impedance, which is the product of the sound velocity and the density of the material. Again, Table 18.1 gives some values.

When a sound wave hits the boundary of two materials, the proportion of its energy which is reflected is a function of the acoustic impedance of the two materials.

$$R = \frac{(W_1 - W_2)^2}{(W_1 + W_2)^2} \tag{19.1}$$

where R is the reflection coefficient and W_1 and W_2 are the acoustic impedances of the two materials.

When the boundary comprises a gas and a solid, reflection within the solid is virtually 100 per cent even if the thickness of the gas normal to the wave is microscopic.

Table 19.1 Acoustic Properties of Some Common Materials

Material	**Acoustic velocity (longitudinal wave)** ($m/s \times 10^3$)	**Acoustic impedance** ($kg/m^2\,s^{-1} \times 10^6$)
Aluminium	6·32	17
Brass	3·83	33
Glass (crown)	5·66	14
Steel	5·90	46·5
ABS	2·0	2·2
Acetal	2·4	3·6
Nylon 66	2·6	2·9
HDPE	2·4	2·3
Polycarbonate	2·0	2·4
PMMA	2·74	3·2
Polypropylene	2·4	2·1
Polystyrene	2·3	2·4
Rigid PVC	2·3	3·2
Poly(4-methyl pentene)	2·0	1·66
PTFE	1·5	3·2

Once the reflected pulse is received from the flaw, it must be displayed to the operator in a convenient form. The methods of achieving this are called scans and the three most common are termed 'A', 'B' and 'C'.

'A' scan is the most simple and commonly used. The received pulse is displayed on a cathode ray tube as an amplitude against time, the time sweep being started by the transmitted pulse. The displacement of the flaw echo from the start of the trace gives the depth of the flaw, and the height of the pulse gives some indication of the size of the flaw normal to the beam. A typical 'A' scan trace is shown in Figure 19.3.

When using 'A' scan, the operator moves the transducer over the component's surface and notes its position when a flaw is found.

'B' and 'C' scans give information on the position of the flaw, other than depth. 'B' scans give a plot of flaw depth against distance from a point on the surface (Figure 19.4). 'C' scan gives a map-view of the component showing flawed areas but gives no information on the depth of the flaw (Figure 19.5). These methods require the transducer to be fitted to a mechanical scanning

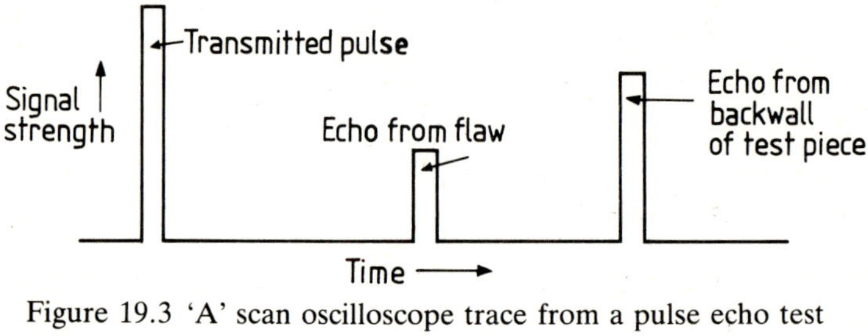

Figure 19.3 'A' scan oscilloscope trace from a pulse echo test

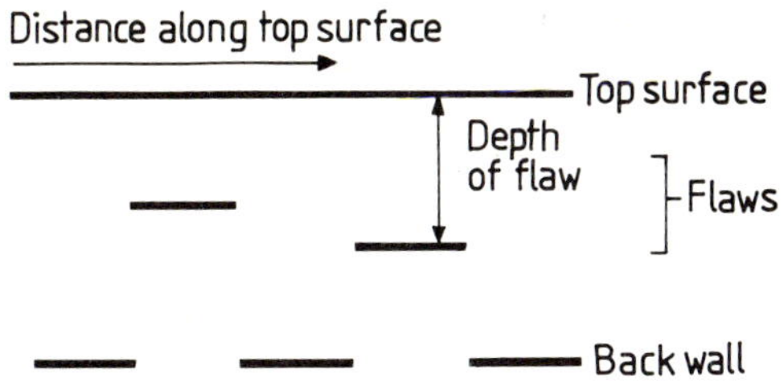

Figure 19.4 'B' scan recorder output from a pulse echo test

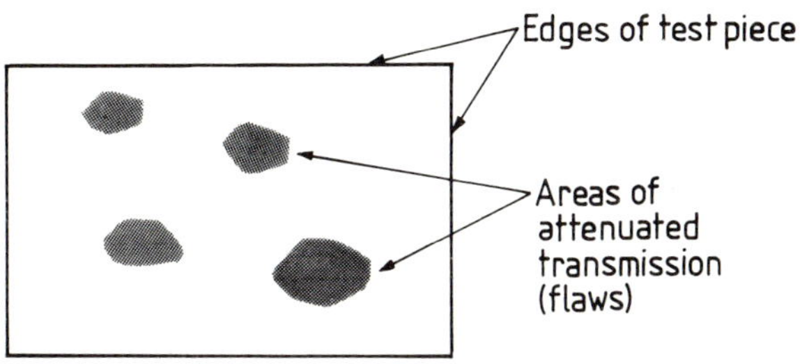

Figure 19.5 'C' scan recorder output from a through transmission test

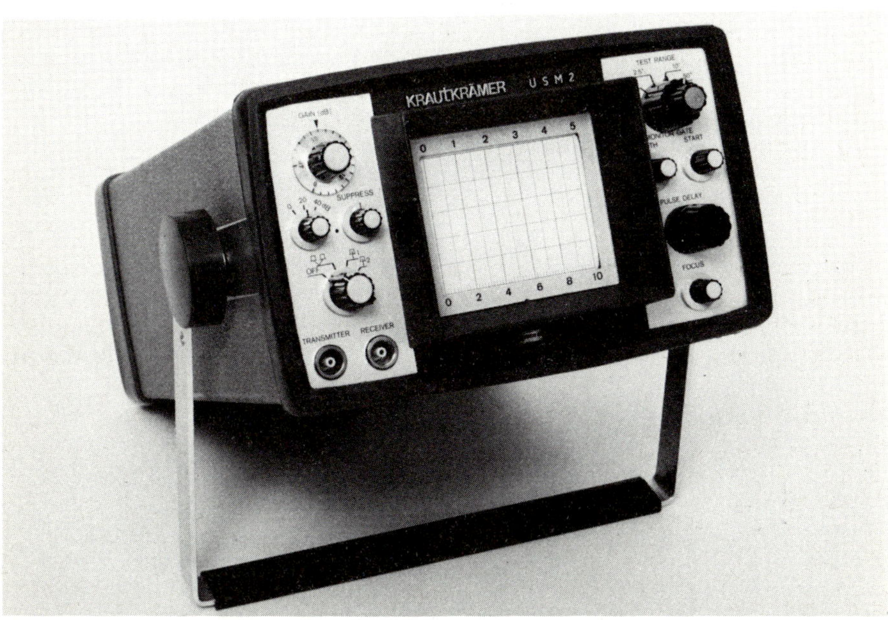

Figure 19.6 Typical ultrasonic pulse echo flow detector with 'A' scan output (*Courtesy Wells-Krautkramer Ltd*)

device and normally present their display on paper. Both 'B' and 'C' scan systems are well suited to automatic inspection methods.

The 'C' scan method has been used successfully to detect areas of voids in fibre-reinforced plastics components.[2–5] For the inspection of welds in thermoplastics, the 'A' scan method has proved to be of value.[6–8]

Ultrasonic spectroscopy, in which the frequency content of the received pulse is examined, has been found to be capable of distinguishing areas of delamination in fibre-reinforced plastics.[9,10]

The velocities at which the longitudinal and transverse waves travel through a material are determined by the elastic constants of that material. Measurement of the velocities allows the calculation of all the moduli and Poisson's ratio.[11] This is of particular importance when dealing with composite materials, as the ability to direct the ultrasonic beam through the material along different axes allows the calculation of the complete set of elastic constants.[12–14] This is also useful in assessing the degree of anisotropy in mouldings.[15] The results obtained from this method are not normally comparable to conventional testing because of the frequency-dependence of the moduli; however, agreement has been found in measurements on epoxy resin.[16]

In conclusion, ultrasonics is a very useful technique for 'looking inside' a component. The high attenuation of many plastics is a major drawback to its usefulness, although increased sensitivity of equipment will continue to alleviate this.

Another problem arises from the fact that, since plastics are easily formed, the component under investigation may be complex in shape, so that the placing of transducers to view suspect areas is difficult. Advances in transducer design will continue to relieve this problem.

It is in the area of processing and displaying the results of ultrasonic inspection that the application of modern electronics is likely to have the most impact, providing equipment that requires less operator skill in its use.

19.3 ACOUSTIC EMISSION

Certain materials, when subjected to stress, exhibit the phenomenon of acoustic or stress wave emission.

This is the release of stress wave pulses, in a broad frequency range, due to deformation and fracture. The frequency range 50 kHz to 1·5 MHz has been found to be the most useful.

Because some deformation must occur to produce acoustic emission, the method is not wholly non-destructive; however, if this deformation and subsequent emission are produced by the first loading of the product, and loads are kept to those used in service, no extra damage will occur.

The stress waves emitted from the component are detected by one or more sensitive piezoelectric transducers. The electrical signal produced is amplified and passed to the signal analysis equipment. The analysis equipment characterises the emission in terms of pulse count, height, length, energy content, frequency content, etc. Details of the systems and analysis techniques have been published.[17–21]

An important factor in the use of acoustic emission is the Kaiser effect, which is simply the observation that loadings of the component which cause no further damage are free from acoustic emission, i.e. when repeated loadings to a fixed stress level are performed, only the first loading will produce real emission. This provides a means of determining whether the emission is a genuine measure of the damage on the component or is a spurious emission due to frictional sources.

Figure 19.7 shows the total stress wave emission count and its associated load curve against time. The total emission begins to rise sharply as the component approaches its critical load point. Finding this point enables a proof load to be determined which, when applied to faulty components, will give rise to characteristic emission evidencing poor quality.

The great majority of acoustic emission studies have been performed on composite materials. The first practical application was pressure testing of GRP Polaris Missile chambers by Green *et al.*[22] Becht *et al.*[23] have used acoustic emission to investigate flaw formation and propagation in GRP pressure tubing, and fracture behaviour of GRP in flexural and fatigue experiments. Raymond[24] has used the method to study the effect of resin content of GRP laminates.

Stress wave emission from adhesive joints have been studied by Curtis[21] and Brockman *et al.*[25] with some success.

Little is known of stress wave emission from unreinforced thermoplastics and thermosets, although Sims[26] reports emission from those that fail through crack propagation.

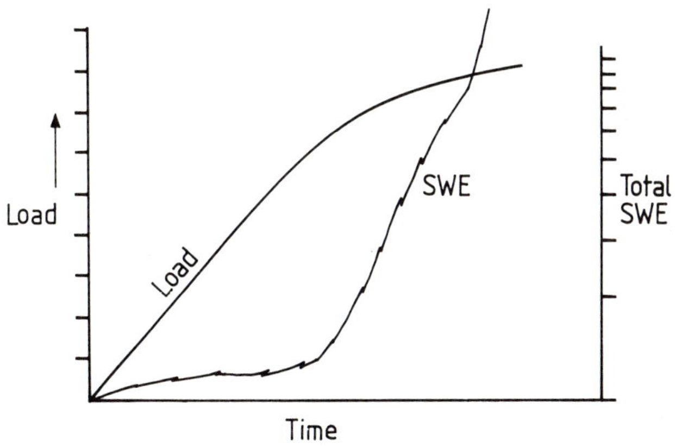

Figure 19.7 Typical acoustic emission output showing load and total stress wave emission versus time

19.4 RADIOGRAPHY

The discovery by Röntgen of X-radiation early in this century has provided non-destructive testing with one of its most useful tools.

Electromagnetic waves in the visual part of the spectrum have little penetration power—hence the opacity of most of the objects about us. The wave energy is reflected from, or absorbed by, the object. However, as we move beyond the visible spectra through the ultraviolet, the energy of the emitted photon increases. After passing through the far ultraviolet we reach the X-ray spectra with wavelengths between 500 and 0·1 Å. These waves are highly penetrative, the shorter wavelength more so than the longer.

The amount of X-ray attenuation within a material is a characteristic of that material. Hence, just as it is possible to see voids within a translucent material, so, with X-rays, it is possible to study the interior structure of an object.

X-rays are produced electrically by means of a tube similar to a common cathode ray tube. Whereas the domestic television set uses accelerating potentials of around 9 kV, X-ray tubes use potentials between 10 and 800 kV to accelerate the electrons. In the television set, the accelerated electrons strike the phosphor of the screen and cause visible light to be emitted; the higher energy electrons in the X-ray tube give rise to the higher energy X-ray when they hit a metal plate called the target. The X-radiation from this target covers a broad band of frequencies. Different frequencies are attenuated differently by a material, and different materials have different attenuation characteristics. For this reason, the choice of accelerating potential is important. Compared with steel, plastics are quite transparent to X-rays. In order to detect fine detail in the structure of plastics components, it is usually necessary to use low energy. To enable a tube to produce these low energy X-rays, it must possess a beryllium 'window' through which the radiation can emerge, since the usual glass tube envelope absorbs these energies. The final selection of accelerating potential is made by trial on the component to be tested with a view to obtaining the most detailed picture possible.

As the human eye is not able to see X-radiation, a means must be provided to view the picture produced. Normally a photographic film is used which is exposed to the X-radiation emerging from the item under test. The disadvantages of film are its high cost and its lengthy processing time. Systems are produced which do not have these problems; the fluorescent screen, which emits visible light when struck by X-rays, is the most popular. Such systems, especially when linked to closed circuit television, provide an efficient and rapid solution, but to date they are unable to provide the quality and detail available from film.

Unlike photography, no lenses are used in radiography, hence the geometry of the test set-up is important. The general form of an X-ray system is shown in Figure 19.8. The X-ray source may be regarded as a point. The X-rays emerge from the window and diverge. The film is placed immediately behind the object. The film-to-source distance (FTSD) is important in determining the resolution of the radiograph. The higher the ratio of FTSD to object thickness, the higher will be the resolution of the radiograph. Ratios of 5:1 are adequate for detecting large voids and for gross inspection, 10:1 is suitable for detecting small voids in mouldings. Ratios of the order of 100:1 are required for optimum resolution of very small flaws in critical components.

A further criterion which determines the X-ray system required is the maximum size of object to be inspected. For large objects, e.g. pressure vessels, chemical plant, aircraft structures, a mobile X-ray source in a shielded room with remote controls is required. For smaller items (up to about 760 mm square) a fixed source in a shielded cabinet will usually suffice. This cabinet-type system has the advantage that it requires little installation, operators require the minimum of training and the units are inherently safe.

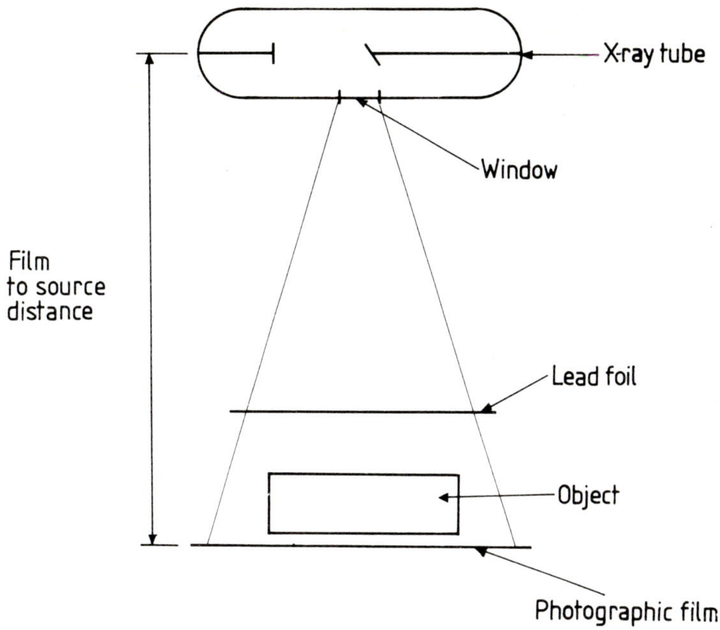

Figure 19.8 General form of an X-ray system

The application of X-radiography to plastics components is limited by cost and the size of flaw to be detected. Voids of less than 1 mm diameter in a plastics mouldings are readily found by means of radiographic film, but are far more difficult to detect if the less costly fluorescent screen technique is used.

The use of gamma-emitting isotope sources as a substitute for conventional X-ray equipment is common in heavy industry, but is unlikely to find application to plastics because of the very high energies of radiation produced. A far more interesting type of radiation is that of the neutron.

A thermal neutron beam is heavily attenuated by materials rich in hydrogen, such as plastics, but is little affected by its passage through many heavier materials, notably steel. This provides a means of investigating plastics components even when they are surrounded by metal. Spowart[27] reviews the technique and gives examples of its use.

The source of the neutron beam is commonly a nuclear reactor, a severe restriction on the technique's use in industry; however, isotope sources are now available providing semiportable facilities. Equipment costs, however, are very high and unlikely to fall.

19.5 OPTICAL METHODS

19.5.1 Visual Inspection

The human eye is the finest item of non-destructive testing apparatus available. Visual inspection of components is a prime requirement for quality control,

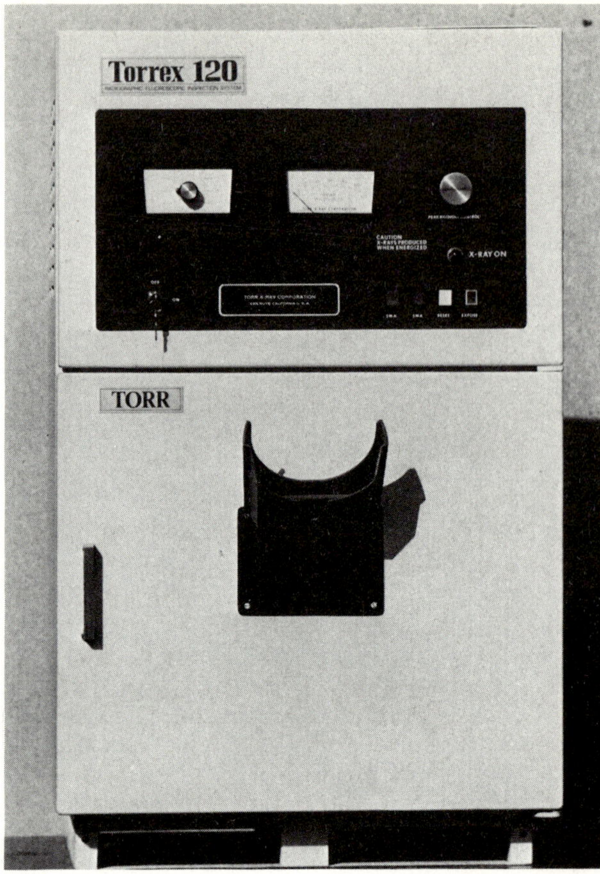

Figure 19.9 Cabinet-style X-ray unit with fluoroscopic viewing system (*Courtesy Wells-Krautkramer Ltd*)

providing an assessment of shape, surface finish and superficial flaws. When assisted by the techniques described in the following sections a comprehensive picture of the quality of a component can be generated. (Instrumental optical techniques, wherein the fundamental optical properties of the material are measured, have been dealt with in Chapter 14).

Simple visual inspection is probably the technique most ignored and taken for granted. The least that is required is that the inspection area is clean, uncluttered and well lit and that the operative is comfortably situated. Failure to provide these basic requirements will result in the acceptance of obviously sub-standard production. Lighting should be strong but should not throw glare at the inspector. A figure of 1500 Lux is recommended for fine inspection. For small components an illuminated magnifier is to be recommended.

Inspection of cavities and bores is aided by the use of borescopes and endoscopes of which a wide range is available.

Where the detection of fine cracks is required the use of a fluorescent dye penetrant is advisable. The dye is introduced on to the surface of the component and then removed. Removal will not be complete within a crack, hence when the component is subsequently viewed under UV lighting the crack will appear as an obvious bright line.

19.5.2 Photoelasticity

As discussed in Chapter 14, many plastics are optically active when stress is applied. When they are examined with polarised light the direction and magnitude of the stresses within the material may be deduced. Apart from the straightforward case of examination of a transparent moulding to assess the residual stresses, a technique has been developed to examine the strain distribution, in use, of opaque and optically inactive materials. A thin coating of optically active material is applied to the component in the unstressed condition, the test loading is applied and the surface coating is examined using reflected light. Instruments called photoelastic reflection polariscopes are available to measure the strains in the structure under test.

19.5.3 Coherent Optical Techniques

The development of the laser allowed the practical use of a number of well known optical phenomena. The ability of the laser to produce an intense, parallel and coherent beam of monochromatic light made feasible the application of interferometry techniques to component testing. The measurement of small displacements using light wave interference, e.g. by the Michelson interferometer, has been known for many years. The technique of holography, the storage of three-dimensional information on photographic medium, was also well known but the lack of a suitable light source prohibited its application. The laser enables these two techniques to be applied together to provide the method of holographic interferometry. Using the optical arrangement shown in Figure 19.10, a photographic plate is partially exposed with the object in the unstressed condition, the component is stressed and the plate further exposed. The plate now contains three-dimensional information of the component in both the stressed and unstressed states. When the hologram is reconstructed using the same optical arrangement but with the component removed, two superimposed images of the component are formed. Where the images are different, i.e., where the stress has caused movement of the surface, an interference fringe pattern is formed, the density of the fringes indicating the amount of displacement. This is known as lapsed time holographic interferometry.

The technique of real time holographic interferometry uses the same basic optical arrangement but the photographic plate is developed after the first exposure and the resulting image is projected onto the component as the stresses are applied, hence the interference pattern is formed in real time.

Both techniques require very stable optical and mechanical arrangements, real time holography being the more critical as gross movement between the components of the system must be limited to $\frac{1}{8}$ of the wavelength of the light. The test stress may be applied mechanically or thermally.

Where there is an interest in the effect of resonant vibration within the component, the technique of time-averaged holographic interferometry, as described by Stetson and Powell[28] and Stetson and Taylor,[29] can be utilised. A further technique makes use of the speckle pattern formed when coherent light is scattered from a surface. When two speckle patterns are superimposed an interference pattern similar to the holographic interference pattern is produced. Its advantage is that the speckle patterns are compatible with the resolution of a television camera, giving the possibility of on-line data processing, and removing problems of film handling.

Systems for both lapsed time holography and speckle pattern interferometry are commercially available.

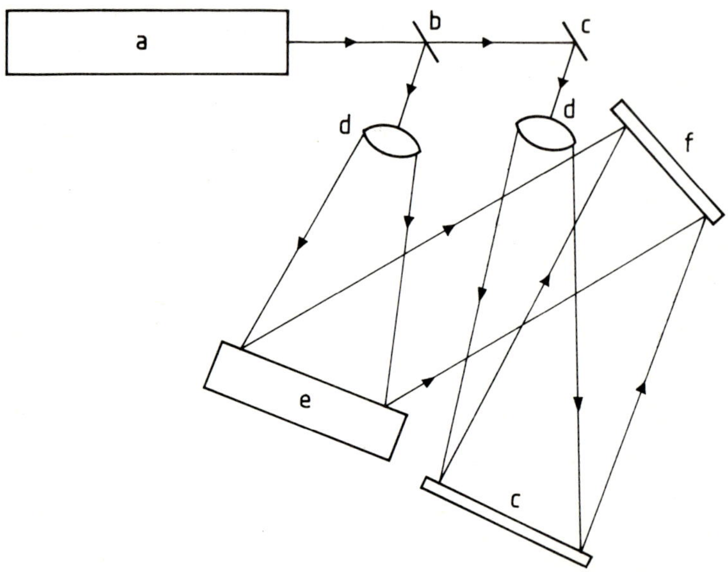

Figure 19.10 Optical arrangement for holographic interferometry
a Laser
b Beam splitter
c Mirror
d Lens
e Object
f Photographic plate

The applications of these techniques are wide ranging. Grunwald *et al.*[30] have detected faults in laminated fuel tanks and measured the deformations under stress of GFR plates.[31] Peterson *et al.*[32] and Krenz[33] have used holographic interferometry to investigate crazing. Archbold and Ennos[34] have used HI to test the components of a plastics pipe system. The technique of speckle photography has been investigated by Archbold and Ennos.[35] Sensitivity is much lower than with interferometry, but large objects can be investigated using lower laser power.

19.6 OTHER NON-DESTRUCTIVE TEST METHODS

Apart from the techniques already described, there is a number of methods which may find application to plastics components.

Infrared radiation techniques using either a scanning sensor or a thermal camera (thermography) have been used to observe the heat flow through a component, an alteration in heat flow being indicative of a change in mechanical properties. Kaplan[36] and McLaughlin[37] describe the techniques and give examples of application.

Beta-radiation gauges are in common use as on-line thickness gauges for some continuous processes. Their ability to provide a measure of mass per unit area enables measurement of porosity, wall thickness, etc., to be made.

Microwave radiation of millimetric wavelength has been used to detect voids in GRP structures.[38] The sensitivity of this technique is limited by the wavelength used, although it would appear useful for large voids and thickness measurements.

Acoustic holography which uses coherent light techniques to display a picture of the internal structure of an object in the path of an ultrasonic beam is finding increasing application. Details and some applications of the technique are given by Brenden and Collins.[39]

Eddy current testing is not normally applicable to plastics, as a conducting material is required; however, with carbon fibre laminates, the conductivity of the fibres is sufficient to allow its use. Owston[40] has assessed the technique's potential to indicate variations in fibre volume fraction and fibre lay-up.

REFERENCES

1. Krautkramer, J. and Krautkramer, H. (1969) *Ultrasonic Testing of Materials*, Springer.
2. Stone, D. E. W. and Clarke, B. (1975) *Non-destructive Testing*, June, 137.
3. Jones, B. R. and Stone, D. E. W. (1976) *Non-destructive Testing*, April, 71.
4. Martin, B. G. (1976) *NDT International*, October, 242.
5. Van Dreumel, W. H. M. (1978) *NDT International*, October, 233.
6. Herrmann, H. (1971) *Kunststoffe*, **61** (November), 839.
7. Herrmann, H. (1973) *Kunststoffe*, **63** (August), 535.
8. Kolb, K. (1976) *Kunststoffe*, **66** (6), 357.
9. Chang, F. H., Yee, W. G. W. and Couchman, J. C. (1974) *Non-destructive Testing*, August, 194.
10. Cousins, R. R. and Markham, M. F. (1977) *Composites*, July, 145.
11. Markham, M. F. (1969) in *Conference, Ultrasonics for Industry*, Iliffe, page 1.
12. Markham, M. F. (1970) *Composites*, **1** (3), 145.
13. Dean, G. D. and Turner, P. (1973) *Composites*, **4,** 174.
14. Dean, G. D. (1974) in Conference, *Composites—Standards, Testing and Design,* National Physical Laboratory, April, page 126.
15. Thomas, K. and Meyer, D. E. (1976) *Plastics and Rubber, Materials and Applications*, **1** (3), 136.
16. Smith, A., Wilkinson, S. J. and Reynolds, W. N. (1974) *Journal of Materials Science*, **9** (4), 547.
17. Arrington, M. (1976) *Metrology and Inspection*, October.

18. Swindlehurst, W. (1973) *Non-destructive Testing*, June, 152.
19. Pollock, A. A. (1973) *Non-destructive Testing*, October, 264.
20. Brindley, B. J., Holt, J. and Palmer, I. G. (1973) *Non-destructive Testing*, December, 299.
21. Curtis, G. (1974) *Non-destructive Testing*, April, 82.
22. Green, A. T., Lockman, C. S. and Steele, R. K. (1974) *Modern Plastics*, **41** (July), 137.
23. Becht, J., Schwalbe, H. J. and Eisenblaetter, J. (1976) *Composites*, October, 245.
24. Raymond, J. A. (1974) *Ninth Biennial International Conference*, British Plastics Federation, Paper 8.
25. Brockman, W., Roeder, E. and Crostack, H. A. (1975) *Kunststoffberäter, Rundschau und Technik*, Part 1, **20** (5), 263; Part 2, **20** (6), 316.
26. Sims, G. D. (1976) *Plastics and Rubber, Materials and Applications*, September, 205.
27. Spowart, A. R. (1972) *Journal of Physics, E*, **5,** 497.
28. Stetson, K. A. and Powell, R. L. (1965) *Journal of the Optical Society of America*, **55,** 1694.
29. Stetson, K. A. and Taylor, P. A. (1972) *Journal of Physics, E,* **5** (9), 923.
30. Grunwald, K. and Fritzsch, W. (1973) *Kunststoff-Rundschau*, **20** (12), 593.
31. Grunwald, K., von Harnier, A., Roth, E. and Fritzsch, W. (1974) *Kunststofftechnik*, **13** (5), 98.
32. Peterson, T L., Ast, D. G. and Kramer, E. J. (1974) *Journal of Applied Physics*, **45** (10), 4220.
33. Krenz, H. G., Kramer, E. J. and Ast, D. G. (1975) *Polymer Letters*, **13,** 583.
34. Archbold, E. and Ennos, A. E. (1974) *Journal of Strain Analysis*, **9** (1), 10.
35. Archbold, E. and Ennos, A. E. (1972) *Optica Acta*, **19,** 253.
36. Kaplan, H. (1972) in *Applied Polymer Symposium No. 19*, Wiley, page 441.
37. McLaughlin, P. V., McAssey, E. V. and Dietrich, R. C. (1980) *NDT International*, April, 56.
38. Botsco, R. J. (1968) *Plastics Design and Processing*, November, 12.
39. Brenden, B. B. and Collins, H. D. (1974) in *Holographic Nondestructive Testing*, R. K. Erf (Ed.), Academic Press, page 405.
40. Owston, C. N. (1972) *Cranfield Memo No. 73,* Cranfield Institute of Technology, June.

Chapter Twenty

Testing Products

20.1 INTRODUCTION

Moulding powders are, of course, products, as are moulded sheet and laminates, but in the context of the bulk of testing subjects considered in this book they are products for which the normal, and the easiest, approach will be to produce standard test pieces. For 'finished' products, it is often desirable to cut or machine standard test pieces by the methods discussed in Chapter 3, rather than specially to mould test pieces or sheet. The reasons are self-evident: there is no chance of the test material being different from that of the product and the properties measured will not be changed by a different processing procedure being used for the test pieces. There are, however, the disadvantages of the extra effort required to produce test pieces and the fact that machining may affect the property in question.

These considerations apart, it is the product which has to perform and the surest way to test for product performance is to forget standard test pieces and test the whole product. This is not always easy or convenient and often judgments on performance will still have to be made on the basis of standard tests. (These tests will of course also continue to be valuable in quality assurance, material evaluation, etc.)

Tests on products are standardised in product specifications, although perhaps not as often as would be desired, but a great many more product tests are invented and used on an *ad hoc* basis. It is not possible in this book to consider every product test known to the authors but it is appropriate that attention is drawn to the types of test which can be carried out, with advantage, on the product itself. Anyone contemplating tests on products is advised to check the ISO and the appropriate national standards yearbook for tests on the same or similar products which have already been standardised.

20.2 SEMI-FINISHED PRODUCTS

Semi-finished products can be taken to include sheet, film, laminates and coated fabrics which will later be used to fabricate other articles. In general, all the usual laboratory tests from tensile strength to gas permeability can be made on test pieces cut from these products, the only deviation from standard procedures perhaps being the need to use test pieces which are the thickness of the product in question rather than that given in the standard method.

In the case of films, test methods for mechanical properties such as tensile strength and impact have been specially formulated and standardised and have been mentioned in earlier chapters. Coated fabrics are treated as being a little apart from either rubbers or plastics and a large number of test methods have evolved specifically for them. In Britain the tests which have been standardised are collected together as BS 3424 (1973).[1] The equivalent ISO standards are published separately as individual test methods.

Decorative and industrial laminates can be tested for many properties in the same manner as sheets by the process of cutting standard test pieces, but they will have particular requirements for properties such as abrasion resistance and the effect of hot objects, for which tests need to be specially formulated. Tests standardised to date can be found in specifications for laminates, for example those of ISO and British Standards.[2–7] For some properties, a variation on the standard methods used for plastics in general is specified for laminates. This may be included in the standard method of test for that property but may be published under a separate number; for example, in the BS 2782[8] series there are separate methods for temperature of deflection under load (121C), crushing strength after heating (131B), resistance to dry heat (132A) and dimensional change (106H).

20.3 FIBRE REINFORCED PLASTICS

The fibre reinforced plastics are really a type of plastics rather than a particular product but it is emphasised that this class of material does sometimes require a different approach to testing. This important class of materials is considered quite briefly here because most of the tests carried out on fibre reinforced materials are in principle the same as those carried out on other plastics. The need for special test piece geometries for such tests as tensile stress–strain properties must be noted; where these requirements have been incorporated into standard methods they have been discussed in Chapter 8.

There is a number of standard test methods specific to fibre reinforced materials. Although it is intended to include tests for tensile properties in ISO 527, currently glass reinforced materials are covered by ISO 3268 (1978)[9] which is discussed in Chapter 8. There is a very similar European Standard EN 61 (1977)[10] which is reproduced as BS 2782 (1977) Method 1003.[11] This proliferation of similar standards (GRP is also covered in BS 2782 Method 220) is unfortunate and it is to be hoped that they will be rationalised in the future.

An unusual technique has been developed particularly for the evaluation of the raw materials of filament-wound structures. It involves the making of NOL (Naval Ordnance Laboratory) rings, which is described in ASTM D2291 (1976).[12] The rings can be made from dry reinforcement or pre-impregnated materials. The winding system for dry reinforcement is shown schematically in Figure 20.1. After having been wound onto a mould or mandrell, the material is oven cured. Three sizes of ring are specified, each 146 mm internal diameter by 6·35 mm wide, the wall thicknesses being in two cases 1·52 mm and the other

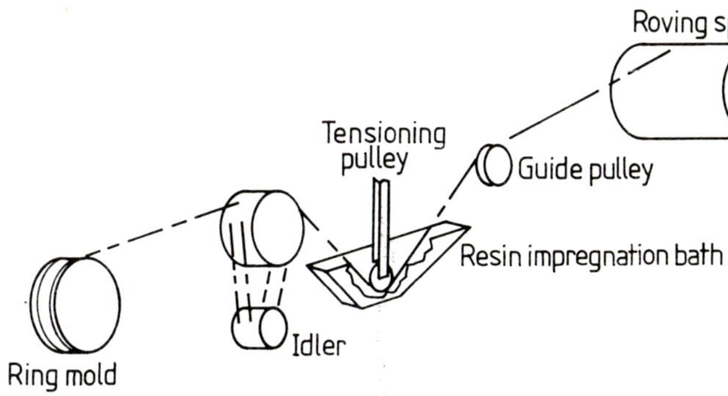

Figure 20.1 Schematic diagram of wet roving ring winding apparatus: ASTM D2291

3·18 mm. The difference between the two apparently identical rings is that one is made over-thick and machined down, whereas the other is moulded to the correct wall thickness directly. The 3·18 mm ring is also moulded over-thick and machined down. The procedure for machining is not given.

Tests that may be applied to these rings include split disc ring tensile strength, ring compression strength, flexural strength and interlaminar shear strength. The interlaminar shear test is given in ASTM D2344 (1976),[13] and is essentially the method given in BS 2782 Method 341A (see Chapter 8), using test pieces cut from the rings. Flexural strength can be measured using segments of the ring and compression strength measured on the complete ring. The split ring tensile test is given in ASTM D2290 (1976)[14] and uses the complete ring. A test fixture to be used for straining the ring is shown in Figure 20.2.

There are two ISO standards for tests on rods made from textile glass rovings, ISO 3597 (1977)[15] for the determination of flexural strength and ISO 3605 (1978)[16] for compressive strength, both of which give details of test piece preparation. ISO 3597 uses a modification of the general flexural strength method, ISO 178, to suit rods, while ISO 3605 simply specifies compression to failure at 1 mm/min using jigs to support the ends of the test piece. There is also a number of standards for testing the textile glass reinforcement but consideration of these is outside the scope of this book. Methods for the preparation of panels of glass reinforced plastics for test purposes have been considered in Chapter 3.

European Standard test methods exist for glass reinforced plastics and these have also been reproduced in the BS 2782 series. EN 59 (1977)[17] covers measurement of hardness using the Barcol impressor (Chapter 8). EN 60 (1977)[18] is substantially in agreement with ISO 1172 (1975)[19] and is intended to give a measure of glass content, although the results will be influenced by the presence of ingredients other than glass and resin. The method for tensile properties, EN 61 (1977)[10] has been mentioned above and is largely in agreement with ISO 3268 (1978).[9] All these European Standards refer to EN 62 (1977),[20] *Standard Atmospheres for Conditioning and Testing*, which is effectively a copy of those internationally standardised atmospheres which are relevant to European climatic conditions. EN 63 (1977)[21] for flexural properties is equivalent to the content of ISO 178 (1975)[22] (see Chapter 8) which is

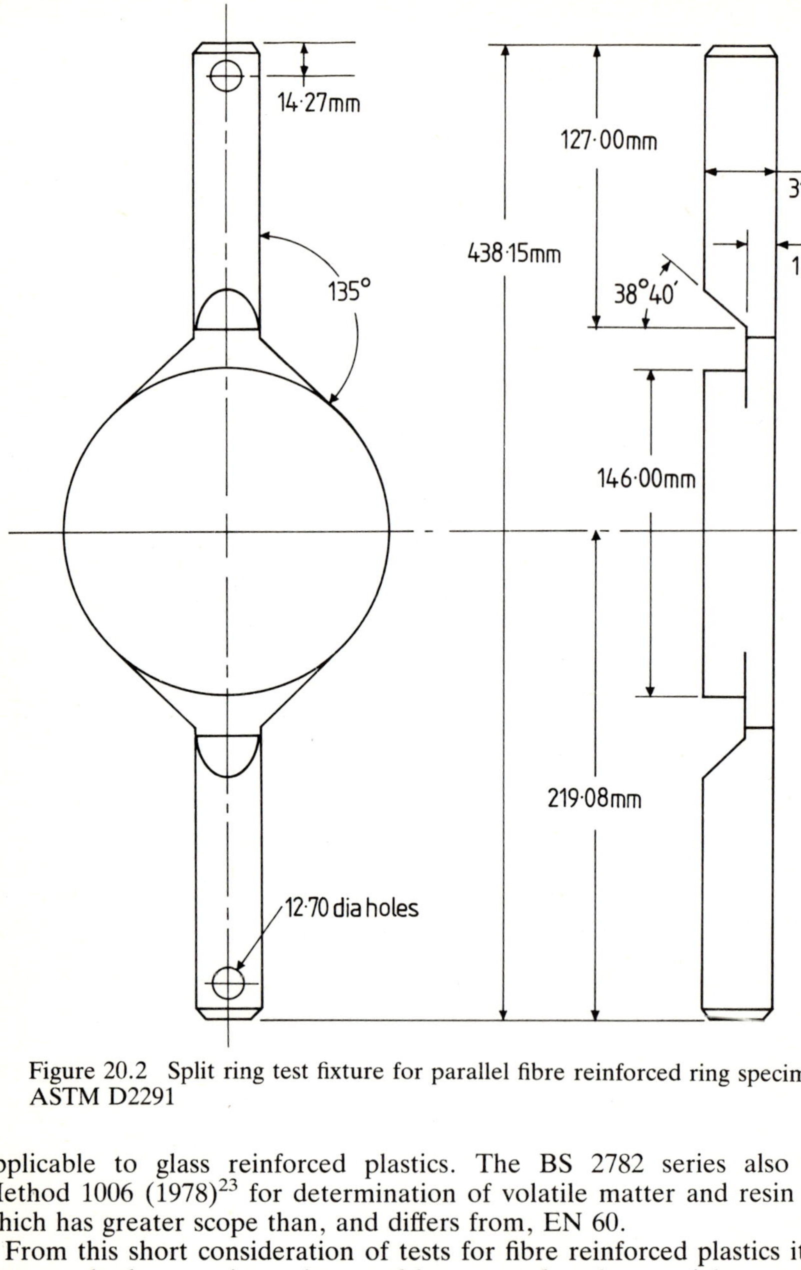

Figure 20.2 Split ring test fixture for parallel fibre reinforced ring specimens: ASTM D2291

applicable to glass reinforced plastics. The BS 2782 series also includes Method 1006 (1978)[23] for determination of volatile matter and resin content, which has greater scope than, and differs from, EN 60.

From this short consideration of tests for fibre reinforced plastics it is clear that standards committees have neither treated such materials as a separate entity nor wholly included them with plastics in general. The result is that a little care is needed to search out the most appropriate standard test method for a given property of the fibre reinforced material. The fact is that, although in most cases the basic type of test will be the same, the conditions needed for fibre reinforced materials will differ from those for other plastics and this will be especially so for highly oriented reinforcement. An outline of the problems encountered with a number of common mechanical tests is given by Ewing[24] and test techniques for filament wound materials are discussed by Bert.[25]

20.4 MECHANICAL PROPERTIES OF PRODUCTS

Some mechanical tests can be readily adapted to test a complete product. The most common is probably impact, which can be applied to dustbins, buckets, pipe and even window frames. Either the product itself can be dropped or a weighted striker can be dropped onto the product; occasionally, a pendulum type of impact machine can be used.

The choice of how to make the impact will depend on the convenience of testing a particular shape of object and on how impacts in service are likely to occur. For a dustbin or a bucket it is sensible both for convenience and for simulating service to drop the product. An example of where such a test has been standardised is BS 4998 (1974)[26] for dustbins. A rig is specified (Figure 20.3) to enable the drop to be made smoothly and to ensure that the bin strikes at the right attitude. Similar tests are used for many containers but often no advice is given on how to align and release the test product. This can be a

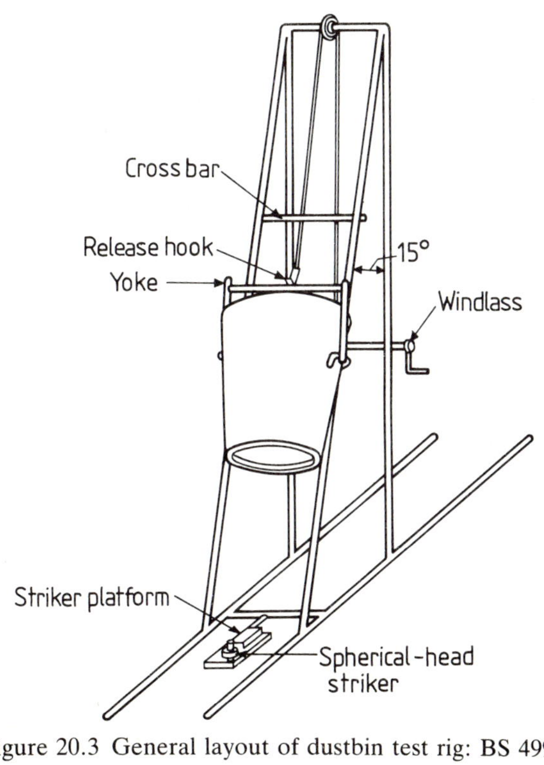

Figure 20.3 General layout of dustbin test rig: BS 4998

little difficult when impacts onto an edge are required, as in the United Nations recommendations for the packaging of dangerous goods.[27]

For pipes it is usual to drop a weight onto the product and the apparatus is exactly the same in principle as that used for laboratory prepared test pieces (see Chapter 8). An example can be found in BS 4991 (1974)[28] where the test is required to be made at 0°C. With a little ingenuity, rigs can be devised to hold a variety of products to be impacted in this manner. Falling weight impact apparatus can be instrumented to measure force and deceleration and this much more complicated and expensive apparatus is used for testing crash helmets.[29]

If a large enough stressing frame is available it is quite feasible to monitor the strains induced in a product as large as a fibre reinforced plastics car body (Figure 20.4) by the use of strain gauges attached to the product. Many similar products can be similarly tested using a more humble universal tensile testing machine. A very simple example is a flexural test suggested for the handles of domestic hollow-ware[30] and many extrusions can readily be subjected to a similar test without test piece preparation.

Pressure testing must be carried out on the product and is often required for pipe, drums, bottles, etc. In the simple forms of test, a proof pressure is applied internally for a given time and there may be both long-term and short-term tests

Figure 20.4 Large stressing frame

as in BS 4991 (1974),[28] for example. The pressurising medium is usually water. By testing at a series of pressures the time to failure can be determined as a function of pressure. For some purposes the pressure may be applied externally, as in a test for leakproofness of pipe joints.[31]

In service, products may be subjected to rapidly varying pressures which will reduce their fatigue life and more advanced pressure testing systems can be programmed to give cyclic pressurisation. The dynamic fatigue of PVC pipe formulations has been considered by Gotham and Hitch,[32] while Stapel[33] has compared the pressure variations occurring in water distribution systems with the fatigue properties of PVC.

20.5 ENVIRONMENTAL RESISTANCE

Testing the environmental resistance of plastics products presents the same difficulties as similar testing on laboratory samples—the long exposure times needed or, alternatively, the uncertainties of using accelerated conditions. The very size and value of some products will add to the difficulty or the cost of testing, and simulation of the stresses realised in service is more complicated than when stressing test pieces.

The environmental stress cracking resistance of products exposed to liquids can be estimated by simply immersing the product, but in most cases it is desirable to apply stress externally. The impressed ball method discussed in Chapter 17 can be applied to almost any object or in many cases some other form of mechanical restraint can be devised. Bottles can very conveniently be tested by applying an internal pressure and such a method has been standardised in ASTM D2561 (1970).[34] Of three methods given in D2561 for polyethylene blow moulded containers, Procedure C involves pressurisation. Test methods for environmental stress cracking, including those on products, have been critically reviewed by Brown.[35]

As the size of the product increases so the difficulty or cost of environmental testing escalates. For exposure to heat, cold or high humidity, conditioned rooms become necessary and very large facilities for artificial weathering and exposure to corrosive atmospheres are not often available. Natural weathering is very time-consuming but product specifications do on occasions require exposure at sites having extreme climatic conditions. One advantage of embarking on such exposure trials before the product is marketed is that any degradation which may occur can be noted before products in service reach the same state.

20.6 SIMULATED SERVICE TESTS

Perhaps the ultimate proof in the laboratory of product performance is a simulated service test, which usually involves some type of test rig designed specifically for that product. The stressing frames and impact tests mentioned in §20.4 (mechanical tests) could be said to be in this category and there is no sharp division between tests covered in that and the environmental section, and those considered here.

It may not be necessary to use any apparatus in the usual sense—for example, to measure the stackability of drums requires only a flat surface, and material to fill the drums. Generally, however, some sort of device is needed to subject the product to the types of stress and strain it will receive in service. Often it is fatigue which worries the designer, in which case product testing rigs need to have a dynamic action. A simple example is the fatigue test specified in BS 4083 (1974)[36] and used for plastics handles for hollow-ware[30] (Figure 20.5). Packaging materials can be subjected to vibration by using an electromagnetic shaker.

When cyclic action has been built into the test rig there arises the question of the effect of the environment on fatigue life and the apparatus is further complicated by housing it in conditioned cabinets or rooms. Ultimately the rig becomes excessively costly as attempts are made to simulate all the conditions which are realised in service. When designing product test rigs, a compromise must always be sought between uneconomic complication and failure to cover factors which may prove to be of overriding importance in service.

In some cases it may be environmental effects rather than direct mechanical actions that are of greatest interest. In pipe systems for water distribution, for example, temperature cycling of the water being conveyed may be more likely to cause failures than any direct mechanical action on the pipe. A BPF publication[37] gives guidance on a test procedure which subjects a pipe system to cycling and fluctuating temperatures.

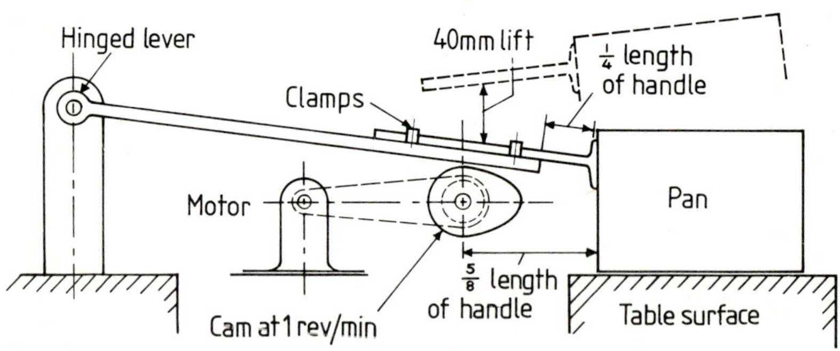

Figure 20.5 Fatigue test for handle and attachment: BS 4083

20.7 WHEN TO TEST PRODUCTS

If our knowledge of the properties and behaviour of plastics and hence our design rules were such that we could predict the performance of the product accurately from tests on laboratory test pieces then product testing would rarely be needed.

The fact is that our understanding of the properties of plastics is simply not good enough to make such predictions reliably in a great many cases and hence we need to test the product, since this is the only way to be sure that it will perform satisfactorily. As we have seen, it can be extremely difficult and/or expensive to devise tests to simulate service adequately and as a matter of expediency we have to resort to less adequate trials on laboratory test pieces. In general, the product should be tested to prove its performance wherever there is doubt as to the reliability of predictions made from tests on test pieces, appreciating that the cost of such product tests may become prohibitive. The more critical the component, the more justification there will be for high expenditure on testing.

The same principle applies to quality control testing, but here there is a much greater probability that the experience gained from proving the product initially will allow the quality of subsequent production to be reliably judged on the basis of tests on test pieces. Sometimes a product test will give more valuable assessment of quality for the same testing cost as needed for test pieces. This would be true, for example, for impact resistance of a bucket because the cost of moulding test pieces would be little different from the value of the bucket and the testing costs would be equal. Impact testing the bucket would actually be cheaper than machining standard impact test pieces from it.

When the value of the product far outweighs the cost of making test pieces it is again necessary to judge whether control on test pieces gives us sufficient confidence to reject the costly alternative of product tests. It is here that non-destructive tests on the product become especially attractive and, not surprisingly, great effort is made to devise such tests, which give more confidence than the use of test pieces and additionally may even be cheaper to carry out.

REFERENCES

1. BS 3424 (1973) *Methods of Test for Coated Fabrics.*
2. ISO 1642 (1975) *Industrial Laminated Sheets Based on Thermosetting Resins—Basis for Specifications.*
3. BS 4965 (1974) *Decorative Laminated Plastics Sheet. Veneered Boards and Panels.*
4. BS 3953 (1976) *Glass Fabric Laminated Sheet.*
5. BS 2572 (1976) *Specification for Phenolic Laminated Sheet.*
6. BS 5102 (1974) *Phenolic Resin Bonded Paper Laminated Sheets for Electrical Applications.*
7. BS 3794 (1973) *Decorative Laminated Plastics Sheet.*
8. BS 2782 *Methods of Test for Plastics.*
9. ISO 3268 (1978) *Glass Reinforced Materials—Determination of Tensile Properties.*
10. EN 61 (1977) *Glass Reinforced Plastics. Determination of Tensile Properties.*
11. BS 2782 Method 1003 (1977) *Glass Reinforced Plastics. Determination of Tensile Properties.*
12. ASTM D2291 (1976) *Fabrication of Ring Test Specimens for Reinforced Plastics.*
13. ASTM D2344 (1976) *Apparent Horizontal Shear Strength of Reinforced Plastics by Short Beam Method.*
14. ASTM D2290 (1976) *Apparent Tensile Strength of Ring or Tubular Plastics by Split Disc Method.*
15. ISO 3597 (1977) *Textile Glass Reinforced Plastics—Composites in the Form of Rods made from Textile Glass Rovings—Determination of Flexural (Cross-breaking) Strength.*
16. ISO 3605 (1978) *Textile Glass Reinforced Plastics—Composites in the form of Rods made from Textile Glass Rovings—Determination of Compressive Strength.*
17. EN 59 and BS 2782 Method 1001 (1977) *Glass Reinforced Plastics. Measurement of Hardness by means of a Barcol Impressor.*

18. EN 60 and BS 2782 Method 1002 (1977) *Glass Reinforced Plastics. Determination of Loss on Ignition.*
19. ISO 1172 (1975) *Textile Glass Reinforced Plastics—Determination of Loss on Ignition.*
20. EN 62 and BS 2782 Method 1004 (1977) *Glass Reinforced Plastics. Standard Atmospheres for Conditioning and Testing.*
21. EN 63 and BS 2782 Method 1005 (1977) *Glass Reinforced Plastics. Determination of Flexural Properties. Three Point Method.*
22. ISO 178 (1975) *Determination of Flexural Properties of Rigid Plastics Materials.*
23. BS 2782 Method 1006 (1978) *Determination of Volatile Matter and Resin Content of Synthetic Resin-impregnated Textile Glass Fabric.*
24. Ewing, P. D. (1974) in *Conference, Composites, Standards Testing and Design*, National Physical Laboratory, April, page 144.
25. Bert, C. W. (1974) *Composites*, **5** (1), 20.
26. BS 4998 (1974) *Moulded Plastics Dustbins.*
27. United Nations (1976) *Transport of Dangerous Goods*, UN Sales No. 76. VIII. 2.
28. BS 4991 (1974) *Propylene Copolymer Pressure Pipe.*
29. BS 5361 (1976) *Specification for Protective Helmets for Vehicle Users.*
30. Healey Mouldings Limited (1978) *Specification for Thermosetting Handles for Hollow-ware*, STPH/78.
31. ISO 3459 (1976) *Polyethylene (PE) Pressure Pipes—Joints Assembled with Mechanical Fittings—Internal Under-pressure Test Method and Requirement.*
32. Gotham, K. V. and Hitch, M. J. (1978) *British Polymer Journal*, **10,** 38.
33. Stapel, J. J. (1977) *Pipes and Pipelines International*, February, 11; April, 33.
34. ASTM D2561 (1970) *Environmental Stress-crack Resistance of Blow-moulded Polyethylene Containers.*
35. Brown, R. P. (1980) *Polymer Testing*, **1** (4), 267.
36. BS 4083 (1974) *Aluminium Hollow-ware Cooking Utensils.*
37. British Plastics Federation (1976) *Guide for the Determination of Performance for Thermoplastics Pipes and Fittings for Use in Hot and Cold Water Applications*, BPF Publication No. 132/1.

Index